Animal Models of Acute Neurological Injuries

Animal Models of Acute Neurological Injuries

Edited by

Jun Chen, M.D.
University of Pittsburg School of Medicine, Pittsburg, PA, USA

Xiao-Ming Xu, M.D., Ph.D.
Indiana University School of Medicine, Indianapolis, IN, USA

Zao C. Xu, M.D., Ph.D.
Indiana University School of Medicine, Indianapolis, IN, USA

John H. Zhang, M.D., Ph.D.
Loma Linda School of Medicine, Loma Linda, CA, USA

Humana Press

Editor
Jun Chen, M.D.
Department of Neurology
University of Pittsburgh School of Medicine
Pittsburgh, PA, USA

Xiao-Ming Xu, M.D., Ph.D.
Spinal Cord and Brain Injury Research Group
Stark Neurosciences Research Institute
Department of Neurological Surgery
Indiana University School of Medicine
Indianapolis, IN, USA

Zao C. Xu, M.D., Ph.D.
Department of Anatomy and Cell Biology
Indiana University School of Medicine
Indianapolis, IN, USA

John H. Zhang, M.D., Ph.D.
Department of Physiology and Pharmacology
Department of Neurosurgery
Department of Anesthesiology
Loma Linda University Medical School
Loma Linda, CA, USA

ISBN: 978-1-60327-184-4 e-ISBN: 978-1-60327-185-1
DOI: 10.1007/978-1-60327-185-1

Library of Congress Control Number: 2008935542

Printed on acid-free paper

9 8 7 6 5 4 3 2 1

springer.com

Foreword

The past 30 years have witnessed an explosion in our understanding of the central nervous system structure and function in both health and disease. This explosion and the excitement it generates have led to a flurry of activity in translational biomedical efforts, with many clinicians and scientists utilizing animal models to dissect important features of specific CNS disease and/or injury processes, not only to better identify potential preclinical therapeutic targets, but also to screen newly-developed therapies as a prelude to their testing in well-controlled human studies. Despite this explosive progress and the launching of numerous large-scale, Phase III human clinical trials, the overall success of these clinical efforts has not been positive. This has been particularly so for those clinical studies targeting acute aspects of CNS injury, including ischemia, subarachnoid and intraparenchymal hemorrhage, and traumatic brain and spinal cord injury. While the failure of these clinical/translational studies has been linked to multiple patient management, center variability, pharmacokinetic, and preclinical laboratory shortcomings, there is also a consensus that animal modeling issues, including investigator non-compliance in faithfully reproducing specific animal models, has also been a significant contributing factor. In this vein, there is a general recognition that although the field of acute neurological injury has embraced state-of-the-art technologies to address relevant questions in terms of CNS pathobiology and therapeutic management, the occasional non-consideration of rigorous animal modeling and appropriate laboratory animal management have led to the generation of less than optimal data. Such non-optimal data have added little to the field and, more importantly, have not allowed for meaningful comparisons to be drawn from data generated by different research groups.

In this book, focusing on acute neurological injuries, the editors Drs. Jun Chen, Xiao-Ming Xu, Zao C. Xu, and John Zhang have provided an important contribution to the field of central nervous system injury via the provision of an outstanding compendium of chapters focusing on a wide array of animal models currently used for assessing acute neurological injury associated with ischemia, subarachnoid and intraparenchymal hemorrhage, and traumatic injury to the brain and spinal cord. In doing so, the editors have assembled an impressive array of senior and mid-level investigators, all of whom are actively involved in the study of acute neurological injury as assessed in established animal models of ischemia, hemorrhage, and CNS injury. In each of these modeling chapters, the authors not only review the relevant history of their respective animal models, but also detail specific steps in animal model preparation and use, thereby providing the reader with a detailed guide on how to reproduce a specific animal model in their own laboratory. Additionally, and of equal importance, these detailed animal modeling chapters are preceded by several chapters focusing on general animal care and related surgical considerations that are integral to the conduct of well-controlled and well-designed animal modeling studies. While this compendium is not entirely complete, given its non-consideration of penetrating and blast injuries and other more generalized acute neurological injuries, this is not seen as a major detracting factor. Rather, as current models of blast and penetrating injuries are neither well developed nor supported by a wide body of literature, it would be premature to so burden this outstanding text.

From the reader's perspective, this book is an important companion to the study of acute CNS injury. It should serve as an invaluable guide for junior as well as seasoned investigators, reminding all of the key steps in animal preparation, injury, and post injury management relevant to the specific research questions being asked in the context of acute CNS injury. The editors are to be congratulated on their efforts in guiding this important collection of chapters, which hopefully will have a positive impact upon those using animal models in the study of acute neurological injury. It is hoped that this text will serve as an invaluable resource to those using animal models in the study of CNS injury, with the expectation that the use of this book will lead to more consistent and reproducible findings as well as more meaningful preclinical therapeutic trials. Hopefully, by more strict adherence to animal modeling, many of the above-identified failures in preclinical/translational studies will be obviated, leading to more appropriately targeted clinical trials for the better care and management of the patients who have sustained CNS injury.

Richmond, VA J.T. Povlishock

Preface

Acute neurological injuries are devastating events that affect the full spectrum of human life. Understanding the mechanisms of cell death and tissue damage will help to develop pharmacological agents and therapeutic interventions to ameliorate or even prevent damage to the nervous system. For the past two decades, numerous new studies have been conducted and exciting discoveries have been reported. However, only limited treatment is available today for the victims of acute neurological injuries. This indicates that more research is needed to explore the roles and underlying mechanisms of acute neurological disorders. To reach this goal, reliable animal models are the key for success in experimental neurobiology. Although many animal models for acute neurological injuries have been developed in recent years, many variations exist in the experimental procedures and outcome measures between laboratories even when nominally using the same animal model. These inconsistent preparations make comparisons nearly impossible and jeopardize the advance of the field. We believe that the large variability existing in the field is largely due to the lack of a standard manual detailing the methodology used in animal models of acute neurological injuries.

In contrast to other excellent scientific books on the market that focus on theoretical details, this book is constructed as a benchwork manual of animal models. We selected the most commonly used animal models of acute neurological injuries that have been proven to be both reliable and reproducible. The contributors of this book are experts in each of these respective fields and have many years of hands-on experience with these models. Many authors are the original developers of these models. Each chapter contains a step-by-step working procedure along with multiple photographs and/or schematic drawings. References and descriptions have been kept to a minimum for the clarity. It is expected that even someone with little experience could use this manual as a guide to produce a reliable animal model for clinical or basic science research. We hope that this book will be useful for students, postdoctoral fellows, and clinical and basic scientists who would like to study acute neurological disorders.

Pittsburg, PA — Jun Chen
Indianapolis, IN — Xiao-Ming Xu, Zao C. Xu
Loma Linda, CA — John H. Zhang

Contents

Section II Focal Cerebral Ischemic Models

Section III Neonatal Hypoxia-Ischemia

Part III Cerebral Hemorrhage Animal Models

Section I Subarachnoid Hemorrhage

Contributors

Faisal Adhami
Department of Pediatrics, Division of Developmental Biology and Neurology, Children's Hospital Medical Center, Cincinnati, OH, USA; Physician Scientist Training Program (PSTP), University of Cincinnati, College of Medicine, Cincinnati, OH, USA

Henry L. Alexander
Department of Critical Care Medicine, Safar Center for Resuscitation Research, Pittsburgh, PA, USA

Robert Ayer
Department of Physiology and Pharmacology, Loma Linda University Medical School, Loma Linda, CA, USA

Ludmila Belayev
Neuroscience Center of Excellence, Louisiana State University Health Science Center, New Orleans, LA, USA

Andrew R. Blickman
Laboratory Animal Resource Center, Indiana University School of Medicine, Indianapolis, IN, USA

Klas Blomgren
Center for Brain Repair and Rehabilitation, Institute of Neuroscience and Physiology, Göteborg University, Göteborg, Sweden
Department of Pediatric Oncology, The Queen Silvia Children's Hospital, Göteborg, Sweden

Lisa J. Brossia
Laboratory Animal Resource Center, Indiana University School of Medicine, Indianapolis, IN, USA

Julian Cahill
Department of Physiology and Pharmacology, Loma Linda University Medical School, Loma Linda, CA, USA

Tarkan Calisaneller
Department of Neuroscience, University of Virginia Health System, Charlottesville, VA, USA
Neurosurgery Department, Baskent University Faculty of Medicine, Ankara, Turkey

Clifton W. Callaway
Department of Emergency Medicine, University of Pittsburgh School of Medicine, Pittsburgh, PA, USA

Pak H. Chan
Department of Neurosurgery, Department of Neurology and Neurological Sciences, and Program in Neurosciences, Stanford University School of Medicine, Stanford, CA, USA

Jun Chen
Department of Neurology, University of Pittsburgh School of Medicine, Pittsburgh, PA, USA

S.T. Chen
Department of Neurology, Chang Gung Memorial Hospital, Taoyuan, Taiwan

W.M. Cheung
Institute of Biomedical Sciences, Academia Sinica, Taipei, Taiwan

Tak-Ho Chu
Department of Anatomy, Li Ka Shing Faculty of Medicine, The University of Hong Kong, Hong Kong

Robert S.B. Clark
Departments of Critical Care Medicine and Pediatrics, Safar Center for Resuscitation Research, Children's Hospital of Pittsburgh, Pittsburgh, PA, USA

Ping Deng
Department of Anatomy & Cell Biology, Indiana University School of Medicine, Indianapolis, IN, USA

Matthew Derrick
Department of Pediatrics, Evanston Northwestern Healthcare and Northwestern University, Evanston, IL, USA

C. Edward Dixon
Departments of Neurosurgery, Physical Medicine & Rehabilitation, Psychology, Center for Neuroscience, Brain Trauma Research Center, Center for the Neural Basis of Cognition, and the Safar Center for Resuscitation Research, University of Pittsburgh, Pittsburgh, PA, USA

Hidenori Endo
Department of Neurosurgery, Department of Neurology and Neurological Sciences, and Program in Neurosciences, Stanford University School of Medicine, Stanford, CA, USA
Department of Neurosurgery, Tohoku University Graduate School of Medicine, Sendai, Japan

Donna M. Ferriero
Department of Neurology and Pediatrics, University of California San Francisco, San Francisco, CA, USA

Mark Fitzgerald
Department of Neuroscience, University of Virginia Health System, Charlottesville, VA, USA

Beth Friedman
Department of Neuroscience, University of California at San Diego, La Jolla, CA, USA

Steven H. Graham
Department of Neurology, University of Pittsburgh School of Medicine, Pittsburgh, PA, USA

Mark Grossetete
Department of Neurology, University of New Mexico Health Sciences Center, Albuquerque, NM, USA

Yuxiang Gu
Department of Neurosurgery, University of Michigan, Ann Arbor, MI, USA

Erdem Gueresir
Department of Neurosurgery, Johann Wolfgang Goethe-University, Frankfurt/Main, Germany

Y.Y. He
Chi-Ching Huang Stroke Research Center, Taipei Medical University, Taipei, Taiwan

Robert W. Hickey
Department of Pediatrics, Division of Pediatric Emergency Medicine, Children's Hospital of Pittsburgh, Pittsburgh, PA, USA

Chung Y. Hsu
Chi-Ching Huang Stroke Research Center, Taipei Medical University, Taipei, Taiwan

Bingren R. Hu
Cerebral Vascular Disease Research Center and Laboratory of Neurochemistry of Brain Injury, Department of Neurology, University of Miami Miller School of Medicine, Miami, FL, USA

Ya Hua
Department of Neurosurgery, University of Michigan, Ann Arbor, MI, USA

Vikram Jadhav
Department of Physiology and Pharmacology, Loma Linda University Medical School, Loma Linda, CA, USA

Lyn B. Jakeman
Department of Physiology and Cell Biology and Department of Neuroscience, The Ohio State University College of Medicine, Columbus, OH, USA

Sarah C. Jernigan
Department of Neurological Surgery, Kentucky Spinal Cord Injury Research Center, University of Louisville School of Medicine, Louisville, KY, USA

Xinhai Ji
Department of Pediatrics, Evanston Northwestern Healthcare, Evanston, IL, USA

Hideyuki Kamii
Department of Neuroanesthesia, Kohnan Hospital, Sendai, Japan

Ayako Kawashima
Division of Neurosurgery and the Keenan Research Centre in the Li Ka Shing Knowledge Institute of St. Michael's Hospital and the Division of Neurosurgery, University of Toronto, Toronto, ON, Canada

Gab Seok Kim
Department of Neurosurgery, Department of Neurology and Neurological Sciences, and Program in Neurosciences, Stanford University School of Medicine, Stanford, CA, USA

Hiroyuki Kinouchi
Department of Neurosurgery, Faculty of Medicine, University of Yamanashi, Tamaho, Yamanashi, Japan

David Kleinfeld
Department of Physics, Graduate Program in Neurosciences, University of California at San Diego, La Jolla, CA, USA

Anthony E. Kline
Departments of Neurosurgery, Physical Medicine & Rehabilitation, Psychology, Center for Neuroscience, Brain Trauma Research Center, Center for the Neural Basis of Cognition, and the Safar Center for Resuscitation Research, University of Pittsburgh, Pittsburgh, PA, USA

Kohei Kokubun
Division of Neurosurgery, Department of Neuro- and Locomotor Science, Akita University School of Medicine, Akita, Japan

Chia-Yi Kuan
Department of Pediatrics, Division of Developmental Biology and Neurology, Children's Hospital Medical Center, Cincinnati, OH, USA

Kevin S. Lee
Department of Neuroscience, University of Virginia Health System, Charlottesville, VA, USA
Department of Neurological Surgery, University of Virginia, Charlottesville, VA, USA

Jonathan Lifshitz
Spinal Cord & Brain Injury Research Center (SCoBIRC), Departments of Anatomy & Neurobiology and Physical Medicine & Rehabilitation, University of Kentucky Chandler Medical Center, Lexington, KY, USA

Chih-Lung Lin
Department of Neuroscience, University of Virginia Health System, Charlottesville, VA, USA
Department of Neurosurgery, Kaohsiung Medical University, Kaohsiung, Taiwan

T.N. Lin
Institute of Biomedical Sciences, Academia Sinica, Taipei, Taiwan

Chunli Liu
Laboratory of Neurochemistry of Brain Injury, Department of Neurology, University of Miami Miller School of Medicine, Miami, FL, USA

Eng H. Lo
Neuroprotection Research Laboratory, Departments of Neurology and Radiology, Massachusetts General Hospital, Harvard Medical School, Charlestown, MA, USA

Eric S. Logue
Department of Emergency Medicine, University of Pittsburgh School of Medicine, Pittsburgh, PA, USA

Patrick D. Lyden
Graduate Program in Neurosciences, Department of Neuroscience, University of California at San Diego, La Jolla, CA, USA

R. Loch Macdonald
Division of Neurosurgery and the Keenan Research Centre in the Li Ka Shing Knowledge Institute of St. Michael's Hospital and the Division of Neurosurgery, University of Toronto, Toronto, ON, Canada

Mioara D. Manole
Department of Pediatrics, Division of Pediatric Emergency Medicine, Safar Center for Resuscitation Research, Children's Hospital of Pittsburgh, Pittsburgh, PA, USA

Ying Mao
Department of Neurosurgery and Institute of Neurosurgery, Huashan Hospital, Fudan University School of Medicine, Shanghai, China

Anthony Marmarou
Department of Neurosurgery, Virginia Commonwealth University Medical College of Virginia Campus, Richmond, VA, USA

Christina R. Marmarou
Department of Neurosurgery, Virginia Commonwealth University Medical College of Virginia Campus, Richmond, VA, USA

Dana M. McTigue
Department of Neuroscience, The Ohio State University College of Medicine, Columbus, OH, USA

Kazuo Mizoi
Division of Neurosurgery, Department of Neuro- and Locomotor Science, Akita University School of Medicine, Akita, Japan

Yoshihiro Murata
Neuroprotection Research Laboratory, Departments of Neurology and Radiology, Massachusetts General Hospital, Harvard Medical School, Charlestown, MA, USA

D. Jeannie Myer
Department of Neurosurgery, Department of Neurology and Neurological Sciences, and Program in Neurosciences, Stanford University School of Medicine, Stanford, CA, USA

Kuniyasu Niizuma
Department of Neurosurgery, Department of Neurology and Neurological Sciences, and Program in Neurosciences, Stanford University School of Medicine, Stanford, CA, USA

Shigeru Nishizawa
Department of Neurosurgery, University of Occupational and Environmental Health, Fukuoka, Japan

Chikako Nito
Department of Neurosurgery, Department of Neurology and Neurological Sciences, and Program in Neurosciences, Stanford University School of Medicine, Stanford, CA, USA

Christine D. Nunn
Kentucky Spinal Cord Injury Research Center & Department of Neurological Surgery, School of Medicine, University of Louisville, Louisville, KY, USA

Masaya Oda
Division of Neurosurgery, Department of Neuro- and Locomotor Science, Akita University School of Medicine, Akita, Japan

Thomas O'Lynnger
Department of Neurosurgery, University of Michigan, Ann Arbor, MI, USA

Stephen M. Onifer
Spinal Cord and Brain Injury Research Center & Department of Anatomy and Neurobiology, College of Medicine, University of Kentucky, Lexington, KY, USA

Peter Poon
Toronto Western Hospital, Toronto, ON, Canada

Ruth Prieto
Department of Neurosurgery, Virginia Commonwealth University Medical College of Virginia Campus, Richmond, VA, USA

Aaron H. Puckett
Kentucky Spinal Cord Injury Research Center & Department of Neurological Surgery, School of Medicine, University of Louisville, Louisville, KY, USA

Ami P. Raval
Cerebral Vascular Disease Research Center, Department of Neurology, University of Miami Miller School of Medicine, Miami, FL, USA

Kelly N. Roberts
Sanders-Brown Center on Aging, University of Kentucky College of Medicine, Lexington, KY, USA

Gary A. Rosenberg
Department of Neurology, University of New Mexico Health Sciences Center, Albuquerque, NM, USA

Laurel Schantz
Indiana University School of Medicine, Indianapolis, IN, USA

Stephen Scheff
Sanders-Brown Center on Aging, University of Kentucky
College of Medicine, Lexington, KY, USA

Volker Seifert
Department of Neurosurgery, Johann Wolfgang Goethe-University,
Frankfurt/Main, Germany

Christopher B. Shields
Department of Neurological Surgery, Kentucky Spinal Cord Injury Research
Center, University of Louisville School of Medicine, Louisville, KY, USA

Lisa B.E. Shields
Department of Neurological Surgery, Kentucky Spinal Cord Injury Research
Center, University of Louisville School of Medicine, Louisville, KY, USA

Andy Y. Shih
Department of Physics, University of California at San Diego, La Jolla,
CA, USA

Bradford T. Stokes
Department of Physiology and Cell Biology, The Ohio State University
College of Medicine, Columbus, OH, USA

Takashi Sugawara
Department of Physiology and Pharmacology, Loma Linda University
Medical School, Loma Linda, CA, USA

Taku Sugawara
Division of Neurosurgery, Department of Neuro- and Locomotor Science,
Akita University School of Medicine, Akita, Japan

Sidhartha Tan
Department of Pediatrics, Evanston Northwestern Healthcare and
Northwestern University, Evanston, IL, USA

Charles H. Tator
Toronto Western Hospital, Toronto, ON, Canada

Keisuke Taya
Department of Neurosurgery, Virginia Commonwealth University
Medical College of Virginia Campus, Richmond, VA, USA

Teiji Tominaga
Department of Neurosurgery, Tohoku University Graduate School of
Medicine, Sendai, Japan

Naoya Ukita
Department of Neuroscience, University of Virginia Health System,
Charlottesville, VA, USA
Department of Neurological Surgery, Okayama University School of
Medicine, Okayama, Japan

Hartmut Vatter
Department of Neurosurgery, Johann Wolfgang Goethe-University,
Frankfurt/Main, Germany

Zinaida S. Vexler
Department of Neurology, University of California San Francisco, San Francisco, CA, USA

Patricia Walters
Department of Physiology and Cell Biology and Department of Neuroscience, The Ohio State University College of Medicine, Columbus, OH, USA

Xiaofei Wang
Spinal Cord and Brain Injury Research Group, Stark Neurosciences Research Institute, Department of Neurological Surgery, Indiana University School of Medicine, Indianapolis, IN, USA

Xiaoyang Wang
Department of Pediatrics, the Third Affiliated Hospital of Zhengzhou University, China
Perinatal Center, Institute of Neuroscience and Physiology, Göteborg University, Göteborg, Sweden

Xiaoying Wang
Neuroprotection Research Laboratory, Departments of Neurology and Radiology, Massachusetts General Hospital, Harvard Medical School, Charlestown, MA, USA

Yi Wang
Department of Neuroscience, University of Virginia Health System, Charlottesville, VA, USA
Department of Neurosurgery, Rui Jin Hospital, Shanghai Jiao Tong University School of Medicine, Shanghai, PRC

Scott R. Whittemore
Department of Neurological Surgery, Kentucky Spinal Cord Injury Research Center, University of Louisville School of Medicine, Louisville, KY, USA

Wutian Wu
Department of Anatomy, Li Ka Shing Faculty of Medicine, The University of Hong Kong, Hong Kong

Guohua Xi
Department of Neurosurgery, University of Michigan, Ann Arbor, MI, USA

Xiao-Ming Xu
Spinal Cord and Brain Injury Research Group, Stark Neurosciences Research Institute, Department of Neurological Surgery, Indiana University School of Medicine, Indianapolis, IN, USA

Zao C. Xu
Department of Anatomy & Cell Biology, Indiana University School of Medicine, Indianapolis, IN, USA

Guo-Yuan Yang
Med-x Research Institute Shanghai JiaoTong University
Shanghai, China
and

The Center for Cerebrovascular Research, Department of Anesthesia and Perioperative Care, University of California San Francisco, San Francisco, CA, USA

Ryo Yoshimura
Department of Neuroscience, University of Virginia Health System, Charlottesville, VA, USA
Department of Neurological Surgery, Wakayama Medical University, Wakayama, Japan

Harold F. Young
Department of Neurosurgery, Virginia Commonwealth University Medical College of Virginia Campus, Richmond, VA, USA

Wise Young
W.M. Keck Center for Collaborative Neuroscience, Rutgers University, Piscataway, NJ, USA

Feng Zhang
Department of Neurology, University of Pittsburgh School of Medicine, Pittsburgh, PA, USA

John H. Zhang
Department of Physiology and Pharmacology, Department of Neurosurgery, Department of Anesthesiology, Loma Linda University Medical School, Loma Linda, CA, USA

Yi Ping Zhang
Department of Neurological Surgery, Kentucky Spinal Cord Injury Research Center, University of Louisville School of Medicine, Louisville, KY, USA

Changlian Zhu
Center for Brain Repair and Rehabilitation, Institute of Neuroscience and Physiology, Göteborg University, Göteborg, Sweden
Department of Pediatrics, the Third Affiliated Hospital of Zhengzhou University, China

Wei Zhu
Department of Neurosurgery and Institute of Neurosurgery, Huashan Hospital, Fudan University School of Medicine, Shanghai, China

Mario Zuccarello
Department of Neurosurgery, The Neuroscience Institute, University of Cincinnati School of Medicine, and Mayfield Clinic, Cincinnati, OH, USA

Part I

Animal Care and Surgery

Chapter 1

Animal Protocol

Laurel D. Schantz

Abstract This chapter outlines the important points that must be considered in developing an animal protocol using a model of acute neurological injury. It highlights the issues that should be considered in designing such a study as well as the issues that must be addressed when writing a protocol for submission to the Institutional Animal Care and Use Committee (IACUC). The key issues for regulatory compliance are addressed.

Keywords: Animal protocol, IACUC, Regulatory compliance

1 Regulatory Oversight

In the United States, the use of animals in research has extensive regulatory oversight. The U.S. Animal Welfare Act (AWA) and the Animal Welfare Regulations (AWRs) are administered by the United States Department of Agriculture (USDA). These regulations cover warm-blooded vertebrates except birds, rats of the genus *Rattus*, mice of the genus *Mus*, and farm animals used in production agriculture.[1] The Public Health Service Policy on Humane Care and Use of Laboratory Animals (PHS Policy) has its statutory mandate in the U.S. Health Research Extension Act (HREA), and is administered by the Office of Laboratory Animal Welfare (OLAW) of the National Institutes of Health (NIH). It applies to all institutions that use live vertebrates in research supported by any component of PHS. Along with the AWA and AWRs, the PHS policy also mandates compliance with the U.S. Government Principles for the Utilization and Care of Vertebrate Animals Used in Testing, Research, and Training (US Government Principles), and the National Research Council's Guide to the Care and Use of Laboratory Animals (The Guide).[2,3] In addition to governmental oversight, many research institutions seek accreditation by the Association for Assessment and Accreditation of Laboratory Animal Care International (AAALAC International). AAALAC International is a private, nonprofit organization that provides a program of voluntary accreditation to institutions that use animals in research. AAALAC International uses The Guide as a primary reference document when assessing animal care programs.[4]

From: *Animal Models of Acute Neurological Injuries.*
Edited by: J. Chen (eds.) New York, Totowa, NJ

The AWA, AWRs, US Government Principles, and The Guide give the responsibility for oversight and evaluation of the animal care program and use of animals in research to the Institutional Animal Care and Use Committee (IACUC). The IACUC is tasked with assessing every protocol involving animal use in research, testing, and education. The topics that it is required to evaluate include the rationale, purpose, and uniqueness of animal use, the justification of the species and number of animals, lack of non-animal alternatives, training of personnel, use of sedation, analgesia, anesthesia, appropriate methods of euthanasia, development of humane end points, conduct of operative procedures, thorough post-procedural monitoring and care of the animals, unusual husbandry or housing requirements, and safety of the working environment.[5]

By its very nature, use of an animal model of acute neurological injury presents a challenge to design studies that achieve the desired experimental outcomes and minimize animal pain and distress. However, this is not only the regulatory mandate; it is the goal of everyone involved in such work. To accomplish this goal, developing a team approach incorporating the investigative staff, veterinary staff, and animal care staff is most useful. Involving all the necessary team members in the design phase of the protocol development usually makes acceptance of the final written animal protocol by the IACUC much more expedient.

2 Designing the Animal Protocol

When using an animal model for the first time, there are several factors that must be considered as having the potential to introduce undesirable variables into the study. Training of personnel, special surgical methods, anesthesia monitoring, postoperative care and monitoring, analgesia, identification of humane end points, method of euthanasia, outcome assessment tool development, and identification of any hazards must all be considered.

Provision must be made for training of individuals participating in the animal work. Particularly, the skills of individual surgeons have a dramatic impact on model outcomes. If a surgeon has not had experience with the procedures to be carried out in the study, then it is important that accommodations be made to provide such experience prior to beginning work on experimental animals. Obviously, initial practice should be given using cadavers and is best guided by a surgeon with experience in using such procedures. If there is no one available with experience in the procedures to provide the "hands-on" training, often arrangements can be made to visit another facility to observe the surgery being performed by experienced surgeons as a survival procedure. Practice on cadavers must be done until the surgeon is very comfortable with the new procedures. Training on live animals as part of non-survival surgeries is the next step. Once live animals are needed even for training purposes, the work must be covered by a written animal protocol approved by the IACUC. Laboratory animal resource centers at some institutions may have animals available for training purposes. Often, the work is then covered under an animal protocol, which is the responsibility of an institutional veterinarian. Sometimes, an amendment must be written to add the actual surgical procedure that will be practiced and/or the name of the surgeon who will participate.

Sometimes, the training animal protocol is broad enough to cover non-survival procedures as long as the veterinarian is present and anesthesia/euthanasia is performed by animal care personnel. If such an institutional animal protocol is not available, then the training on live animals will need to be included in the written animal protocol covering the experimental work.

The surgical procedure itself must be described in detail in the animal protocol. For survival surgeries, particular attention must be given to describing how the requirements for establishing and maintaining asepsis during surgery will be met. If the procedure will not be done in a designated operating room, the area where surgery will be done will need to be inspected by representatives of the IACUC prior to performing the work.[5] It is a good idea, when possible, to arrange for this inspection prior to submitting the animal protocol. During the inspection, a thorough discussion of not only the steps taken to insure asepsis but also those concerning animal welfare during transportation and/or recovery can lead to improving the perioperative plan for animal care. If a procedure will require special consideration due to equipment that cannot be moved or equipment that cannot be sterilized by conventional means, then justification for the exceptions to required asepsis should be provided. If a procedure is new to an institution and there will be a need to break the aseptic technique due to specialized equipment or techniques, then a pilot study may be required with veterinary observation and/or outcome reporting to the IACUC to allow evaluation of the impact on animal welfare of the deviation from aseptic practices. The IACUC inspection team can be a valuable resource in the design of such a pilot study. The pilot study would then need to be described in the animal protocol.

Anesthetic choice will be dictated by the needs of the study. If the anesthesia chosen is not one commonly used in standard veterinary care for the species involved, it may require justification in the written animal protocol. Regardless of anesthetic choice, anesthetic monitoring will need to be considered. Proper monitoring of an anesthetized animal is often critical to successful outcomes of surgical procedures. The parameters that will be monitored need to be determined on the basis of animal species, anesthetic choice, and expected effects of the surgical manipulation. Availability of the equipment necessary for monitoring and record keeping should be considered. For prolonged studies done on anesthetized animals, the means necessary to monitor and maintain normal physiologic functions within the study parameters must be considered. The veterinarian can be helpful in forming this plan and providing training, should it be needed. Because proper monitoring and care of the anesthetized animal ensures that the animal does not experience pain or distress during the surgical procedure, it must be described in detail in the animal protocol.

The plan for postoperative care and monitoring should also be considered in the study design. Animals recovering from acute neurological injury must often be considered "critical care" patients, especially in the immediate postoperative period. Frequency of monitoring, parameters to be monitored, and any special equipment required must be considered. The veterinarian can be helpful in developing this part of the plan and in training personnel in recognition of clinical signs or critical values that require immediate action. Attention should be paid to maintaining body temperature and hydration status as well as other physiologic parameters. The use of specific analgesics, sedatives, and other drugs should be considered. Those drugs that would be helpful to the animal and not interfere with the experimental outcomes should be identified. Having a well-thoughtout plan for addressing potential postoperative complications is not only beneficial to enhancing better study outcomes, but is also required as part of the written animal protocol, as it can have a major impact on animal welfare.

The chronic care of the animals following injury must also be considered carefully. There may be housing or husbandry accommodations that could be made to enhance the animals' well being without compromising experimental goals. Again, the veterinarian and animal care staff can be valuable resources for identifying potential enhancements that might apply to a specific model and species. Some examples might include providing softer and/or deeper bedding, offering moistened chow in a floor feeder, or making water more easily accessible from the cage floor. If appetite suppression is anticipated, then highly palatable diets could be provided to help prevent weight loss. Increased or decreased amounts of environmental enrichment could be provided. Decreased exposure to loud sounds or bright lights might help in some cases.[4] It is usually a good idea to initiate special diets or environmental changes prior to creating the injury model in an animal. Most animals benefit from the opportunity to acclimate to these changes prior to injury. Prior acclimation usually speeds up acceptance of the changes and maximizes their benefit to the animals.[6] Providing such special care and describing it in the animal protocol indicate the importance of maximizing animal welfare while conducting the study.

Animal monitoring protocols must be developed to provide chronic monitoring of experimental and health issues in the animals. These could be a combination of clinical observations, physical measurements, and behavioral measurements. The parameters measured should be specific to the species of animal used and the expected experimental outcomes. Some examples of parameters to measure might include body temperature, body weight, body condition score, respiratory rate, heart rate, normal/abnormal posturing, normal/abnormal gait, home cage activity, open field activity, coat/skin condition, skin turgor, food consumption, urine output, fecal output, approach response, touch response, sound response, etc. Specific monitoring forms may be developed for the study to include the most applicable parameters for assessing the experimental goals and animal health. Established behavioral species-specific screens may be used and/or modified appropriately for the study. Monitoring should begin prior to injury induction. Baseline measurements should be established in all parameters that will be evaluated in the post-injury period. Since many of the observations could be subjective, identifying and training the personnel involved in the animal monitoring prior to beginning the study is important. Research, animal care, and veterinary staff can all be involved in this phase of the study. It is valuable to standardize observations as much as possible between observers through group discussions and training sessions prior to beginning experimental work.[4] Monitoring frequency should be planned to increase, if needed, during periods of expected increase in severity of clinical symptoms from the injury model being studied. Thorough record keeping is also important to document animal monitoring. Finally, potential complications should be considered and interventions planned that would be in the best interest of both the experiment and the animal. Contact information for the vital research staff and investigator should be available to facilitate communication when critical decisions about animal welfare and experimental needs must be made. The plan for health and experimental monitoring should be described in detail in the written animal protocol. Inclusion of a sample monitoring form is often helpful.

Development of experimental end points is inherent in study design. These end points must also take into account the animal's welfare as much as possible. The earliest possible experimental end points should be chosen to limit animal pain and distress. Humane end points also must be defined clearly in the animal protocol. These are end points that are developed and used to reduce the severity and duration of pain and/or distress.[7] Often, these are defined using clinical, behavioral, and physiological signs that are believed to indicate a level of animal pain or distress that is unacceptable and should be relieved by euthanasia. A study that does not have humane end points and requires that death of the animals is the final data point must provide stringent scientific justification for this necessity.[8] To be convincing, the justification should include data from a pilot study indicating that no identifiable clinical sign – or group of signs – could be found that were irrevocably linked to death.[9] If such a study has not been done, then the animal protocol would do well to include provisions to perform this pilot study to determine if humane end points can be identified using the monitoring form already developed for the study.

The AWRs and The Guide state that the method of euthanasia must be consistent with the current version of the Report of the AVMA Panel on Euthanasia unless a deviation is justified for scientific or medical reasons.[1,5] The method of euthanasia must also be chosen to provide the least interference with experimental data collection. The Report of the AVMA Panel on Euthanasia recommends anesthetizing rodents prior to euthanasia by physical means such as decapitation or cervical dislocation.[10] Thus, if use of anesthetics would interfere with experimental end points, euthanasia of rodents by physical means alone must be rigorously justified in the written animal protocol. If scientific justification is not available but interference is suspected, a pilot study should be designed and included in the animal protocol to determine whether or not anesthetics used in euthanasia interferes with experimental data. Training of personnel who will perform euthanasia is paramount – especially if euthanasia will be performed by physical means without anesthesia.[10] If the personnel involved require training, practice should be done first using cadavers and then deeply anesthetized animals until the personnel are completely comfortable with the technique. Veterinary staff should be able to provide training in means of euthanasia should it be required. Again, the experience of personnel with these techniques should be documented in the animal protocol.

In determining the experimental data points, the best paradigms for outcome assessment are chosen by the investigator as part of the study design. These could include quantitative or observational behavioral screening tests as discussed for animal monitoring. It could also include performance of behavioral tasks, blood or urine sample collection, physiologic measurements, tissue collection, etc. The data collection may require specialized equipment such as photocell arrays or implanted mini transmitters, or specialized environment such as metabolic caging or reverse light cycle housing. Data collection in some species may require a form of prolonged restraint. All these raise their own set of concerns for animal welfare that will need to be addressed in the study design period and in the written animal protocol.

In general, with any experimental manipulation that has the potential to cause more than momentary pain or distress to an animal, there is some necessary information that must be provided in the written animal protocol. Thus, it must be considered in designing the study. First, scientific justification for the need to use the manipulation must be provided. Second, a thorough and detailed description of what happens to the animal during the manipulation must be included. If an apparatus other than the animal's home cage is used, that apparatus should be described including dimensions of the primary enclosure. Unless scientific justification is provided for an exception, the surfaces of a primary enclosure should be smooth and sealed to allow sanitization and prevent injury.[5] Third, a description of monitoring the animal during the manipulation and how it will be documented should be provided. Fourth, the plan for stopping or intervening in the manipulation should the animal show evidence of unacceptable pain or distress should be described. The parameters that would lead to recognition of unacceptable pain or distress and the interventions that would be taken on the animal's behalf should be included. Finally, personnel performing the task must have sufficient experience with it or have received training. Developing this information during the design phase should allow the investigator to provide the IACUC with the information required to properly evaluate the animal welfare concerns involved in the animal protocol.

The IACUC is also tasked with evaluating hazards posed to personnel working with the animals or other animals in the institution as part of its review of animal protocols. There are some species that pose inherent zoonotic risks, such as working with nonhuman primates. All animals can pose some physical risks to personnel. If a study involves working with a species that will be new to the investigator, the animal care staff and veterinarian can provide valuable training in proper handling and treatment procedures to the study participants to help limit risks to both personnel and animals. It is best to seek this training as part of the study design phase. Experience with the species in question will need to be documented in the animal protocol. If a study involves a chemical, infectious, and/or radiation hazard, it must be identified in the animal protocol.[11] The institution usually has other committees that must also review the use of hazards in research and are required to approve such use before experimental work can begin. Once the hazards are identified in the study design process, application for their use should be made to the appropriate committees. Personnel participating in the animal work will probably need to be enrolled in the institution's Occupational Health program if this has not already been done. This will help prevent a delay in starting experimental work.

3 Writing the Animal Protocol

Each IACUC will have developed an application form to guide investigators in writing a protocol for animal use in research at that institution. All the applicable information discussed in the design phase will be requested on this form. In addition to this information, a few more areas of concern must be addressed.

There will be a requirement to provide evidence that the proposed work will not duplicate any previously performed studies and that alternatives are not available to the use of animals or painful procedures.[1,5] These justifications are most often required in the form of a recent and thorough literature search coupled with the assurance of the investigator based on his expertise in the field. (It is of note that, when performing a literature search for alternatives to animal use or use of painful procedures, having the word "alternative" as a keyword potentially makes the search more productive.)

Because IACUCs are required to have nonscientist members, a statement explaining the purpose, benefit, and brief overview of the work should be included using terminology that can be easily understood by the lay public. Brief but sufficient background information should also be provided for a scientist to understand the justification for the approach taken and the study design. It is easy for the investigator who is intimately involved with a study to assume that reviewers also have an in-depth understanding of the subject. Unfortunately, this is not always the case when a large institution is involved. It is very helpful for the investigator to take a step-by-step approach when explaining what will happen to the animals in a study and when each step will happen in the study timeline. Details such as dose, approximate volume, route of administration, and frequency of treatment should be provided for any substance given to the animals. Likewise, if body fluid samples are to be taken, the method of collection, volume collected, and frequency of collection should be specified. These details allow the IACUC members to determine if standard veterinary methods are being followed or if exceptions need to be considered.

The choice of appropriate species, strain, and genotype of animal for use in the study must be explained. Some strain choices might be dictated by ability to study genetically manipulated animals to further refine the experimental design. The number of animals needed for the work must also be explained. The US Government principles state that investigators should use the minimum number of animals required to obtain valid results.[3] The Guide states that "Whenever possible the number of animals requested should be justified statistically."[5] If statistical justification cannot be performed, a pilot study might be necessary to provide sufficient information to complete statistical calculations. Alternatively, documentation of published works using the same model could be provided and sample group size choice for the study based on group sizes that provided statistical significance in those studies might be considered sufficient justification for the requested animal numbers. Previously published works using the same model would at least provide a good sample-size estimate for a pilot study if such is required.

All animals that might be needed for successful completion of the study must be included in the total number of animals requested for the study. An estimate of the percentage amount of expected attrition among experimental animals should be available for an established model and must be included in the overall number of animals requested for the study. Attrition would include animals that must be euthanized or removed from the study prior to final data collection owing to humane concerns or incorrect model development within those animals. Beyond experimental animal needs, there might be the need to include training animals in the study if another animal protocol is not available for this purpose. An investigator or veterinarian experienced in the procedures being developed should be able to provide a justifiable number of animals needed to become competent in the techniques involved. Finally, if a breeding colony must be maintained to produce the experimental animals – such as genetically manipulated rodents – those animals that are born as part of the colony but not used experimentally will also need to be counted.

Writing an animal protocol for a study of acute neurological injury can be challenging. However, involving veterinary staff, animal care staff, IACUC staff, and other appropriate institutional committees, if necessary, in the planning process, should help to make the process more efficient. Addressing all the key issues and providing necessary detail in describing everything done with the animals will also help the IACUC in its review process. It may be necessary to perform one or more pilot studies as part of initiating a new model at an institution. These could be described in a single animal protocol or written as separate protocols to facilitate work on gathering the pilot information. With everyone working together, the goal of meeting the highest scientific standards and the highest animal welfare standards should be achieved.

References

1. CFR (Code of Federal Regulations). 1985. Title 9: Animals and Animal Products, Chapter 1, Subchapter A: Animal Welfare. Washington, DC: Office of Federal Register
2. PHS (Public Health Service). 1996. Public Health Service Policy on Humane Care and Use of Laboratory Animals. Washington, DC: U.S. Department of Health and Human Services, 28 pp. [PL 99–158, Health Research Extension Act, 1985]
3. IRAC (Interagency Research Animal Committee). 1985 US Government Principles for the Utilization and Care of Vertebrate Animals Used in Testing, Research, and Training. Washington, DC: Office of Science and Technology Policy
4. NRC (National Research Council). 2003. Guidelines for the Care and Use of Mammals in Neuroscience and Behavioral Research. Washington, DC: National Academies Press
5. NRC (National Research Council). 1996. The Guide for the Care and Use of Laboratory Animals. Washington, DC: National Academies Press
6. NIH (National Institutes of Health). 2002. Methods and Welfare Considerations in Behavioral Research with Animals: Report of a National Institutes of Health Workshop. Morrison, A.R., Evans, H.L., Ator, N.A., Nakamura, R.K. (eds). NIH Publication No. 02–5083. Washington, DC: US Government Printing Office
7. Stokes, W.S. Reducing unrelieved pain and distress in laboratory animals using humane endpoints. ILAR J 2000; 41(2):59–61
8. ARENA-OLAW (Applied Research Ethics National Association – Office of Laboratory Animal Welfare). 2002. Institutional Animal Care and Use Committee Guidebook (2nd ed). Washington, DC: US Government Printing Press
9. Morton, D.B. A systematic approach for establishing humane endpoints. ILAR J 2000; 41(2):80–86
10. AVMA (American Veterinary Medical Association). 2000 Report of the AVMA panel on euthanasia. J Am Vet Med Assoc. 2001; 218(5):669–696
11. NRC (National Research Council). 1997. Occupational Health and Safety in the Care and Use of Research Animals. Washington, DC: National Academies Press

Chapter 2

Anesthesia and Analgesia for Research Animals

Andrew R. Blickman and Lisa J. Brossia

Abstract This chapter provides a general overview of anesthesia and analgesia for animals used in acute neurologic surgical models. Injectable and inhalant anesthetic agents are discussed, as well as analgesia to control postoperative pain. Considerations in choosing the appropriate agent, anesthetic adjuncts, and suggestions for monitoring are also covered in this chapter. Common agents and doses are listed by species in Tables 1 and 2.

Keywords: Animal models, Anesthesia, Analgesia, Monitoring

1 Introduction

The provision of proper anesthesia and analgesia to research animals is a legal and moral obligation and ensures animal welfare. The optimal anesthetic and analgesic will vary depending on the surgical or experimental procedure, the species and strain/breed of animal. Some anesthetics and analgesics may alter experimental outcomes.

2 Anesthetic Considerations

Anesthesia is neuroprotective because it reduces the central nervous system (CNS) metabolic rate and oxygen requirement.[1] Cerebral blood flow (CBF) is affected by arterial O_2 and CO_2, mean arterial pressure, and venous flow. Anesthesia affects all those parameters. Increased CBF causes increased intracranial pressure (ICP), which may have adverse consequences.

3 Injectable Anesthetics

3.1 Barbiturates

Barbiturates include pentobarbital, thiopental, and thiobarbital. They inhibit the release of acetylcholine, norepinephrine, and glutamate, and alter γ-aminobutyric acid (GABA). At high doses, barbiturates inhibit the uptake

From: *Animal Models of Acute Neurological Injuries.*
Edited by: J. Chen (eds.) New York, Totowa, NJ

Table 1 Common anesthetic drug protocols for induction and maintenance[5].

Species	Drug (dDosage) and route
Mouse	• Xylazine (5–16 mg/kg) + ketamine (100–200 mg/kg) given IP • Acepromazine (2–5 mg/kg) + ketamine (100 mg/kg) given IP or IM • Pentobarbital (50–90 mg/kg) IP or IV injection • Halothane or Isoflurane inhaled, to effect • Avertin (125–250 mg/kg)-for single survival anesthesia only; 1.2% solution; made fresh with sterile saline, given IP • Acepromazine + butorphanol (sedation only) 50:50 mixture of 10 mg/ml drugs dosed at 0.005–0.01 ml/g of body weight, given IM or SQ
Rat	• Xylazine (5–13 mg/kg) + ketamine (40–90 mg/kg) given IP • Acepromazine (2.5 mg/kg) + ketamine (75–80 mg/kg) given IP or IM • Pentobarbital (40–60 mg/kg) given IP or IV • Halothane or Isoflurane inhaled, to effect • Acepromazine + butorphanol (sedation only) 50:50 mixture of 10 mg/ml drugs dosed at 0.005–0.01 ml/g of body weight, given IM or SQ
Rabbit	• Xylazine (3 mg/kg) + ketamine (35 mg/kg) given IM • Halothane or Isoflurane inhaled, to effect • Acepromazine + butorphanol: (sedation only) 50:50 mixture of 10 mg/ml drugs dosed at 1 mg/kg; given IM or SQ
Cat	• Atropine 0.01–0.02 mg/kg IM or SC • Acepromazine (0.025–0.1 mg/kg) + butorphanol (0.2–0.4 mg/kg) IM as preemptive sedative/analgesic • Induce with Isoflurane 3–5% then maintain through ET tube at 1.5–3%. • Other induction agents after preemptive sedation: Thiopental (2–10 mg/kg IV); Propofol (1–3 mg/kg IV)
Dog	• Acepromazine (0.1–0.5 mg/kg) given IM + atropine (0.04 mg/kg) given IM or IV as a pre-anesthetic followed by thiopental (22 mg/kg) given IV for induction + 1–3% Isoflurane by inhalation through ET tube to effect for maintenance • Pentobarbital (20–30 mg/kg) given IV
Swine	• TKX (500 mgTelazol + 250 mgketamine + 250 mgxylazine; 1 cc per 50 lbs; IM) + thiopental (10–20 mg/kg) if needed for intubation + 1–3% isoflurane by inhalation through ET tube, to effect • Ketamine–xylazine (1 g bottle ketamine + 142 mgxylazine of the 100 mg/ml concentration of xylazine, 0.2 cc per kg, IM split into two site if > 5 ml total dose)
Macaque	• Ketamine (5–20 mg/kg; IM) for induction + thiopental (5–7 mg/kg; IV) if needed for intubation intubation + 1–3% Isoflurane, inhaled to effect
All species	• Local anesthetic total maximum safe dose: lidocaine 4 mg/kg; bupivacaine 1–2 mg/kg

IM intramuscular; *IP* intraperitoneal; *SC* subcutaneous; *IV* intravenous; *ET* endotracheal; *PO* per os (orally)

of calcium at nerve endings. They have no intrinsic analgesic properties. Barbiturates are highly lipid-soluble, and patient fat content alters their distribution in tissues. They enter the CNS quickly but then are redistributed to other tissues. Barbiturates are CNS depressants but may cause an excitatory reaction in some species, such as cats. The long half-life of barbiturates may result in a prolonged anesthetic recovery period. Residual barbiturates cause CNS depression, which may obscure postoperative evaluations. In most species (especially cats), barbiturates cause respiratory depression and reduce intestinal tract motility.[2]

Thiopental is an ultra-short-acting barbiturate, with onset of effect within 1 min of IV administration. It enters the CNS very rapidly, but it is quickly

Table 2 Suggested analgesic doses[5].

Species	Drug	Dose and route of administration
Mouse	Aspirin	120 mg/kg PO
	Buprenorphine	0.05–0.10 mg/kg SQ every 8–12 h
	Morphine	2–5 mg/kg SQ every 4 h
Mouse or Rat	Butorphanol	1–2 mg/kg SQ or IM every 4 h
	Carprofen	5 mg/kg PO of SC daily
	Acetaminophen	200 mg/kg PO daily
Rat	Aspirin	100 mg/kg PO
	Buprenorphine	0.01–0.05 mg/kg SQ every 8–12 h
	Morphine	2 mg/kg IM or SQ every 4h
	Meloxicam	1.0 mg/kg PO or SC daily
	Ketoprofen	5 mg/kg SC or PO daily
Rabbit	Aspirin	100 mg/kg PO
	Buprenorphine	0.01–0.05 mg/kg SC or IM every 6–12 h
	Carprofen	4 mg/kg SC daily
	Meloxicam	0.2 mg/kg SC daily up to 3 days
	Ketoprofen	3 mg/kg SC daily
Cat	Buprenorphine	0.005–0.02 mg/kg IM or SC every 8–12 h
	Butorphanol	0.2–0.5 mg/kg IM or SC every 4 h
	Meloxicam	0.1 mg/kg SC or PO for 3 days then 0.1 mg total dose PI daily every 1–3 days
Dog	Aspirin	10–25 mg/kg PO every 8–12 h
	Buprenorphine	0.005–0.02 mg/kg SQ, IM or IV every 8–12 h
	Carprofen	4.4 mg/kg SC or PO single dose or divided twice daily; 4.4 mg/kg SC or PO once daily (may be well-tolerated long term)
	Ketoprofen	2 mg/kg SC daily for up to 3 days; 1 mg/kg daily orally up to 5 days
	Meloxicam	0.2 mg/kg SC or PO single dose; 0.2 mg/kg SC or PO on day 1, then 0.1 mg/kg SC or PO daily (may be well-tolerated long term)
Swine	Aspirin	10 mg/kg PO daily
	Buprenorphine	0.005–0.05 mg/kg IM every 8–12 h
	Carprofen	2–4 mg/kg SC daily
	Flunixin	1–2 mg/kg SC daily
	Ketoprofen	3 mg/kg IM once
Monkey	Aspirin	20 mg/kg PO every 6–8 h
	Buprenorphine	0.005–0.01 mg/kg SQ or IM every 6–12 h
	Carprofen	3–4 mg/kg SC daily for up to 3 days
	Ketoprofen	2 mg/kg SC daily

IM intramuscular; *IP* intraperitoneal; *SC* subcutaneous; *IV* intravenous; *PO* per os (orally)

redistributed to the muscle and adipose tissue, resulting in a short duration of action. Thiopental is frequently used to induce anesthesia in large animals (cats, dogs, primates, and swine). It is rapidly metabolized by the liver. It must be given intravenously, preferably via an intravenous catheter, as extravasation results in tissue necrosis. Inactin (a thiobarbiturate) is occasionally used in acute rodent procedures. Pentobarbital is considered a short-acting barbiturate, but it is less lipophilic and has a longer duration of action than thiopental (up to 3–4 h). Pentobarbital has cardiovascular effects, such as tachycardia, decreased myocardial contractility, and stroke volume, and may cause hypotension in dogs.[2,3]

3.2 Ketamine

Ketamine is a dissociative anesthetic that activates the limbic system and suppresses the thalamoneocortical system. It increases sympathetic tone, thereby increasing cardiac output, heart rate, and other cardiovascular parameters. It may cause pain and necrosis of muscle cells upon intramuscular injection.[4] It increases CBF, oxygen requirement, ICP, and EEG activity. It may cause convulsions and seizures. Ketamine has some analgesic properties.[3] It is commonly used alone as an immobilizing agent for nonhuman primate handling. Ketamine causes hypersalivation in primates, and therefore an anticholinergic should be given 30 min prior to ketamine. Ketamine is routinely administered with other drugs, including xylazine (for rodent anesthesia), valium (for dogs and cats), and Telazol with zolazepam (as an induction agent for swine anesthesia).[5]

3.3 Propofol

Propofol is a short-acting hypnotic used as an induction agent or for maintenance via constant rate infusion in large animals. Recovery is smooth.[3]

3.4 Local Anesthetics

Local anesthetics block innervation to specific regions when infused at the site of pain or at the nerve root. They can also be administered epidurally. Local anesthetics include lidocaine and bupivacaine. Of these, bupivacaine has the longest duration of action. It may be administered as a local infiltration of the incision site or as a nerve block. The reader is referred to Hall et al.[6] for more information on nerve blocks. Lidocaine can be mixed with sodium bicarbonate to reduce the pain associated with injection in conscious patients. Local anesthetic doses should be carefully calculated on the basis of body weight.

4 Gaseous Agents

Inhalant agents produce a generalized reversible depression of the CNS via unknown mechanisms. The solubility of the agent determines the rate of induction and recovery; those with low blood/gas partition coefficients have the most rapid inductions and recoveries. Inhalants produce respiratory and cardiovascular depression, both of which may be offset by sympathetic stimulation during surgery.[1] If the surgical site is the neck or skull, inhalant anesthetics may require a special apparatus to maintain a patent airway and to allow access to the incision site. A stereotaxic apparatus may have an integrated nose cone for anesthetic gas delivery built onto the incisor bar (Fig. 1). Waste gas management must be addressed to reduce occupational exposure.

4.1 Isoflurane

Isoflurane is one of the most commonly used inhalant anesthetics in veterinary medicine. Its high volatility and low solubility allow rapid induction and recovery, as well as rapid responses to changes in isoflurane concentration during maintenance anesthesia.[6] Isoflurane produces a dose-dependent depression of cardiovascular parameters, but less so than halothane. Unlike halothane,

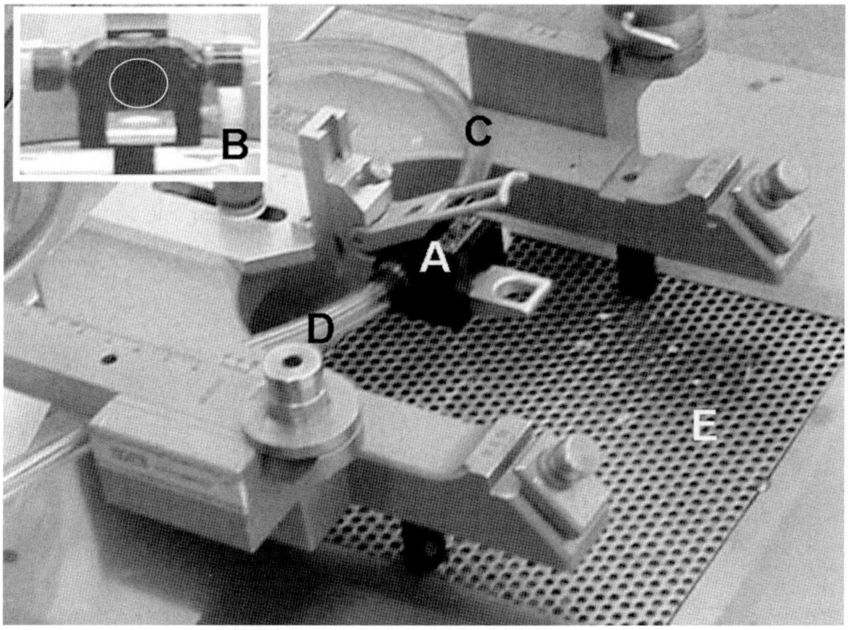

Fig. 1 A modified stereotaxic apparatus with anesthesia mask. (**A**) A plastic nose cone is mounted proximal to the incisor bar. (**B**) Inset: front view close-up of (**A**), showing where the nose is inserted for gas delivery. (**C**) Anesthetic gas is supplied via polyethylene tubing from the right and (**D**) waste gas is scavenged to the left. (**E**) A down-draft surface collects all leaked gases to protect the surgeon from occupational waste gas exposure. Photo courtesy of Dr. Eric Engleman, Department of Psychiatry Indiana University School of Medicine

isoflurane reduces brain oxygen requirement by reducing the CNS metabolic rate. Isoflurane suppresses burst activity on electroencephalograms, interfering with diagnostic testing.[1]

4.2 Halothane

Halothane produces a rapid induction and recovery and has high potency, and there are few adverse effects. It causes vasodilation and increases CBF by 200%, which may increase ICP.[1] It also causes a dose-dependent myocardial depression. Halothane predisposes to catecholamine-induced arrhythmias.[6] Halothane can alter research outcomes (affect cortical morphology)[7] and may cause hepatotoxicity in rats (where the P450 enzyme has been preinduced).[6]

4.3 Other Inhalants

Nitrous oxide is an adjunct anesthetic. Veterinary applications are limited. It has strong analgesic properties. It increases CBF and oxygen requirement.[1] Desflurane has a very low solubility in blood, resulting in very quick induction and recovery. It requires a special temperature-regulated vaporizer. Sevoflurane has similar characteristics as isoflurane (rapid induction and recovery, less cardiovascular depression than halothane).[6]

5 Hypothermia

Hypothermia has been used as anesthesia for neonatal rats. It can be induced by submersion of the rat in a protective latex sleeve into ice water. It can be an effective immobilization agent and may provide some analgesia, but the process of cooling is considered painful.[8] Also, it may alter brain morphology and behavior.[9] Alternative effective anesthetics exist and should be used whenever possible for animal welfare reasons.

6 Analgesia

It is clearly stated in applicable regulations and guidelines that animals must not experience pain unless it is scientifically justified and approved in the institutional animal care and use committee (IACUC) protocol.[10] The rule of thumb is that if a procedure is considered painful in a human, then it is painful in an animal.[11] Justification for withholding an analgesic should include objective data from a pilot study, published adverse effects of analgesia on experimental data, plans for monitoring for excessive pain, and criteria for withdrawal from study or euthanasia when pain is excessive.

Pain and analgesics may alter interpretation of experimental data. By withholding analgesics, the physiologic and behavioral effects of pain will need to be appropriately interpreted in the context of the experiment. And, just as pain can affect the experiment, analgesics can also interfere with experimental outcomes.[12]

When considering analgesic regimens, it is important to consider the intensity of tissue manipulation during surgery.[13] The surgical site is also important, as cervical spine surgery is more painful than thoracic or lumbar spinal surgery.[1] Opioid analgesics are considered optimal for moderate to severe pain,[3] whereas nonsteroidal anti-inflammatory drugs (NSAIDs) may be more appropriate for mild to moderate pain. In many cases, a combination approach (utilizing both NSAIDs and opioids) is most efficacious and should be considered.

6.1 Opioids

Opiate receptors are found in the brain, gastrointestinal tract, and other smooth muscle. Opioids are categorized according to what receptor they target as well as what their action is at that receptor. Opiate receptors include mu (found in pain-regulating areas in the brain), kappa (deep cerebral cortex and spine), and delta (in the limbic areas).[2] Opiates are agonists, partial agonists, or antagonists at these receptors.

Morphine is the standard opiate to which all others are compared. The primary effects of morphine and other mu-receptor agonists are analgesia, cough suppression, respiratory depression, sedation, emesis, and dependence. Secondarily, they may cause euphoria, bradycardia, and hypotension. Morphine is used for short-term analgesia, sedation, and as an anesthetic adjunct. It can be given epidurally. Some species may react differently to morphine. Dogs and humans develop miosis, while cats develop mydriasis.[2] Cats and swine may develop excitement after morphine administration, whereas dogs may become depressed.[3]

Buprenorphine is 25–50 times more potent than morphine. It is a partial mu-receptor agonist and kappa-receptor antagonist. It has a high affinity for the mu-receptors of the CNS,[3] allowing for a relatively long duration of action (8–12 h in rodents).[5] It decreases blood pressure and heart rate. It may cause less respiratory depression than other opiates.

Butorphanol is 4–7 times more potent than morphine. It is a kappa-receptor agonist and a weak mu-receptor antagonist, causing more sedation and less analgesia than mu-receptor agonists.[3] Its analgesic effect is short lived, and lasts about 1 h in dogs and 4 h in cats.[5]

Fentanyl may be conveniently administered via transdermal patch, and can be an effective method of pain control.[3] For example, a 20-kg pig would need a 50 μg h^{-1} patch, placed 12–24 h before surgery, and typical duration is 72 h.[3]

6.2 Nonsteroidal Anti-Inflammatory Drugs

NSAIDs reduce inflammation by inhibiting the synthesis of prostaglandins by binding directly to cyclo-oxygenase (COX). The two isoforms of COX (COX-1 and COX-2) play different roles. COX-1 not only produces prostaglandins but is also important in gastrointestinal and renal health, and platelet function. COX-2 is more specific for the production of inflammatory prostaglandins that cause fever and pain.[3]

Aspirin, the prototype anti-inflammatory, is not COX selective, although it may bind COX-1 selectively at lower doses. Aspirin is also an antiplatelet drug and prevents thromboemboli formation.[2] Carprofen is relatively COX-1 sparing compared to aspirin. Carprofen is primarily prescribed to treat musculoskeletal pain and the acute pain of surgery.[3] Its duration of action is up to 24 h. Ketoprofen inhibits both COX isoforms, and it is effective in treating moderate pain and inflammation. It has a long duration of action, which may be up to 24 h.[14]

Newer NSAIDs are COX-2 selective or more COX-1 sparing than older NSAIDs. Meloxicam is an example of a newer NSAID. Meloxicam has less COX-1 activity; therefore it can be used chronically with fewer adverse effects on the gastrointestinal tract and kidneys.[3]

Acetaminophen is an analgesic and antipyretic with few anti-inflammatory effects. It is a relatively weak analgesic. It cannot be used in cats.[2]

7 Monitoring

Monitoring should start during the preoperative period. Baseline physiologic parameters should be obtained prior to induction of anesthesia. Parameters may include body weight and temperature, heart rate, and blood pressure, as well as respiratory rate and quality. If a pain scale will be used to detect postoperative pain, a preoperative baseline evaluation may be useful.

Intraoperative monitoring parameters may include temperature, heart rate, blood pressure, electrocardiogram (ECG), blood gas analysis, and biphasal EEG activity. As the animal emerges from anesthesia, temperature, heart rate, respiratory rate, and behaviors that may indicate pain should be recorded. Adverse events should be identified and addressed quickly and recorded. Chronic postoperative records may include pain scale scores, body weight, food and water consumption, incision integrity, and other observations related to the procedure performed.

References

1. Harvey RC, Paddleford RR. Anesthesia for the Central Nervous Systems and Ophthalmic Surgery. In: Slatter DS, ed. Textbook of Small Animal Surgery, 2nd edn., vol. II. Philadelphia: W.B. Saunders, 1993: 2271–2278
2. Plumb DC. Veterinary Drug Handbook. 3rd edn. Ames, IO: Blackwell Publishing, 1999
3. Papich MG. Saunders Handbook of Veterinary Drugs. 2nd edn. St Louis, MO: W.B. Saunders, 2007
4. Smiler KL, Stein S, Hrapkiewicz KL, Hiben JR. Tissue response to intramuscular and intraperitoneal injections of ketamine and xylazine in rats. Lab Anim Sci 1990; 40(1):60–64
5. Flecknell P, Waterman-Pearson A. Pain Management in Animals. Edinburgh, UK: W.B. Saunders, 2001
6. Hall LW, Clarke KW, Trim CM. Veterinary Anesthesia, 10th edn. London, UK: W.B. Saunders; 2001
7. Nunez JL, Juraska JM. Neonatal halothane anesthesia affects cortical morphology. Brain Res. Dev Brain Res 2000; 124(1–2):121–124
8. Clowry GJ, Flecknell PA. The successful use of fentanyl/fluanisone ("Hypnorm") as an anaesthetic for intracranial surgery in neonatal rats. Lab Anim 2000; 34:260–264
9. Nunez JL, Koss WA, Juraska JM. Hippocampal anatomy and water maze performance are affected by neonatal cryanesthesia in rats of both sexes. Horm Behav 2000; 37(3):169–178
10. Animal Welfare Regulations, 9 C.F.R. Sect. 2.31 (2005)
11. U.S. Government Principles for the Utilization and Care of Vertebrate Animals Used in Testing, Research, and Training. Interagency Research Animal Committee (1985)
12. Ferland C, Veilleux-Lemieux D, Vachon P. Effects of buprenorphine on intracerebral collagenase-induced hematoma in Sprague-Dawley rats. J Am Assoc Lab Anim Sci 2007; 46(3):13–16
13. Liles JH, Flecknell PA. The effects of surgical stimulus on the rat and the influence of analgesic treatment. Brit Vet J 1993; 149(6):515–525
14. Roughan JV, Flecknell PA. Behaviour-based assessment of the duration of laparotomy-induced abdominal pain and the analgesic effects of carprofen and burprenorphine in rats. Behav Pharmacol 2004; 15(7):461–472

Chapter 3

Rodent Surgical Procedures and Tissue Collection

Robert Ayer, Julian Cahill, Takashi Sugawara, Vikram Jadhav, and John H. Zhang

Abstract The use of animal models for neurological investigation often requires a specific skill set that allows the investigator to adequately prepare the animal for experimental surgery, correctly collect their brain samples from the specimen, and prepare them for subsequent analysis. This chapter will provide detailed explanations for methods of rodent intubation, brain perfusion and removal, as well as the separation of the brain into anatomical sections and triphenyltetrazolium chloride (TCC) staining. Knowledge of these procedures is often necessary regardless of the specific disease processes you are modeling in the rodent: subarachnoid hemorrhage, middle cerebral artery occlusion, intracerebral hemorrhage, etc., and will therefore be useful to anyone working with the rodent models regardless of their specific area of investigation.

Keywords: Rodent model, Ventilation, Trachestomy, Rodent model, Intubation, Brain, Perfusion, Fixation, Brain dissection, Triphenyltetrazolium chloride (TCC) staining

1 Airway Management in the Rodent Model

For many years the intubation of the rodent has been problematic, largely because of the small size of the subject and the lack of specialized equipment available. However, as research advanced and more complicated operations were being attempted in the rodent model, secure intubation and ventilation has become more and more important. Early attempts required expensive instruments to achieve intubation, such as laryngoscopes, fiber optic light sources, and microscopes. A myriad of methods have been described over the years including blind intubation, intubation under direct vision, tracheostomy, and transillumination techniques. In this section a tracheostomy and the intubation of a rat will be described.

The main difficulty with intubating the rat or, indeed, a mouse has been its size. Larger animals such as the canine model or even rabbits can be successfully intubated using standard pediatric equipment.[1] Because of this difficulty, there are many described techniques for ventilating rodents, ranging from face

From: *Animal Models of Acute Neurological Injuries.*
Edited by: J. Chen (eds.) New York, Totowa, NJ

masks[2] to intubation. With regard to intubation there are two techniques to choose from: blind intubation[3] and intubation under direct vision. According to Stark et al., blind intubations require three or four attempts and results in success in 90% of cases. With regard to direct vision intubation, there are a number of methods to choose from, including transillumination of the neck,[4,5] microscopes,[6] laryngoscopes,[7] and various head or mirrored light source techniques.[8] Many of these methods are difficult to master, or require expensive specialized equipment.

By far the simplest method to intubate the rodent is using the Seldinger technique under direct vision with the use of an otoscope. With minimal practice, this technique can be performed quickly, cheaply, in one attempt, and with a success rate of close to 100%.[1] Many descriptions of this technique elaborate on the use of inclined plastic or metal plates in order to secure the animal; however, the method described in this chapter mounts the otoscope instead of the animal.[1,9] Major advantages of this method of intubation include the following:

(a) It provides a safe and secure airway
(b) It is easy and cheap to perform, without the need of expensive equipment
(c) It causes minimal damage to the upper respiratory passages of the animal
(d) Extubation is quick and easy
(e) It does not require extensive surgery and is well tolerated by the animal

A tracheostomy in the rat is relatively easy to perform, but can be difficult to maintain or close. Given the relative ease of intubation techniques now available, tracheostomy has become less popular. However, in some instances it is still indicated, such as in cardiothoracic research.[10,11] The longer a tracheostomy remains in place, the more difficult it is to maintain. Suturing the tracheostomy in place is often required to minimize the chances that the animal will dislodge it. Additionally, secretions can be difficult to combat and suctioning on a regular basis is recommended. The use of antimuscarinic drugs can minimize these secretions. Despite these precautions, it is still possible for animals to asphyxiate. Major advantages of a tracheostomy include the following:

(a) It is suitable for some types of specialist surgery, e.g., cardiothoracic surgery
(b) It works well as an alternative to intubation
(c) It is easy to perform and can be closed postoperatively
(d) It is reasonably well tolerated in the short term but may require suction

1.1 Materials

1. Induction chamber (Braintree Scientific, PO Box 850929, Braintree, MA 02185-0929, USA)
2. Otoscope and a 4-mm viewing cone (Welch Allyn, Inc. 4341 State Street Road, Skaneateles Falls, New York 13153, USA)
3. Castaloy three-prong extension clamp
4. Bougee: a metal guide wire (gauge 0.0889 cm)
5. An aluminum 15 G blunt needle over which the plastic catheter from a 14 G (Insyte) intravenous catheter has been cannulated
6. Small-animal volume controlled ventilator, model 683 (Harvard Apparatus, 84 October Hill Road, Holliston, Massachusetts 01746 USA)
7. Syringe (20 ml)

8. Hair clippers (with #40 blade) and skin disinfectant, e.g., iodophors (Betadine) or chlorhexidine (Novasan)
9. Sterile surgical drapes, gown, and gloves
10. Operating surface (nonporous, reusable, and cleanable), well-ventilated and well-lit working area
11. Operating microscope (Kent Scientific)
12. MV-9 electro surgical unit (Kent Scientific)
13. Standard operating surgical instruments and micro instruments if available. This should include a scalpel (disposable), fine forceps, vascular clamps, needle holder, retractors, and microscissors
14. Suction
15. Sterile sutures, 4/0 nylon monofilament suture, 10/0 silk ties and 3/0 silk suture
16. PE 250 tubing with an obliquely fashioned tip

1.2 Procedures

1.2.1 Intubation

1. Prior to intubation the animal has to be anesthetized. If a gas anesthetic is to be used, the rat is placed in a gas anesthetizing chamber (Braintree Scientific) and can be induced using 1-chloro-2,2,2-trifluoroethyl difluromethyl ether (Isoflurane) 4% with a mixture of 40% oxygen and 60% medical air. If an injection method is to be used, DecapiCones (Braintree Scientific) are ideal for restraining the animal for injection.
2. Once the animal is suitably anesthetized, the oral cavity is swabbed out. The animal's head is held in the left hand. Using an otoscope (Fig. 1), which is mounted on a Castaloy three-prong extension clamp, a 4-mm viewing cone is inserted into the oral cavity and the epiglottis can be visualized (Fig. 2). In order to do this, it is important to hold the tongue of the animal to one side. Using the tip of the viewing cone, push the epiglottis forward; this allows you to insert a guide wire (gauge 0.0889 cm) past the vocal cords and into the upper trachea (Fig. 2). This helps avoiding the obvious danger of placing the endotracheal tube in the esophagus. The view achieved using the otoscope can be seen in Fig. 3.
3. Each endotracheal tube is made using an aluminum 15 G blunt needle over which a plastic catheter from a 14 G (Insyte) intravenous catheter is placed and taped securely (Fig. 1). The endotracheal tube is then placed over the guide wire and the animal is intubated. The guide wire is then removed, and the animal is attached to a small-animal volume controlled ventilator (Harvard Apparatus, model 683).
4. Maintenance of the anesthetic can be achieved with 2–2.5% isoflurane and a mixture of 40% oxygen and 60% medical air, or a long-acting injection anesthetic. As with any intubation, the operator must make sure that the animal is properly intubated. Techniques include looking at the respiratory effort, checking for condensation in the tube, feeling air being exhaled from the tube, or measuring the carbon dioxide output using a capnograph.

1.2.2 Tracheostomy

1. A vertical incision should be made in the midline of the animal's neck extending from the thyroid cartilage to just above the sternum. The underlying subcutaneous tissue, salivary glands, and lymphoid tissue are all reflected

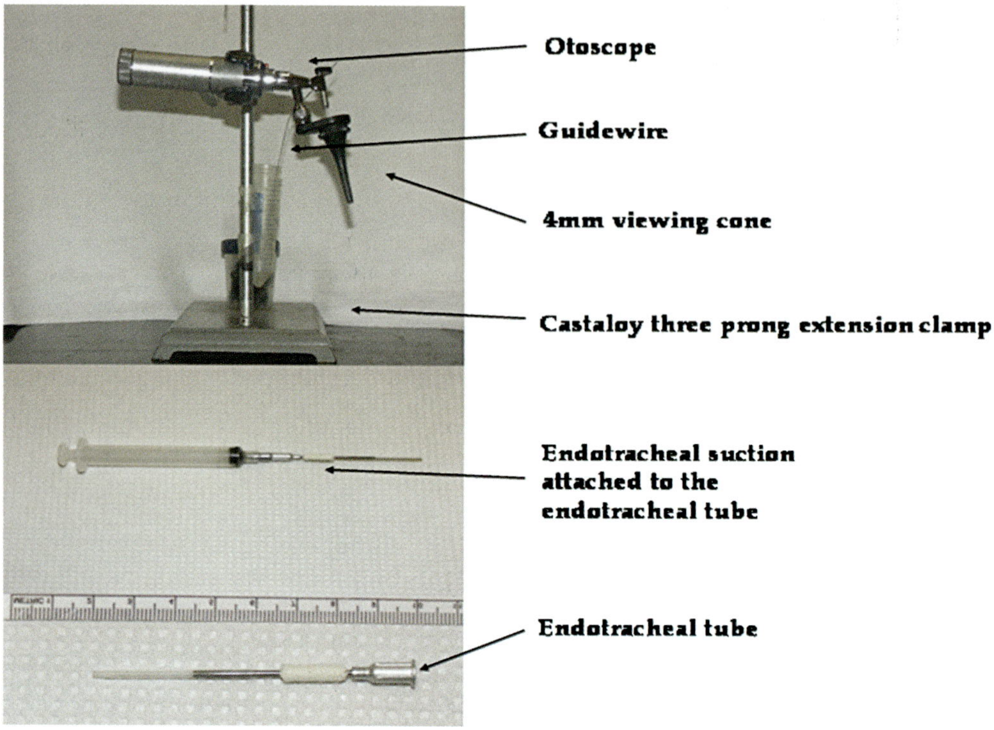

Fig. 1 *Rat intubation equipment*. A simple otoscope is used to intubate the rat using an endotracheal tube, made from an aluminum 15 G blunt needle over which a plastic catheter from a 14 G intravenous catheter is placed. After placing a tape around the endotracheal tube, it is then secured by suturing the tape to the animal's lip

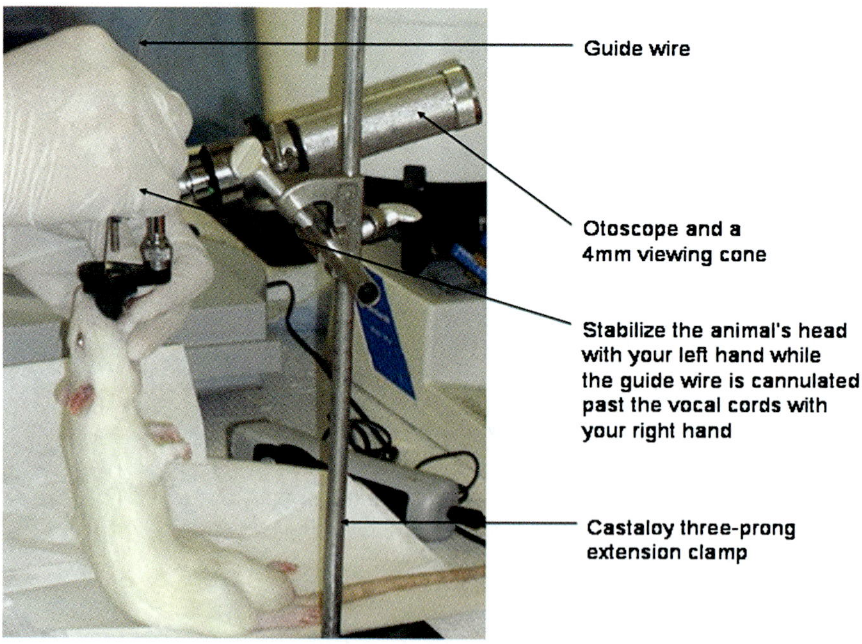

Fig. 2 *Intubating the rodent*. By placing the 4-mm viewing cone into the oral cavity of the rodent, the vocal cords can be visualized. The head of the rodent is stabilized with the left hand, while the guide wire is passed into the trachea with the right hand

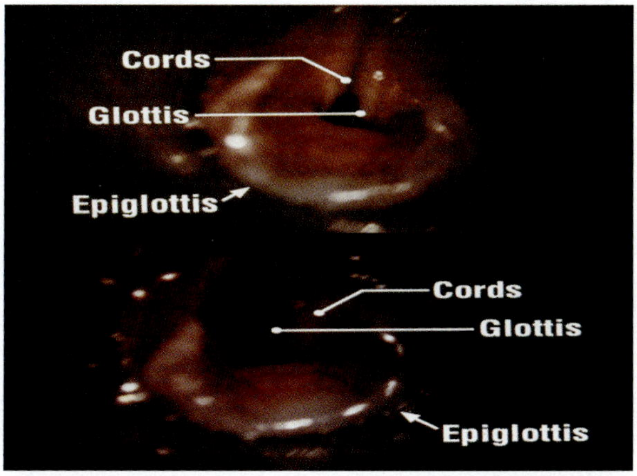

Fig. 3 *View of the rat epiglottis*. The epiglottis of the rat is visualized using an otoscope, which allows the insertion of a guide wire over which an endotracheal tube is inserted. The upper photograph shows the glottis in a closed position, while the lower photograph demonstrates the open position

laterally. The trachea, covered by the sternohyoid muscle, should now be visible in the midline.

2. Using blunt dissection, the sternohyoid muscle is split in the midline and reflected to either side of the trachea. By using blunt dissection there should be little or no bleeding. The thyroid and cricoid cartilages should now be visible superiorly, and the tracheal rings are seen extending inferiorly into the sternum (Fig. 4a).
3. The second or third tracheal ring is chosen for the insertion of the tracheostomy tube. Between each tracheal ring lies the annular ligament (ligamenta anularia tracheae), which carries small branches of the inferior thyroid artery. These small arteries should be diathermied prior to making an incision. A small incision is then made through the annular ligament and any bleeding is diathermied before continuing (Fig. 4b). At this point it is often useful to suction the trachea prior to advancing the polyethylene tubing.
4. Using the obliquely cut tip of the PE250 tubing, the trachea is cannulated and the tubing should be rotated as it is advanced, which allows the tubing to be inserted easily (Fig. 4c). Make sure that the trachea is supported using a broad forceps to prevent tearing the tissue as the tube is inserted. Once inserted, the tube should be suctioned.
5. Once the tube is in place, the skin is suture-closed around the tube using a 3/0 silk suture. The tracheal tube should also be tied into the wound closure and cut to fit the animal's neck. Alternatively, if the tracheostomy is being used only for the duration of the surgery, the stoma can be closed using a 10/0 monofilament suture after surgery (Fig. 4d). The wound is then closed as before.

1.3 Applications, Complications, and Limitations

1. It is important to ensure that all the equipment to be used in this procedure is available and working before beginning. Once the animal is anesthetized, you will not have time to go looking for equipment.

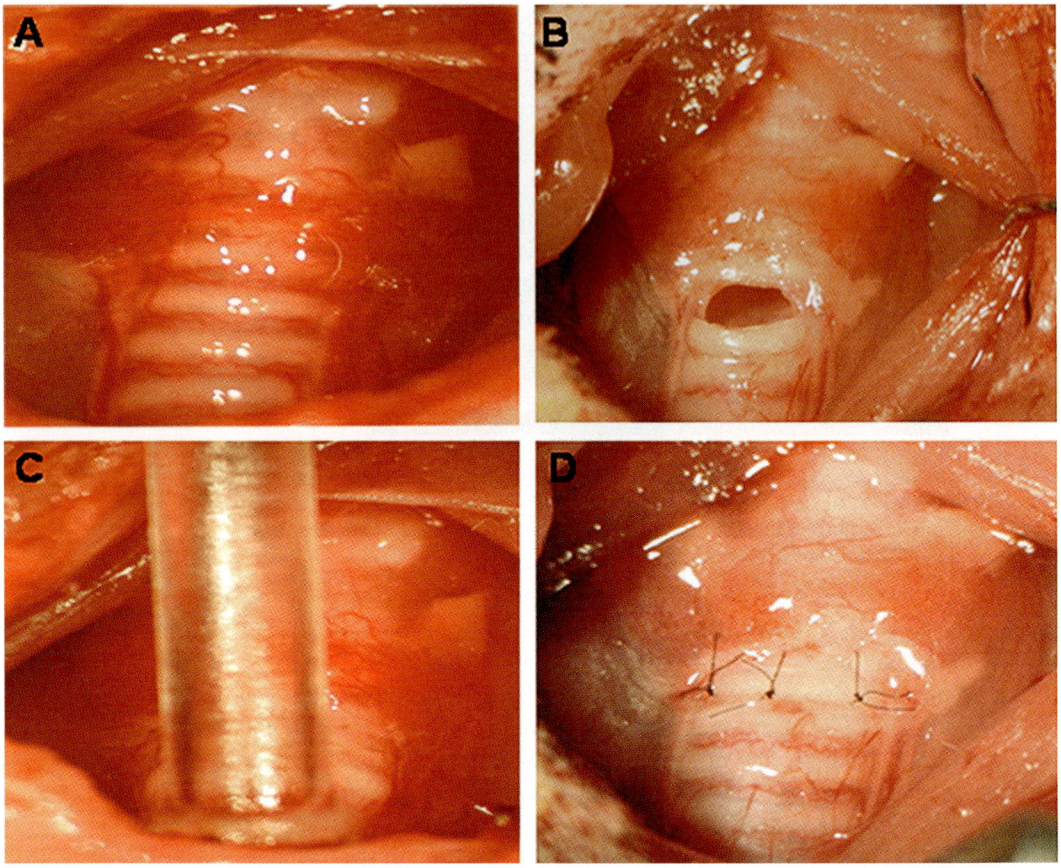

Fig. 4 Insertion of the tracheostomy tube. (**A**) After splitting the sternohyoid muscle, the thyroid and cricoid cartilages come into view. The tracheal rings can be seen inferiorly. (**B**) It is important to ensure that the edges of the tracheostomy are dry prior to inserting the tube (**C**). (**D**) The tracheostomy can be closed using a 10/0 monofilament suture

2. A plastic bag with a corner cut off is a cheap alternative for restraining animals as compared to DecapiCones (Braintree Scientific), and works just as well.
3. An alternative to the otoscope is the top of a 3-ml plastic syringe, which is cut obliquely to create a wedge.[12,13] This method lacks a light source and therefore can be more difficult.
4. The use of the aluminum blunt needle simply gives the endotracheal tube more stability; however, the use of an Insyte catheter on its own has been described.[1]
5. It is a good idea to suture the endotracheal tube in place, using a 3/0 silk suture. The suture can be placed in the tape around the endotracheal tube (Fig. 1) and secured to the animal's lip.
6. The settings for the ventilation of any animal depend largely on the animal being used and size of the animal. For a Sprague-Dawley rat between 280 and 320 g, tidal volumes of approximately 2.5–3 ml stroke^{-1} is suitable and the respiratory rate is kept between 45–55 breaths per minute, with a mean respiratory pressure of between 9 and 12 mmHg. The response of the animal

to the settings chosen on the ventilator and continuous blood gas analysis will allow you to maintain ventilation for long periods.

7. It is important to diathermy the small branches of the inferior thyroid artery within the annular ligament prior to making your incision. This is to prevent a surprising degree of bleeding from these small vessels, blood from which can be easily aspirated by the animal.
8. In our experience, for an SD rat with weights of 280 and 320 g, PE250 tubing is the most suitable. Smaller bore tubing leads to the escape of air around the tubing and larger bore tubing leads to an unacceptable degree of trauma while inserting the tubing into the trachea. If larger or smaller animals are being used, the size of the tubing must be changed accordingly.

2 Transcardiac Brain Perfusion and Fixation

Perfusion and fixation of the brain is often required for the preparation of samples before testing. Fixation is necessary for maintaining the structural integrity of the brain, stabilizing proteins, and allowing antibody binding. There are a number of described perfusion methods; below is a technical description of perfusing and fixing the brain via the vascular system from a transcardiac approach. The first step, perfusion, is performed with solution that has a physiologic osmolarity and pH (~7.4) to prevent the lysis of cells or denaturing of proteins. The purpose of perfusion is to remove the blood from the vascular system. Phosphate buffers are typically selected for this purpose, and volumes of 6–8 times the blood volume of rats, BV (ml) = (0.06 × BW (g) + 0.77),[14] are usually perfused over several minutes. The next step, fixation, is the procedure that actually maintains the cellular and structural integrity of the tissue. No fixative is ideal,[15] but aldehyde fixatives are typically used in brain collection because of its good penetration and ability to fix tissues via cross-linking with minimal disruption of the structural integrity of the proteins.[16] There are numerous fixatives to select from, and specific solutions may have stronger indications depending on the analysis you are going to perform. Extensive information on the topics of fixation and tissue preservation are available, and will not be discussed here.[15–18]

2.1 Materials

1. One 15 G × 1.5 in. aluminum hub blunt needle
2. Three-way valve
3. Syringe (60 ml for perfusion under manual pressure)
4. Operating surface located next to a large sink that is lit
5. Standard operating surgical instruments and micro instruments if available. This should include forceps, vascular clamps, Kelly clamps, scissors, and microscissors

2.2 Procedures

2.2.1 Transcardiac Perfusion

1. After sedating the animal, the procedure begins by gaining access to the abdominal cavity 1–1.5 in. below the xiphoid process. Lift the skin with a pair of forces to prevent cutting intrabdominal viscera, and gain access to

the abdominal cavity with a pair of sharp scissors. Once access is obtained, extend the incision laterally toward the axilla on both sides. Stop the cuts just below the diaphragm.

2. Take down any abdominal organs that may be obscuring your view of the diaphragm (usually the liver). Through the diaphram you are able to see the lungs expand with each breath (Fig. 5a). Once the diaphragm is disrupted and air enters the pleural spaces, the animal will no longer be able to draw breath into its lungs. At this point it is important to move quickly so that tissue hypoxia does not affect your results.
3. Cut the diaphragm along the costal margins. Then cut the ribs along the mid-axillary lines bilaterally. Raise the ribcage superiorly. It is recommended that you grab the xiphoid process with a Kelly clamp to perform this maneuver. The heart is now clearly exposed and still beating (Fig. 5b).
4. Carefully dissect the thymus from the root of the aorta. Use a vascular clamp or Kelly to clamp the inferior vena cava (Fig. 5b) as well as the descending aorta which runs along the spinal cord behind the left lung. Clamping these vessels prevents the infusion from reaching the lower half of the animal and reduces the amount of solution required for adequate perfusion (Fig. 6).
5. Next, cut the apex of the heart to create an access point to the left ventricle and ascending aorta. This can be done with dissecting scissors or an 11 scalpel blade. Insert the 15 G × 1.5 in. aluminum hub blunt needle that is already attached to the 60 cc syringe and hanging bag of fluid to be perfused, or the perfusion pump, into the left ventricle. Direct it into the ascending aorta. Make sure there are no air bubbles in any of the lines. Visualize the tip of the needle in the aorta and use a Kelly clamp to lock it in place (Fig. 7).
6. Once the syringe is in place, cut the auricle of the right atrium (Fig. 7).

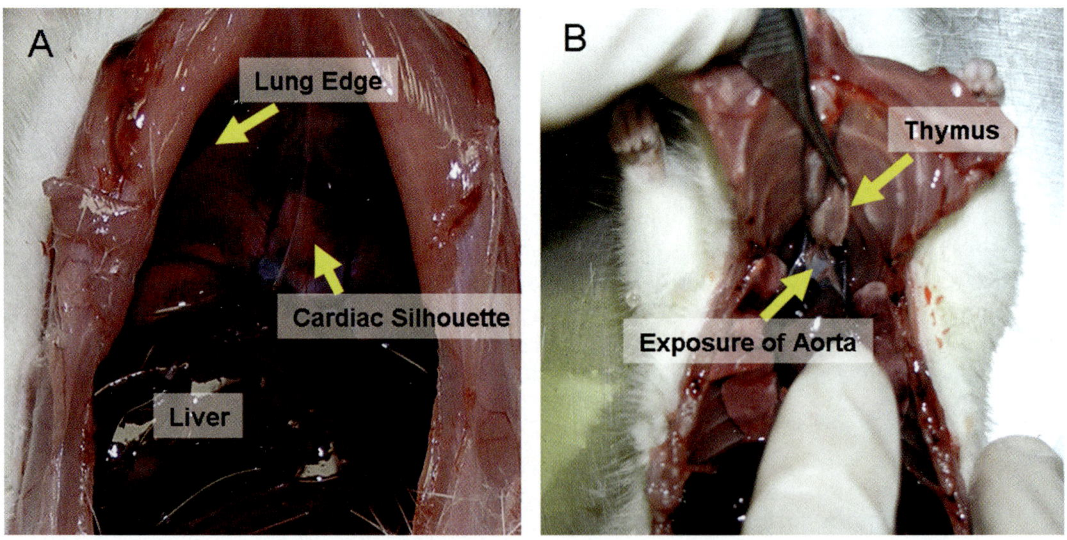

Fig. 5 Transcardiac Perfusion (**A**) Visualization of the thoracic organs through the diaphragm prior to opening the chest cavity. Once the diaphragm is penetrated, the transcardiac perfusion must be performed promptly. (**B**) Exposure of the ascending aortal following the opening of the thoracic cavity and dissection of the thymus

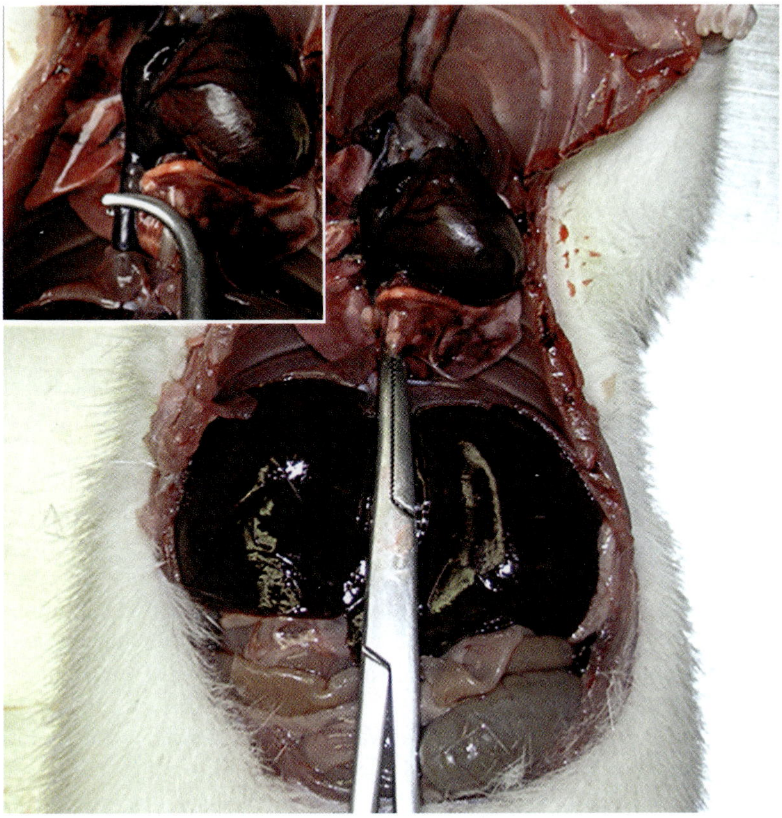

Fig. 6 Clamping of the inferior vena cava prior to perfusion. A small clamp is on the descending aorta, but is covered by the left lung

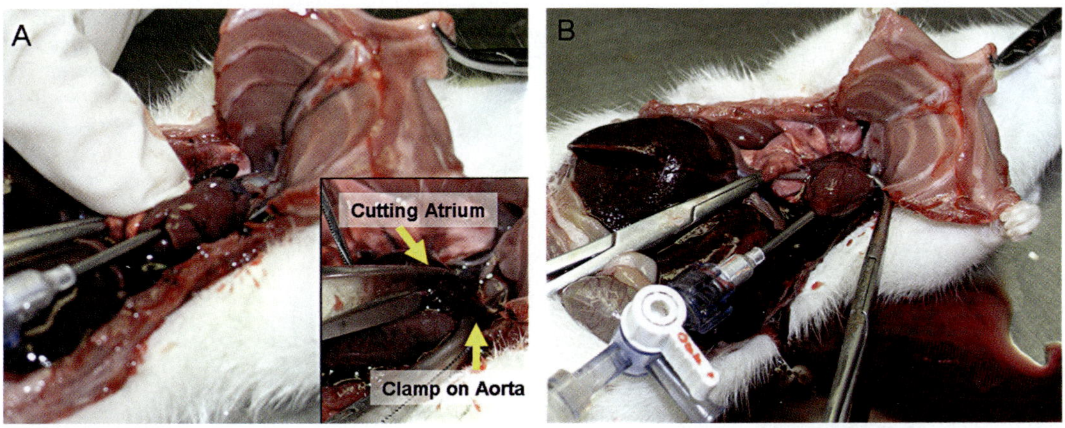

Fig. 7 Transcardiac Perfusion (**A**) After making an incision in the apex of the heart, the 15 G × 1.5 in. aluminum hub blunt needle is placed into the left ventricle and inserted into the aorta. Once the hub needle is in the aorta, clamp it in place, and cut the right atrium. (**B**) Fluid flows from the atrium after returning from the cerebral circulation during perfusion. Note: Vessels around the sternum being cleared of blood (A vs. B) indicates a good perfusion

7. Once you begin the perfusion, you should see the perfusion fluid begin to flow out of the incision in the right atrium. You may also see the blood in the vasculature surrounding the sternum being flushed out (Fig. 7b). Perfusion is usually performed with 120–180 cc of buffer solution for a 250–350 g rat when flushing with manual pressure. This step can also be done with the use of gravity (see below) or a perfusion pump.
8. Once you have completed your perfusion, remove the needle and clamps from the animal. The animal is now ready for extraction of the brain. Successful perfusion will result in the "whitening" of the rat's eyes.

2.3 Applications, Complications, and Limitations

1. It is better to perform this procedure while the animal is still alive with an intact circulation, so that there is no stagnant blood in the arterial system that may prevent adequate perfusion.
2. Whether or not to perfuse and fix the tissues under gentle manual pressure or under controlled pressure with the use of a perfusion pump or gravity is at the discretion of the investigator. In studies in which cerebral vascular tissues cannot be exposed to high pressures in order to preserve accurate measurements of vascular diameter, or the endothelial cell layer cannot be exposed to greater than normal shear forces, perfusion should be done at pressures under normal physiologic values: systolic pressure 120; diastolic 80 mmHg; mean arterial pressure of 100 mmHg for the rat. Below is an illustration of a simple and effective way to perfuse animals while monitoring pressure (Fig. 8). This method uses standard IV tubing, and is set up to allow gravity to perform the perfusion. Two IV lines, one with formalin and one with phosphate-buffered saline (PBS), run separately before joining

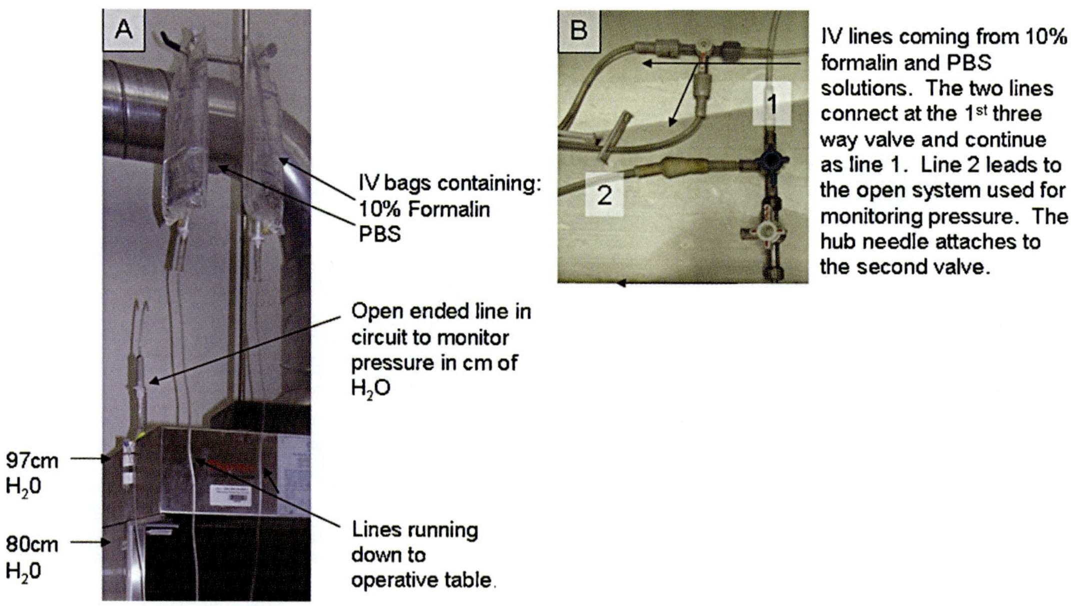

Fig. 8 (**A**) Tubing system and pressure monitor. Two IV bags, one with formalin and the other with PBS, run down to the animal to be perfused. The open end of the system, which monitors the perfusion pressure, is marked at 80 and 97 cm of H_2O. (**B**) Formalin and PBS join a common line, and the system is connected to an open circuit to monitor perfusion pressure

a common line (Fig. 8). The operator uses the three-way valve to change between formalin and PBS perfusions. The circuit then connects to an open line that runs upward toward the IV bags (Fig. 8b). This open circuit is used to measure the pressure of the perfusion. The pressure is adjusted by changing the flow rates in the IV tubing using the drip rate adjustment gauges that come standard with IV tubing. Our lab has had success in perfusing brains at pressures between 80 and 97 cm H_2O marks (58.8 and 71.3 mmHg, respectively) over a period of 15 min. The cm of H_2O markings shown in Fig. 8a are determined by measuring the distance above the surface where the animals are being perfused. Similar setups have been used in the perfusion of mice.[19]

3 Methods of Brain Removal

Extracting the brain from the skull quickly and in a manner that minimizes disruption of the brain's structures is ideal. There are many methods for removing the brain from the skull. This section provides illustrations for one effective method of brain removal. Following removal, the brain is often stored in 30% sucrose solutions prior to storage to prevent freezing artifacts during sectioning.[20,21]

3.1 Materials

1. Guillotine
2. Scalpel handle (F.S.T. 10003-12) and scalpel blades (F.S.T. 10015-00)
3. Forceps (Adson, F.S.T. 11027-12)
4. Curved Kelly hemostat (F.S.T. 13019-14)
5. Spring scissors (F.S.T. 15005-08)
6. Friedman rongeur (F.S.T. 16000-14)
7. Hippocampus tool (F.S.T. Cat # 10099-15)
8. Sucrose solution (30%)

3.2 Procedures

3.2.1 Brain Removal

1. Deeply anesthetize the animal before sacrifice. Decapitation of the rat is performed with a guillotine at a level just above the shoulder blades.
2. Once the head is removed, the first step is to make a midline incision starting in the parietal region and extend it to the end of the nasal bone. Then turn the skull 180° and extend the incision in the opposite direction to the end of the caudal tissues. Remove the periostium from the parietal, frontal, and nasal bones by scraping the surface of the skull with the scalpel blade (Fig. 9a).
3. Turn the skull over and make a forceful incision between the superior incisor teeth. This cut should split the nasal bone. A couple of passes may be required to ensure a clean cut (Fig. 9b).
4. Return the skull to the upright position. Take a curved Kelly clamp and insert the curved end into the split in the nasal bones. Slide the tips as caudally as possible along the incision, as this will ensure a clean midline split in the skull when the Kelly clamp is opened.

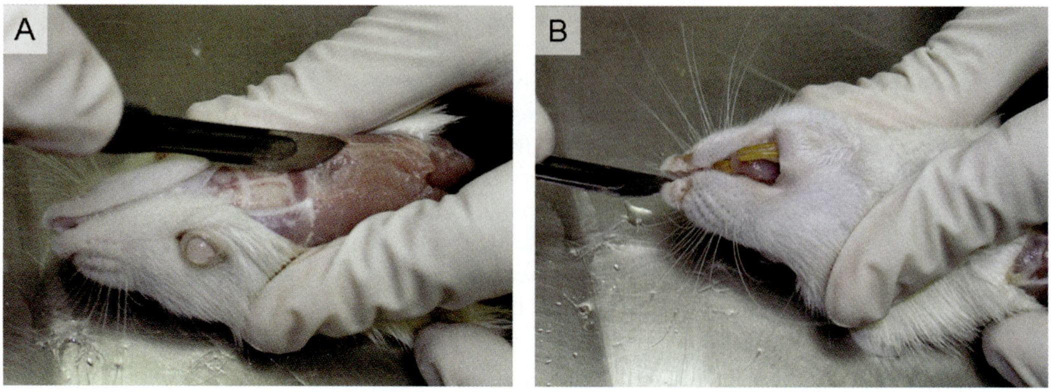

Fig. 9 (**A**) Rostral and caudal incisions along the superior nasal bone and skull following by removal of the periostium from the skull. (**B**) Splitting the nasal bone by making an incision in between the incisors

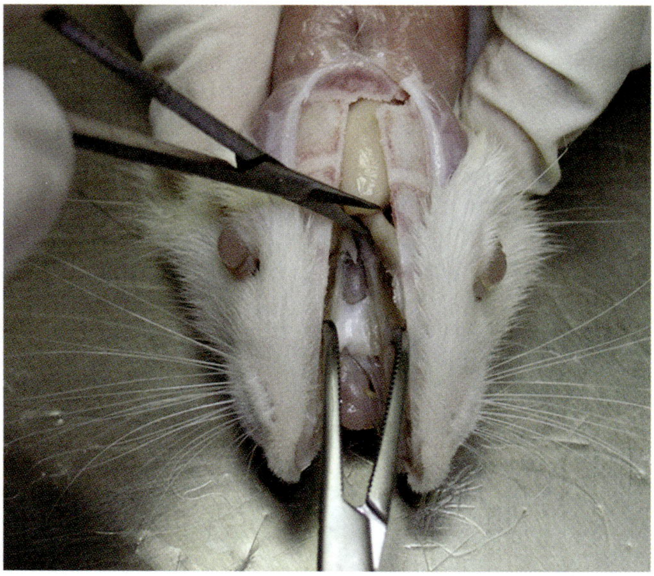

Fig. 10 Splitting of the nasal and frontal bones along the midline. Here one of the olfactory bulbs is also being cut

5. Open the Kelly clamps and spilt the skull along the midline. This should create a midline split in the nasal and frontal bones that extends to the lambdoid suture proximal to the interparietal bone. As the olfactory bulbs become exposed, cut them so that they do not tear and damage the brain (Fig. 10).
6. The superior portions of the skull can be dissected from the brain with your hands as you extend the nasal bones laterally. This step should create a specimen in which the rat's brain appears to be resting on the tongue (Fig. 11).
7. Use a rongeur to remove the interparietal bone, taking care not to tear the underlying brain. Cut the tentorium before removing the temporal bones.

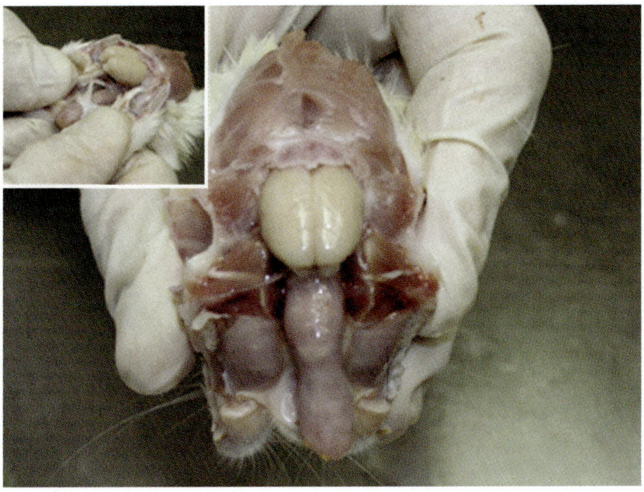

Fig. 11 Opening the skull to expose the superior portions of the hemispheres

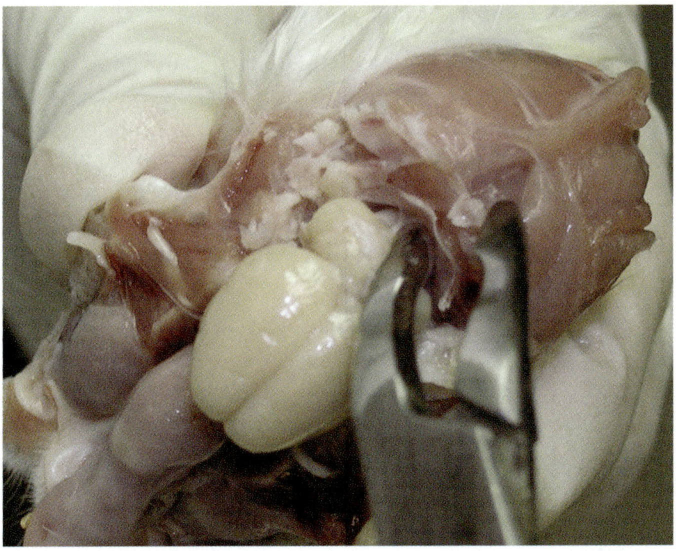

Fig. 12 Using a bone rongeur to remove the interparietal bone and expose the cerebellum

If you try to remove the temporal bones without cutting this membrane, it may tear the brain as you dissect (Fig. 12).

8. After removing the occipital and temporal bones in order to free the cerebellum, lift the brain from the skull base. The pituitary stalk is often left in the pituitary fossa with this maneuver. As you lift the brain, cut away any cranial nerves and the vertebral arteries (Fig. 13).
9. Once the underside of the brain is free, cut the spinal cord. The brain should easily separate from the skull base (Fig. 13c).

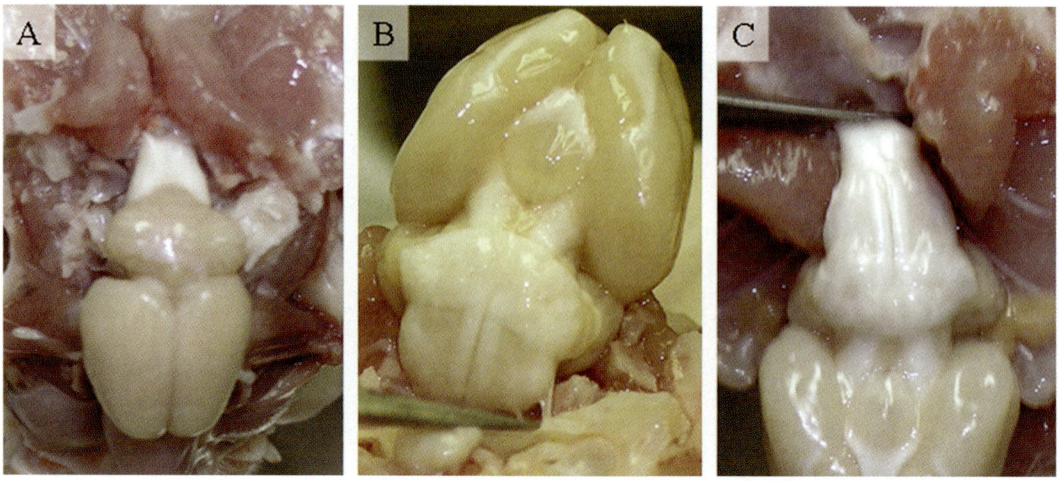

Fig. 13 (**A**) Complete exposure of the superior aspect of the brain. (**B**) Removing the brain from the skull base and cutting cranial nerves. (**C**) Cutting of the spinal cord

3.3 Applications, Complications, and Limitations

1. It is important to note that in un-perfused animals the process of decapitation can create subarachnoid hemorrhage in the spinal cord that extends rostrally into the brain stem. If these brain stem tissues are relevant to your particular experiment, it may be important to keep blood away from these structures. Therefore, we recommend a more caudal decapitation that leaves a significant portion of the spinal cord extending from the brain stem. A decapitation at the level of just above the ribs should suffice.

4 Separation of the Brain into Anatomical Sections

After the brain is removed, the investigator may want to separate the brain into several different sections: cortex, basal ganglia, hippocampus, cerebellum, and brain sstem. This is frequently done because different regions of the brain have variable responses to injury and treatment.

4.1 Materials

1. Well lit working area
2. Standard operating surgical instruments and micro instruments if available. For this procedure it should include micro spatula (F.S.T. 10091-12) or hippocampus tool (F.S.T. 10099-15), microscissors, and fine forceps

4.2 Procedures

4.2.1 Dissection

1. In this example the brain was perfused with 0.01 M PBS before removal and dissection. Once the brain has been removed from the cranial vault, separate the brain into the two hemispheres, cerebellum, and brain stem. Remove the cerebellum from the brain stem by gentle dissection from the caudal end.

Make the separating cut at the cerebellar peduncles (Fig. 14a). Split the hemispheres at the interhemispheric fissure (Fig. 14b).

2. One hemisphere at a time, separate the basal ganglion from the surrounding cortex. Blunt dissection should separate the two parts completely except for the white matter tracks, which can be cut with scissors (Fig. 15).

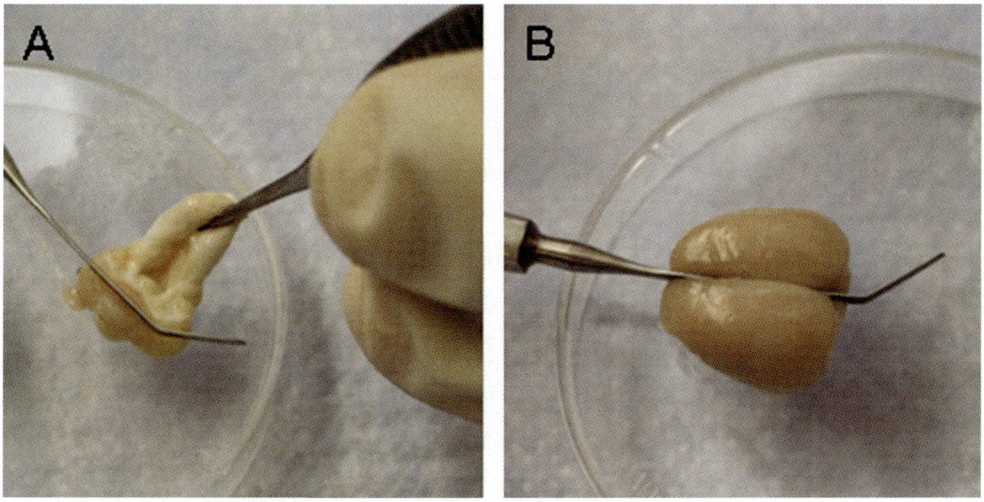

Fig. 14 (**A**) Separation of the cerebellum from brainstem. (**B**) Separation of the hemispheres

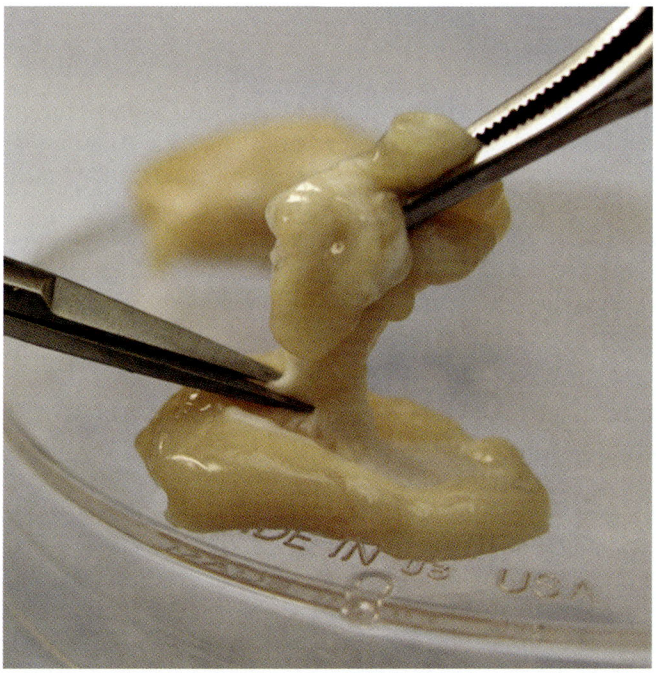

Fig. 15 Removal of the basal ganglia from the cortex

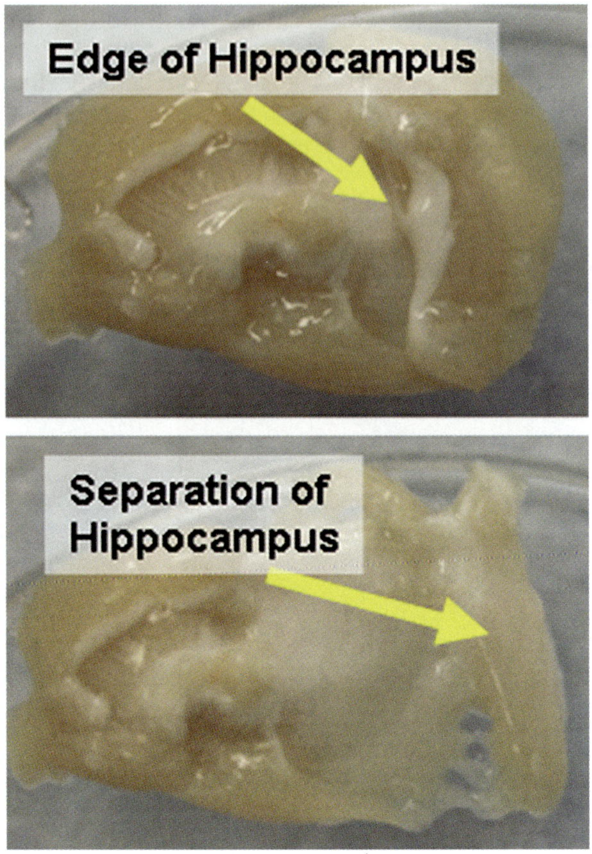

Fig. 16 Separation of the hippocampus from the cortex

3. The hippocampus brain region can now be separated from the cortex. The hippocampus resides at the caudal end of the cortex. The free edge of the hippocampus can be gently lifted from the interior of the removed cortex. The two pieces can be separated by gentle dissection (Fig. 16).
4. Repeat this procedure for the opposite hemisphere. Once completed, the brain should be separated into the following sections: right and left cortical regions, right and left basal ganglion, right and left hippocampus regions, cerebellum, and brain stem.

5 Preparation of Brain Slices for Triphenyltetrazolium Chloride Staining

Another common test utilized immediately following brain removal is TCC staining. Staining with TTC was developed to be a rapid, convenient, inexpensive, and reliable method for the detection and quantification of cerebral infarction in rats 24 h after the onset of ischemia.[22] The stain identifies viable mitochondria in brain tissues. Infarcted areas in the brain that do not take up the TTC (red color) stain are identified after slicing the brain into 1-mm thick sections and immersing in 2% TTC solution.

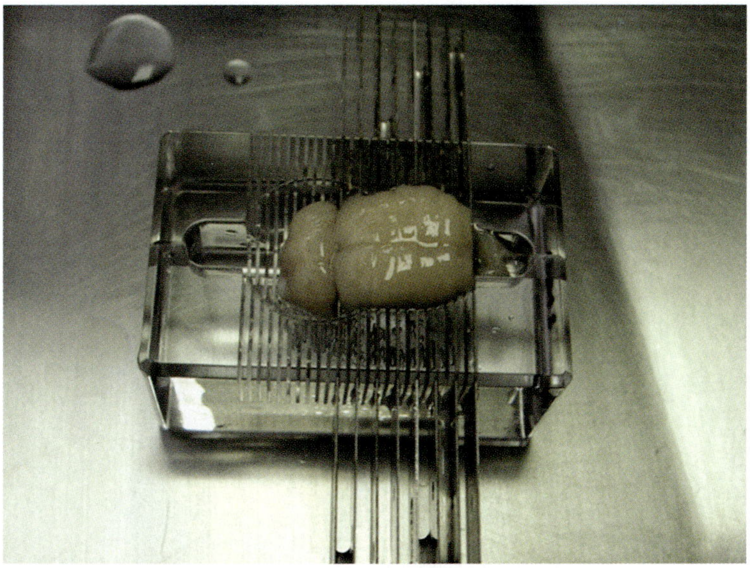

Fig. 17 Brain in matrix with razor blades dividing it into 1-mm sections

5.1 Materials

1. TCC solution (2%, Sigma T – 8877)
2. PBS (Sigma P-3563)
3. Razor blades
4. Brain matrix (Kent Scientific RBMA-300C)
5. Petri dish
6. Formalin (10%)

5.2 Procedures

5.2.1 TCC Staining

1. Post sacrifice, the brain is removed carefully and immersed in ice-cold PBS. This helps to make the brain firm, making it easier to cut sections.
2. Brain is cut into 1-mm thick sections in the brain matrix using razor blades (Fig. 17).
3. The razor blades are lifted and separated carefully as the tissue tends to stick to the edges of the blades. To prevent any damage to the sections, the edge of the blade with the brain section adhering to it is immersed in the 2% TTC solution. The TTC solution is light/heat sensitive and should be kept covered in aluminum foil till just before use. The brain slices are allowed to stay immersed in TTC solution and the change in color is continuously monitored. This process can be expedited by providing heat, e.g., incubator/oven at 37°C.
4. When the surface of the brain sections turns pinkish, the sections are flipped over. The change in color is an indication of the TTC staining the viable mitochondria in the brain tissue. The infarcted regions that lack viable mitochondria do not change color and remain white (Fig. 18).

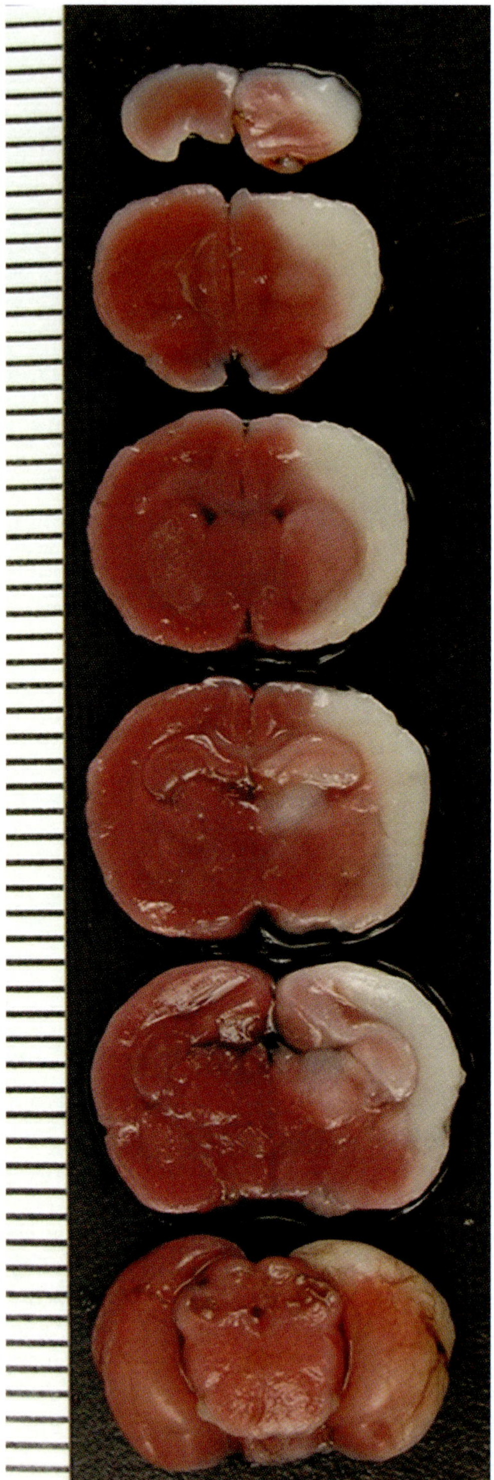

Fig. 18 TCC staining showing infarction after middle cerebral artery occlusion in an adult male Sprague-Dawley rat

5. When the sections have turned light red, the reaction is stopped by washing off the TTC with PBS.
6. The brain slices need to be photographed to quantify the infarct volume. However, the sections can be fixed by immersing in formalin overnight. This will harden the sections and make it easy to take pictures of both surfaces.

5.3 Applications, Complications, and Limitations

1. TTC staining is a standard procedure performed in research laboratories for calculation of infarct volume.
2. The TTC solution is light sensitive and should be stored away from light.
3. Perfusion or the lack of perfusion with PBS prior to preparing the brain for TCC staining has not been found to affect our results.

Acknowledgments: The authors wish to thank Suzzanne Marcantonio, Norihito Shimamura and Gen Kusaka for all their help. This study is partially supported by grants from NIH NS54680, NS53407, NS45694, and NS43338 to JHZ.

Reference

1. Kastl S, Kotschenreuther U, Hille B et al. Simplification of rat intubation on inclined metal plate. Adv Physiol Educ 2004;28:29–32
2. Johns TN, Olson BJ. Experimental myocardial infarction. I. A method of coronary occlusion in small animals. Ann Surg 1954;140:675–682
3. Stark RA, Nahrwold ML, Cohen PJ. Blind oral tracheal intubation of rats. J Appl Physiol 1981;51:1355–1356
4. Cambron H, Latulippe JF, Nguyen T, Cartier R. Orotracheal intubation of rats by transillumination. Lab Anim Sci 1995;45:303–304
5. Yasaki S, Dyck PJ. A simple method for rat endotracheal intubation. Lab Anim Sci 1991;41:620–622
6. Pena H, Cabrera C. Improved endotracheal intubation technique in the rat. Lab Anim Sci 1980;30:712–713
7. Costa DL, Lehmann JR, Harold WM, Drew RT. Transoral tracheal intubation of rodents using a fiberoptic laryngoscope. Lab Anim Sci 1986;36:256–261
8. Alpert M, Goldstein D, Triner L. Technique of endotracheal intubation in rats. Lab Anim Sci 1982;32:78–79
9. Rivera B, Miller S, Brown E, Price R. A novel method for endotracheal intubation of mice and rats used in imaging studies. Contemp Top Lab Anim Sci 2005;44:52–55
10. Boyer S, Faure K, Ader F et al. Chronic pneumonia with Pseudomonas aeruginosa and impaired alveolar fluid clearance. Respir Res 2005;6:17
11. Yaghi A, Bradbury JA, Zeldin DC et al. Pulmonary cytochrome P-450 2J4 is reduced in a rat model of acute Pseudomonas pneumonia. Am J Physiol Lung Cell Mol Physiol 2003;285:L1099–L1105
12. Jou IM, Tsai YT, Tsai CL et al. Simplified rat intubation using a new oropharyngeal intubation wedge. J Appl Physiol 2000;89:1766–1770
13. Jou IM, Tsai YT, Tsai CL et al. Simplified rat intubation using a new oropharyngeal intubation wedge. J Appl Physiol 2000;89:1766–1770
14. Lee HB, Blaufox MD. Blood volume in the rat. J Nucl Med 1985;26:72–76
15. Bancroft JD, Stevens A. Theory and Practice of Histological Techniques. 766 edn. Edinburgh: Churchill Livingstone; 2007
16. Srinivasan M, Sedmak D, Jewell S. Effect of fixatives and tissue processing on the content and integrity of nucleic acids. Am J Pathol 2002;161:1961–1971

17. O'Leary TJ. Standardization in immunohistochemistry. Appl Immunohistochem Mol Morphol 2001;9:3–8
18. Shi SR, Cote RJ, Taylor CR. Antigen retrieval techniques: current perspectives. J Histochem Cytochem 2001;49:931–937
19. Parra A, McGirt MJ, Sheng H et al. Mouse model of subarachnoid hemorrhage associated cerebral vasospasm: methodological analysis. Neurol Res 2002;24:510–516
20. Rosene DL, Mesulam MM. Fixation variables in horseradish peroxidase neurohistochemistry. I. The effect of fixation time and perfusion procedures upon enzyme activity. J Histochem Cytochem 1978;26:28–39
21. Ebbeson SOE. The selective silver-impregnation of degenerating axons and their synaptic endings in nonmammalina species. *Contemporary Research Methods in Neuroanatomy*. Berlin: Springer; 1970:132–162
22. Bederson JB, Pitts LH, Germano SM et al. Evaluation of 2,3,5-triphenyltetrazolium chloride as a stain for detection and quantification of experimental cerebral infarction in rats. Stroke 1986;17:1304–1308

Chapter 4

Brain Monitoring

Julian Cahill and John H. Zhang

Abstract Invasive measurement of cerebral blood flow (CBF) and blood pressure (BP) are important adjuncts to many different types of animal surgery. CBF can be measured by a number of different methods, including invasive methods such as placing probes into the subarachnoid space (Am. J. Physiol. Heart Circ. Physiol. 281:H308–H315, 2001), which can lead to brain injury, and noninvasive methods such as rheoencephalography (Physiol. Meas. 25:1371–1384, 2004) and radiography (Magn. Reson. Med. 49:479–487, 2003), which are expensive. This section will describe a relatively simple method to measure CBF that is cheap, easy to replicate, and does not result in damage to the parenchyma.

BP monitoring is used commonly in most types of surgery. There are a number of noninvasive machines available that are easy to use and quite reliable. However, the invasive measurement of BP is more accurate. The placement of a femoral line will be described in detail here.

Keywords: Cerebral blood flow, Doppler, Blood pressure, Femoral line

1 Model Selection

The measurement of cerebral blood flow (CBF) is useful in a number of experimental models, including the effect of drugs on CBF,[4] head injuries,[5] cerebral ischemia,[6] and seizures,[7] to name but a few. CBF measurement allows confirmation of the injury, and allows you to confirm the period of time in which there is reduced or no blood flow, this being particularly important in stroke models. Although the measurement of CBF cannot indicate the extent or the severity of the injury accurately, particularly in subarachnoid hemorrhage (SAH) models, it will allow you to plot the various groups in your study against each other and, therefore, demonstrate that the treatment and the nontreatment groups are similar with regard to the injury. It is important to point out that there is no "normal" value for CBF. There are too many variables to choose from, such as site, condition of the animal, anatomical variations in the rat, etc. For this reason it is better to describe data not in absolute figures

From: *Animal Models of Acute Neurological Injuries.*
Edited by: J. Chen (eds.) New York, Totowa, NJ

but in terms of percentage of normal for each rat in the study. Major advantages of this model include the following:

1. Measurement of CBF using the Doppler technique is reliable and reproducible
2. By placing the CBF probe on the surface of the dura, the brain is not injured during the study
3. The CBF probe can be left in place for repeated measurements
4. CBF measurement is useful for SAH, stroke, and head injury studies
5. A femoral line can be used for heart rate assessment as well as the administration of drugs and fluids and the measurement of blood constituents

2 Materials

(a) Anesthetic equipment, including intubation equipment. Alternatives include a tracheostomy or simply supplemental oxygen, but this is not recommended
(b) Hair clippers (with #40 blade) and skin disinfectant, e.g., iodophors (Betadine) or chlorhexidine (Novasan)
(c) Sterile surgical drapes, gown and gloves
(d) Benchmark sterotactic frame with 45° ear bars (Kent Scientific, 1116 Litchfield Street, Torrington, CT 06790, USA)
(e) MV-9 electro surgical unit (Kent Scientific)
(f) DRL235 slender 1/2 in. diameter hand piece with a 3/32 in. drill bit (Kent Scientific)
(g) Normal saline (0.9%)
(h) Small, straight Periflux probe (407), probe holder (PH 07 series), Periflux System 5000 (Periflux PF 5010 Laser Doppler Perfusion Monitoring Unit), laptop computer, and Perisoft for windows version 2.10 (Perimed, Inc, 4873 Princeton Dr., North Royalton, OH 44133, USA)
(i) Tissue adhesive, 3/0 silk ties, 4/0 viacryl and 3/0 nylon sutures
(j) Standard operating surgical instruments and micro instruments if available. This should include a scalpel (disposable), fine forceps, needle holder, retractors, and microscissors
(k) Suction, if available, or Q-tips (preferred)
(l) PE50 and PE240 polyethylene tubing and a three-way stopcock

3 Procedures

1. The animal should be prepared in advance, including shaving the head and the use of skin disinfectants.
2. The animal can now be turned prone and placed in a head clamp. A Benchmark sterotactic frame with 45° ear bars (Kent Scientific) is used to secure the animal's head firmly in place (Fig. 2a).
3. A midline sagittal incision is then made from the frontal to the occipital bone. The MV-9 electrosurgical unit (Kent Scientific) can be used for diathermy as needed.
4. The skin and subcutaneous tissues are then retracted laterally on the right-hand side and the underlying periosteum is lifted. This is a very thin, transparent membrane on the surface of the skull and is usually removed with the subcutaneous tissue in this model. This allows good exposure of the midline, coronal suture, and the right parietal bone (Fig. 1b).

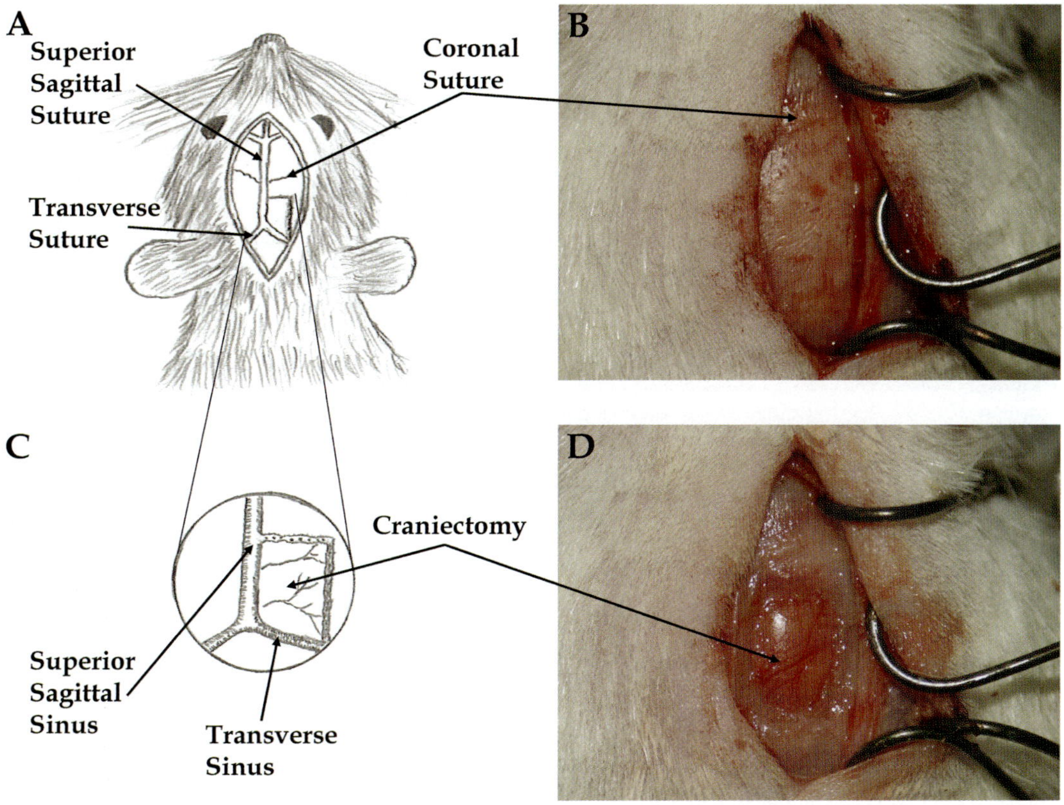

Fig. 1 Cartoon and intraoperative pictures of a craniectomy in the rodent. Cartoon and intraoperative pictures showing a midline incision in the skull of a rat (**A**, **B**), followed by a small craniectomy (**C**). The craniectomy should be located superior to the transverse sinus and lateral to the superior sagittal sinus. After the craniectomy, the dura can be seen, to which a Doppler probe is later attached (**D**)

5. A 1-cm^2 bony window is then fashioned in the parietal bone of the skull, superior to the transverse sinus, lateral to the superior sagittal sinus, and inferior to the coronal suture using a DRL235 slender 1/2 in.-diameter hand piece with a 3/32 in. drill bit (Kent Scientific), cooled with 0.9% normal saline (Fig. 1d).
6. A Periflux probe holder is then glued over the craniectomy site and held firmly in place using tissue glue (Fig. 2). There are a number of different probe holders and probes to choose from; the method described here is for a straight probe.
7. The overlying subcutaneous tissues and skin are then suture-closed using a 4/0 viacryl and a 3/0 nylon suture, respectively; alternatively you can simply use the tissue adhesive to close the skin (Fig. 2d).
8. The periflux laser probe, which generates a laser beam at a wavelength of 780 nm to a depth of approximately 1 mm, is then inserted into the probe holder, which in turn is connected to the Periflux PF 5010 Laser Doppler perfusion monitoring unit. This system is then connected to a laptop computer where the readings can be analyzed using Perisoft for windows version 2.10.

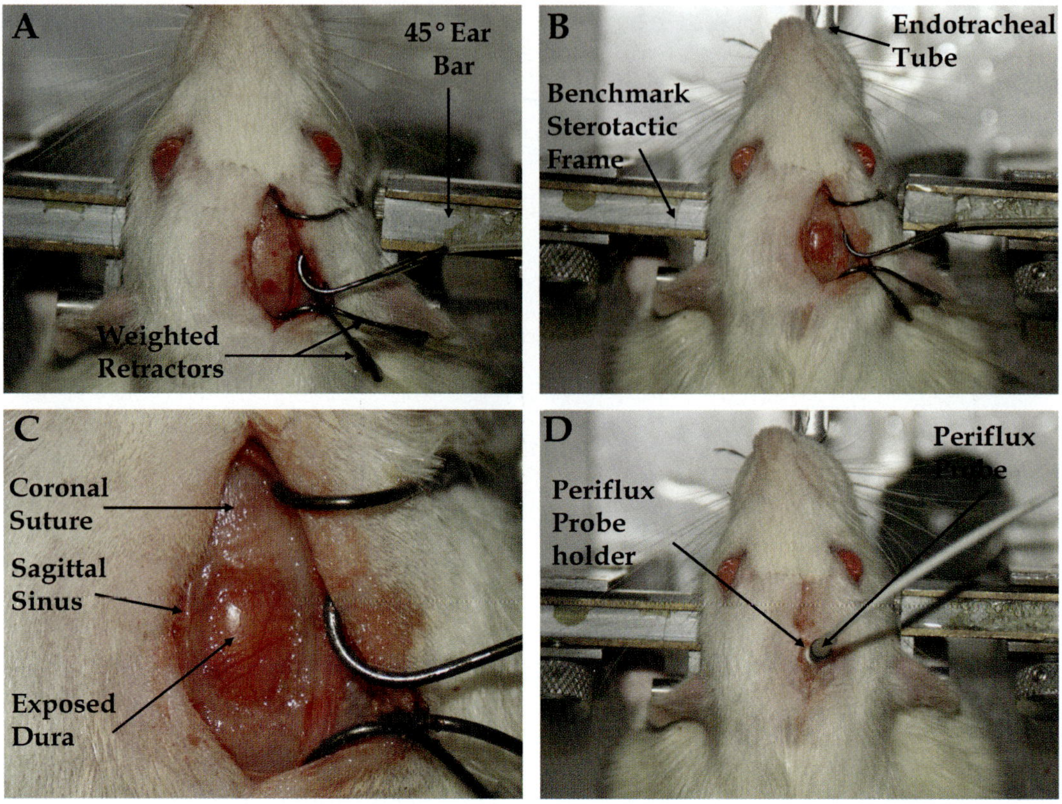

Fig. 2 A Benchmark sterotactic frame with 45° ear bars and a Periflux Doppler probe. A Periflux probe holder is glued over the craniectomy site (**B**, **C**) and held firmly in place using tissue glue (**D**). The periflux laser probe, which generates a laser beam at a wavelength of 780 nm to a depth of approximately 1 mm, is inserted into the probe holder, which in turn is connected to the Periflux System 5000. This system is then connected to a laptop computer where the readings are analyzed using Perisoft for windows version 2.10

9. The Periflux System 5000 can be seen in Fig. 3c. In addition, examples of the traces from different animals can be seen (Fig. 3b). Note the different base line values for each animal. However, the overall data can be shown as a percentage of baseline for the group. An example of such a graph can be seen in Fig. 3b. In addition, there is also a picture of a Benchmark sterotactic frame (Kent Scientific), which is very useful for this procedure.
10. In order to place a femoral line, the right femoral area is shaved and painted with an iodine solution. This is allowed to dry for 10 min before continuing. The animal is then placed supine and a 1-cm incision is made overlying the right femoral artery (Fig. 4a).
11. Throughout their course, both the femoral artery and vein are encapsulated within a common perivascular sheath. It has only one branch, the epigastric artery, which branches just before the division. Once the incision is made, the inguinal fat pad comes into view, which is then reflected laterally after incising through its medial, superior, and inferior borders. Hemostasis is achieved using a MV-9 electrosurgical unit (Kent Scientific). The distal ends of the epigastric vessels can be seen entering the fat pad, and they are preserved.

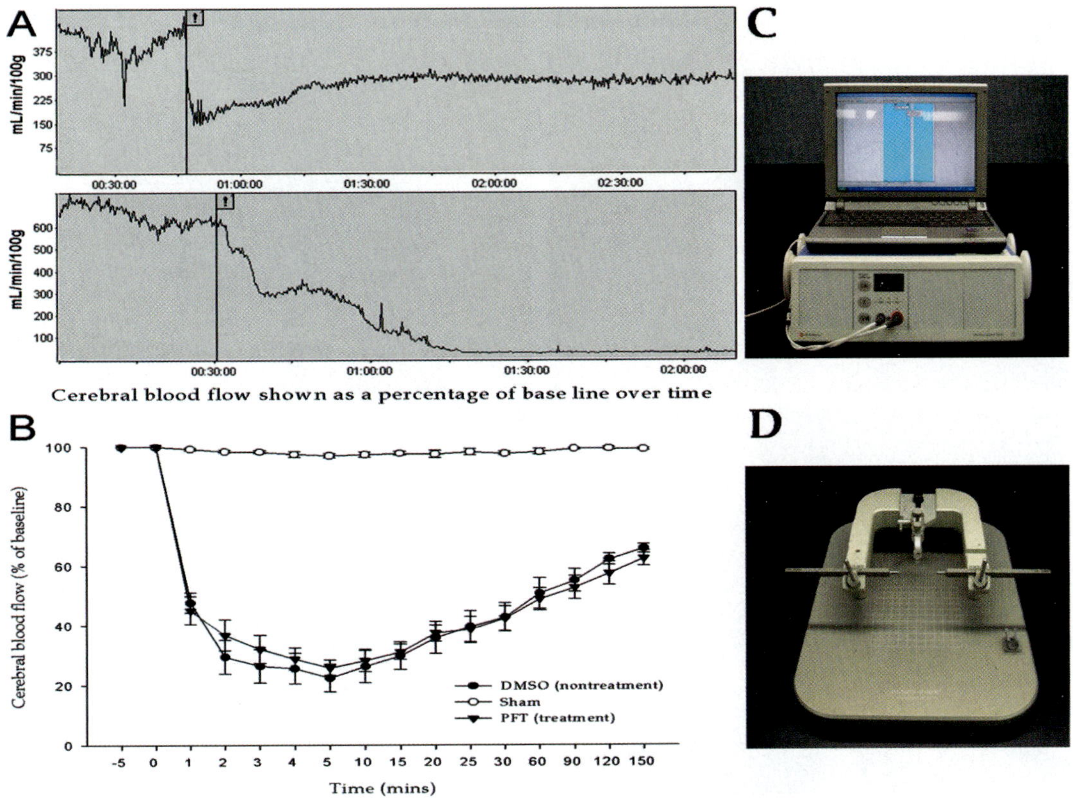

Fig. 3 Examples of data acquired and some of the equipment used. **A**) Shows two examples of CBF tracings in rats experiencing a SAH (the lower tracing was fatal). The exclamation mark indicates the puncture. (**B**) Actual published data using the described technique.[8] The data is displayed as a percentage of baseline. (**C**) The Periflux System 5000. (**D**) A Benchmark sterotactic frame with 45° ear bars (Kent Scientific)

12. The common femoral vessels should now be in view. To completely expose the length of the vessel, the abdominal wall muscles are bluntly reflected medially until the inguinal ligament comes into view. Once the whole of the perivascular sheath is exposed, it is incised parallel to the femoral vessels.
13. The artery is then separated and isolated from the vascular sheath. The distal end of the artery is then ligated using a 3/0 silk tie (Fig. 4b). A second tie is placed at the proximal end but this is left loose and attached to an artery clip.
14. Using the microscissors a hole is made in the vessel. The proximal loose tie is used to elevate the vessel slightly, which will prevent bleeding (Fig. 4b). The femoral artery is then cannulated with polyethylene tubing (Fig. 4c). For the femoral artery PE50 tubing is used, which is attached to PE240 tubing, which in turn is attached to a three-way stopcock. Once the vessel has been cannulated, the line can be advanced by approximately 3 cm, before the line is securely fastened with the second tie (Fig. 4d).
15. The wound is lightly covered for the duration of the surgical procedure. Postoperatively the line must be removed. An additional tie is placed around the line, which is then removed and the femoral artery is tied off. The skin is then suture-closed.

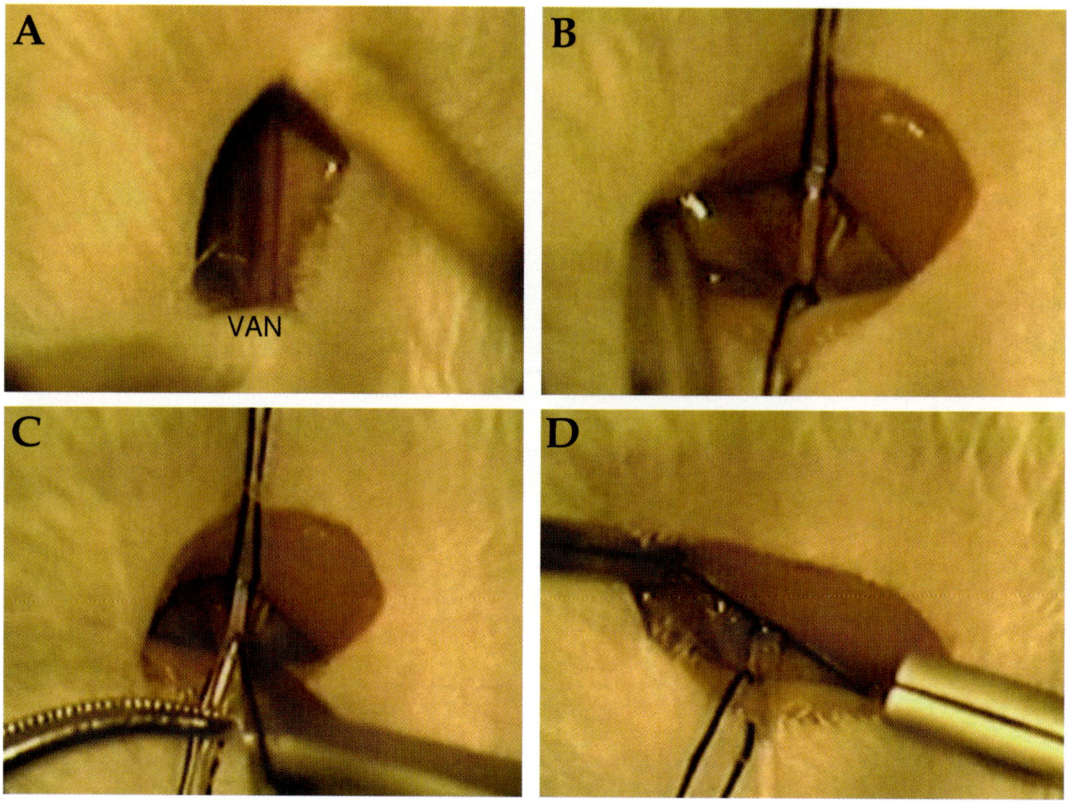

Fig. 4 Femoral Artery dissection and catheter insertion. (**A**) The anatomical position of the femoral vessels as classically described, vein, artery, and nerve (VAN). (**B**) After the dissection of the femoral artery, the vessel is tied distally and the proximal ligature is left loose. (**C**) A catheter is then inserted into the femoral artery for blood pressure measurements. (**D**) Once complete, the catheter is secured proximally

4 Applications, Complications, and Limitations

1. When preparing the skin, care is required to ensure that the eyes of the animal are not damaged.
2. The Benchmark sterotactic frame allows you to secure the animal's mouth and, in particular, endotracheal tube, which is always secured first. The ear bars can then be applied. The standard ear bars have a 45° taper into the ear, which avoids puncturing the tympanic membrane.
3. When you are using the drill, the use of normal saline is very important to cool it. A continuous stream should be deployed to avoid overheating and subsequent damage to the brain. Care must be taken to ensure that the dura remains intact.
4. A bony window of $1\,cm^2$ is bigger than that classically described. However, using a larger window allows a suitable site for the probe holder to be chosen.
5. In practice, the Doppler probe and the probe holder are held over the craniectomy site first. This allows you to choose the best site for the probe holder. If the probe is not lying over a vessel, you will not get a good signal. Move the probe (in the holder) around the craniectomy site until a good signal is found. Once a site has been chosen, the probe holder is glued to place. Adequate time must be given for the tissue adhesive to harden. Make sure that your forceps are not glued to the holder.

6. After fixing the probe in place, baseline readings are usually taken from the territory of the right middle cerebral artery (MCA).[9] In the case of SAH, the contralateral hemisphere is usually chosen for the insertion of the Doppler probe to avoid the documented brief decrease in CBF, following the filament insertion.[10] If you are measuring CBF for other reasons, e.g., stroke, it may be suitable to measure the CBF on the ipsilateral side to the injury.
7. In the rat, the femoral artery is about 1.5 cm long, commencing at the mid-point of the inguinal ligament and ending at its division into the superficial femoral artery and the deep branches.
8. When the perivascular sheath is incised, the vasa vasorum may be cut. Any bleeding is usually self-limiting, but for persistent hemorrhage diathermy can be used to achieve hemostasis.
9. The femoral line should be kept free of blood throughout the procedure. Normal saline can be used to flush the line. The use of heparin-based solutions can be considered depending on the surgery; however, it is unnecessary as saline works perfectly well.
10. Although this procedure results in the ligation of the femoral artery, the collateral flow is excellent in the rat and so an ischemic limb is avoided.

References

1. Hong, K. W., Shin, H. K., Kim, C. D., Lee, W. S., & Rhim, B. Y. (2001). Restoration of vasodilation and CBF autoregulation by genistein in rat pial artery after brain injury. *Am J Physiol Heart Circ Physiol* **281**, H308–H315
2. Bodo, M., Pearce, F. J., and Armonda, R. A. (2004). Cerebrovascular reactivity: rat studies in rheoencephalography. *Physiol Meas* **25**, 1371–1384
3. Tailor, D. R., Roy, A., Regatte, R. R., Charagundla, S. R., McLaughlin, A. C., Leigh, J. S., and Reddy, R. (2003). Indirect 17(O)-magnetic resonance imaging of cerebral blood flow in the rat. *Magn Reson Med* **49**, 479–487
4. Gao, F., Sugita, M., and Nukui, H. (2005). Phosphodiesterase 5 inhibitor, zaprinast, selectively increases cerebral blood flow in the ischemic penumbra in the rat brain. *Neurol Res* **27**, 638–643
5. Cherian, L., Chacko, G., Goodman, J. C., and Robertson, C. S. (1999). Cerebral hemodynamic effects of phenylephrine and L-arginine after cortical impact injury. *Crit Care Med* **27**, 2512–2517
6. Lauer, K. K., Shen, H., Stein, E. A., Ho, K. C., Kampine, J. P., and Hudetz, A. G. (2002). Focal cerebral ischemia in rats produced by intracarotid embolization with viscous silicone. *Neurol Res* **24**, 181–190
7. Sato, T., Takeda, Y., Hagioka, S., Zhang, S., and Hirakawa, M. (2001). Changes in nitric oxide production and cerebral blood flow before development of hyperbaric oxygen-induced seizures in rats. *Brain Res* **918**, 131–140
8. Cahill, J., Calvert, J. W., Solaroglu, I., and Zhang, J. H. (2006). Vasospasm and p53-induced apoptosis in an experimental model of subarachnoid hemorrhage. *Stroke* **37**, 1868–1874
9. Schmid-Elsaesser, R., Zausinger, S., Hungerhuber, E., Baethmann, A., and Reulen, H. J. (1998). A critical reevaluation of the intraluminal thread model of focal cerebral ischemia: evidence of inadvertent premature reperfusion and subarachnoid hemorrhage in rats by laser-Doppler flowmetry. *Stroke* **29**, 2162–2170
10. Bederson, J. B., Germano, I. M., and Guarino, L. (1995). Cortical blood flow and cerebral perfusion pressure in a new noncraniotomy model of subarachnoid hemorrhage in the rat. *Stroke* **26**, 1086–1091

Chapter 5

Veterinary Care Methods for Rats and Mice in Experimental Spinal Cord Injury Studies

Aaron H. Puckett, Christine D. Nunn, and Stephen M. Onifer

Abstract The incidences and causes of traumatic injury to the adult human spinal cord vary throughout the world. Regardless, spinal cord injury (SCI) similarly can result in devastating dysfunctions that lead to tremendous and long-term physical, emotional, and financial hardships not only for the survivor but also for family and friends. For these reasons, both clinical trials[1,2] and experimental studies of SCI treatments are taking place. The treatments that are being investigated include ones to protect spinal cord cells or replace them, sprout or regenerate axons, improve endogenous repair mechanisms, and rehabilitate.[3]

Advances in acute and chronic medical care have markedly improved the quality and length of the lives of persons with traumatic SCI.[4–7] We feel that veterinary care of spinal cord injured animals should have no less of a goal. Moreover, spinal cord injured animals in optimal acute and chronic health are essential for successful experimental studies of SCI treatments. In this chapter, we will report veterinary care methods to optimize the health of experimental animals, particularly rats and mice, with tetraplegia or paraplegia after cervical or thoracic SCI.

Keywords: Care, Cervical spinal cord injury, Handling, Housing, Morbidity, Mortality, Pain, Paraplegia, Tetraplegia, Thoracic spinal cord injury

1 Model Selection

There are four types of traumatic injury to the adult human spinal cord: (1) contusion evolving to cavity formation, (2) massive compression, (3) laceration, and (4) solid cord.[8,9] These can be anatomically and functionally complete or incomplete. Monkeys, dogs, cats, opossums, and rodents are being used for mammalian models of the different types of traumatic adult human spinal cord injury (SCI) in experimental studies.[10–16] Since damage processes and dysfunctions in spinal cord injured rats and mice are similar to those that occur in humans after SCI,[17–21] they are frequently used.[22–30]

Veterinary care methods for rats and monkeys with paraplegia after SCI have been described.[31] Additionally, methods for rats and mice are included

From: *Animal Models of Acute Neurological Injuries.*
Edited by: J. Chen (eds.) New York, Totowa, NJ

in articles about experimental SCI studies. We will describe veterinary care methods developed and utilized at the Department of Physiology, The Miami Project to Cure Paralysis, the Kentucky Spinal Cord Injury Research Center, and the Spinal Cord and Brain Injury Research Center as well as by the faculty, veterinarians, and their technical staffs at Ohio State University. These methods were approved by the Institutional Animal Care and Use Committees of the SUNY Health Science Center, University of Miami, University of Louisville, University of Kentucky, and Ohio State University. They are in accordance with the Public Health Service Policy on Humane Care and Use of Laboratory Animals and with the Policies on the Use of Animals and Humans in Neuroscience Research.

2 Materials and Methods

2.1 Before Surgery and SCI

2.1.1 Housing

Keeping the animal's distress to a minimum will lead to optimal performance throughout an experimental study and both less morbidity and mortality. This can start upon the arrival of the animals at the research institution. The American Association for Laboratory Animal Science suggests at least 48 h for animals to recover from their trip and to acclimate to new housing conditions.[32] Housing is typically done in and under the direction of the research institution's veterinary care facility. Throughout the study, rats and mice are housed in separate rooms. After weaning, males and females are housed separately. The home cage (SE Labs, Napa, CA; Lab Products, Inc., Seaford, DE; Alternative Design Manufacturing, Siloam Springs, AR; TECNIPLAST, Exton, PA) typically consists of a three tier system: a cage, a lid to hold food (Harlan TEKLAD, Madison, WI; PMI Nutrition International, Brentwood, MO) and a water bottle for ad libitum eating and drinking, and a micro-filter top that fits over the lid. Various types of cage bedding are available including corn-cob (The Andersons, Mauinee, OH), wood chips (P. J. Murphy Forest Products, Montville, NJ), and moisture-absorbent Alpha-dri (Shepherd Specialty Papers, Kalamazoo, MI). The dimensions of the cage depend upon the storage rack and the space available in the room for the maximum number of cages. The maximum number of animals housed in a cage is dictated by their sizes and guidelines for the minimal amount of cage floor space required by that size of animal.[33] The housing density to that maximum number will be dictated by the experimental study. The cages are stored on racks with or without water systems and air systems (Lab Products, Inc., Seaford, DE; Alternative Design Manufacturing, Siloam Springs, AR; TECNIPLAST, Exton, PA). Typically, the cages are changed at least twice a week. The rooms are usually on a 12 or 10 and 14 h dark/light cycle, at 75°F, and under positive or negative air pressure. All experimental and veterinary care methods typically are performed during the light cycle.

2.1.2 Physical Examination

Pathogen-free animals are purchased from commercial vendors. It is also recommended that the animals undergo a physical examination during the acclimation period. The results should be recorded (Table 1). Veterinarians and the vendors should be consulted with before using an animal that exhibits any complications.

Table 1 Physical examination checklist.

Project Name:

ID#: Strain: Sex: Date of Birth: Arrival Date:

Examination Date: Examiner: Body Weight (gms):

Check for the following health issues. Circle signs that are present. If an animal exhibits any of these signs, seek veterinarians and vendors assistance to decide whether to remove the animal from the study.

- ❑ Discharges from the eyes, ears, nose, and mouth
- ❑ Dehydration (lift the skin over the back and determine whether it quickly returns to its original position)
- ❑ Constipation or diarrhea
- ❑ Fractures or amputations
- ❑ Bites, scratches, wounds, abscesses, or calluses
- ❑ Hair loss or overgrooming
- ❑ Labored breathing

2.1.3 Acclimation to Handling and Apparatuses

Experimental SCI studies typically require more frequent handling of the animals for daily veterinary care and functional assessments than which routinely occurs at the vendors and animal care facilities. Handling and being out of their home cage are distressing to the animals. As a result, scratching and biting can occur. This can be reduced by acclimation. First aid for bites and scratches, and the possible allergic reactions to them, should be readily available. Regarding handling, it is strongly recommended that this be done frequently till the animals do not vocalize. The animals should not be lifted or transferred by their tails, especially after SCI. This should be done firmly but not impair breathing. Frequently transferring the animals to the apparatuses used for behavior or electrophysiological training and assessments so that they can explore them are also highly recommended.

2.1.3.1 Rats can be handled the following ways:

2.1.3.1.1 Gently approach the rat while it rests on bedding. Place thumb and index finger of a gloved hand over right and left sides of the body. Gently pick up the rat. It will brace itself with the forepaws on handler's thumb. Be cautious of the rat's ability to maneuver itself and bite.

2.1.3.1.2 Gently approach the rat while it rests on bedding. Cover it from head to base of tail with a cloth or paper towel leaving excess cloth over its head like a hood. Place the gloved thumb and index finger over the towel and the right and left sides of the rat's body. The thumb will be under its forepaws. The excess towel and thumb position will decrease the likelihood of the rat biting the handler. Rats that are accustomed to being handled do not require the towel.

2.1.3.1.3 Gently approach the rat while it rests on bedding. When stationary, place the rat's head between the gloved index and middle fingers. Lift rat and apply gentle pressure to be certain it does not wiggle excessively. Use the opposite hand to support hindquarters and tail. This "V" method is useful when giving intramuscular injections in the hindquarters.

2.1.3.2 Mice can be handled the following way:

2.1.3.2.1 Wait until the mouse is in a corner and at rest. Gently approach it while cupping both gloved hands. If the mouse is not in a corner, slowly move it towards one with cupped hands. Wait for the mouse to move onto the hands then close them. If it does not move onto the hands, gently close them around it. Transport the mouse out of the cage and place it on top of a wire cage top. Gently grip the tip of its tail with one hand. The mouse will grasp the cage top with both forelimbs. At this time, use the thumb and index finger of the opposite hand to grasp the skin between the mouse's shoulder blades. To avoid excessive movement, gently place the tail between the ring and last fingers.

2.1.4 Food Restriction

Optimal training for behavior assessments that involve obtaining food or a food reward require that the animal be fasted overnight or every other overnight or continuously food restricted throughout the training period.[34,35] For continuous food restriction, it is recommended that the animals be housed individually and that their body weights be measured every day and maintained between 80 and 90% of littermates housed individually and eating ad libitum. It is also recommended that ad libitum eating occur for at least 1 week before surgery and SCI. Food restriction after SCI may be detrimental especially to animals with severe cervical SCI. Moreover, we found that ad libitum eating did not alter the food retrieval performances of sham-operated rats compared to their baseline performances.[34]

2.2 Surgery and SCI

Surgery can be performed once the animals are acclimated to housing and handling and once the necessary baseline assessments are performed. Specific survival surgery methods are described in the other chapters. In general, survival surgery requires that aseptic methods be used to prevent infections. These include using sterile instruments, sterile materials, and sterile drapes for the surgery sites as well as sterile gloves, masks, caps, clean labcoats, and disinfected surgery areas. Instruments, materials, and gloves should be changed when moving from skin to muscle, bone, and nervous system tissues. Anesthetics, analgesics, antibiotics, ointments/topicals, and fluids should be used (Table 2). Normothermia should be maintained throughout surgery.

2.2.1 Sterilization

2.2.1.1 According to manufacturer's instructions (Fine Science Tools, Inc., Foster City, CA; Roboz Surgical Instrument Co., Inc., Gaithersburg, MD), instruments should be cleaned with a neutral pH detergent in distilled or deionized water and a soft nylon toothbrush or ultrasonic cleaner. They should be thoroughly rinsed with distilled or deionized water and dried.

2.2.1.2 Surgery instruments and materials should be heat sterilized in pouches with an autoclave, gas sterilized, or cold sterilized with a disinfectant not caustic to animal tissues. After cleaning and rinsing with sterile distilled or deionized water, instruments should be heat sterilized with a glass bead sterilizer between animals on the same day.

2.2.1.3 Many of the surgery materials indicated above and below can be obtained from Henry Schein, Inc. (Melville, NY).

Table 2 Products names and manufacturers information.

	Product name	Manufacturer information
Anesthetics/tranquilizers	Isoflurane	Baxter Healthcare Corporation, Deerfield, IL
	Halothane	Halocarbon Laboratories, River Edge, NJ
	Nembutal® Sodium Solution	Abbott Laboratories, North, Chicago, IL
	Ketaset®	Fort Dodge Animal Health, Fort Dodge, IA
	Acepromazine Maleate	Boehringer Ingelheim Vetmedica, Inc., St Joseph, MO
Analgesics	Xylazine	IVX Animal Health, Inc., St Joseph, MO
	Buprenorphine	Reckitt Benckiser Healthcare, Ltd., Hull, England
Antibiotics	Gentozen® (Gentamicin Sulfate Solution)	Phoenix Scientific, Inc., St Joseph, MO
	Penicillin G Procaine	Bimeda, Inc., Irwindale, CA
	Baytril	Bayer Healthcare LLC, Shawnee Mission, KS
	Clavamox®	Pfizer, New York, NY
Ointments/topicals	Betadine®	Purdue Products L.P., Stamford, CT
	Lacri-Lube®	Allergan, Irvine, CA
	Bacitracin Zinc Ointment USP	Fougera, Melville, NY
Fluids	0.9% Sodium Chloride Solution (Saline)	Hospira, Inc., Lake Forest, IL
	Lactated Ringers Solution	B. Braun Medical, Inc., Irvine, CA
	5% Dextrose in Lactated Ringers Solution	B. Braun Medical, Inc., Irvine, CA

2.2.2 Preparation

2.2.2.1 It is recommended that these methods be performed in an area that is different than the surgery area. The animal should be gently weighed and this, as well as other surgical information should be recorded (Table 3).

2.2.2.2 Anesthetics are required for preparation and surgery. A Drug Enforcement Agency License is necessary for their purchase (for example from Henry Schein, Inc., Melville, NY). Inhalational (INH) gas anesthetics (Isoflurane, Halothane) mixed with oxygen and nitrous oxide can be delivered to a covered induction chamber containing the animal. For this and for surgery, safe gas delivery and evacuation are required (Smiths Medical PM, Inc., Waukesha, WI). Sterile chemical anesthetics (Nembutal® Sodium Solution, Ketaset®) can be injected intraperitoneally (IP) or intramuscularly (IM). Sterile chemical analgesics (Xylazine) and tranquilizers (Acepromazine Maleate) can be mixed with the chemical anesthetics. Antibiotics (Gentozen®) should also be administered at this time. The animal should be placed in a covered and empty cage to prevent aspiration of bedding or food. To prevent hypothermia, the induction chamber and cage should be placed on a cloth towel over a water circulating pad set at 37.5°C or an electric heating pad at the lowest setting. The absence of the blink reflex to touching the cornea or limb withdrawal to toes or digit pinch should be used to ensure that the animal is not lightly anesthetized before proceeding and during surgery. Respiratory rate and mucous membrane or paw pads color should also be monitored for the depth of anesthesia. Gray or blue coloration of the mucous membranes around the eyes and mouth or paw pads indicates inadequate tissue oxygenation. If necessary, "mouth-to-mouth" resuscitation can be given to rats using a 10 or 20 mL syringe with the plunger removed.

Table 3 Surgery sheet.

Project Name: ____

ID#: ____________ **Date:** ____________ **Surgeon:** __________________ **Investigator:** _____________

Strain: _________________ **Sex:** _________ **Starting Weight:** _________g

Procedure:

____

Anesthesia: **IP** **IM** **SQ** **IV** **INH** **Additional:** ______________mL
_______________mL

Pentobarbital: ____________mL

Ketamine/Xylazine (70%/30%): ______________mL

Isoflurane Halothane (0.5 –4.0): ______________v (NO2 70% – O2 30%)

Gentamicin IM: **0.1 mL** **Fluids:** _____cc **SQ** **L. Ringer** **Dex./LR** **Saline**

Topical: Bacitracin – Incision site ________ **Lacri-lube – Eyes** ________

Comments:____

____

____

____

Complications:____

____

____

Illustration of procedure done or damage that may have occurred:

Additional Procedures / Surgeries:

Date: _______ **Weight:** _______g **P / KZ / Hal.** ________ **Procedure:** _______

Date: _______ **Weight:** _______g **P / KZ / Hal.** ________ **Procedure:** _______

Date: _______ **Weight:** _______g **P / KZ / Hal.** ________ **Procedure:** _______

Date: _______ **Weight:** _______g **P / KZ / Hal.** ________ **Procedure:** _______

If the animal sounds and/or feels raspy or is choking, pull the tongue to one side of the mouth. Use a gavage feeding needle (Fine Science Tools, Inc.) to carefully suction fluids or objects that are obstructing the airway. Do not attempt this on an unanesthetized animal.

2.2.2.3 Hair at and around the surgery site should be removed by careful shaving and vacuuming. Plucking the hair or a depilatory cream also can be used. Shaving the abdomen and hindquarters of animals can be done to prevent the skin from being damaged by urine-soaked hair.

2.2.2.4 Beginning at the surgery site then moving outward in increasingly larger circles, the skin should be scrubbed with gauze or a cotton-tipped applicator containing povidone-iodine microbiocide (Betadine®, Table 2) then 70% ethanol. This should be repeated and end with the microbiocide.

2.2.2.5 Ophthalmic ointment (Lacri-Lube®, Table 2) should be applied to the eyes to prevent the corneas from drying. It is recommended that the ointment be put on to a clean cotton-tipped applicator then the eye to prevent contamination.

2.2.2.6 One-half the final volume of sterile, core temperature fluids (0.9% sodium chloride or saline solution, lactated Ringers solution, Table 2) can be injected subcutaneously (SQ) to prevent dehydration and to account for blood loss during surgery. It is recommended that the fluids be injected away from the surgery site at two locations adjacent to the body's midline.

2.2.2.7 An identification system should be administered at this time. Permanent markers can be used to put numbers or codes on the tail. This needs to be repeated throughout the experimental study.

2.2.2.8 The animal's core temperature should be maintained at 37.5°C during surgery. This can be accomplished with a homeothermic blanket system (Harvard Appartus, Inc., Holliston, MA). A water circulating heating pad (Gaymar Industries Inc., Orchard Park, NY) and thermometer with a flexible rectal probe (Physitemp Instruments, Inc., Clifton, NJ) are also recommended. Electric heating pads, warmers (Grabber Performance Group, Grand Rapids, MI), and heat lamps can be used with extreme caution. A towel or drape should be between the animal and the blanket, pads, or warmers.

2.2.2.9 A sterile drape with a previously made hole in it should be placed over the animal. This will expose the surgery site and prevent hair from entering it.

2.2.3 Surgery and SCI

2.2.3.1 It is essential that the animal's depth of anesthesia and core temperature be monitored as described above. Additionally, monitoring of blood pressure, heart rate, blood flow, oxygen saturation, respiratory rate, and expired carbon dioxide is suggested.

2.2.3.2 The surgery site should be flushed with sterile, core temperature saline to prevent dehydration of the exposed tissues.

2.2.3.3 Immediately following contusive cervical SCI and during incision closure, we have found that inhalation of 100% oxygen can be used to compensate for potential respiratory dysfunction.[36]

2.2.3.4 Muscle and skin incisions should be sutured in layers. Absorbable materials, such as chromic gut, and round needles can be used to suture muscles. Skin can be sutured with stainless steel wound clips or staples. Non-absorbable materials, such as nylon, and cutting needles can also be used. Cyanoacrylate adhesives are also available but should be used with caution and for small incisions.

2.2.3.5 Ointments containing antibiotics (Bacitracin Zinc Ointment USP, Table 2) should be applied to sutured skin incisions.

2.2.3.6 The remaining one-half of the sterile, core temperature fluids (saline solution, lactated Ringers solution with or without 5% dextrose, Table 2) can be injected subcutaneously as described above.

2.3 Acutely and the First Week after Surgery and SCI

2.3.1 Housing

2.3.1.1 After manually expressing their bladders (see below), rats and mice should be placed on a clean paper towel or drape in a clean, empty, and covered home cage in a recovery room or their housing rooms. The animals should not be handled by the tail. They should lie on their side and the head should face toward the cage front. The cage should be placed on a cloth towel over a water-circulating pad set at 37.5°C or an electric heating pad at the lowest setting. Importantly, systemic hypothermia[37] or hyperthermia[38] after SCI can be either therapeutic or harmful. Should a recovery room be used, cages of mice and rats should be separated from each other and the environmental conditions should be the same as those in the housing rooms.

2.3.1.2 The animals should be watched closely and turned until in sternal recumbency. This may not occur in animals with cervical SCI. Instead, the blink reflex should be monitored. When sternal recumbency or a blink reflex is observed, the animal should be transferred to a home cage prepared as described above. Information about the surgery should be listed on the cage card for identification and for suture or wound clip removal.

2.3.1.3 Importantly, the number of animals in a cage, the cage size, and whether to add objects (tunnels, running wheels, etc.) to the cage must be carefully considered given the equivocal results thus far for the effects of environmental enrichment after SCI[39–46] and the experimental study's goals. Animals that are housed together should be separated when mutilation and extreme domination behaviors occur.

2.3.1.4 We recommend that the soft and moisture absorbent Alpha-dri bedding be used. One half of the cage should be on a water-circulating or an electric heating pad so that the animal can move away from it.

2.3.2 Veterinary Care

2.3.2.1 It is our contention that spinal cord injured animals should be treated no differently than spinal cord injured humans. Thus, the animals should undergo physical examination and veterinary care once daily or more times as necessary. The frequency directly depends on the site and severity of SCI. This should also be done for animals that undergo sham methods. An Animal Care Sheet (Table 4) is of utmost benefit after surgery and SCI as it can be used to document the day-to-day health status (activity/lethargy, body weight, hydration, bowel, bladder, incisions, scratches, abscesses, callouses, autophagia, etc.) and veterinary care that is being provided (food supplements, fluids, bladder expression, antibiotics, analgesics, etc.) as well as aid the veterinarians when consultations are required.

2.3.2.2 Hydration should be assessed daily in the morning. This can be done as described above and by presenting the spout of a water bottle to the animal. A water bottle with a curved spout should be used after SCI. An additional water bottle can be placed on the bedding. Sterile, core temperature saline solution or lactated Ringers solution (3–5 mL SQ for rats and 0.5–1 mL SQ for mice, Table 2) can be injected subcutaneously as described above. The bladder status (see below) should also guide fluid administration.

Table 4 Animal care sheet.

Month/Year:

- ❑ Date, AM/PM:

Rat/Mouse ID #:

Surgery/Surgeon's initials/Surgery date:

- ❑ Activity: None, Hypo, OK
- ❑ Forelimb/Hindlimb flaccid paralysis:
- ❑ Forelimb/Hindlimb support:
- ❑ Locomotion:
- ❑ Nares: Blood, OK
- ❑ Ears/Paws: Pale color, OK
- ❑ Eyes: Ptosis, crusty, OK
- ❑ L,R Forepaw nails or digits 1–5 chewing:
- ❑ L,R Hindpaw nails or digits 1–5 chewing:
- ❑ L,R Hindpaw callouses:
- ❑ L,R Forelimb hair loss:
- ❑ L,R Hindlimb hair loss:
- ❑ Body weight (gm):
- ❑ Feces: None, solid, moist, dry, runny
- ❑ Bladder: Voiding or S/M/L
- ❑ Urine: Clear, concentrated, bloody, cloudy, crystals, on abdomen
- ❑ Thirsty/Dehydrated:
- ❑ Fluids: Dextrose-L. ringers, L. ringers, saline
- ❑ Incision: Infected, OK
- ❑ Suture/Clips: Off @ 7–10 days post-op
- ❑ Implant: Infected, OK
- ❑ Scratching: Location
- ❑ Bacitracin: SID, PRN
- ❑ Gentamicin: 0.1 mL IM every 48 h × 5 days
- ❑ Baytril: 0.1 mL IM Daily for bloody urine
- ❑ Buprenorphine: IM daily for 48 h
- ❑ Comments:

2.3.2.3 Body weight should be measured daily in the morning. Weight loss after surgery and SCI naturally occurs. Food and cereal can be broken into bite-size pieces and scattered on the bedding. When weight loss persists, poor hydration and/or nutrition may be the causes. In addition to the fluids indicated above, commercially available, core temperature liquid diets and precision pellets can be used (BIO-SERV, Frenchtown, NJ). A mash of food soaked in water can be placed on a paper towel or in glass dishes. Apples and potatoes can also be used for both nutrition and hydration. The mash, fruits, and vegetables should be examined daily and changed when contaminated. Since these will increase the moisture in the covered cage, more frequent cage changes are recommended. Even though they are initially hesistant, we have also found that animals that are severe tetraplegics or lethargic will drink core temperature ENSURE® from a 1 mL syringe or glass dish.[36] Resistance by the animal indicates when this should stop.

2.3.2.4 Bladder function and urine should be evaluated at least once daily in the morning to avoid urinary tract infections, renal complications, and urine scald. Not only do these negatively affect the health of the animal, they also could alter behavior. It is very helpful if a bladder and its location are seen in an euthanized animal before proceeding with this method.

2.3.2.4.1 The bladder should be located and gently palpated with the thumb and fingers while the animal is on the bedding or being handled. Its size and firmness should be recorded as well as whether urine is on the hair and skin. If the bladder contains urine, the animal should be handled over an empty cage. Typically, the fingers and/or thumb of the free hand are used to gently press the bladder ventral-to-dorsal or to gently squeeze it lateral-to-lateral. Return the animal to its home cage when the bladder is empty. Tremendous care must be taken, especially in males where there is more resistance to urine flow, to not fatally rupture the bladder. When expression is difficult, the bladder can be massaged, the urethra should be checked for plugs or blood clots, or the animal returned to its home cage before trying again.

2.3.2.4.2 Voided or expressed urine should be examined for color, clarity, and odor. It is normally dilute and free of cloudiness, crystals, puss, blood, and a strong odor. Any change of the urine should be aggressively treated. Fluid administration is recommended especially for concentrated urine and the dehydration this suggests. Urine that is cloudy and has crystals or a strong odor can be treated with Gentozin® (50 mg mL^{-1} for rats, 0.2 mg mL^{-1} for mice, 0.1 mL IM every other day for 5 days, Table 2). Core temperature Clavamox® can also be used for rats (62.5 mg mL^{-1}, 0.1 mL orally from a 1 mL syringe for 5 days, Table 2). Blood may be in the urine initially after SCI. Should this stop then reappear, more gentle and frequent expression should be tried to reduce bladder damage. Baytril can also be used (22.7 mg mL^{-1} for rats, 5 mg mL^{-1} for mice, 0.1 mL IM daily for 3 days, Table 2). Vitamin C tablets can be administered to acidify the urine. Consult with the veterinarian should these conditions persist beyond a couple of days.

2.3.2.4.3 When crystals appear on the hair, treatment should be as described above. Urine scald occurs when urine persistently keeps the hair and skin moist and hindlimb weight support is lim ited. Moreover, this can lead to a vagina or penis infection. We have found that covering the shaved abdomen with diaper rash ointment or Desitin Ointment (Pfizer Consumer Healthcare, Morris Plains, NJ) is helpful. The affected area should be washed with warm soapy water and thoroughly dried. More frequent bladder expression and bedding changes are also recommended.

2.3.2.5 Bowel function and feces should be examined at least once daily in the morning. Feces should be solid and moist or dry. While infrequent, no new feces suggest constipation. Abdomen palpation with the fingers and hydration should ensue. The anus should be checked for plugs. We have found that a core temperature mixture of sterile water and Epsom Salts (2 cups/gallon of water, Cumberland Swan, Smyrna, TN) administered orally from a 1 mL syringe to be useful. Very moist feces indicate diarrhea and that fluid intake and bladder status should be managed. Discolored feces typically are due to nutritional supplements. Blood may infrequently be seen on the feces. Consult with the veterinarians should these conditions persist beyond a couple of days.

2.3.2.6 Antibiotics should be delivered, but cautiously, during this time. The dosage and administration days depend on which antibiotics are being used. For example, Gentozin® (50 mg mL^{-1} for rats, 0.2 mg mL^{-1} for mice, 0.1 mL IM, Table 2) can be delivered every other day for 5 days. Diarrhea can occur due to changes of intestinal flora. Antibiotic ointment can also be applied daily to skin incisions.

2.3.2.7 Analgesics can be delivered at this time if necessary and appropriate. We administer the mixed opioid agonist-antagonist buprenorphine (0.05 mg mL^{-1} for rats, 0.015 mg mL^{-1} for mice, 0.1 mL SQ, Table 2) 12–24 h after surgery, so as to not fatally depress the respiratory rate, and as needed.

2.3.2.8 If skin incisions are healing, wound clips and sutures should be removed at 7–10 days. Afterwards, they can be a source of irritation, scratching, and infection. This can be done without anesthesia in gentled animals or with gas anesthesia.

2.3.2.9 Issues that also may arise during this time include excessive grooming, scratches, wounds, and autophagia. Bitter ChewGuard (Summit Hill Laboratories, Navesink, NJ) can be sprayed when excessive grooming occurs before the skin is opened. Gumabones (BIO-SERV) can also be an effective deterrent. Scratches and superficial wounds can be cleansed with 3% hydrogen peroxide and thoroughly rinsed with sterile water. New-Skin® liquid bandage (Medtech, Jackson, WY) can be applied to the treated scratch or wound. This approach can also be used for callouses that are infrequently found on the hindpaw heels. We have found that topical application of New-Skin® and metronidazol is effective for preventing and abolishing autophagia.[47] Clean bedding and applying antibiotic ointment then wrapping with gauze, Vetrap™ (3M, St. Paul, MN), and tape are approaches that we have found to be effective for deep wounds caused by autophagia. The bandages and bedding should be changed frequently. Digits and paws swelling is also an infrequent occurrence. They should be soaked twice a day for 10–15 min in a mixture of warm water and Epsom Salts until the swelling subsides. Ulcers at injection sites or from pressure at bony prominences and abscesses at incision sites are infrequently seen. Animals may contort their body due to them. Callouses can be infrequently found on the hindpaws. In consultation with the veterinarians, antibiotics, such as Penicillin G Procaine (300,000 units mL^{-1}, 0.1 mL IM daily for 3–5 days, Table 2) and analgesics, such as buprenorphine, may need to be administered when these and the other issues arise. When severe, euthanasia should be performed.

2.4 Chronically after Surgery and SCI

2.4.1 The cages should be removed from the water-circulating or electric heating pads when the animals spend little time over them. This typically happens after the first week post-surgery or –SCI. Additionally, the core temperature can be checked. The pads can be removed when the animals are normothermic. Should an animal be found to be cold to the touch at anytime or shivering, the pads should be used.

2.4.2 The physical examination and all of the veterinary care methods described above should continue to be used daily. Hydration and nutrition, bladder and bowel function, and skin condition should be paid particular attention. Even though reflexive voiding may occur, it is recommended that the decision to stop expressing the bladder be very conservative.

Acknowledgements: A.H.P., C.D.N., and S.M.O. were supported by NIH/NCRR/COBRE P20 RR15576 (Core C: S.M.O.; S.R. Whittemore, Kentucky Spinal Cord Injury Research Center, University of Louisville). S.M.O. was supported by NIH/NINDS R01 NS40411 and P30 NS051220 (E.D. Hall, Spinal Cord and Brain Injury Research Center, University of Kentucky). Sincere appreciation is extended to Dr. Russell G. Durkovic, Deyanira Santiago, Dr. Frances M. Doepel, Dr. William W. King, Dr. Jeffrey Smiley, the MASCIS Group of Investigators (*Unpublished Document*), as well as the faculty, veterinarians, and their technical staffs at Ohio State University (Basso DM, Beattie MS, Bergdall V, Bresnahan JC, Buford JA, Jakeman LB, Kostyk SK, McTigue DM, and Popovich, PG, *Unpublished Document*, Spinal Cord Injury Training Manual, NINDS, Facilities of Research Excellence in Spinal Cord Injury, Copyright Ohio State University, 2004–2008) for their guidance in providing the highest quality of veterinary care to spinal cord injured animals.

References

1. Blight AR, Tuszynski MH. Clinical trials in spinal cord injury. J Neurotrauma 2006;23:586–593
2. http://clinicaltrials.gov/search/term = Spinal%20Cord%20Injury
3. Kleitman N. Keeping promises: translating basic research into new spinal cord injury therapies. J Spinal Cord Med 2004;27:311–318
4. Chiodo AE, Scelza WM, Kirshblum SC, Wuermser LA, Ho CH, Priebe MM. Spinal cord injury medicine. 5. Long-term medical issues and health maintenance. Arch Phys Med Rehabil 2007;88:S76–S83
5. Cortez R, Levi AD. Acute spinal cord injury. Curr Treat Options Neurol 2007;9:115–125
6. Strauss DJ, Devivo MJ, Paculdo DR, Shavelle RM. Trends in life expectancy after spinal cord injury. Arch Phys Med Rehabil 2006;87:1079–1085
7. Wuermser LA, Ho CH, Chiodo AE, Priebe MM, Kirshblum SC, Scelza WM. Spinal cord injury medicine. 2. Acute care management of traumatic and nontraumatic injury. Arch Phys Med Rehabil 2007;88:S55–S61
8. Bunge RP, Puckett WR, Becerra JL, Marcillo A, Quencer RM. Observations on the pathology of human spinal cord injury. A review and classification of 22 new cases with details from a case of chronic cord compression with extensive focal demyelination. Adv Neurol 1993;59:75–89
9. Bunge RP, Puckett WR, Hiester ED. Observations on the pathology of several types of human spinal cord injury, with emphasis on the astrocyte response to penetrating injuries. Adv Neurol 1997;72:305–315
10. Xu XM, Martin GF. The response of rubrospinal neurons to axotomy in the adult opossum, Didelphis virginiana. Exp Neurol 1990;108:46–54
11. Blight AR. Morphometric analysis of a model of spinal cord injury in guinea pigs, with behavioral evidence of delayed secondary pathology. J Neurol Sci 1991;103:156–171

12. Khan T, Havey RM, Sayers ST, Patwardhan A, King WW. Animal models of spinal cord contusion injuries. Lab Anim Sci 1999;49:161–172
13. Fry EJ, Saunders NR. Spinal repair in immature animals: a novel approach using the South American opossum Monodelphis domestica. Clin Exp Pharmacol Physiol 2000;27:542–547
14. Rossignol S, Chau C, Giroux N, Brustein E, Bouyer L, Marcoux J, Langlet C, Barthelemy D, Provencher J, Leblond H, Barbeau H, Reader TA. The cat model of spinal injury. Prog Brain Res 2002;137:151–168
15. Jeffery ND, Smith PM, Lakatos A, Ibanez C, Ito D, Franklin RJ. Clinical canine spinal cord injury provides an opportunity to examine the issues in translating laboratory techniques into practical therapy. Spinal Cord 2006;44:584–593
16. Courtine G, Bunge MB, Fawcett JW, Grossman RG, Kaas JH, Lemon R, Maier I, Martin J, Nudo RJ, Ramon-Cueto A, Rouiller EM, Schnell L, Wannier T, Schwab ME, Edgerton VR. Can experiments in nonhuman primates expedite the translation of treatments for spinal cord injury in humans? Nat Med 2007;13:561–566
17. Beattie MS, Bresnahan JC, Komon J, Tovar CA, Van Meter M, Anderson DK, Faden AI, Hsu CY, Noble LJ, Salzman S, Young W. Endogenous repair after spinal cord contusion injuries in the rat. Exp Neurol 1997;148:453–463
18. Popovich PG, Wei P, Stokes BT. Cellular inflammatory response after spinal cord injury in Sprague-Dawley and Lewis rats. J Comp Neurol 1997;377:443–464
19. Metz GA, Curt A, van de Meent H, Klusman I, Schwab ME, Dietz V. Validation of the weight-drop contusion model in rats: a comparative study of human spinal cord injury. J Neurotrauma 2000;17:1–17
20. Norenberg MD, Smith J, Marcillo A. The pathology of human spinal cord injury: defining the problems. J Neurotrauma 2004;21:429–440
21. Fleming JC, Norenberg MD, Ramsay DA, Dekaban GA, Marcillo AE, Saenz AD, Pasquale-Styles M, Dietrich WD, Weaver LC. The cellular inflammatory response in human spinal cords after injury. Brain 2006;129:3249–3269
22. Kuhn PL, Wrathall JR. A mouse model of graded contusive spinal cord injury. J Neurotrauma 1998;15:125–140
23. Farooque M. Spinal cord compression injury in the mouse: presentation of a model including assessment of motor dysfunction. Acta Neuropathol (Berl) 2000;100:13–22
24. Jakeman LB, Guan Z, Wei P, Ponnappan R, Dzwonczyk R, Popovich PG, Stokes BT. Traumatic spinal cord injury produced by controlled contusion in mouse. J Neurotrauma 2000;17:299–319
25. Jacob JE, Pniak A, Weaver LC, Brown A. Autonomic dysreflexia in a mouse model of spinal cord injury. Neuroscience 2001;108:687–693
26. Inman D, Guth L, Steward O. Genetic influences on secondary degeneration and wound healing following spinal cord injury in various strains of mice. J Comp Neurol 2002;451:225–235
27. Joshi M, Fehlings MG. Development and characterization of a novel, graded model of clip compressive spinal cord injury in the mouse: Part 1. Clip design, behavioral outcomes, and histopathology. J Neurotrauma 2002;19:175–190
28. Joshi M, Fehlings MG. Development and characterization of a novel, graded model of clip compressive spinal cord injury in the mouse: Part 2. Quantitative neuroanatomical assessment and analysis of the relationships between axonal tracts, residual tissue, and locomotor recovery. J Neurotrauma 2002;19:191–203
29. Zheng B, Lee JK, Xie F. Genetic mouse models for studying inhibitors of spinal axon regeneration. Trends Neurosci 2006;29:640–646
30. Onifer SM, Rabchevsky AG, Scheff SW. Rat models of traumatic spinal cord injury to assess motor recovery. ILAR J 2007;48(4):385–395
31. Santos-Benito FF, Munoz-Quiles C, Ramon-Cueto A. Long-term care of paraplegic laboratory mammals. J Neurotrauma 2006;23:521–536
32. Laboratory Animal Technician Training Manual. Animal Health Maintenance. Memphis, TN: Sheridan Books; 2005:88–89

33. National Research Council. Guide for the care and use of laboratory animals. Washington (DC): National Academy Press; 1996:22–28
34. Onifer SM, Rodríquez JF, Santiago DI, Benitez JC, Kim DT, Brunschwig J-P, Pacheco JT, Perrone JV, Llorente O, Martinez-Arizala A. Cervical spinal cord injury in the adult rat: assessment of forelimb dysfunction. Restor Neurol Neurosci 1997;11:211–223
35. Titsworth WL, Onifer SM, Liu N, Xu X-M. Focal phospholipases A2 group III injections induce cervical white matter injury and functional deficits with delayed recovery concomitant with Schwann cell remyelination. Exp Neurol 2007;207(1):150–162
36. Onifer SM, Nunn CD, Decker JA, Payne BN, Wagoner MR, Puckett AH, Massey JM, Armstrong J, Kaddumi EG, Fentress KG, Wells MJ, Calloway CC, Schnell JT, Whitaker CM, Burke DA, Hubscher CH. Loss and spontaneous recovery of forelimb evoked potentials in both the adult rat cuneate nucleus and somatosensory cortex following contusive cervical spinal cord injury. Exp Neurol 2007;207(2):238–247
37. Yu CG, Jimenez O, Marcillo AE, Weider B, Bangerter K, Dietrich WD, Castro S, Yezierski RP. Beneficial effects of modest systemic hypothermia on locomotor function and histopathological damage following contusion-induced spinal cord injury in rats. J Neurosurg 2000;93:85–93
38. Yu CG, Jagid J, Ruenes G, Dietrich WD, Marcillo AE, Yezierski RP. Detrimental effects of systemic hyperthermia on locomotor function and histopathological outcome after traumatic spinal cord injury in the rat. Neurosurgery 2001;49:152–158
39. Lankhorst AJ, ter Laak MP, van Laar TJ, van Meeteren NL, de Groot JC, Schrama LH, Hamers FP, Gispen WH. Effects of enriched housing on functional recovery after spinal cord contusive injury in the adult rat. J Neurotrauma 2001;18:203–215
40. Woerly S, Doan VD, Evans-Martin F, Paramore CG, Peduzzi JD. Spinal cord reconstruction using NeuroGel implants and functional recovery after chronic injury. J Neurosci Res 2001;66:1187–1197
41. Van Meeteren NL, Eggers R, Lankhorst AJ, Gispen WH, Hamers FP. Locomotor recovery after spinal cord contusion injury in rats is improved by spontaneous exercise. J Neurotrauma 2003;20:1029–1037
42. Koopmans GC, Deumens R, Honig WM, Hamers FP, Steinbusch HW, Joosten EA. The assessment of locomotor function in spinal cord injured rats: the importance of objective analysis of coordination. J Neurotrauma 2005;22:214–225
43. Erschbamer MK, Pham TM, Zwart MC, Baumans V, Olson L. Neither environmental enrichment nor voluntary wheel running enhances recovery from incomplete spinal cord injury in rats. Exp Neurol 2006;201:154–164
44. Moon LD, Leasure JL, Gage FH, Bunge MB. Motor enrichment sustains hindlimb movement recovered after spinal cord injury and glial transplantation. Restor Neurol Neurosci 2006;24:147–161
45. Burke DA, Magnuson DS, Nunn CD, Fentress KG, Wilson ML, Shum-Siu AH, Moore MC, Turner LE, King WW, Onifer SM. Use of environmentally enriched housing for rats with spinal cord injury: the need for standardization. J Am Assoc Lab Anim Sci 2007;46:34–41
46. Fischer FR, Peduzzi JD. Functional recovery in rats with chronic spinal cord injuries after exposure to an enriched environment. J Spinal Cord Med 2007;30:147–155
47. Zhang YP, Onifer SM, Burke DA, Shields CB. A topical mixture for preventing, abolishing, and treating autophagia and self-mutilation in laboratory rats. Contemp Top Lab Anim Sci 2001;40:35–36

Part II

Cerebral Ischemic Animal Models

Section I

Global Cerebral Ischemic Models

Chapter 6

Four-Vessel Occlusion Model in Rats

Ping Deng and Zao C. Xu

Abstract Four-vessel occlusion model in rat induces transient forebrain ischemia that resembles cardiac arrest in clinical situations. In this model, two vertebral arteries are coagulated and two common carotid arteries are reversibly occluded. This model produces reliable outcome with selective, delayed cell death. For example, CA1 pyramidal neurons in the hippocampus die 2–3 days after ischemia, whereas CA3 neurons survive the same insult. The complications of this model include mortality and seizure. The methods of quantitative analysis of ischemic outcome are introduced. The advantages and limitations of this model are discussed.

Keywords: Wistar rat, Transient global ischemia, Hippocampus, Striatum, Ischemic depolarization

1 Model Selection

Four-vessel occlusion (4-VO) rat model has been widely employed to investigate the mechanisms underlying brain damage after transient global ischemia, which mimics the pathological conditions of cardiac arrest. 4-VO is produced by permanent coagulation of two vertebral arteries and temporary ligation of two common carotid arteries. This model induces transient forebrain ischemia because the blood supply to the hindbrain remains relatively intact.[1] Such insults cause severe, reversible forebrain ischemia with consistent neuronal damage in a species that is large enough to allow physiological monitoring and maintenance with relatively low costs. The original model involves a 2-day procedure inducing an ischemic insult in awake animals. Several modified 4-VO models have been developed in the past decades: for example, to perform the 2-day procedure in anesthetized animals[2]; to perform all procedures on the same day in anesthetized rats while maintaining brain temperature and recording of ischemic depolarization (ID)[3]; to coagulate vertebral arteries from ventral approach,[4] etc.

From: *Animal Models of Acute Neurological Injuries.*
Edited by: J. Chen (eds.) New York, Totowa, NJ

2 Materials

2.1 Animal

The choice of animal for 4-VO model is male adult Wistar rats. The strain is an important factor affecting the success rate. 4-VO induces severe ischemia in ~80% of Wistar rats,[1] whereas only 50–60% of Sprague-Dawley rats exhibit successful ischemia.[5] It is worthwhile to point out that even Wistar rats from different suppliers may exhibit different outcomes. It is therefore necessary to purchase rats with a high success rate from the same supplier. The age of the animal is another factor influencing the outcome. Neuronal sensitivity to ischemia decreases significantly in neonatal rats because of their developmental nature. The success rate of complete ischemia is significantly reduced in aged rats because of the increase of blood vessel collaterals to the brain. The rats with 200–300 gm body weight have higher success rates and more consistent outcomes for the 4-VO model.

2.2 Anesthesia

Anesthetics used for 4-VO models include ether, barbiturates, isoflurane, etc. Each anesthetic has its side effects, such as lowering the body temperature and metabolism, and blocking certain glutamate receptors or ion channels, which might complicate the ischemic pathophysiology and therefore the data interpretation. Isoflurane is recommended because of its relatively fewer side effects and rapid recovery.

2.3 Equipment

2.3.1 Anesthesia

Isoflurane mixer (Matrx, VIP 3000)*
Nasal masks (Fig. 1)

2.3.2 Surgery

Surgical instruments (scissors, forceps, knives, etc.)
Stereotaxic device (KOPF 1430)
Dissecting microscope (Nikon, SMZ-1B)
Drill (Dremel, Multipro, model 395, type 5)
Vessel cauterizer (cordless soldering iron, ISOTIP, model 7700 with microtip)
Buttons (made of Teflon, Fig. 2)
Plastic tube (made of 1 cc syringe, Fig. 2)
Laboratory tubing (SILASTIC, CAT NO. 508-003, OD 1.19 mm (0.025 in.), ID 0.64 mm (0.047 in.)
Hook for occlusion (Ligation aid, model 18062, Fine Science Tools)
Temperature control unit (Temperature controller and alarm, TCAT-1A)
Thermometer (Physitemp, BAT-10)
Temperature probe: (Harvard) Rectal: BS4 52-1526; brain: BS4 52-1732 (OD, 0.64 mm)

2.3.3 Recording of Ischemic Depolarization

Amplifier (Neuroprobe, model 1600)
Electrode puller (KOPF model 750)
Electrode (Thin-walled glass capillary with filament, OD 2.0 mm, ID 1.5 mm)

---

*Equipment in parenthesis is the one used in our lab.

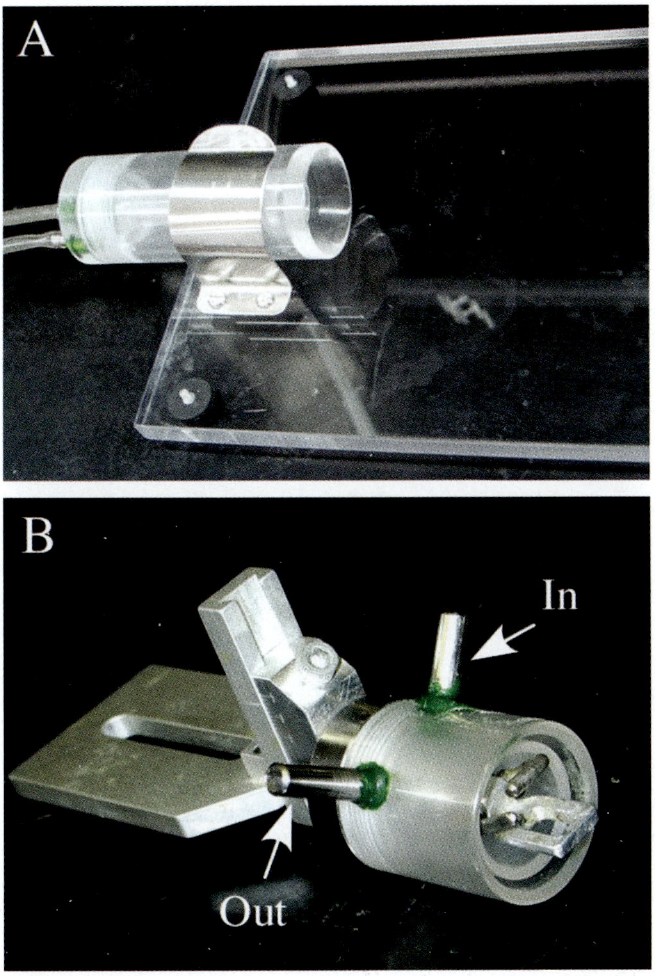

Fig. 1 (**A**) Nasal mask for surgery on ventral neck. (**B**) Nasal mask for stereotaxic apparatus. Both masks have an inlet for anesthesia and an outlet for retrieval of anesthesia by connecting to a vaccum

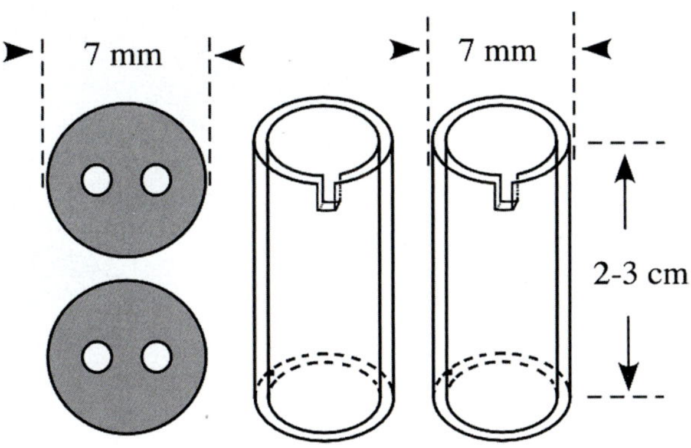

Fig. 2 Drawings of buttons and plastic tube for occlusion

3 Procedures

3.1 Original 4-VO Model

Rats are anesthetized with isoflurane (a mixture of 1–2% isoflurane in 33% O_2 and 66% N_2), and placed on an operating table. Anesthesia is maintained with a nasal mask.

3.1.1 Placement of Occluding Device Around Carotid Artery

1. Make a ventral midline cervical incision to open the skin.
2. Separate the muscles covering the ventral neck with two small curved forceps to expose the common carotid arteries.
3. Isolate the common carotid arteries from the vagas nerves and the cervical sympathetic chains.
4. Place the occluding device, a piece of silicone tubing (0.025 in. ID, 0.047 in. OD, 10–15 cm length), loosely around each common carotid artery and pass through the two holes in a small Teflon button (Fig. 3a).
5. Pass the two ends of the silicone tubing through the plastic tube and tie the tubing with a knot to form a loop. The length of the loop should be shorter than the length of the plastic tube that will be used as the supporting apparatus for occlusion. So, when the silicone tubing loop is pulled through the plastic tube, it will press the common carotid artery against the button and stop the blood flow (Fig. 3c).
6. Close the wound with 1–2 sutures.

3.1.2 Electrocauterization of Vertebral Artery

1. Place the rat on a stereotaxic frame. A dorsal neck incision is made from the occipital bone to the second cervical vertebra (about 1 cm in length). The paraspinal muscles are separated to expose the alar foramina of the first cervical vertebrae. To improve the exposure of the alar foramina, the cervical spine should be gently extended with a rubber band tied on the rat's tail.
2. Insert the tip of the soldering iron into the alar foramen to permanently cauterize the vertebral arteries (Fig. 4). It should be mentioned that the cauterization is crucial for successful ischemia. Heating for too long will

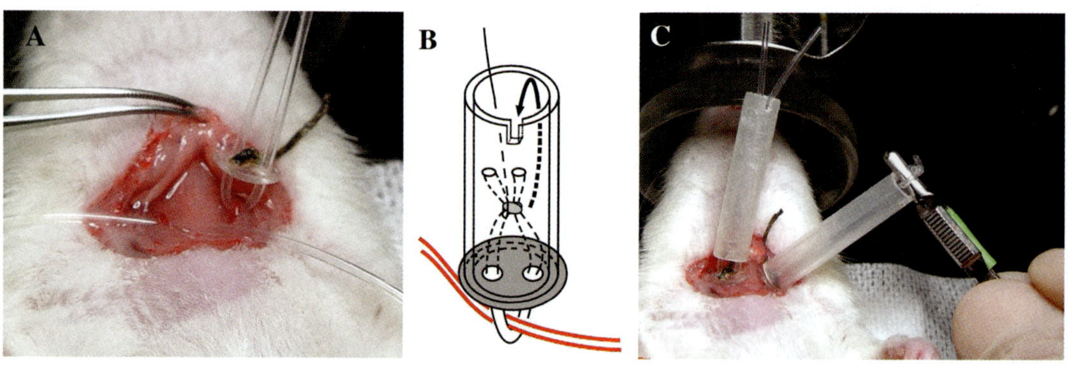

Fig. 3 (**A**, **C**) Photomicrograph of the preparation of the occluding device. (**B**) Schematic drawing of occluding devices for 4-VO

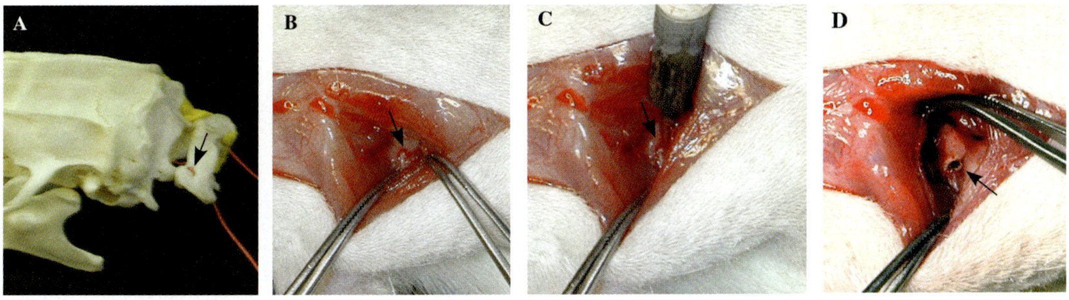

Fig. 4 (**A**) Skeletal specimen showing the alar foramen and vessels (*arrow*). (**B**) Exposure of alar foramen (*arrow*). (**C**) Electrocauterization of vertebrate artery. (**D**) Photomicrograph showing the alar foramen after electrocauterization

damage the pons underneath the cervical vertebrae and increase the mortality, but heating for too short a time will not coagulate the vertebral arteries, resulting in incomplete ischemia. There is no perfect method to detect whether the vertebral arteries are permanently occluded. In our experience, a small probe with a sharp tip, such as a syringe needle, is inserted into the alar foramina and gently twisted. If bleeding occurs, another cauterization is needed.

3. Close the wounds with surgical clips and return the animal to its cage for recovery.

3.1.3 Occlusion of Bilateral Common Carotid Arteries

1. On day 2, the awakened rat (fasted overnight to provide uniform plasma glucose levels) is gently held by one experimenter and put upside down to expose the ventral neck (a rat adapter could be used to secure the head and better expose the operating area).
2. Another experimenter removes the sutures on the ventral neck, pulls the silicone tubing loop through the plastic tube, and secures the knot to the edge of the tube to completely occlude carotid blood flow (Fig. 5). After occlusion of both common carotid arteries, the rat should become unresponsive and lose its righting reflex within 1 min if the ischemia is completed. The core body temperature should be maintained at 37°C during occlusion, via a rectal probe coupled to a heating lamp.
3. Release the knot to restore blood flow after occlusion for a certain period of time, which depends on the brain region of interest. The animal should recover shortly.

3.2 One-Day 4-VO Model (with ID Recording)

Brain temperature is the most important factor influencing the severity of cell death after ischemia. Another potential problem is the collaterals of blood vessels from the skull and neck muscles resulting in incomplete ischemia. This modified 4-VO model resolves these problems by maintaining the brain temperature at 37°C and using ID as an indication of complete ischemia.[6]

3.2.1 Surgical Preparation

1. The procedures for placement of occluding devices on common carotid arteries and electrocauterization of vertebral arteries are the same as

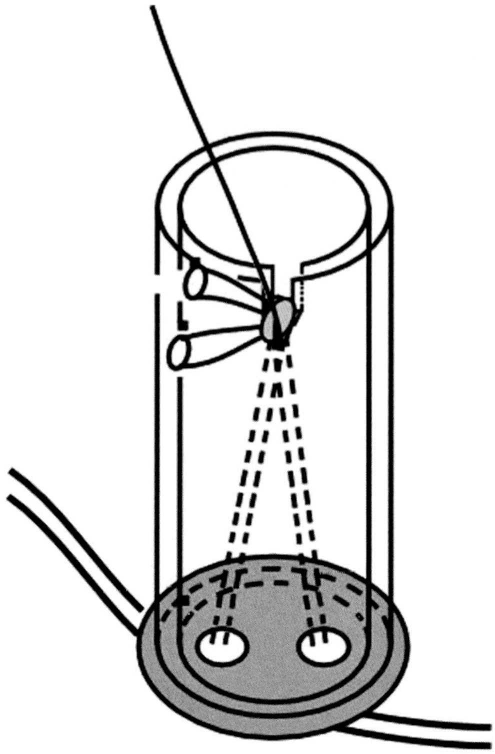

Fig. 5 Schematic drawing of the occlusion of the common carotid artery

described earlier, except that a surgical suture thread is tied to the knot of the tubing.

2. Drill a burr hole on one side of the skull for placement of a temperature probe.
3. Gently insert a tiny temperature probe (0.025 in. in diameter) underneath the skull in the extradural space, and maintain brain temperature at 37°C with a heating lamp, using a temperature control system (Fig. 6).
4. Drill another hole on the other side of the skull above the brain region of interest. The coordinates for hippocampus are AP 5.0 mm, ML 2.5 mm; the coordinates for striatum are: AP 9.5 mm, ML 3.0 mm (Interaural).
5. Peal off the dura matter with a sharp-tipped forceps and keep the brain surface moist with saline.
6. Put a reference electrode (silver wire) underneath the skin at the opening of the dorsal neck.

3.2.2 ID Recording

1. Prepare a glass microelectrode with electrode puller and break the tip of the electrode to yield a tip diameter of 5–10 μm. Fill the electrode with 2 M NaCl.
2. Slowly advance the microelectrode to the region (hippocampus DV 2.5 mm; striatum DV 3.0 mm).
3. Connect the recording electrode and reference electrode to the amplifier. Adjust the DC potential to 0 mV.

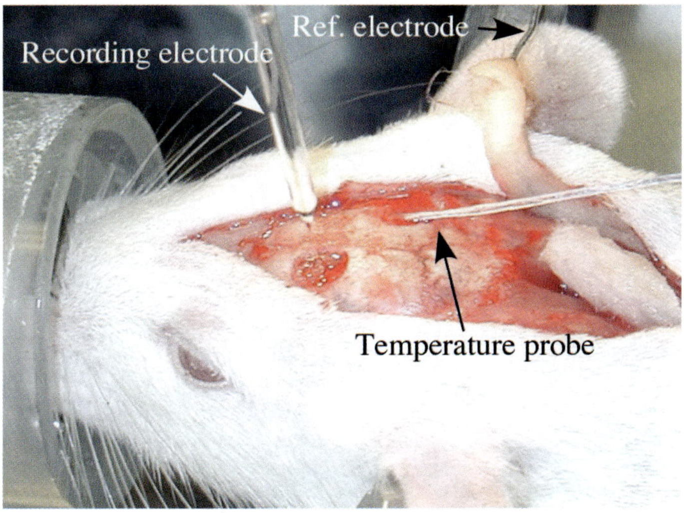

Fig. 6 The operation site of ID recording and temperature monitoring

4. Following baseline recording for 1–2 min, occlude the common carotid arteries bilaterally by pulling the thread, and secure the knot to the notch of the plastic tube. A sudden drop of DC potential from 0 to ~–20 mV should occur 2–3 min after occlusion if complete ischemia is achieved (Fig. 7).
5. Release the occlusion after ID has lasted for 10 min for studying the hippocampus and 20 min for studying the striatum. The DC potential should return to 0 mV in 1–2 min.
6. Remove the electrodes and close the wound with surgical clips.

4 Outcome Evaluation

Several end points, including behavioral, physiological, and histological features, could be used to assess the brain damage and validate the success of the animal model. Among them, histological changes in ischemia-vulnerable regions are the most reliable measurement. Hematoxylin–eosin (HE) staining of a paraffin section is a standard pathological technique and has been widely used to evaluate cell death after transient global ischemia.

4.1 Ischemic Cell Death in the Hippocampus

The CA1 pyramidal neurons in the hippocampus are the neurons most vulnerable to ischemia. Neurons in the dorsal hippocampus are more sensitive to ischemia than those in the ventral hippocampus. Therefore only sections from the dorsal hippocampus are used for evaluation (Fig. 8). CA1 neurons start to show morphological signs of cell death, i.e., cell body swollen or shrunk, and nuclear condensation 2–3 days after ischemia. The cell death begins from the medial part of the CA1 zone, gradually progresses towards the lateral part of the CA1 zone and stops at the junction of the CA3 zone. If the ischemia is relatively mild, the cell death stops somewhere in the middle of the CA1 zone.

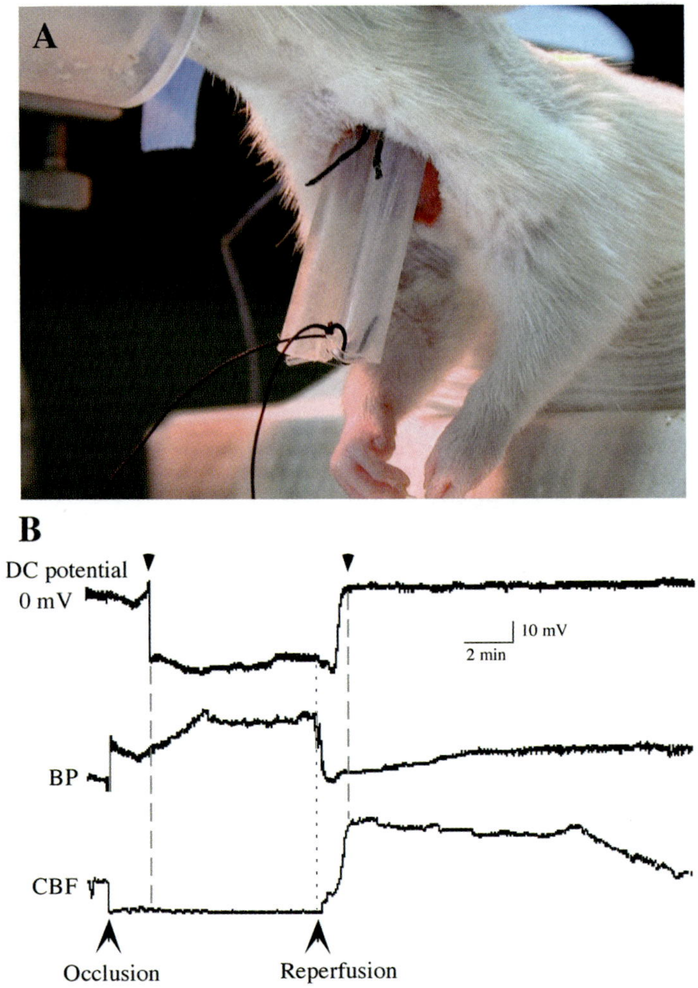

Fig. 7 (**A**) Occlusion of common carotid arteries. (**B**) Traces showing the changes of cerebral blood flow (CBF), arterial blood pressure (BP), and DC potential after ischemia. After occlusion, the blood pressure immediately increases and the DC potential drops to −20 mV (ID). After reperfusion, the blood flow significantly increases (hyperperfusion), and the blood pressure and DC potential return to control levels. The ID duration is defined as the distance between the onset of potential drop and the return to 0 mV (*arrow heads*)

In contrast, the CA3 pyramidal neurons, dentate granule cells, and interneurons in the CA1 region survive the ischemic insult. Another intriguing phenomenon in hippocampus is delayed cell death after ischemia. Approximately 2 days after 10–15 min global ischemia, CA1 neurons start to die and will reach the maximal cell death in 7 days.

There are many ways to quantify ischemic cell death. The grading method[7] is relatively simple, and the results are well correlated to the severity of ischemic insult:

1. Perfuse animal with 4% paraformaldelhyde 7 days after ischemia
2. Embed brain blot containing hippocampus into paraffin

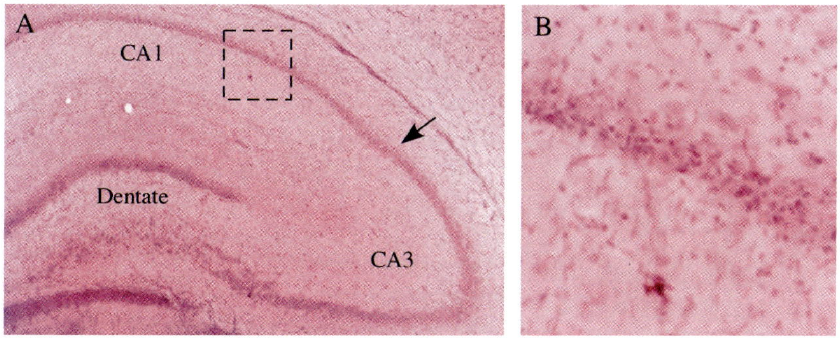

Fig. 8 Photomicrograph of paraffin sections showing the hematoxylin–eosin (HE) staining of the hippocampus 7 days after 10 min forebrain ischemia. (**A**) Most of CA1 pyramidal neurons have degenerated by this time. The *arrow* indicates the boundary of CA1 and CA3 zones where the cell death stops. (**B**) Higher magnification picture of the square area in (**A**) showing the morphological features of cell death in CA1 region after ischemia

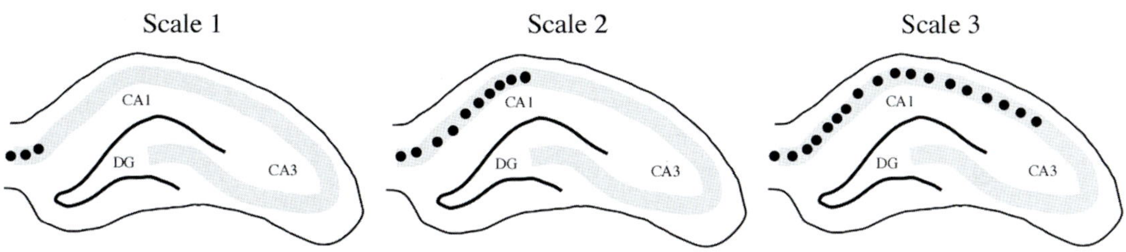

Fig. 9 Schematic drawing showing the grade scale method to quantify the cell death in the hippocampus after transient global ischemia

3. Cut coronal sections of 8 μm thickness of hippocampus and stain with hematoxylin–eosin

   (The details of the above procedure can be found in any histology manual)

4. Select sections from the dorsal hippocampus (between AP 4.8–5.8 mm, interaural or AP −3.3 to −4.3 mm, Bregma) and grade the damage on the following scale

   No damage (control) = 0
   A few neurons damaged = 1
   Many neurons damaged = 2
   Majority of neurons damaged = 3

According to the natural curves in the coronal plane of the CA1 cell layer, the cell death within the medial portion (subiculum) is defined as scale 1. The cell death reaching the middle part of the CA1 zone (the peak of the dorsal curve) is defined as scale 2. The cell death reaching the junction of CA3 zone is defined as scale 3 (Fig. 9). Cell death progressing between these landmarks could be defined as 1.5 or 2.5. This method could obtain reliable results across the experimental groups.[8]

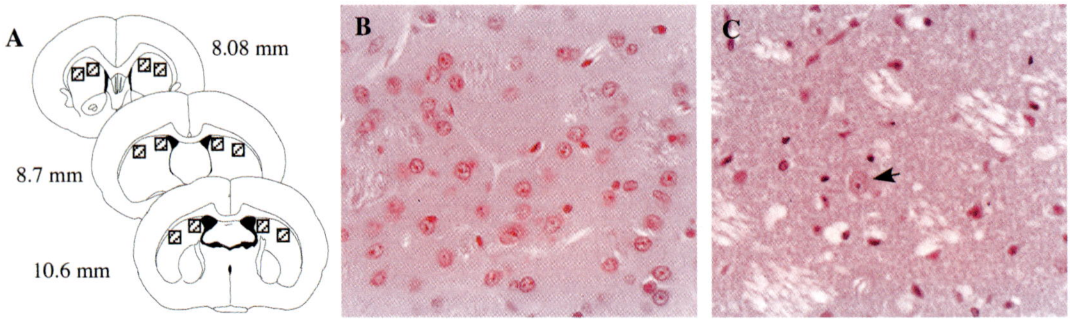

Fig. 10 (**A**) Schematic drawing showing the section plane and selected region (*square*) for cell counting. (**B**) HE staining of striatal section of a control rat. (**C**) HE staining of section 1 day after 20 min ischemia. Moat of the cells have died but some interneurons (*arrow*) remain viable

5. Evaluate at least four sections from each hemisphere. Get average grading from two hemispheres of each animal and use it as the score of that animal.
6. Compare the data between groups using nonparametric Mann–Whitney *U*-test.

4.2 Ischemic Cell Death in the Striatum

In the striatum, small- to medium-sized spiny neurons are highly sensitive to transient global ischemia. It requires ~25 min ischemia to induce cell death in the dorsal striatum. The cell death starts about 6–8 h after reperfusion and most cells die within 24 h. The neurons in the dorsolateral striatum are more vulnerable than those in the ventrolateral striatum. Interneurons in the striatum are resistant to ischemia (Fig. 10b, c). Quantification of cell death in the striatum is achieved by counting the number of survived neurons in a given area according to the procedure described.

1. Prepare paraffin sections as described above.
2. Select four sections at the anterior, middle, and posterior plans (AP 10.6, 8.7, and 8.8 mm, interaural). Identify a region in the dorsomedial and a region in the dorsolateral striatum with the same size and approximately similar location (Fig. 10a).
3. Count the number of survived neurons in the selected region.
4. Compare the data between groups using nonparametric Mann–Whitney *U*-test.
5. Get average grading from two hemispheres of each animal and use it as the score of that animal.

5 Advantages, Complications, and Limitations

5.1 Advantages

1. Rats are widely used in neuroscience research, so the results can be compared with previous studies
2. 4-VO model in rat is well established and yields reproducible neurophathological results
3. The species is large enough to allow relatively easy surgery, physiological monitoring, and maintenance
4. It involves low costs in purchase and maintenance

5.2 Complications

Developing seizures is a common complication after cerebral ischemia. The incidence of seizures is positively correlated to the severity and duration of ischemic insult. In Wistar rats, a 20 min ischemia might induce seizures in 20–30% of the animals. The brain damage caused by seizures is different from that caused by ischemia. Therefore, the animals with seizures after ischemia must be identified and excluded from the study. One of the signs of seizures is the spreading of the flooring materials around the cage because these materials are kicked out of the cage during seizures.

5.3 Limitations

5.3.1 Technical Difficulties

In some animals, the alar foramina might deform or disappear, making electrocauterization impossible. The more difficult task is the successful electrocauterization of the vertebral arteries that are hidden underneath the alar foramina. Unsuccessful electrocauterization is the major cause of incomplete ischemia. Therefore it is essential to make sure that the vertebral arteries are cauterized and the blood flow is completely stopped using the methods described earlier.

Recording of DC potential change during ischemia might be another difficulty for someone without experience in electrophysiology. Actually, it is not as difficult as it sounds. Recording of ID only involves a very basic electrophysiological technique and can be easily learned and mastered.

5.3.2 Mortality

Approximately 10% of the rats will die either during ischemia or shortly after reperfusion even in experienced hands. The respiratory failure during the first 2–3 min of occlusion is a major complication that contributes to the animal's death during ischemia. Immediate resuscitation normally rescues the animals. Death after reperfusion may stem from severe ischemia and cerebral vesicular blood clot, combined with surgical trauma in the neck and head. The mortality in strains other than Wistar rats might be even higher.

5.3.3 Variation

Despite the reproducible outcomes, the results of 4-VO might vary among animals. Even in animals with successful ischemia, the severity of cell death between the two hemispheres might be different. Many factors, including cerebrovascular collaterals, brain temperature, brain glucose, anesthesia, oxygen and carbon dioxide tension, etc., influence the ischemic outcome.[9] It is important to strictly control the variables and optimize the experimental conditions to obtain consistent results.

References

1. Pulsinelli WA, Brierley JB. A new model of bilateral hemispheric ischemia in the unanesthetized rat. Stroke 1979;10(3):267–72
2. Schmidt-Kastner R, Paschen W, Ophoff BG, Hossmann KA. A modified four-vessel occlusion model for inducing incomplete forebrain ischemia in rats. Stroke 1989;20(7):938–46
3. Xu ZC, Pulsinelli WA. Responses of CA1 pyramidal neurons in rat hippocampus to transient forebrain ischemia: an in vivo intracellular recording study. Neurosci Lett 1994;171(1–2):187–91

4. Yamaguchi M, Calvert JW, Kusaka G, Zhang JH. One-stage anterior approach for four-vessel occlusion in rat. Stroke 2005;36(10):2212–4
5. Ginsberg MD, Busto R. Rodent models of cerebral ischemia. Stroke 1989;20(12): 1627–42
6. Xu ZC. Neurophysiological changes of spiny neurons in rat neostriatum after transient forebrain ischemia: an in vivo intracellular recording and staining study. Neuroscience 1995;67(4):823–36
7. Pulsinelli WA, Brierley JB, Plum F. Temporal profile of neuronal damage in a model of transient forebrain ischemia. Ann Neurol 1982;11(5):491–8
8. Xu ZC, Gao TM, Ren Y. Neurophysiological changes associated with selective neuronal damage in hippocampus following transient forebrain ischemia. Biol Signals Recept 1999;8(4–5):294–308
9. Pulsinelli WA, Jacewicz M. Animal models of brain ischemia. In: Barnett HJM, Stein BM, Mohr JP, Yatsu FM, eds. Stroke: Pathophysiology, Diagnosis, and Management. New York: Churchill Livingstone; 1992:49–64

Chapter 7

Rat Model of Global Cerebral Ischemia: The Two-Vessel Occlusion (2VO) Model of Forebrain Ischemia

Ami P. Raval, Chunli Liu, and Bingren R. Hu

Abstract Animal models of brain ischemia can be divided into focal and global brain ischemia. Among global (often referred to as forebrain) ischemia animal models, the 2-vessel occlusion (2VO) and 4-vessel occlusion (4VO) models in rats are most frequently employed in the studies of molecular mechanisms of neuronal damage after brain ischemia followed by reperfusion. This chapter discusses the rat 2VO model. In this model, reversible forebrain ischemia is produced by bilateral common carotid artery (CCA) occlusion combined with systemic hypotension (50 mmHg) sufficient to induce brain ischemia. This ischemia-and-reperfusion model affects widespread areas of the forebrain. There are two chief histopathological changes in this 2VO ischemia model: (1) selective neuronal vulnerability; typically CA1 pyramidal neurons of the hippocampus are most vulnerable, followed by dorsoventral striatal small- and medium-sized neurons, and pyramidal neurons in the layers 3–4 of the neocortex; and (2) delayed neuronal death, i.e., neuronal death does not occur immediately after transient ischemic episode, but takes place after 2–3 days of reperfusion. During the surgical procedure, this model requires artificial maintenance of (1) rectal and brain temperatures at 37°C, (2) the mean arterial blood pressure (MABP), and (3) blood gases. This 2VO ischemia model has the advantage of a one-step surgical procedure that produces high-grade forebrain ischemia. The reproducibility of the ischemic histopathological damage of this model is more than 90%. This animal model is also suitable for molecular, biochemical, and physiological studies, as well as for evaluation of neuroprotective procedures and agents.

Keywords: Transient cerebral ischemia, 2-Vessel occlusion, Reperfusion, Delayed neuronal death, Selective vulnerability, Hypotension, Survival studies, Hippocampus

From: *Animal Models of Acute Neurological Injuries.*
Edited by: J. Chen (eds.) New York, Totowa, NJ

1 Introduction

Human brain ischemia is enormously diverse in causes, manifestations, and anatomic region. For instance, ischemic stroke is often caused by focal brain ischemia, whereas global brain ischemia likely results from cardiac arrest or heart surgeries. For a scientific study of molecular mechanisms, treatments, and prevention of brain ischemia, it is essential to employ physiologically controlled and reproducible in vivo animal models. In this context, rodent models of cerebral ischemia have been widely employed, owing to small animal size convenient for handling, low cost of animals, simple surgical procedures, and relative homogeneity within strains because of inbreeding.

The aim of this chapter is to describe the two-vessel occlusion (2VO) model of global ischemia. Because global ischemic insult does not damage the cerebellum, this model is also referred to as the forebrain ischemia or cerebral ischemia model. This model was first developed in Dr Siesjö's laboratory in Lund University, Sweden, more than 30 years ago.[1] During the ischemic period, mean arterial blood pressure (MABP) is reduced to 50 mmHg by withdrawing blood from either the jugular vein or femoral artery, immediately followed by occlusion of bilateral common carotid arteries, in order to produce brain ischemia to a degree that causes neuronal damage. The withdrawn blood is returned to the animal at the end of the ischemic period. This model produces a "square-wave" ischemic insult in which the transition from non-ischemia to ischemia and its subsequent reversal are abrupt.[1–3] This model gives rise to delayed neuronal injury selectively in vulnerable brain regions, including the CA1 pyramidal neurons of the hippocampus, the dorsoventral striatum, and neocortex.[1] In this model, ischemic duration of more than 4 min in Wistar rats is required to produce consistent neuronal damage; animals can survive an ischemic duration for as long as 20–30 min.

2 Materials, Equipment, and Supplies

2.1 Animals

Male Wistar rats at 280–300 g body weight[1] are mostly employed in studies using this 2VO model. Although challenging technically, mice are also used for this 2VO model.[4] Selection of an appropriate species is important owing to genetic variants that affect the ischemic outcome. Because of the known effects of estrogen on the ischemic outcome, male rats are most frequently used in this 2VO model.[5] Similarly, animal age also affects the ischemic outcome. Therefore, when using this 2VO model, animal strain and species, sex, and age must be taken into consideration.

2.2 Equipment

Figure 1 illustrates typical surgical equipment required for performing the rat 2VO ischemia model:

(A) Anesthesia box (Harward apparatus, Holliston, MA, USA)
(B) Anesthesia machine (Matrx Medical Inc., Orchard Park, NY, USA)
(C) Respiratory ventilator (Columbus Instruments, Columbus, OH, USA)
(D and E) Computerized Powerlab/400 for MABP and electroencephalogram (EEG) measurements (AD Instruments, Colorado Springs, CO, USA)

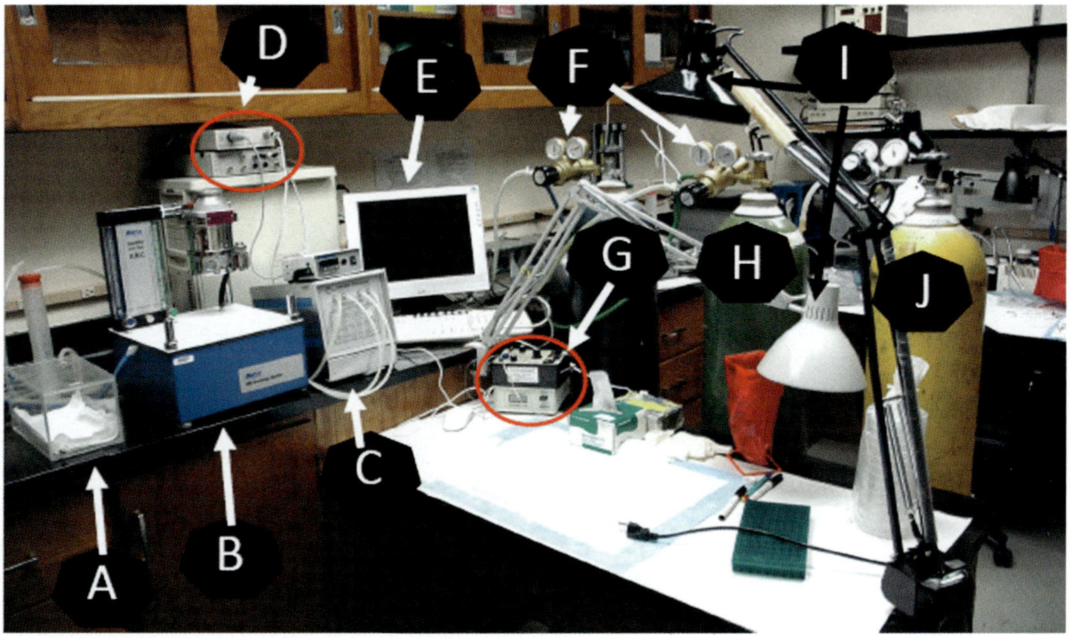

Fig. 1 A surgical equipment table for performing the rat 2VO ischemic model. (**A**) Anesthesia box, (**B**) Anesthesia machine, (**C**) Respiration ventilator, (**D**, **E**) Computerized Powerlab/400, (**F**) Gas regulators, (**G**) Temperature controller, (**H**) Oxygen (O_2) tank, (**I**) Lamps, (**J**) Nitrous oxide (NO) tank

(F) Gas regulators (VWR, West Chester, PA, USA)
(G) Temperature controller (Omega, Stamford, CT, USA)
(H) Oxygen (O_2) tank
(I) Lamp (Luxo corporation, New York, NY, USA)
(J) Nitrous oxide (NO) tank

2.3 Anesthesia Agent

Isoflurane and halothane are commonly used to anesthetize animals during surgical procedures of the 2VO forebrain ischemic model. Because global brain ischemia may affect respiratory system function, ventilator-controlled respiration with a rodent ventilator is required.[1]

2.4 Surgery Instruments and Supplies

These include a pair of serrated, slightly curved forceps (Roboze surgical instrument company, Maryland, USA; Cat # RS-8254), microdissecting spring scissors (Roboze, Cat #RS-5658), operating scissors (Roboze; Cat #RS-6951), a pair of microdissecting forceps (Roboze; Cat #RS-5244), a pair of delicate hemostatic forceps (Roboze; Cat #RS-7111), microdissecting tweezers (Roboze; Cat #RS-5058), and a bulldog-type serrated fine clamp (Fine Science Tools, Foster City, CA, USA; Cat # 18374-43).

Also required are syringes (1 and 10ml; BD Biosciences, San Jose, CA, USA), syringe adaptors (23 and 20G; BD Biosciences), intubation tube of 14GA, 1.75 in., 2.1 × 45mm (BD biosciences; cat # 381467), polyethylene tubing(PE-50,

BD Biosciences), heparinized capillaries (Radiometer Copehagen, cat # D941G-240–85), silastic tubing (inner diameter 0.76 mm and outer diameter 1.65 mm; Dow Corning Corporation, Midland, MI, USA), rectal temperature probe (Omega, Stamford, CT, USA; Cat # HYPO-33–1-T-G-60-SMP-M), and head temperature probe (YSI, Dayton, OH, USA).

3 Surgical Procedure

All procedures are performed under aseptic conditions for survival studies. The person performing the surgery should wear a mask, clean gown, and sterile gloves. All surgical instruments are sterilized before surgical procedures and maintained aseptic in a solution, such as hexazol solution (Burns Veterinary Supply, Inc., Westbury, NY, USA), between the animal procedures.

A Male Wistar rat weighing 250–300 g is fasted overnight. The rat is anesthetized by placing it in an anesthesia box (Fig. 1a) supplied with 4% isoflurane in a 70/30% nitrous oxide/oxygen (NO/O_2) gas mixture for 5–8 min until the rat becomes numb and reaches a state of apathy. Subsequently, the rat is endotracheally intubated with a plastic tube by means of an otoscope (Fig. 2). In this technique, an otoscope is fixed on a frame (Fig. 2a). The anesthetized rat is positioned such that its mouth can be introduced over the otoscope tip. Through the otoscope, the surgeon can see the vocal cords in the trachea (Fig. 2b). A flexible guide wire with a blunt and smooth end (Fig. 2a) is gently introduced between the vocal cords into the trachea through the otoscope. The rat, together with the guide wire, is then pulled from the otoscope. A plastic intubation tube (BD Biosciences) can now be inserted via the guide wire into the trachea. There is usually a slight resistance as the intubation tube passes the vocal cords. After removing the guide wire, the intubation tube is fixed with suture onto the mouth skin and then connected to the ventilator. The rat is artificially ventilated at a respiration rate of 60 min^{-1} with 1.5 ± 0.5% of isoflurane and 70/30% nitrous oxide/oxygen (NO/O_2). The circuit to connect the ventilator to the rat is illustrated in Fig. 3.

A B C

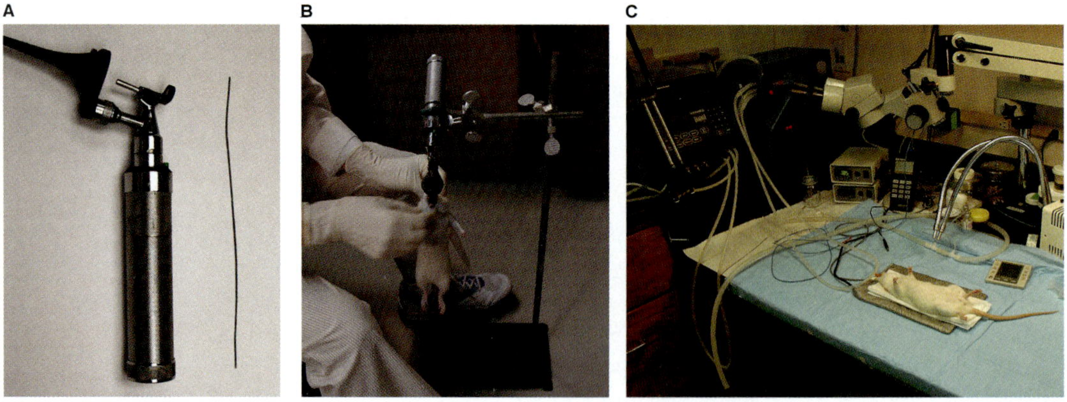

Fig. 2 Otoscope intubation method. (**A**) A guide wire and otoscope. (**B**) Endotracheal intubation of rat with otoscope assistance. (**C**) Connecting the endotracheal-intubated rat with a ventilator

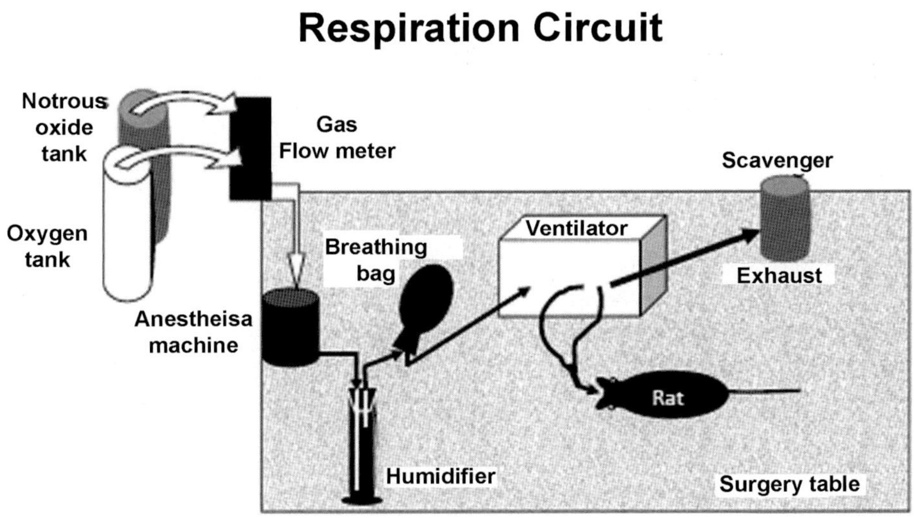

Fig. 3 Schematic drawing of a respiratory circuit

The rats skin is shaved and scrubbed with povidone iodine (10% solution, Purdue Products, Stamford, CT, USA) before any incision is made. Incisions are made to expose a tail artery and the right femoral artery and vein. They are then catheterized with single lumen polyethylene tubes (PE-50; BD Biosciences). The tail artery catheter is for monitoring the MABP and arterial sampling of blood gases and plasma glucose. Two needle electrodes are inserted through small skin incisions into the temporalis muscle bilaterally to obtain an interhemispheric EEG. MABP and EEG are monitored with a Powerlab system (AD Instruments, Inc., Colorado Springs, CO, USA) throughout the experiment. The femoral artery and vein catheters are for withdrawing the blood during ischemia and intravenous infusion of drugs such as heparin and pancuronium bromide, respectively (see below). Heparin at a dose of 0.2 ml (1,000 USP units ml^{-1}) is slowly injected through the femoral vein catheter to heparinize the animal. Arterial blood (0.1 ml) is sampled with a heparinized glass tube (Radiometer Copehagen, Brønshøj, Denmark) through the tail artery catheter for measuring arterial blood gases and plasma glucose levels. An arterial blood gas analyzer (Fig. 4a; Model: ABL 50; Radiometer Copehagen, Brønshøj, Denmark) and plasma glucose level detector (Fig. 4b; Model: YSI-2300 STAT plus; YSI, Dayton, OH, USA) are used (Fig. 4). Arterial blood gases and plasma glucose levels are measured 15 min before and 15 min after the ischemic period, as well as intermittently throughout the experiment whenever required. Blood gases are maintained in the physiological range throughout the experiment (Table 1).

An incision is made in the ventral neck skin of the rat. Both common carotid arteries are gently dissected free of the surrounding tissues and nerve fibers. Thick sutures are loosely placed under the arteries for pulling them out when clamping during induction of ischemia (see below).

A rectal temperature probe (Omega, Stamford, CT, USA) is gently inserted into the rectum to monitor body temperature. A thermocouple 33G needle temperature probe (YSI, Dayton, Ohio, USA) is implanted into the temporalis

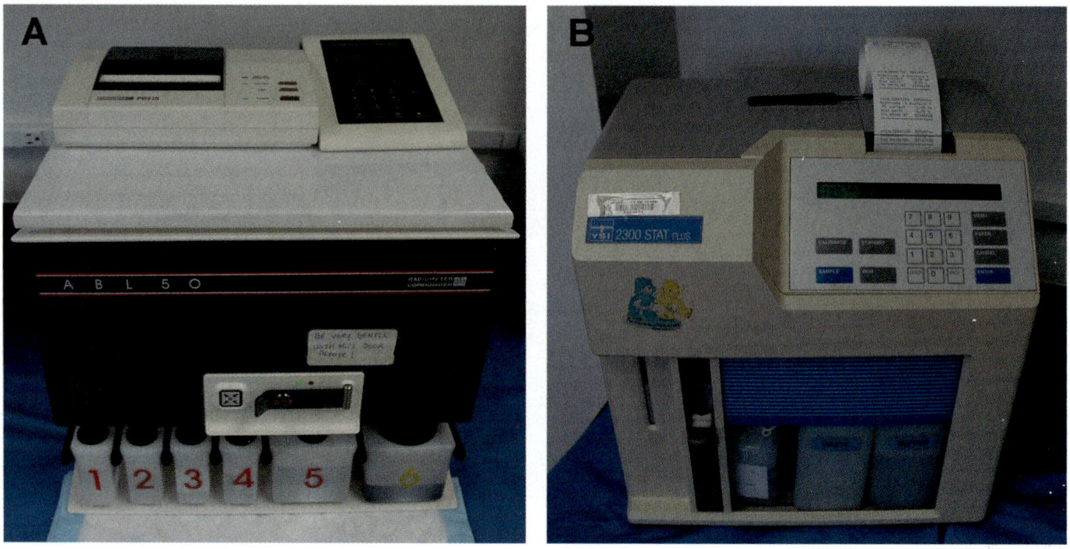

Fig. 4 Photographs of (**A**) blood gas analyzer and (**B**) blood glucose analyzer

Table 1 Physiological parameters.

Variables	Before ischemia	During ischemia	After ischemia
Body weight (g)	300 ± 20	–	–
pH	7.4 ± 0.04	7.4 ± 0.08	7.4 ± 0.04
pCO_2 (mmHg)	37 ± 3	39 ± 3	37 ± 3
pO_2 (mmHg)	120 ± 10	110 ± 10	120 ± 10
MABP (mmHg)	120 ± 20	50 ± 2	120 ± 20
Plasma glucose (mg /ml^{-1})	120 ± 10	–	120 ± 10

muscle for monitoring head temperature. Body and head temperatures are maintained at 37 ± 0.2°C throughout the experiment with the assistance of lamps placed above the animal's body and head. When the surgical preparation is ready for induction of ischemia, pancuronium bromide (2 mg kg^{-1}) is injected (intravenously) to immobilize the rat while the isoflurane concentration is maintained at 1.0%.

4 Induction of Cerebral Ischemia

Fifteen minutes before induction of ischemia, blood gases, MABP, EEG, and rectal and head temperatures are monitored and maintained within the physiological range (see Table 1). Immediately before induction of ischemia, blood is gradually withdrawn from the femoral artery with a heparinized syringe (10 ml; adaptor 20G, BD Biosciences) to reduce systemic blood pressure to 50 mmHg. Cerebral ischemia is then induced by clamping both common carotid arteries with surgical clamps (bulldog type serrated fine clamp; Fine Science Tools, Foster, CA, USA). After bilateral occlusion of the common carotid arteries,

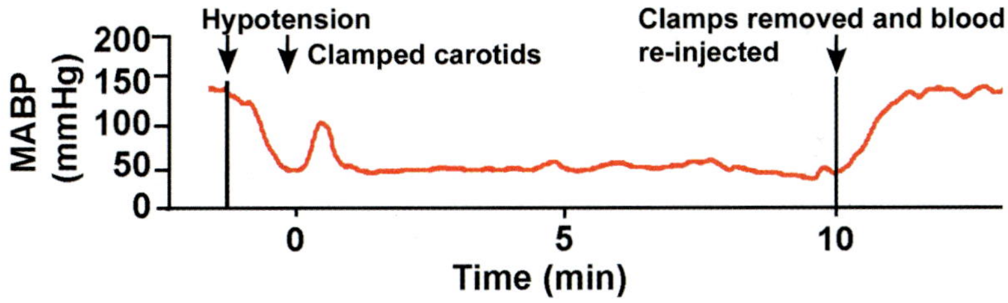

Fig. 5 Representative curve of mean arterial blood pressure measurement before, during, and after the ischemia period

MABP usually increases sharply (Fig. 5), which must be reduced instantly by withdrawal of blood and controlled at 50 mmHg by reinfusion and withdrawal of blood through the femoral artery catheter during the ischemic period. The onset of cerebral ischemia is defined as the time of cessation of EEG activity, usually within 1 min after clamping of arteries. Cerebral ischemic durations of 10–15 min are mostly used according to the literature.

To end the ischemic episode, the common carotid artery clamps are removed first to allow post-ischemic reperfusion, and then the withdrawn blood is returned immediately through the femoral artery at a slow rate. The re-infusion of blood usually restores MABP to a range of 120–140 mmHg. After about a 15–30 min stabilizing period and measurement of blood gases (see above), the catheters can be removed, arteries and vein closed with sutures, and anesthesia stopped. However, the rat still must be continuously ventilated with 70/30% nitrous oxide/oxygen (NO/O_2, and the head and rectal temperatures maintained at 37°C ± 0.2 until the onset of spontaneous respiration during recovery from anesthesia, which usually takes about 30–60 min depending on ischemic duration). The ventilator can now be disconnected, whereas the endotracheal tube is kept in place. After removing the head and rectal temperature probes, the rat can be returned to its cage. The rat is monitored closely until spontaneous respiration smoothly recovers before removing the endotracheal tube. After 1–2 h monitoring, the animal can be transferred to the housing room with food and water supplied.

5 Outcome Evaluation

Ischemia duration of 10–15 min in this 2VO model leads to histopathological damage within selectively vulnerable structures, including neurons in the hippocampal CA1, striatum, and neocortical regions.[3] Histopathological changes are therefore reliable indicators of ischemic damage in this model. Because neuronal death takes place in a delayed fashion, within 2–3 days of reperfusion, animals are usually perfusion-fixed for 7 days for histopathological examination (see below).

6 Histopathology

After 7 days of reperfusion following an episode of cerebral ischemia, the rat is anesthetized in an anesthesia box supplied with 4% isoflurane in 70/30% nitrous oxide/oxygen, intubated through tracheotomy, and ventilated with 2% isoflurane in 70/30% nitrous oxide/oxygen. The chest case of the rat is opened

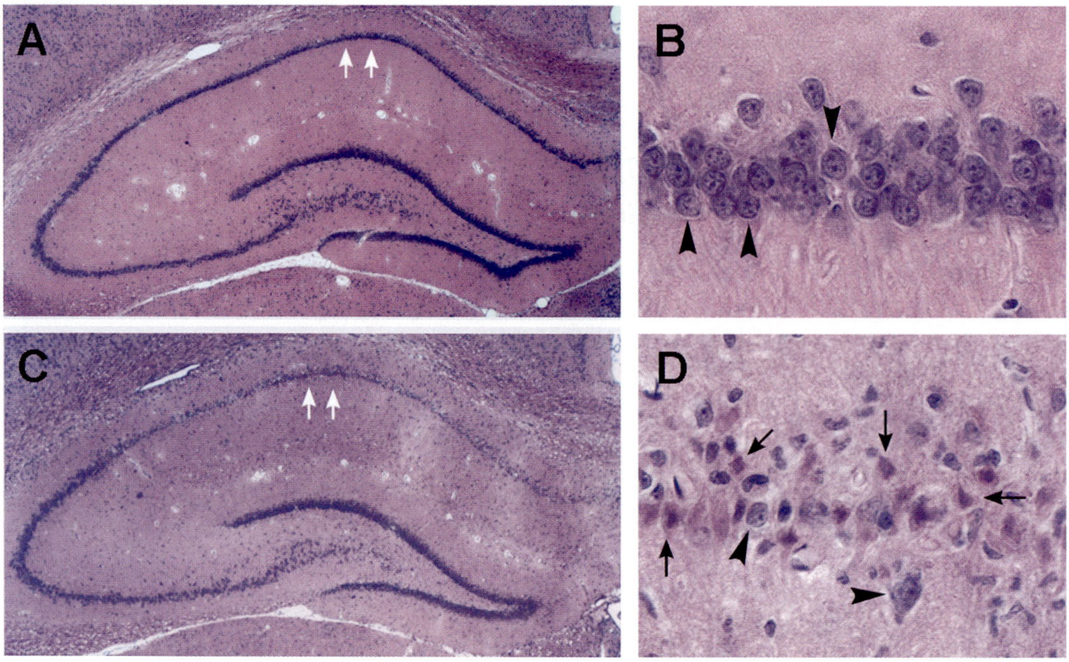

Fig. 6 Histopathology changes in the hippocampus after 10 min of cerebral ischemia in the 2VO model. (**A**) A hippocampal section from a sham-operated rat. (**B**) A higher magnification of the CA1 region indicated by *white arrows* in the sham-operated rat (**A**). (**C**) A hippocampal section from a rat subjected to 10 min of ischemia followed by 24 h of reperfusion. (**D**) A higher magnification of the CA1 region indicated by *white arrows* in the post-ischemic rat (**C**). *Arrowheads* in (**B**) indicate normal neurons. *Black arrows* in (**D**) point to ischemic dead neurons that are identified by shrunken cells and nucleus as well as acidophilically stained cytoplasm

surgically, and a catheter is inserted into the ascending aorta through the left ventricle. After clamping the descending aorta with a surgical clamp, the rat's upper body is immediately perfused through an ascending aorta catheter initially with saline for 1–2 min, and then with FAM (a mixture of 40% formaldehyde, glacial acetic acid, and methanol, 1:1:8 by volume) for 10 min at a constant pressure of 110–120 mmHg. The rat is then decapitated. The brain is removed from the skull, and immersed in FAM at 4°C for 1 day. Coronal brain blocks of about 3–4 mm are cut and embedded in paraffin; coronal sections of 5–10 μm are sectioned with a microtome (Leica Microsystems GmbH, Ernst-Leitz-Strasse 17–3735578 Wetzlar) and stained with hematoxylin and eosin. Brain sections at different levels can be examined.

Delayed neuronal death after transient cerebral ischemia under light microscopy results in the following changes: (1) cells shrunken to about half or one-third of normal neuronal sizes; (2) shrunken, darkly stained, and polygonal-shaped nuclei; and (3) eosinophilic or acidophilic cytoplasm (Fig. 6).

7 Quantification of Neuronal Death

The number of dead neurons in defined brain regions after transient cerebral ischemia in this 2VO model can be quantified. Quantification is commonly done by counting all the normal and dead neurons (see above) at the same anatomic levels of neocortical, striatal, and hippocampal brain sections from both sham-operated control and post-ischemic subjects under light microscopy.[1] The extent of injury is usually expressed as the percentage of dead neurons of the total neurons in the defined regions examined. For statistical comparisons of neuronal death between experimental groups, the Kruskal–Wallis test followed by the Mann–Whitney *U*-test is commonly used. Typically, in this 2VO model, more than 4 min. of transient ischemia is required to produce consistent neuronal death. Ten min. of ischemia induces delayed neuronal death selectively in about 80% of CA1 pyramidal neurons, and a few neocortical neurons. Fifteen min. of ischemia damages more than 90% CA1 neurons, the dorsolateral striatal neurons, and about 5–10% of neocortical neurons, whereas 20 min of ischemia leads to delayed neuronal death in about 30% of neocortical neurons and also extends delayed death to neurons in the CA3 and DG regions.

Figure 6 illustrates typical histopathological changes in the hippocampus after transient cerebral ischemia in this 2VO model. Male Wistar rats were subjected to either sham surgery without ischemia, or 10 min of ischemia followed by 7 days of reperfusion. Paraffin-embedded coronal sections (10 μm) were photographed by light microscopy. Selective neuronal death is evident in more than 80% of pyramidal neurons in the CA1 region of the hippocampus after 10 min of cerebral ischemia in the 2VO model.

8 Advantages and Limitations

The advantages of this 2VO transient cerebral ischemia model include (1) one-stage surgical procedure; (2) highly reproducible ischemic damage; (3) high animal survival rate for studies of reperfusion injury; and (4) suitability for studies of physiological, molecular, biochemical, and behavioral alterations during the post-ischemic phase and for testing the efficacy of neuroprotective agents.

The limitations of this 2VO transient cerebral ischemia model are that (1) brain ischemia cannot be induced in awake rats; and (2) it requires systemic control of hypotension.

9 Potential Technical Difficulties and Pitfalls

Because global ischemia may affect respiration in this 2VO model, mechanical maintenance of respiration and blood gases with a ventilator is essential during and after ischemia.[1] Endotracheal intubation of rats and mice for connecting a ventilator can be technically challenging and often requires training and practice for several weeks. There are two methods commonly used for endotracheal intubation: (1) with direct vision of the trachea through the animal's mouth under local illumination[1]; and (2) with otoscopic vision of the tracheal vocal cords as described earlier. In comparison, the otoscopic method is easier but requires an otoscope.

The following is vital for reproducing the same degree of histopathological damage in this 2VO transient cerebral ischemia model: (1) Both rectal and brain temperatures must be controlled constantly at 37 ± 0.2°C during the procedure. The brain temperature is particularly important because cutting off blood supply to the brain during the ischemic period can lead to reduction of brain temperature. Small changes in brain temperature markedly alter the ischemic histopathological outcome. (2) It is critical to withdraw food (fast) overnight to maintain a constant blood glucose level. A high blood glucose level before ischemia often leads to very severe ischemic neuronal damage. (3) It is also essential to control MABP as much as possible to 50 mmHg in order to induce the same degree of brain ischemia during the ischemic period.

10 Mortality

In our laboratory, the success rate of this model is more than 90%. The mortality is mostly due to postoperative complications. Inappropriate intubation can damage the lung and lead to swelling of the trachea, causing respiratory difficulties during the post-ischemic period. A long duration of surgery period and careless surgical damage can lead to poor recovery. In addition, an excessive dose of anesthesia can cause animal death during surgery.

References

1. Smith ML, Bendek G, Dahlgren N, Rosen I, Wieloch T, Siesjo BK. Models for studying long-term recovery following forebrain ischemia in the rat. 2. A 2-vessel occlusion model. Acta Neurol Scand 1984;69(6):385–401
2. Siesjo BK, Katsura K, Zhao Q, Folbergrova J, Pahlmark K, Siesjo P, Smith ML. Mechanisms of secondary brain damage in global and focal ischemia: a speculative synthesis. J Neurotrauma 1995;12(5):943–956
3. Smith ML, Auer RN, Siesjo BK. The density and distribution of ischemic brain injury in the rat following 2–10 min of forebrain ischemia. Acta Neuropathol (Berl) 1984;64(4):319–332
4. Wellons JC III, Sheng H, Laskowitz DT, Burkhard Mackensen G, Pearlstein RD, Warner DS. A comparison of strain-related susceptibility in two murine recovery models of global cerebral ischemia. Brain Res 2000;868(1):14–21
5. Bramlett HM. Sex differences and the effect of hormonal therapy on ischemic brain injury. Pathophysiology 2005;12(1):17–27

Chapter 8

Asphyxial Cardiac Arrest

Clifton W. Callaway and Eric S. Logue

Abstract Sudden cardiac arrest is a significant source of morbidity and mortality, with neurological injury affecting the majority of persons who initially have restoration of pulses. The rat asphyxial cardiac arrest is an established model for neurological injury after cardiac arrest. This model reproduces many aspects of neurological injury observed in humans – transient coma, evolving motor deficits and persistent sensorimotor deficits – as well as the systemic metabolic disruptions that attend the clinical disease. Rats are anesthetized, intubated and placed on mechanical ventilation. After establishing femoral arterial and venous access, rats are chemically paralyzed, and the ventilator turned off. Hypoxemia leads to highly reproducible brady-asystolic circulatory arrest. Cardiac arrest is reversed after 8 min by resuming mechanical ventilation, chest compressions and intravenous administration of epinephrine. After restoring circulation, rats are supported on the ventilator for 1–2 h and then extubated. After extubation, rats require fluid supplementation for several days until they are able to resume feeding. Temperature during the first day of recovery strongly influences outcome, making control of temperature essential during experiments. Neurological deficits can be measured clinically, and histological damage to the brain becomes clear after several days.

Keywords: Heart arrest, Ischemia, Hypoxemia, Asphyxia, Rat, Coma

1 Model Selection

Sudden cardiac arrest is a major cause of mortality in developing countries. For persons who survive cardiac arrest, neurological deficits are common and can negatively impact subsequent quality of life. The rat cardiac arrest model provides a reproducible model of post-cardiac arrest neurological injury. Unlike arterial occlusion models, in which forebrain ischemia predominates or in which trickle flow may persist during the injury, cardiac arrest results in true global ischemia involving both brain and other organs. Likewise, the model has face validity for the human disease in that it produces severe neurological deficits that resolve partially or completely over a period of days.

From: *Animal Models of Acute Neurological Injuries.*
Edited by: J. Chen (eds.) New York, Totowa, NJ

The most established model for rat cardiac arrest is asphyxia.[1,2] This technique relies on chemical paralysis and cessation of ventilation to induce circulatory arrest, which is more reliable and reproducible than induction of ventricular fibrillation in rats. Refinements in the model have included meticulous post-resuscitation monitoring of temperature,[3] improved neurological scoring and longer-term quantification of neuronal loss.[4]

2 Materials

2.1 Animal

The asphyxia model uses male adult Sprague-Dawley rats. Wistar rats have less neurological injury with survivable cardiac arrest.[5] Rats from 17 days[6] to 120 days in age[3] survive the cardiac arrest well and exhibit similar behavioral changes and neuropathology. For adult rats, surgical preparation is easily performed when rats weigh 300–350 gm.

2.2 Anesthesia

Inhalational anesthesia has the dual advantages of being easily titrated, and being easy to terminate after the cardiac arrest. Asphyxial cardiac arrest has been performed using 0.8% halothane in oxygen or using isoflurane in N2O:oxygen (70%:30%). Anesthesia is induced in a chamber or using a nosecone, and then maintained via the ventilator circuit after rats are intubated (Fig. 1).

2.3 Equipment

2.3.1 Anesthesia

Halothane or Isofluorane nebulizer (Summit Hill Laboratories, Navesink, NJ)
Oxygen tank with regulator
Supply tubing with connectors
14 gauge × 2 in. intravenous catheter (Angiocath)
Small animal ventilator (CWE, Ardmore, PA)
Laryngoscope with Miller 1 or Miller 0 blade (neonatal laryngoscope)

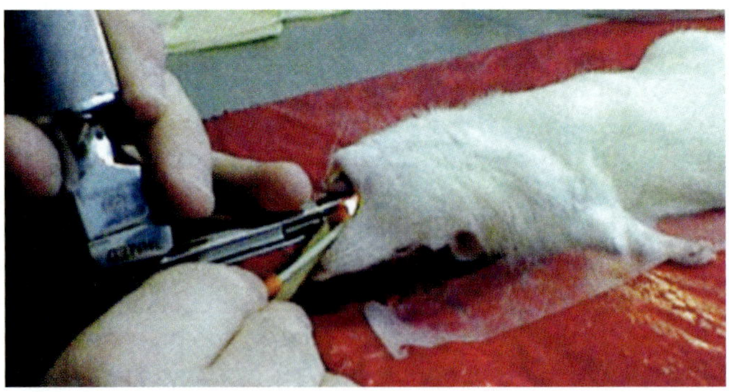

Fig. 1 Endotracheal intubation

2.3.2 Surgery

Instruments (scissors, foreceps, hemostats)
2–0 silk sutures (Sofsilk)
3–0 sewing suture (Sofsilk)
Stereotaxic device (Kopf)
Dental cement (Stoetling)
Stainless-steel skull screws (MX-0080-02F-C, Small Parts, Inc.)
Stainless-steel guide tubing (HTX-20R-36, Small Parts, Inc.)
Stainless-steel wire (GWX-0220-30, Small Parts, Inc.)
Pin Vise hand-drills (240A, 240B, Starrett)
Hand-drill bits (CHSD-50-1, CHSD-3/64-01, Small Parts, Inc.)
Hair-clipper (Wahl)
Povidone-Iodine surgical prep (Betadine)
Intramedic PE-50 polyethylene tubing (BD)
23 gauge needles (fits PE-50 tubing)
1 mL syringes
Three-way stopcocks
Sterile cotton swabs and gauze

2.3.3 Monitoring

ECG electrodes (Kendall Meditrace 530)
Digital multichannel physiological recorder/amplifier (Powerlab)
XM-FH-BP wireless brain temperature probes (Mini Mitter)
TR-3000 receiver for wireless temperature probes (Mini Mitter)
Infrared heating lamp (available in pet stores for reptile vivariums)
5-in. fan
Computer-driven relays to control heating lamps and fans (MiniMitter)
Blood-gas analyzer (ISTAT)
Glucometer (Accu-Chek)

2.3.4 Drugs

Halothane
Vecuronium Bromide
Epinephrine
Sodium bicarbonate
Ampicillin

3 Procedures

3.1 Preparation for Surgery

3.1.1 Temperature Probe Calibration

1. Place temperature probe sensors in a stirred water bath with a precise mercury thermometer (0.1°C gradations). Start with warm water that will cool over time.
2. Record the transmitted frequency from the temperature probes at multiple temperatures that span the range to be used in the physiological experiment (e.g., 33–38°C).

3.1.2 Prepare Guide Cannula and Catheters

1. Using a grinding wheel, cut 25 gauge stainless-steel tubing into lengths 1 mm shorter than the brain-temperature probes. For example, cut 7 mm cannulae if the brain probes are 8 mm long. Smooth the cut edges and be certain that the bore of the cannula is patent.
2. Place a bead of silver solder onto the mid-shaft of each cannula. This bead will allow the dental cement to firmly grip the cannula. Liquid flux facilitates soldering.
3. Cut 18 cm lengths of PE-50 tubing for arterial and venous catheters. With permanent ink, mark the arterial catheter at 2 cm and the venous cathether at 8 cm as a guide for depth of insertion. The tips of these catheters should be beveled.
4. Heat or chemically sterilize the guide cannulae and catheters.

3.2 Implantation of Temperature Monitors

Rats are anesthetized with halothane and placed into the stereotaxic device. Anesthesia is maintained by connecting the halothane delivery tubing to a canopy around the nose of the rat. This open-circuit requires scavenging of halothane with some arrangement of vacuum in order to prevent exposure of the surgeon.

3.2.1 Implantation of Guide Cannula

1. Place anesthetized rat into stereotaxic frame. Shave the top surface of the head and prepare it with povidone-iodine.
2. Make a midline incision through scalp and subcutaneous tissues with a scalpel from neck muscles to a line between the eyes. Bluntly separate the wound edges using cotton swabs. Hemostats can be used to retract the wound edges. The skull surface should be exposed, clean and dry from 4–5 mm anterior to bregma to lambda. Swabbing the surface with topical epinephrine can help establish a dry surface.
3. Place three skull-screws. Starting holes for the screws should be made with a hand drill in positions away from the target location for the guide cannula. For example, if the guide cannula will be placed in the right frontal bone, then one screw can be placed in each of the left frontal, right and left parietal bones. If two guide cannulae will be placed, then one of the screws should be repositioned. Using fewer than three screws increases the chance for the head-cap to become dislodged.
4. Make a hole for placement of the guide cannula. Measure the location first with the stereotaxic device, and then drill the hole using the hand-drill. Placing the guide cannula 2 mm lateral and 2 mm anterior to bregma (level-skull) will aim the cannula at the lateral ventricle.
5. Lower the guide cannula to skull surface. Lower the cannula 1 mm into the skull. This allows the tip of the cannula to sit adjacent to the inner table of the skull.
6. Cover the skull screws and skull surface with dental cement. Build the cement high enough to cover the solder bead on the side of the guide cannula. If desired, a protective cap can be placed in the cement around the screws guide cannula to protect thermistors or other implants. The hollow portion of a 5 mL syringe can be used for this application.

7. When cement is dry, remove stereotaxic carrier. Place a wire stylet, created to fit the length of the guide cannula, into the cannula, and place the animal back in a holding cage to recover. At least 3 days recovery from anesthesia is recommended prior to cardiac arrest.

3.3 Asphyxial Cardiac Arrest

During the entire interval after induction of anesthesia, the procedures are conducted on the temperature monitoring platform, with heating lamp and fans controlled to regulate temperature at 37°C. On the day of surgery, remove the stylet from the guide cannula and place a brain temperature probe into the cannula. The length of the probe should be ~1.0 mm longer than the cannula, such that the thermistor tip sits in the epidural space outside the brain and below the inner table of the skull.

3.3.1 Anesthesia

1. Induce deep halothane anesthesia in a chamber. In order to intubate the rat, it is necessary for anesthesia to be sufficient to cause some respiratory depression. Breathing 4% halothane for 5 min usually suffices.
2. Place the rat supine, and secure the front limbs to the table-top using tape.
3. Hold the upper incisors down with a rubber band attached to the table with tape.
4. Pull the tongue out using forceps and insert a laryngoscope blade into the mouth to visualize the vocal cords.
5. Insert the 14-gauge catheter (needle replaced with blunt stylet) into the vocal cords.
6. Attach the catheter to the ventilator circuit and secure the catheter to the chin using a suture.
7. Ventilate with 1.0–2.0% halothane in oxygen at 9 mL kg^{-1}, 40 breathes per minute and PEEP of 3–5 cm water. Adjust halothane to a surgical plane of anesthesia with a goal of 0.8% halothane at the completion of surgery.
8. Place ECG electrodes on each forelimb and the right hindlimb.

3.3.2 Placement of Intravenous and Arterial Catheters

1. Place the rat supine, and secure the hindlimbs and forelimbs to the table using tape.
2. Shave the left femoral area and prepare with povidone-iodine.
3. Make a 2 cm incision with scissors in the skin parallel to the inguinal crease, about 0.5 cm distal to the inguinal crease. Bluntly dissect towards and below the inguinal ligament using cotton swabs. This dissection should expose the femoral sheath with vein, artery and nerve.
4. Using blunt-tipped forceps, carefully dissect apart the femoral artery and vein.
5. Place two 2–0 silk ties, one proximal and one distal, looped loosely around each of the vessels. These ties allow for control of the vessel.
6. Clean the adventitia off of the vessels between the ties. At this point, the distal ties can be cinched down.
7. Provide gentle traction with hemostats on both proximal and distal ties sufficient to occlude blood flow. Make a small incision transversely into the artery using iris scissors or a similar fine instrument. The incision should not extend more than about 1/3 of the circumference of the artery in order to prevent the artery from tearing.

8. Insert the arterial catheter into the artery, relax the proximal tie, and advance the catheter gently ~2 cm into the artery. The distal tie can be used for traction, but advancement should be halted if there is any resistance. Insertion of the catheter can be facilitated by inserting a hook or fine forceps into the lumen of the artery to spread the arteriotomy.
9. Secure the arterial catheter by cinching down the proximal tie, and by placing a loop in the distal tie over the catheter.
10. Repeat steps 7–9 with the venous catheter to secure the venous access.
11. If chronic access or monitoring is desired, the arterial and venous catheters can be tunneled under the skin to emerge from the back between the scapulae. For acute preparations, the lines can be left emerging from the incision.
12. Cover wound opening with sterile gauze soaked with sterile saline.

3.3.3 Induction of Cardiac Arrest

1. Obtain baseline blood gas, chemistry and hemodynamic measurements. Adjust the ventilator as needed to establish desired starting conditions.
2. Washout the anesthetic for 2–5 min. Turn the halothane to 0%. As emergence from anesthesia requires 5–10 min, this washout will not allow the rat to become conscious prior to cardiac arrest. However, minimizing the concentration of the cardiodepressant anesthetics will greatly increase the success of resuscitation.
3. Slowly infuse 2 mg kg^{-1} vecuronium intravenously over the course of the washout period in order to paralyze the rat.
4. At 2 min prior to asphyxia, disconnect the oxygen source, and allow ventilation with room air.
5. Begin asphyxia by turning off the ventilator (or disconnecting the circuit) at end-expiration.
6. Blood pressure and heart rate will begin to decrease within 30 s (Fig. 2). Simultaneous measurement of venous and arterial pressures indicates that there is no pressure gradient when arterial pressure falls to near 20 mmHg or when pulsatile signals are absent from the arterial tracing. This occurs reliably around 180 s.
7. The ECG may reveal slowing supraventricular complexes, ventricular escape beats or even asystole (Fig. 3). Appearance of ventricular fibrillation is very rare, and portends a difficult resuscitation.

3.3.4 Cardiopulmonary Resuscitation

1. While the animal is disconnected from the ventilator during the cardiac arrest, increase the rate to 60 breathes per minute and reestablish the oxygen supply. Place syringes with resuscitation drugs (0.005 mg kg^{-1} epinephrine and 1.0 mEq kg^{-1} bicarbonate, each diluted to 1 mL kg^{-1}) on the stopcock connected to the venous catheter.
2. After 8 min (or desired duration of asphyxia), reconnect the ventilator.
3. Rapidly infuse the resuscitation drugs. Doses are based on prior dose–response studies.[7]
4. Deliver continuous chest compressions with two fingers tapping the middle of the sternum at a rate of 200–300 beats per minute (Fig. 4). Compressions should be sharp and forceful. The surgeon's other hand can be used to grasp and stabilize the rib cage to improve the chest recoil. Adequate chest compressions are easily recognized by the development of pressure waves

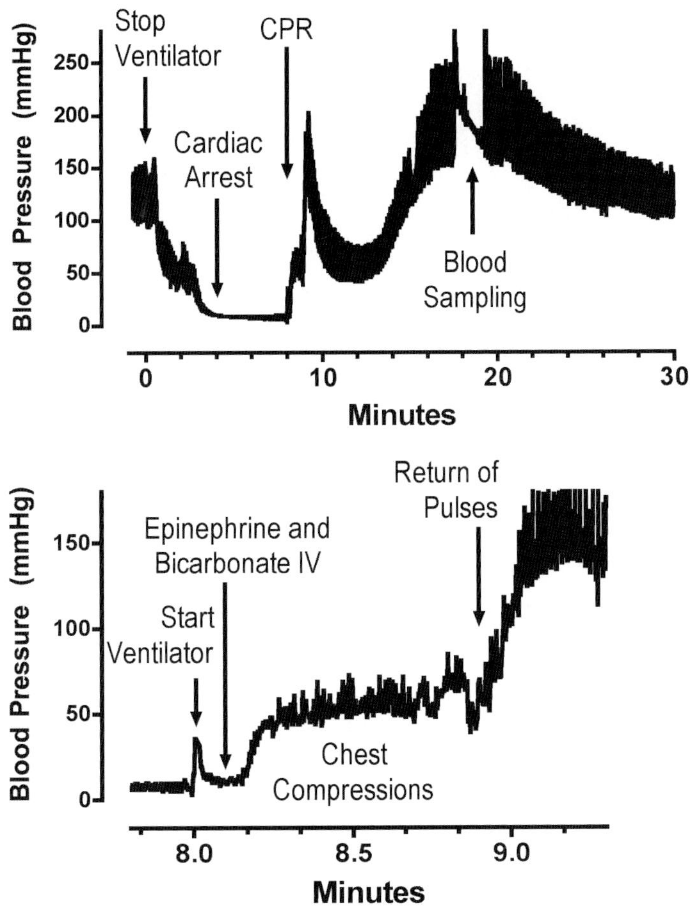

Fig. 2 Overview of blood pressure changes during procedure. *Bottom* panel is expanded in time scale

in the arterial catheter. Mean pressures around 40–50 mmHg are required to restore pulses.

5. Continuous chest compressions will generate a diastolic pressure that increases over 30–60 s. Spontaneous cardiac activity usually begins to appear when diastolic pressures near 40–50 mmHg are achieved (45–75 s of CPR). Chest compressions should continue until spontaneous arterial pressure waves are observed and ECG is detected.

3.3.5 Extubation and Wound Closure

1. Rats will require ventilatory support for at least 60 min. The ventilator rate can be gradually decreased during this time until the animal nears the baseline rate. Blood gases are typically drawn at 10, 30 and 60 min to ensure proper ventilation and oxygenation (Fig. 5).
2. After 60 min, the ventilator circuit can be disconnected from the endotracheal catheter to test whether the rat will spontaneously breathe. If the rat does not take adequate breathes within 30 s, the ventilator should be reconnected for an additional period of support.

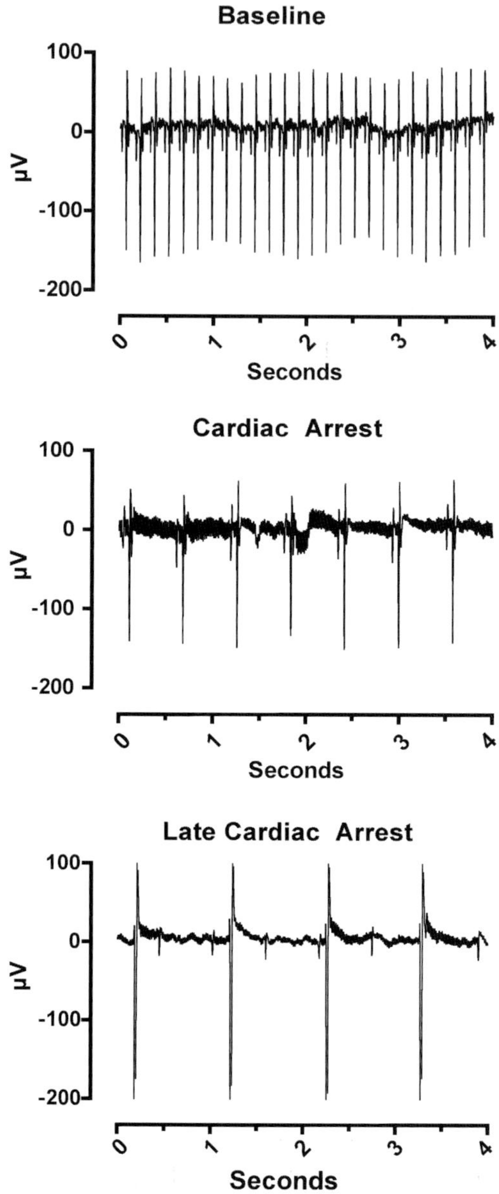

Fig. 3 ECG changes during cardiac arrest

3. When rats breathe well on their own, the suture holding the endotracheal tube can be cut and the tube withdrawn.
4. After extubation, gently suction the interior of the mouth with PE-50 tubing on a syringe to remove any secretions that may have accumulated during the arrest.
5. For survival studies that do not require venous and arterial catheters, these catheters can be withdrawn slowly from the vessel.

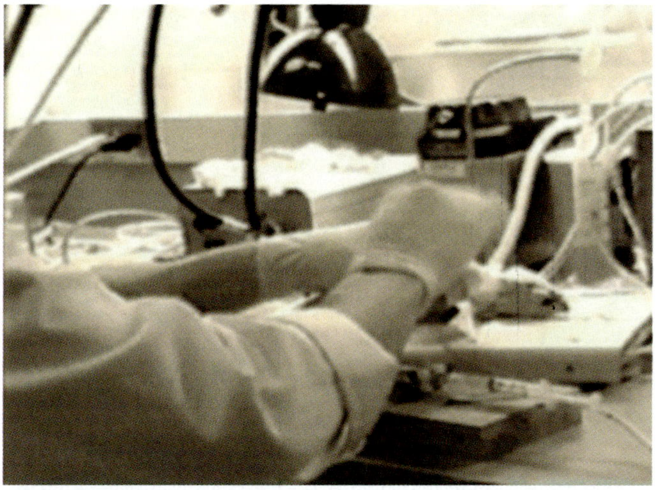

Fig. 4 Technique for chest compressions

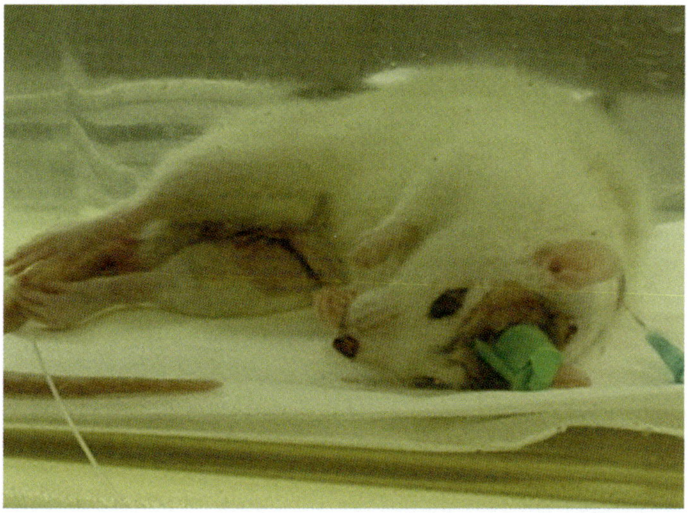

Fig. 5 Recovering rat with venous access and temperature probes in place

6. As the tip of the catheter passes the proximal tie, this tie can be cinched tight onto the vessel, ligating the vessel. Collaterals in the rat hindlimb are sufficient to prevent leg ischemia after this ligation.
7. Femoral wounds can be repaired using running 2–0 nylon suture or wound clips.

3.3.6 Post-Cardiac Arrest Care

1. Rats will not regulate body temperature adequately for extended periods after cardiac arrest,[8] and post-cardiac arrest temperature dramatically affects neurological recovery.[3,4,9] Therefore, thermostatic control of the temperature is absolutely necessary for 24–48 h after resuscitation (Fig. 6).

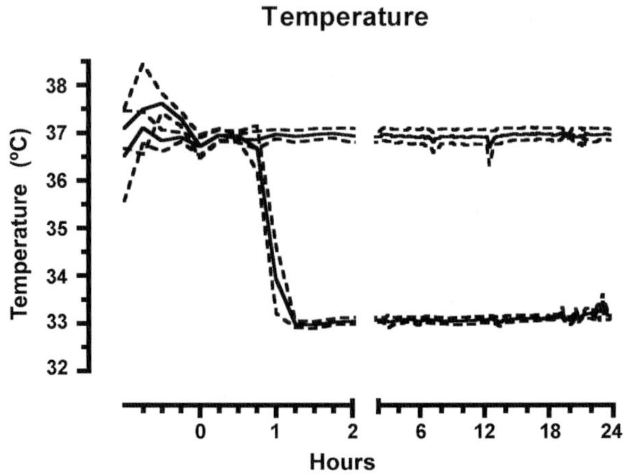

Fig. 6 Tight control of cranial temperature during recovery. Measurements from two rats illustrate normothermia and hypothermia regimens

2. Rats will not feed properly and may be in a coma for several days after cardiac arrest. To prevent dehydration, subcutaneous fluids (e.g., 5% dextrose in 0.9% saline) can be administered (20–40 mL kg^{-1} day^{-1}). Soft mushes of food can be provided to aid feeding, and food and water can be provided to rats with impaired mobility by hand.

4 Outcome Evaluations

4.1 Physiological Measures

1. Blood chemistry and blood gases normalize over the first 60 min after cardiac arrest (Fig. 7). Initial metabolic acidosis usually resolves without supplemental bicarbonate administration. Hyperglycemia is common immediately after cardiac arrest, but also declines over the 60 min after cardiac arrest.
2. Rats exhibit weight loss during the first week after cardiac arrest. This weight loss occurs even when rats appear to have recovered feeding activity. During the second week of recovery, surviving rats typically resume weight gain.
3. Some rats will die at delayed times after cardiac arrest. Death occurs suddenly in rats with the most severe neurological deficits. These deaths may result from airway obstruction or respiratory failure. Setting benchmark neurological scores for certain days after recovery and sacrificing rats that do not achieve the benchmarks will minimize unexpected death. For example, we sacrifice rats that do not reach a neurological score of 5, 10 and 15 on days 3, 5 and 7 respectively.

4.2 Behavioral Recovery

After cardiac arrest, rats exhibit coma for varying lengths of time. Recovery of neurological and behavioral function occurs over several days, similar to the recovery observed in humans.. Therefore, behavioral assessment should be

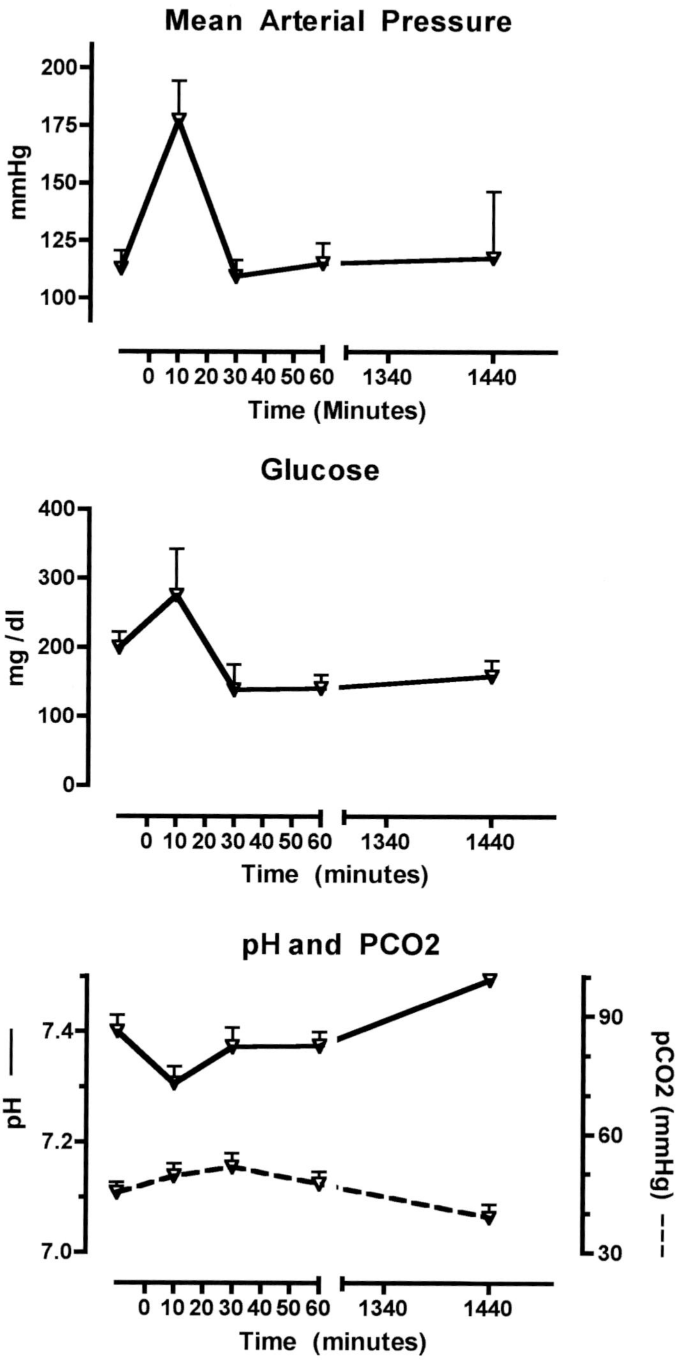

Fig. 7 Mean (and SD) changes in physiological variables in a series of rats during recovery after cardiac arrest illustrates stabilization after 60 min

repeated over time. Most neurological assessment has focused on emergence from coma and severe disability rather than attempting to measure specific memory or learning impairments.

1. Neurological Deficit Score. The original neurological deficit score was adapted from the scores used in dogs and swine after cardiac arrest.[2] The score ranges from 0 (brain dead) to 100 (normal behavior). This score has a positive correlation with summary scores of histological damage in multiple brain regions rather than any specific brain region.[10]
2. Modified Neurological Score. After several years experience with the cardiac arrest model, the neurological score was modified to eliminate items that were never abnormal and to increase inter-rater reliability.[7] The Modified Neurological Score ranges from 0 (no neurological activity) to 26 (normal). Points are awarded for individual behaviors. Testing can be repeated as often as required.
3. Other Behavioral Tests. The time required for a rat to remove a piece of tape from its paw has been used to quantify post-cardiac arrest behavior.[11] Rats that recover to near-normal neurological scores (24–26) may be able to participate in tests of memory and spatial learning. These tests are not appropriate for rats with more severe deficits. For example, water maze protocols are well established, and deficits have been reported in the post-cardiac arrest rat.[12,13]

4.3 Histology

Rat brain is collected for histology in a standard fashion similar to other rat models. Transcardiac perfusion is performed with saline (100 mL) followed by 10% formalin (100 mL). Brains are post-fixed in formalin and frozen sections (20–30 μm) are collected for staining. Results are similar using paraffin-embedded sections (6–10 μm)

1. Hematoxylin and Eosin stains. Cardiac arrest results in the appearance of eosinophilic neurons in many brain regions.[10] This histological sign of degeneration first appears 1–2 days after resuscitation and continues for at least 6 weeks.[8] Therefore, at least 3 days of post-cardiac arrest survival, and preferably more, are required to see histological signs of brain injury.
2. Neuronal stains. Because degenerating neurons continue to appear over weeks after resuscitation, it is desirable to quantify total neuronal loss. Immunohistochemical staining with antibodies to neuronal nuclear antigen (anti-NeuN, Chemicon) will visualize all surviving neurons (Fig. 8). Standard immunostaining kits can be used (e.g., Vector ABC). Two weeks after 8 min of asphyxia, the number of surviving neurons in the CA1 region declines by 50%.[4,9]

5 Advantages, Complications and Limitations

5.1 Advantages

1. A wide variety of neuroscience tools are available for the study of the rat. There is extensive literature about the neurobiology of the rat.
2. With an experienced surgeon, this model provides near 100% return of spontaneous circulation, very reliable injury and minimal duration of ventilator care.

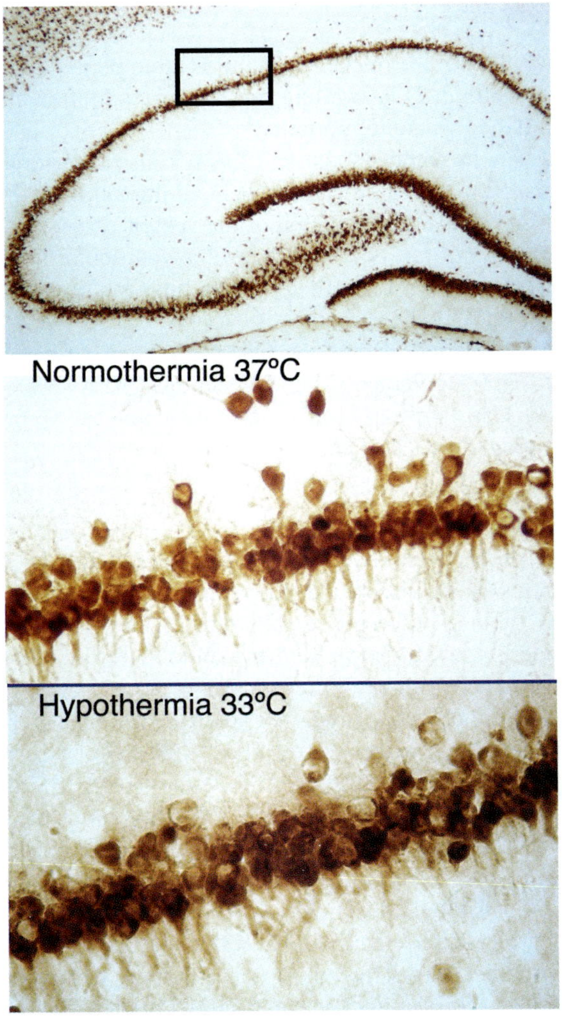

Fig. 8 Anti-NeuN staining of hippocampal CA1 region 14 days after cardiac arrest illustrates loss of neurons with normothermia, but preservation of neurons with hypothermia regimen

3. The cardiac arrest model produces complete ischemia of the whole brain, in contrast to arterial occlusion models, which injure forebrain more than hindbrain, and which may allow "trickle flow" during the period of ischemia.
4. The cardiac arrest model produces the transient systemic physiological changes that are associated with the clinical scenario of global brain ischemia, including hypoxemia, acidosis, systemic inflammation, hypercortisolemia, hyperglycemia.
5. The cost of supplies and resources for the model is low, although the performance of the model is labor-intensive.
6. The arrangement of the brain temperature probe allows intracranial drug injections if desired.[4] The thermistor acts as a stylet, which can be removed, allowing insertion of an injector needle through the guide-cannula. After an injection, the thermistor is replaced as a stylet.

5.2 Complications

1. Difficult intubation of the trachea or multiple attempted intubations can produce airway swelling. This swelling can impair respiration after extubation and potentially contribute to mortality.
2. Arterial and venous catheters can tear the vessels or perforate the far side of the vessels. Slow bleeding can result in additional mortality, but can be avoided by very gentle manipulation of the vessels.
3. A small proportion of rats (<1%) will develop ventricular fibrillation during cardiac arrest. Without a defibrillator, these rats usually cannot be resuscitated using CPR.
4. All rats will lose weight during the first week after cardiac arrest, despite hand feeding and supplemental fluid injections. With severe neurological injury, weight loss can be profound and rats losing >20% of body weight may require premature sacrifice.

5.3 Limitations

1. The surgical preparation requires multiple steps that may be performed slightly differently by individual surgeons. While a single surgeon can produce very reliable and reproducible injury, there is consistent between-surgeon variability in outcomes. Studies should therefore be conducted with a single surgeon performing all preparations.
2. The cardiac arrest procedure requires about 3h of intensive care per rat, with 2–5 days of post-surgery care. Thus, this model is less suitable for high-throughput studies.
3. Acute mortality is rare, and subacute mortality after the procedure is directly related to insult duration and temperature control. Most deaths occur between 1 and 5 days after cardiac arrest and death is directly related to severity of neurological deficits. Severely injured rats appear to have difficulty clearing secretions. In addition the absence of normal movement may result in atelectasis. Thus, most deaths appear to result from respiratory failure. We minimize unexpected death by euthanizing rats that do not achieve benchmark neurological scores on days 3, 5 and 7.

References

1. Hendrickx HH, Rao GR, Safar P, Gisvold SE. Asphyxia, cardiac arrest and resuscitation in rats. I. Short term recovery. Resuscitation 1984;12:97–116
2. Katz L, Ebmeyer U, Safar P, Radovsky A, Neumar R. Outcome model of asphyxial cardiac arrest in rats. J Cereb Blood Flow Metab 1995;15 1032–1039
3. Hicks SD, DeFranco DB, Callaway CW. Hypothermia during reperfusion after asphyxial cardiac arrest improves functional recovery and selectively alters stress-induced protein expression. J Cereb Blood Flow Metab 2000;20:520–530
4. D'Cruz BJ, Logue ES, Falke E, DeFranco DB, Callaway CW. Hypothermia and ERK activation after cardiac arrest. Brain Res 2005;1064:108–118
5. Ebmeyer U, Keilhoff G, Wolf G, Rose W. Strain specific differences in a cardio-pulmonary resuscitation rat model. Resuscitation 2002;53:189–200
6. Fink EL, Alexander H, Marco CD, Dixon CE, Kochanek PM, Jenkins LW, Lai Y, Donovan HA, Hickey RW, Clark RS. Experimental model of pediatric asphyxial cardiopulmonary arrest in rats. Pediatr Crit Care Med 2004;5:139–144

7. Neumar RW, Bircher NG, Sim KM, Xiao F, Zadach KS, Radovsky A, Katz L, Ebmeyer E, Safar P. Epinephrine and sodium bicarbonate during CPR following asphyxial cardiac arrest in rats. Resuscitation 1995;29:249–263
8. Hickey RW, Kochanek PM, Ferimer H, Alexander HL, Garman RH, Graham SH. Induced hyperthermia exacerbates neurologic neuronal histologic damage after asphyxial cardiac arrest in rats. Crit Care Med 2003;31:531–535
9. Logue ES, McMichael MJ, Callaway CW. Comparison of the effects of hypothermia at 33 degrees C or 35 degrees C after cardiac arrest in rats. Acad Emerg Med 2007;14:293–300
10. Radovsky A, Katz L, Ebmeyer U, Safar P. Ischemic neurons in rat brains after 6, 8, or 10 minutes of transient hypoxic ischemia. Toxicol Pathol. 1997;25:500–505
11. Albertsmeier M, Teschendorf P, Popp E, Galmbacher R, Vogel P, Bottiger BW. Evaluation of a tape removal test to assess neurological deficit after cardiac arrest in rats. Resuscitation 2007;74:552–558
12. Katz LM, Young A, Frank JE, Wang Y, Park K. Neurotensin-induced hypothermia improves neurologic outcome after hypoxic-ischemia. Crit Care Med 2004;32:806–810
13. Hickey RW, Akino M, Strausbaugh S, De Courten-Myers GM. Use of the Morris water maze and acoustic startle chamber to evaluate neurologic injury after asphyxial arrest in rats. Pediatr Res 1996;39:77–84

Chapter 9

Rodent Model of Pediatric Asphyxial Cardiac Arrest

Mioara D. Manole, Robert W. Hickey, Henry L. Alexander, and Robert S.B. Clark

Abstract A model of asphyxial cardiac arrest in 17-day old rats is described. This clinically relevant model includes a period of hypoxemia followed by ischemia and resuscitation. Continuous physiologic monitoring is performed before, during, and after the insult. Graded insults produce consistent and dose-dependent brain injury with histological damage and behavioral impairment. The details of the procedures, along with outcome assessments and applications, are discussed.

Keywords: Asphyxia, Cardiac arrest, Cerebral blood flow, Neurodegeneration, Sprague-Dawley rat

1 Model Selection

Cardiac arrest (CA) in infants and children is secondary to asphyxia in most cases.[1] This differs from adults where CA is typically caused by cardiac arrhythmias such as ventricular fibrillation (VF) or ventricular tachycardia (VT). Asphyxial CA differs in physiology from VF- or potassium chloride-induced cardiac arrest because a period of hypoxemia, hypercarbia, and hypotension precedes cardiovascular collapse and circulatory arrest. Regardless of the underlying etiology, all CAs have a period of total body ischemia that can cause impaired and inter-related post-ischemic organ dysfunction. This multisystem organ dysfunction is reproduced in animal models of CA (VF or asphyxial), but not in models of isolated brain ischemia, mimicking the clinical condition.

To model pediatric asphyxial CA, our laboratory modified an established protocol of CA in adult rats[2] and applied it to immature, postnatal day (PND) 17 rats.[3] PND 17 rats resemble a 1-to-4-year-old human child in terms of a number of developmental parameters.[4] The model of pediatric asphyxial CA described in this chapter is therefore especially relevant to scientists whose research interest involves CA in children.

From: *Animal Models of Acute Neurological Injuries.*
Edited by: J. Chen (eds.) New York, Totowa, NJ

2 Materials

2.1 Animal Selection

The use of all photographs has been preapproved by the Institutional Animal Care and Use Committee at the University of Pittsburgh.

We use Sprague-Dawley rat pups, PND 16–18. Pups are breastfed at this age and are therefore housed with a post-partum female rat, allowing them full access to nutrition prior to the experiment. Litters of one mother with eight pups are ordered to obtain rats with similar weights. It is our experience that pups below 30 g have lower survival rates post-resuscitation. Therefore, we typically use PND 16–18 rats weighing 30–40 g on the day of the experiment. Normal cardiovascular physiologic parameters of PND 17 rats under anesthesia with fentanyl or isoflurane are a heart rate (HR) of 350–450 beats min^{-1} and a mean arterial pressure (MAP) or 65–80 mmHg.

2.2 Anesthesia

In this model, we anesthetize the animals with either isoflurane or fentanyl. The choice of anesthetic depends on the parameter studied. Isoflurane provides an excellent plane of anesthesia and stable HR and MAP. Anesthesia with isoflurane has the advantages of being easier to titrate and more predictable. The recovery of the animal is rapid after isoflurane anesthesia; however, isoflurane markedly increases cerebral blood flow (CBF) as compared with fentanyl.[5] Therefore, in experiments where measurement of CBF is important, fentanyl is a more desirable anesthetic choice than isoflurane.

2.3 Equipment

2.3.1 Surgery

- Surgical instruments (Fig. 1a): Roboz by Dumont 45° angle forceps; straight Dumont forceps; Roboz RS-5602 iris scissors; Roboz RS-5137 forceps 45° angle; scissors; curved hemostat; straight forceps
- Dissecting microscope
- Femoral arterial and venous catheters: PE 10 and PE 50 polyethylene tubing (Becton Dickinson; Fig. 1b)
- Glue: 3M Scotch Weld Instant Adhesive CA 4OH
- Syringes
- 6.0 silk sutures
- Short blunted needle, 23G
- Heparinized saline 1 unit ml^{-1}
- Heating pad
- Cotton tip applicators
- Iodine
- Lidocaine solution

2.3.2 Anesthesia and Intubation

- Isoflurane
- Plexiglas chamber
- Laryngoscope with a modified Miller 0 blade, machine-filed from base to tip to reduce the width of the blade (Fig. 1c)

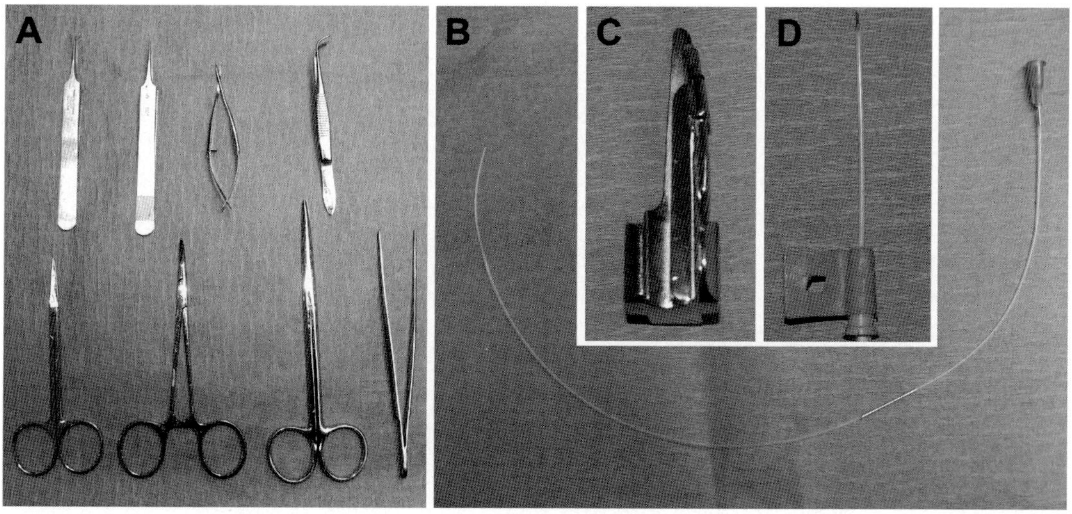

Fig. 1 Examples of equipment necessary for the pediatric asphyxial CA model. (**A**) Surgical instruments (from *top left*): Roboz by Dumont 45° angle forceps, straight Dumont forceps, Iris scissors (Roboz RS-5602), 45° angle forceps (Roboz RS-5137), various scissors, straight forceps; (**B**) Vascular catheter (observe the natural curvature of the catheter, which is placed in the vessel with the concavity cephalad); (**C**) Modified Miller 0 straight blade used for endotracheal intubation: the straight Miller 0 blade is filed from the base of the blade to the tip, resulting in a narrower, triangular shape; (**D**) Endotracheal tube with stylet: the endotracheal tube is an 18G, 2 in. angiocatheter with the needle removed. The stylet is a blunt vascular catheter introducer, and is advanced 2 mm past the tip of the angiocatheter. The stylet permits intubation before glottic spasm occurs, and reduces trauma to the vocal cords from the sharper angiocatheter. The hub of the angiocatheter has a piece of tape with an orifice to secure the catheter after placement

- Teflon angiocatheter 18G, 2 in. long, modified with the hub of the catheter surrounded by tape with a 0.5 cm edge and a central orifice, necessary for securing the angiocatheter after placement (Fig. 1d)
- Stylet custom-made from a central line introducer passed through the angiocatheter and advanced 2 mm beyond the tip. This facilitates advancement of the angiocatheter in the trachea with less trauma. The stylet can be stabilized at the hub of the catheter by taping the stylet to a cotton tip applicator
- Forceps, 45° angle, Roboz RS-5137
- Rubber band
- Rodent mechanical ventilator

2.3.3 Physiologic Monitoring

- Arterial pressure transducer
- Electroencephalogram (EEG) recorder with scalp electrodes
- Electrocardiogram (ECG) electrodes
- Temperature probe

3 Procedures

3.1 Intubation

The rat is anesthetized with isoflurane (mixture of 3% in 50% N_2O/balance O_2) in a Plexiglas chamber until unconscious. The rat is then placed supine on the anesthesia table and isoflurane is provided via a nasal cone made from the barrel of a 20-ml syringe. The front legs and tail are secured with tape to prevent the rat lifting off the table during intubation. A rubber band secured to the anesthesia table is placed over the upper incisors of the rat to stabilize the head during intubation, as shown in Fig. 2a. The tongue is lifted outside the mouth with forceps, then held between the index and thumb of the operator's left hand and pulled toward the left as shown in Fig. 2a. The laryngoscope blade is then introduced with the right hand until resistance is felt (approximately 2 cm in a 30–40 g rat). The tongue is then released, and the operator switches

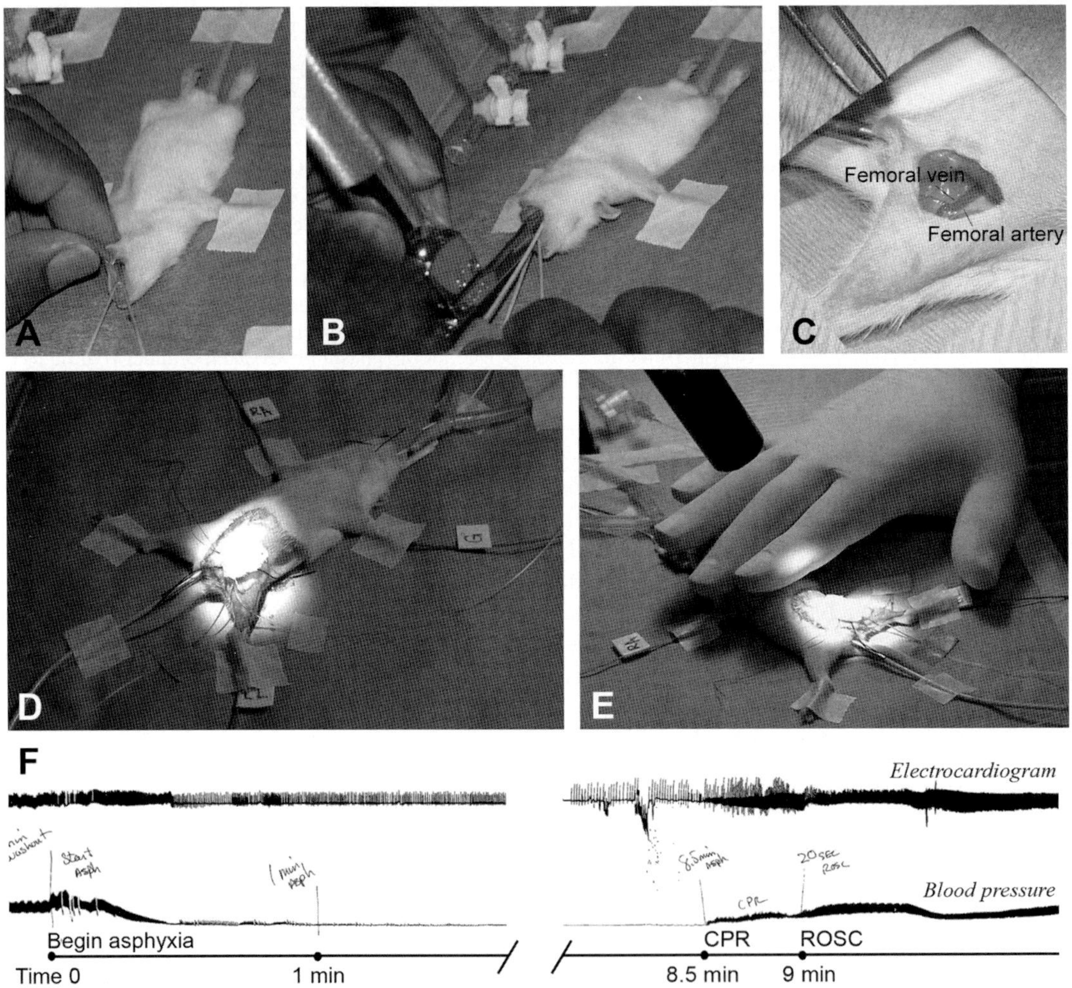

Fig. 2 Asphyxial CA in PND 17 rats. (**A**) The limbs are taped to the operating table and the upper incisors are secured with a rubber band to prevent inadvertent lifting of the head during endotracheal intubation. (**B**) Endotracheal intubation: the tongue is pulled to the left side and the custom-modified laryngoscope blade is used to facilitate placement of the 18G angiocatheter into the trachea. (**C**) Isolation and identification of femoral vein and femoral artery: the artery lies laterally to the vein. (**D**) Proper positioning of the rat and securing of invasive lines, monitors, and endotracheal tube. (**E**) Cardiopulmonary resuscitation (CPR) includes manual chest compressions. (**F**) Physiologic monitoring includes ECG and MAP tracings. Note pulseless electrical activity followed by return of spontaneous circulation (ROSC) after CPR

the laryngoscope handle to the left hand. The operator places the endotracheal tube (18G angiocatheter) in the right hand. The laryngoscope is then retracted until the glottic opening is seen. This is seen as a small (1 mm) orifice that opens and closes with breathing. The 18G catheter with introducer is passed into the trachea as shown in Fig. 2b. The introducer is then removed, the ventilator is connected, and the depth of the tube is adjusted to ensure equal chest expansion. The endotracheal tube is secured in place with a suture to the skin of the anterior neck of the rat.

It is our experience that multiple intubation attempts in these small rats will produce edema of the airway and decreased survival after extubation. When more than one attempt is necessary for intubation of the trachea, we stop the procedure and return the rat to the dam after recovery from anesthesia, and a different rat is used. The rat unsuccessfully intubated can be used on the following day.

3.2 Catheterization of Vessels

- Prepare the intravascular catheters. The catheters have a natural curvature (Fig. 1b). Maintain the normal curvature of the catheters, and orient the curvature towards the rats' head. Cut 25 cm of PE 50 catheter. Connect a 23G blunted needle to one end. Cut 4 cm of PE 10 catheter. Thread the PE 10 into the PE 50 catheter and glue the catheters together. Cut the tip of the PE 10 catheter at a 45° angle bevel, ensuring that a clean cut with no burrs is obtained. Fill catheter with heparinized saline solution and attach a 1-ml syringe to the needle hub.
- Shave the femoral area and clean the area with iodine using aseptic technique.
- While holding the skin with small, smooth tissue forceps, make an oblique incision over the femoral area with iris scissors.
- Under the magnifying microscope, bluntly dissect the tissues to expose the femoral vessels with a cotton tip applicator (Fig. 2c).
- Identify the femoral artery and vein (the vein is medial to the artery).
- With the Dumont 45° angle forceps and straight Dumont forceps, separate the femoral vessels.
- Owing to manipulation and drying, the vessels will contract. To prevent this, place two drops of lidocaine on the surgical area to dilate the vessels.
- Place one silk suture proximally and one distally on the isolated vessel to cannulate.
- Tie the distal suture permanently; the extremity will get blood supply through collateral vessels.
- Make a loose knot with the proximal suture, which will be tightened after the catheter is in place.
- Place a hemostat over the proximal suture ends and apply gentle tension for temporary vessel occlusion.
- Make a small incision in the lower portion of the vessel with the iris scissors (Roboz RS 5602).
- With the left hand, stabilize the proximal side of the incision in the vessel with a Roboz RS 5137 90° forceps.
- With the right hand, pick up the catheter a few millimeters from the tip with a 45° angle forceps. The catheter should be held with the curvature toward the head of the rat, and with the bevel cephalad.

- Introduce the catheter up to the level of the proximal ligature, which is held under tension with the suture and the hemostat. Release the tension from the hemostat, and advance the catheter no more than 5 mm. At this point, the tip of the catheter is not visible. Advancing the catheter further may damage the vessel wall.
- Tie the proximal suture. Align the catheter with the vessel and tie the catheter with the ends of the distal suture to secure the femoral line.
- Repeat the procedure for the other vessel to obtain both arterial and venous access.
- Connect the arterial transducer to the arterial line.
- Close the skin incision.

3.3 Asphyxial Cardiac Arrest

The rats are ventilated with a tidal volume of 0.6 ml and a rate of 35–45 breaths min^{-1} using a Harvard Model 683 Rodent Ventilator (Holliston, MA). Arterial blood gas parameters are measured before the asphyxial arrest and the ventilator is adjusted if necessary to achieve a $PaCO_2$ between 35 and 45 mmHg. Pre-arrest, the temperature is monitored and maintained at 37°C with a heating pad or a warming light. Vecuronium at a dose of 1 mg kg^{-1} is administered intravenously 10 min before CA to prevent spontaneous respirations and allow for apnea.[6] Two minutes before asphyxia, the isoflurane is discontinued to allow anesthetic washout to reduce the potential confounding effects of inhaled anesthetics on neurological injury.[7] Also, anesthetics can cause myocardial depression and hypotension, making it more difficult to successfully resuscitate the rats. Previous pilot experiments testing the anesthetic regimen in uninjured rats show that EEG activity begins to recover in amplitude and frequency after 2 min of anesthetic washout. One minute before asphyxia, the FiO_2 is reduced to 0.21 by disconnecting the oxygen tube from the ventilator, to avoid hyperoxygenation pre-arrest. After the anesthetic washout, the ventilator is disconnected at the endotracheal tube for the desired duration of asphyxia, typically 8–12 min. During CA, the ECG tracing progresses from normal sinus rhythm, to bradycardia, then to pulseless electrical activity, and rarely to asystole or VF. Typically, CA occurs 2–4 min after disconnecting the ventilator, and is defined as occurring when there are no pulsations in the arterial line tracing, i.e., pulseless electrical activity (Fig. 2f).

After the desired duration of asphyxia, epinephrine 0.005 mg kg^{-1} and sodium bicarbonate 1 m eq kg^{-1} are administered intravenously. The sequence of events is as follows: at time −1 min before resuscitation the arterial line is flushed to ensure patency (an accurate arterial pressure wave is essential for monitoring successful return of circulation), and then the medications are administered in quick succession via the venous line, followed by a 10 ml kg^{-1} intravenous fluid bolus of normal saline. At time −30 s before resuscitation, the endotracheal tube is reconnected to the ventilator and the rats are mechanically ventilated at a rate of 10 breaths min^{-1} above baseline using an FiO_2 of 1.0. At time 0, rapid manual chest compressions are performed with two fingers on the mid-chest at a rate of approximately 300 min^{-1}, until return of spontaneous circulation, which typically occurs in less than 1 min. During the CA the rat is kept on a heating pad. Usually, the temperature will spontaneously decrease during CA and the rat will spontaneously re-warm after return of spontaneous circulation. Figure 2f depicts a typical tracing of the HR and MAP at the start of asphyxia through return of spontaneous circulation.

3.4 Post-Resuscitation Care

Arterial blood gas values are measured 15–30 min after CA and the ventilator rate is adjusted if necessary to maintain a $PaCO_2$ of 35–45 mmHg. The FiO_2 is maintained at 1.0, though this could be reduced if desired. The rat is kept on the ventilator until 1 h after return of spontaneous circulation, at which time the arterial and venous lines are removed and the incision is closed with 6.0 silk sutures. The endotracheal tube is disconnected from the ventilator and the ability of the rat to breathe spontaneously through the endotracheal tube is observed for 5 min prior to extubation. The rat is placed on supplemental O_2 (100%) via a nose cone for 5 min after extubation. Afterwards, the rat is given a fluid bolus of D5W 1/2 NS, 20 cc kg^{-1} subcutaneously in the posterior mid-thoracic region because typically the rats are comatose between 12 and 48 h depending upon the duration of asphyxia and therefore do not eat or drink. The rat is then placed in an incubator with supplemental oxygen for 1 h, and then returned to the dam and littermates. The rats typically emerge from coma and are able to feed between 12 and 48 h depending upon the duration of asphyxia. For longer insults (>10 min), the rats are observed overnight in the incubator, re-examined in 24 h, and given another fluid bolus of D5W 1/2½ NS, 20 cc kg^{-1} subcutaneously. Longer insults are possible; however, it is our experience that after 12 min of asphyxia, the rats cannot be weaned from the ventilator and would require prolonged ventilation and post-resuscitative care.

4 Outcome Evaluation

Several end points can be used to evaluate neurological outcome. These include both functional outcome and histological assessment. The duration of coma and spontaneous activity can be used as gross functional assessments, as the duration of coma is directly proportional to the duration of asphyxia.

Vestibulomotor function can be evaluated using a number of quantitative tests. The beam balance test employs a narrow wooden beam (1.5 cm wide), and the duration the rat remains on the beam is measured (for up to 60 s). Prominent beam balance deficits are detected after asphyxia for up to 3 days after 8 min of asphyxia.[3,8] Gross motor function can also be evaluated using an inclined-plane test. This test involves a flat, rubber-covered surface that is raised at 5° angle increments until the rat can no longer maintain its position on the surface. Prominent inclined-plane deficits are detected after asphyxia for up to 2 days after 8 min of asphyxia.[3,8]

Cognitive deficits can be evaluated using the Morris water maze, although there are unique considerations for younger rats. The Morris water maze employs a 180-cm diameter and 60-cm high pool filled with water to a depth of 28 cm and kept at 20–22°C. A platform 10 cm in diameter and positioned 2 cm below the water's surface is used as the hidden goal platform in a room with constant extra-maze cues. Since spatial memory acquisition is not felt to be developed in rats until PND 24, testing begins at least 7 days after asphyxia in PND 17 rats. Because only moderate Morris water maze deficits are seen after 8 min of asphyxia,[3] increasing durations of asphyxia were tested.

However, injury effects remained difficult to detect, primarily because of the poor performance of sham rats (relative to adult rats). The Morris water maze paradigm has since been adjusted to include a 3-day acclimation period, using a visible platform testing, followed by a 4-day hidden platform testing period. Using this modified paradigm, robust injury effects are observed.

Histological assessment has been performed at different time points after different durations of asphyxial CA. Hematoxilin and eosin staining show eosinophilic, pyknotic neurons in CA1 hippocampal neurons and layer 5 cortical neurons after 8 min of asphyxial CA.[3] These cells are also Flurojade B positive. Fluorojade B staining identifies neurons that are undergoing degeneration.[9] Terminal deoxynucleotidyl transferase-mediated biotin dUTP nick end labeling (TUNEL) identifies neurons with DNA damage including those undergoing apoptosis,[10] and is detected in CA1 hippocampal neurons but not cortical neurons after 8 min of asphyxia.[3] CA1 hippocampal neurons are readily quantifiable, and a 40–60% reduction in CA1 hippocampal neuron survival is seen after 8 min of asphyxia.[3] Remarkably, neurodegeneration as assessed by using Fluorojade B is detectable up to 5 weeks after asphyxial CA.[8]

While after 8 min of asphyxia neuronal damage appears to be restricted to layer 5 cortical and CA1 hippocampal neurons, longer durations of asphyxia result in extension of neuronal damage to other brain regions. Figure 3 shows Fluorojade B staining 48 h after 8.5 and 9 min of asphyxial CA. After 8.5 min of asphyxia, neurodegeneration extends to the subiculum and CA2 hippocampal regions (Fig. 3b) and beyond cortical layer 5 (Fig. 3e). After 9 min of asphyxia, neurodegeneration extends to CA3, CA4, dentate gyrus, and hilar hippocampal regions, throughout the cerebral cortex and thalamus, and Purkinje cells in the cerebellum. Thus, asphyxia-dose-dependent increases in neurodegeneration are observed in this model.

5 Advantages and Applications

This model has many advantages, including the capability to (1) perform invasive physiologic monitoring and acute resuscitation that closely mimic guidelines used in pediatric patients; (2) assess acute and long-term functional outcome using standardized behavioral tools; (3) assess acute and long-term biochemical and cellular markers of cell death and molecular mechanisms; and (4) test clinically relevant and novel interventions and treatments for efficacy after asphyxial CA in the developing brain. One example of the latter is therapeutic hypothermia. While hypothermia is now recommended after VF-induced cardiac arrest in adult patients,[11] its efficacy after pediatric asphyxial arrest is less clear. Using this model, brief, induced post-resuscitative hypothermia (32°C for 1 h followed by a 1-h re-warming period) improved motor function and provided long-lasting neuroprotection compared with normothermic controls.[8]

This model can also be used to evaluate CBF patterns after CA in immature rats, information that is currently extrapolated from models of CA in adult animals. To measure CBF, arterial spin labeling (ASL) magnetic resonance imaging (MRI) was used to allow noninvasive serial measurement of regional CBF. This method uses the magnetic signal generated by protons in arterial blood as a tracer. To facilitate this in this model, the anesthetic regimen was

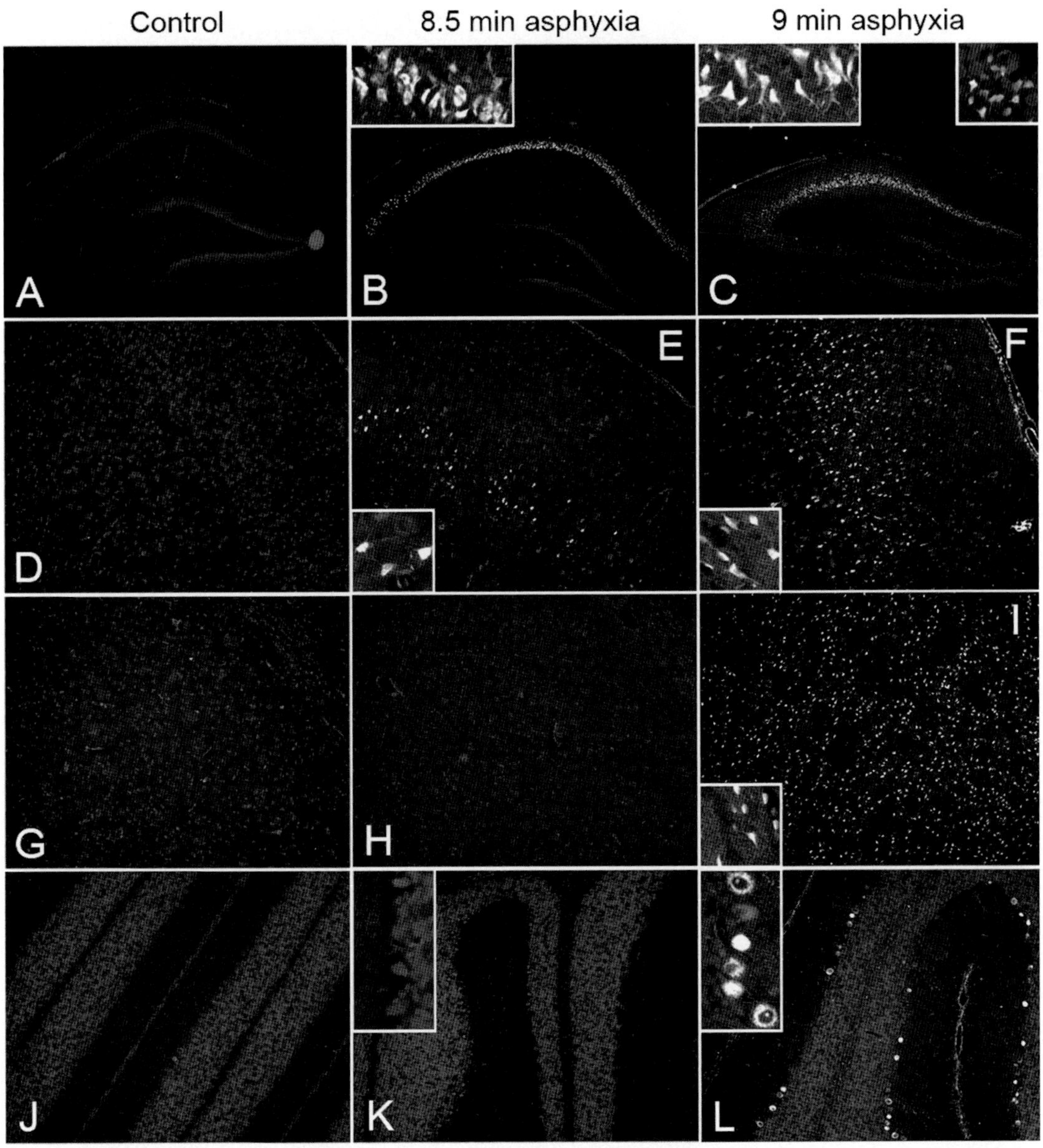

Fig. 3 Asphyxia-duration-dependent increase in neurodegeneration as assessed by Fluorojade B staining 48 h after 8.5 (**B**, **E**, **H**, **K**) and 9 min (**C**, **F**, **I**, **l**) vs. naive control (**A**, **D**, **G**, **J**). Fluorojade positive cells are *bright white* in this black and white image. Note the expansion of Flurojade B staining in regions of the hippocampus (**A**–**C**), cortex (**D**–**F**), thalamus (**G**–**I**), and cerebellum (**J**–**L**) with increasing durations of asphyxia

modified. Anesthesia is initially achieved with isoflurane until arterial and venous lines are placed. Afterwards, the isoflurane is discontinued and fentanyl is administered at a dose of 50 μg $kg^{-1}h^{-1}$ for anesthesia. A continuous vecuronium infusion at a dose of 5 mg $kg^{-1}h^{-1}$ is used to prevent motion during MRI. Figure 4 shows serial CBF measurements in a sham rat and a rat after 8 min of asphyxial CA. CBF was measured using the ASL method and a 7 T Bruker Biospec MRI instrument. After 8 min of asphyxia, prominent hyperemia

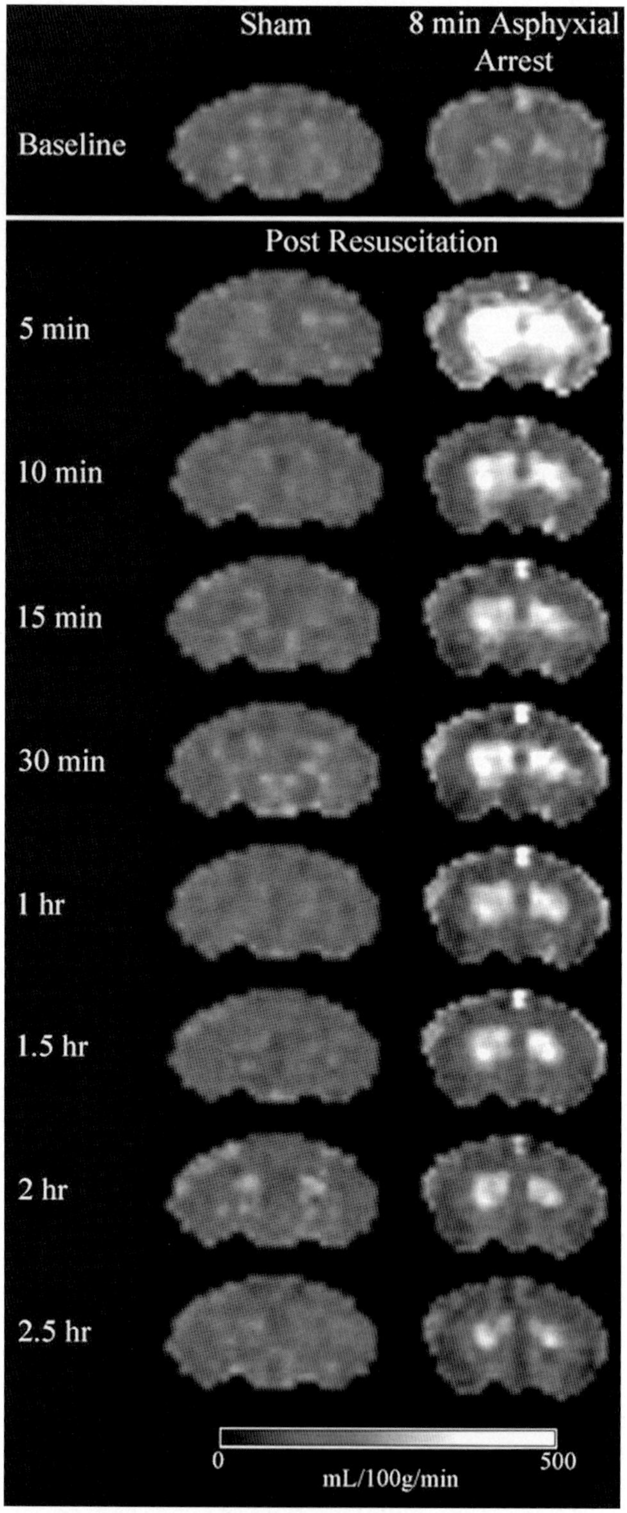

Fig. 4 Regional CBF maps measured serially at baseline and after 8 min of asphyxia. CBF was measured using ASL in a 7 T Bruker Biospec MRI. Note post-arrest hyperemia (*bright areas* in this black and white image) after asphyxia vs. sham

is observed in subcortical brain structures. Therefore, this model, coupled with ASL-MRI, will permit a comprehensive evaluation of cerebrovascular disturbances after asphyxial CA in the developing brain.

6 Conclusions

We have described a clinically relevant model of pediatric asphyxial CA. This model can be used to characterize cerebrovascular disturbances and molecular mechanisms relevant to infant and childhood victims of respiratory arrest, for example, those due to respiratory infections or near-drowning accidents. This model also has the capacity to evaluate potential therapies aimed at improving outcome and reducing the devastating effects of hypoxic-ischemic encephalopathy after CA in infants and children.

References

1. Young KD, Gausche-Hill M, McClung CD, Lewis RJ. A prospective, population-based study of the epidemiology and outcome of out-of-hospital pediatric cardiopulmonary arrest. Pediatrics 2004;114(1):157–64
2. Katz L, Ebmeyer U, Safar P, Radovsky A, Neumar R. Outcome model of asphyxial cardiac arrest in rats. J Cereb Blood Flow Metab 1995;15(6):1032–9
3. Fink EL, Alexander H, Marco CD, et al. Experimental model of pediatric asphyxial cardiopulmonary arrest in rats. Pediatr Crit Care Med 2004;5(2):139–44
4. Rice D, Barone S Jr. Critical periods of vulnerability for the developing nervous system: evidence from humans and animal models. Environ Health Perspect 2000;108(Suppl 3):511–33
5. Hendrich KS, Kochanek PM, Melick JA, et al. Cerebral perfusion during anesthesia with fentanyl, isoflurane, or pentobarbital in normal rats studied by arterial spin-labeled MRI. Magn Reson Med 2001;46(1):202–6
6. Manole MD, Hickey RW, Momoi N, et al. Preterminal gasping during hypoxic cardiac arrest increases cardiac function in immature rats. Pediatr Res 2006;60(2):174–9
7. Statler KD, Kochanek PM, Dixon CE, et al. Isoflurane improves long-term neurologic outcome versus fentanyl after traumatic brain injury in rats. J Neurotrauma 2000;17(12):1179–89
8. Fink EL, Marco CD, Donovan HA, et al. Brief induced hypothermia improves outcome after asphyxial cardiopulmonary arrest in juvenile rats. Dev Neurosci 2005;27(2–4):191–9
9. Schmued LC, Hopkins KJ. Fluoro-Jade B: a high affinity fluorescent marker for the localization of neuronal degeneration. Brain Res 2000;874(2):123–30
10. Loo DT, Rillema JR. Measurement of cell death. Methods Cell Biol 1998;57:251–64
11. 2005 International Consensus on Cardiopulmonary Resuscitation and Emergency Cardiovascular Care Science with Treatment Recommendations. Part 6: Paediatric basic and advanced life support. Resuscitation 2005;67(2–3):271–91

Chapter 10

Transient Global Cerebral Ischemia Model in Mice

Kuniyasu Niizuma, Hidenori Endo, Chikako Nito, D. Jeannie Myer, Gab Seok Kim, Teiji Tominaga, and Pak H. Chan

Abstract The transient global ischemia model is used to analyze selective neuronal death in vulnerable regions including the CA1 subregion of the hippocampus and in certain cortical neurons. In this mouse global ischemia model, bilateral common carotid arteries are reversibly occluded, and the individual anatomical backgrounds are normalized by evaluating the patency of the posterior communicating artery. Using this evaluation, this model might be used for mouse strains with various genetic backgrounds.

Keywords: Blood flow, Delayed neuronal death, Global cerebral ischemia model, Hippocampus, Ischemic brain injury, Mice

1 Model Selection

Transient global cerebral ischemia (tGCI) causes selective neuronal death in vulnerable regions such as the hippocampal CA1 subregion and in some areas of the cerebral cortex.[1–3] The tGCI model has been widely used for studies of neuronal injury. In numerous studies, researchers have attempted to use a representative model of tGCI; however, most have used gerbil or rat models.

Recent advances in transgene and gene knockout technology have enabled us to investigate the precise molecular mechanism of ischemic brain damage. At this time, the predominant species used for the study of genetic mutations is the mouse, and therefore the importance of the mouse tGCI model is significant.

Over the past 10 years, some authors have established mouse tGCI models using two-vessel occlusion,[4–9] two-vessel occlusion with hypotension caused by withdrawing blood,[8,10] three-vessel occlusion,[11,12] and cardiac arrest.[13,14] The tGCI models based on two-vessel occlusion are relatively simple, but result in inconsistent degrees of ischemic damage due to individual differences in collateral blood flow through the Circle of Willis.[5–8,10] The three-vessel

From: *Animal Models of Acute Neurological Injuries.*
Edited by: J. Chen (eds.) New York, Totowa, NJ

occlusion model originally described by Panahian et al.[11] is reproducible; however, the low survival rate may limit its popular use. Yonekura et al.[12] have established a modified three-vessel occlusion model that has a relatively low mortality rate and is highly reproducible, although technically very difficult. Cardiac arrest models are also complicated. A preferable choice for achieving a reproducible but simple tGCI model is a two-vessel occlusion model with careful evaluation of the patency of the posterior communicating artery (PcomA). This evaluation can normalize the anatomical background, as previously described by our group.[7]

2 Materials

2.1 Animals

The patency of the PcomA is regarded as a crucial factor in tGCI, on the basis of the two-vessel occlusion technique. In mice, strain differences have been observed relative to the patency of the PcomA at the Circle of Willis.[4,15] Using a strain that has poor development of the Circle of Willis, such as C57BL/6 or BALB/c, may increase the rate of significant ischemic damage. However, any mouse strain can be used for the tGCI model, with evaluation of the PcomA as described above. We used CD1 mice (35–40 g), taking advantage of their larger size and the availability of several lineages of transgenic and knockout mice that were originally bred or crossbred with a CD1 background.

2.2 Anesthesia

The anesthetics used for the mouse tGCI model are chloral hydrate (CH) and xylazine with an animal ventilator (Harvard Apparatus). CH i.p. anesthesia using CH can induce slight hypotension, which reduces the collateral blood flow compared with inhalation anesthetics including isoflurane and halothane. However, CH causes hypercapnia derived from respiratory disturbance. Therefore, controlled ventilation is needed. This combination of CH anesthesia and controlled ventilation can induce significantly lower blood pressure than isoflurane and decrease $PaCO_2$ to the same level as the isoflurane anesthesia previously described.[7]

2.3 Equipment

- Animal respirator (Harvard Apparatus)
- Arterial catheter for blood pressure monitoring and arterial blood gas analysis (PE-10; Intramedic)
- Blood pressure monitoring unit (Pressure Transducer; Harvard Apparatus)
- Forceps, scissors, etc. (Hegar, Micro dissecting blade, Pattern #5 forceps; Roboz)
- Homeothermic blanket with rectal probe (Harvard Apparatus)
- Lacrilube ophthalmic ointment (Allergan Pharmaceuticals)
- Laser Doppler flowmeter for relative regional cerebral blood flow (rCBF) (Laserflo; Vasamedics)
- Microsurgical clips and clip-applying forceps (Zen temporary clip; Oowatsusho, Tokyo, Japan)
- Suture to close skin (5–0 Dermalon; Tyco Healthcare)

3 Procedures

3.1 Surgical Procedure

1. Anesthetize mouse with CH (350 mg kg^{-1}) and xylazine (4 mg kg^{-1}) i.p.
2. Shave hair around the surgical site in the hood
3. Apply Lacrilube ophthalmic ointment to prevent eyes from drying
4. Place the animal on a homeothermic blanket and insert a rectal probe to maintain body temperature at 37 ± 0.5°C
5. Use Betadine to sterilize the skin over the surgical site
6. Make a skin incision on the midline of the ventral neck (Fig. 1a)
7. Expose the trachea, then insert a tracheal tube
8. Start controlled ventilation using an animal ventilator with a volume of 0.5 ml and a respiratory rate of 120 min^{-1}
9. Expose bilateral common carotid artery bifurcations; then isolate from the sympathetic nerve and the vagus nerve (Fig. 1b)
10. Five minutes after the start of controlled ventilation, apply microsurgical clips to the bifurcations to independently occlude both the external and internal carotid arteries (Fig. 1c)
11. After the desired duration of global ischemia, remove clips and suture the skin using the 5-0 Dermalon suture. The duration of ischemia remarkably affects the level of neuronal damage and mortality. Usually, 10–15 min of ischemia are allowed

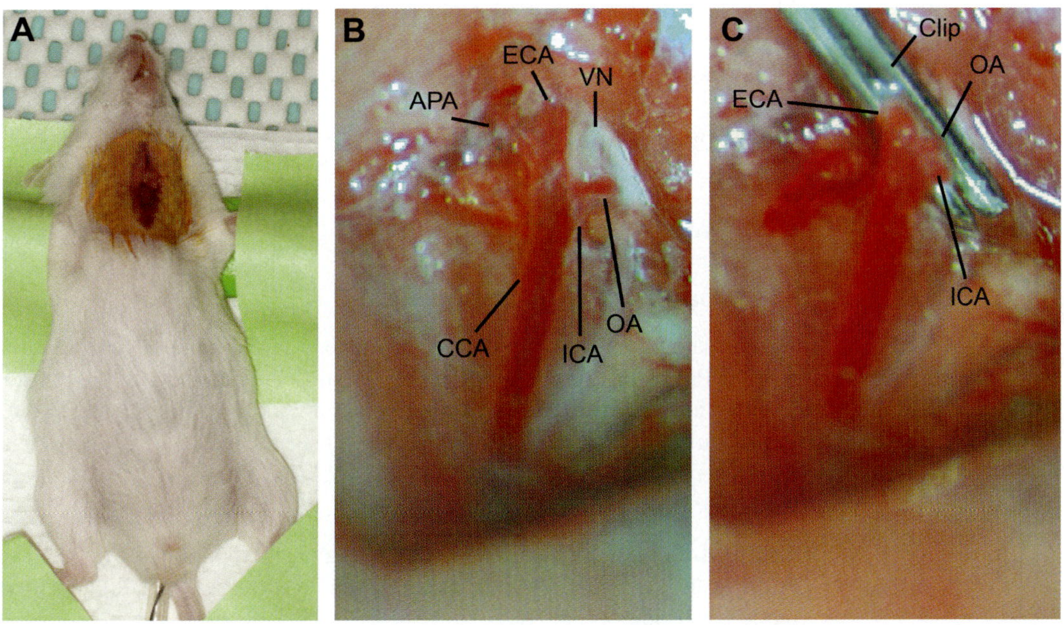

Fig. 1 Step-by-step procedures: (**A**) Skin incision. (**B**) Isolation of left common carotid artery (CCA) bifurcation. (**C**) Application of microsurgical clip to the bifurcation. Clips must be applied to independently occlude both the external and internal carotid arteries. APA, ascending pharyngeal artery; ECA, external carotid artery; ICA, internal carotid artery; OA, occipital artery; VN, vagus nerve

3.2 Monitoring of Physiological Parameters

- Relative rCBF
  Make a small skin incision near the bregma, using glue attach the laser Doppler probe to the intact skull over the ischemic cortex 3.5 mm lateral to the bregma, and then continuously monitor changes in cortical perfusion. There are no differences in the laser Doppler flowmeter readings if the skull is removed or kept intact.[4]
  rCBF is the most critical factor for regulating the level of neuronal damage after tGCI. A reduction in rCBF to <10% correlated with damage in the CA1 subregion after tGCI.[9]

- Mean arterial blood pressure
  Connect the pressure transducer to a PE-10 catheter. Expose the femoral artery and catheterize, then start blood pressure monitoring.
  A high level of mean arterial blood pressure increases cerebral perfusion pressure and results in an increase in rCBF.

- Arterial blood gas
  Blood gas is usually analyzed before, during, and about 15 min after ischemia using a blood gas analyzer.
  High levels of $PaCO_2$ cause vasodilatation to compensate the cerebral blood flow as described above.

- Body temperature
  Insert a rectal probe and monitor body temperature.
  Hyperthermia results in aggravation of brain damage.[16] In contrast, hypothermia, not only during ischemia but also after ischemia, is neuroprotective.[16] Therefore, the mouse's temperature should be controlled at around 37°C during and after ischemia. Post-ischemic hyperthermia has not been seen in the mouse model of tGCI; however, hypothermia has been noted and could be injurious.[9] An incubator should be used to avoid hypothermia after surgery.

4 Outcome Evaluations

4.1 Evaluation of PcomA Plasticity

Since the patency of the PcomA will influence the outcome of tGCI, evaluation of the PcomA has great importance. A significant decrease in rCBF can be seen in the hypoplastic PcomA group compared with the normoplastic PcomA group during bilateral common carotid artery occlusion (Fig. 2). Moreover, ischemic damage is more severe in the hypoplastic PcomA group compared with the normoplastic PcomA group (Fig. 3):

1. Anesthetize mouse with ketamine (200 mg kg^{-1}) and xylazine (10 mg kg^{-1}) i.p.
2. Perform thoracotomy and introduce a cannula into the ascending aorta via the left ventricle.
3. Perform transcardial perfusion fixation with 200 ml, 10 U ml^{-1} heparin in 0.9% saline and 200 ml of 3.7% formaldehyde in phosphate-buffered saline.

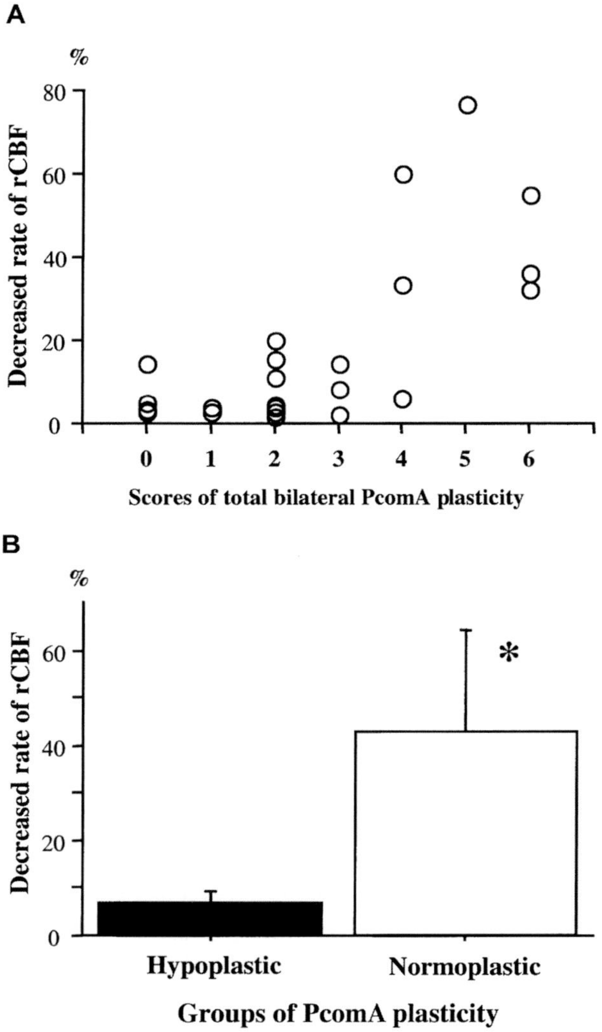

Fig. 2 Decreased rates of rCBF during bilateral common carotid artery occlusion. (**A**) Relationship between rCBF and the total score of plasticity of bilateral PcomAs. Each *open circle* indicates the decreased rate of rCBF in the posterolateral cortex of the brain, with the scores of PcomA plasticity totaling 0–6. (**B**) Brains were classified into two groups on the basis of total scores of bilateral PcomA plasticity; the score was <3 in the hypoplastic PcomA group and >3 in the normoplastic group. A significant difference in the decreased rate of rCBF during ischemia was observed between these groups. $*P < 0.01$ (cited from Ref. 17)

4. Inject 0.2 ml of carbon black in an equal volume of 20% gelatin in H_2O through the cannula.
5. Remove the brain and fix in 3.7% formaldehyde overnight at 4°C.
6. Grade the plasticity of the PcomA on a qualitative scale of 0–3, with 0 = no anastomosis between posterior cerebral artery and superior cerebellar artery; 1 = anastomosis between posterior cerebral artery and superior cerebellar artery in capillary phase; 2 = small truncal PcomA; 3 = truncal

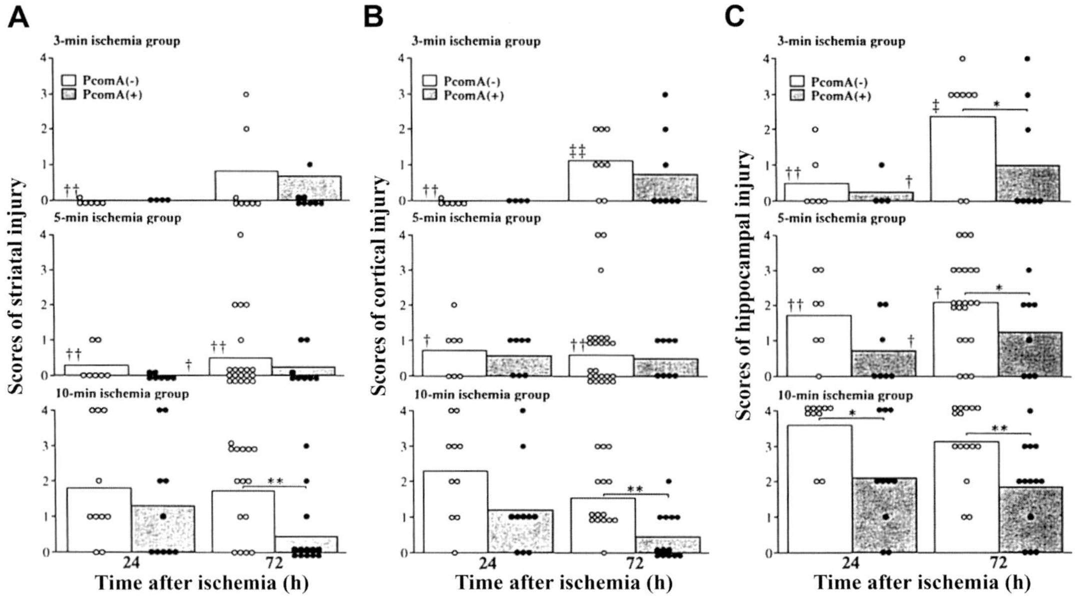

Fig. 3 Qualitative evaluation of neuronal injury in the striatum (**A**), cortex (**B**), and hippocampus (**C**) following tGCI. PcomA (−) means grades 0 and 1, and PcomA (+) means grades 2 and 3. $*P < 0.05$; $**P < 0.01$, PcomA (−) vs. PcomA (+); $^{\dagger}P < 0.05$; $^{\dagger\dagger}P < 0.01$ vs. 10-min ischemia group; $^{\ddagger}P < 0.05$; $^{\ddagger\ddagger}P < 0.01$ vs. 24 h (cited from Ref. 7)

PcomA, under a dissecting microscope (Fig. 4). Then, total the plasticity scores of the PcomAs in both hemispheres. The brains with scores of 0–3 are classified as the hypoplastic PcomA group, and the brains with sums >3 are classified as the normal PcomA group. Hippocampal damage is significantly severe in the hypoplastic PcomA group compared to the normoplastic PcomA group.[7,17]

This grading scale can be used for applications without transcardial perfusion fixation, such as Western blotting and other biochemical assays. After taking the brain, observe the Circle of Willis directly under the microscope and grade the PcomA plasticity.

4.2 Histological Analysis of Ischemic Neuronal Damage

Hematoxylin and eosin staining or cresyl violet staining is widely used to evaluate ischemic neuronal damage. Grading on the 5-point scale below can be used to evaluate the severity of ischemic damage. Typical patterns of damage in the hypoplastic PcomA group are seen in Fig. 5.

1. Perform PcomA evaluation as described above.
2. Using a vibratome cut coronal sections 50 μm thick.
3. Select the section of the striatum (0.5 mm anterior to the bregma) and hippocampus (2.0 mm posterior to the bregma).
4. Stain selected sections with hematoxylin and eosin or cresyl violet.
5. For the striatal and cortical lesions, grade the neuronal damage qualitatively on a scale of 0 to 4, with 0 = normal; 1 = 0–25% neurons dam-

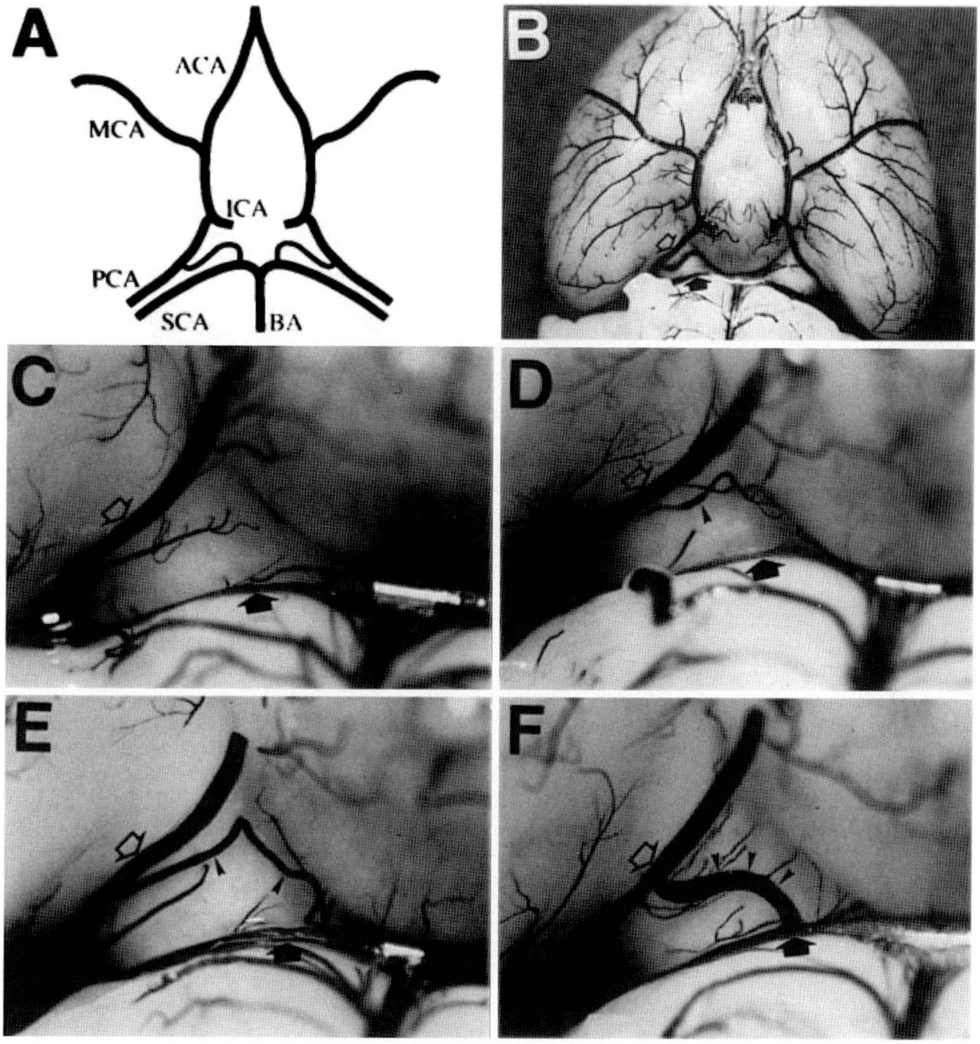

Fig. 4 (**A**) Scheme showing major blood vessels in the mouse brain, including the anterior cerebral artery (ACA), middle cerebral artery (MCA), internal carotid artery (ICA), posterior cerebral artery (PCA), superior cerebellar artery (SCA), and basilar artery (BA). Note that the PCA originates from the ICA in the mouse. (**B**) Photomicrograph showing the major blood vessels. (**C–D**) Photomicrographs showing the PcomA graded with scores 0–3. (**C**) Score 0, (**D**) Score 1, (**E**) Score 2, (**F**) Score 3. *Arrowheads*: PcomA; *open arrows*: PCA; *closed arrows*: SCA (adapted from Ref. 7)

aged; 2 = 26–50% neurons damaged; 3 = 50–75% neurons damaged; and 4 = 76–100% neurons damaged.[18] For the hippocampal lesion, grade the neuronal damage on a scale of 0–4, with 0 = no damage to any hippocampal subregion; 1 = scattered ischemic neurons in CA1 subregion; 2 = moderate ischemic damage in CA1 subregion; 3 = whole pyramidal cells damaged in CA1 subregion; and 4 = extensive cell damage in all hippocampal subregions.[7]

6. Neuronal damage can also be evaluated quantitatively by directly using the percentage of damaged neurons.

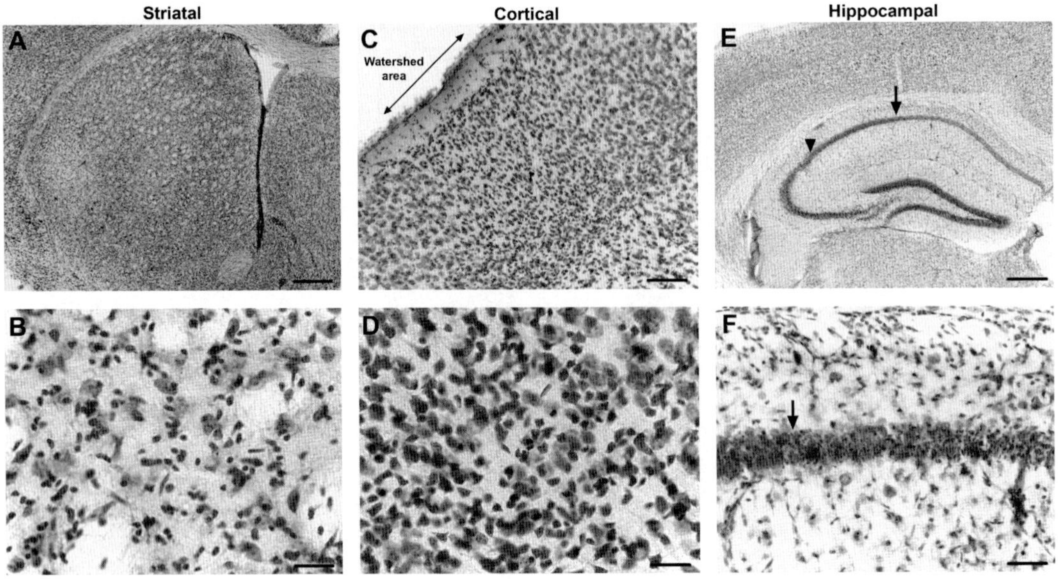

Fig. 5 Representative photomicrographs of cresyl violet staining showing striatal (**A**, **B**), cortical (**C**, **D**), and hippocampal (**E**, **F**) damage 3 days after 15 min of tGCI. (**A**) Damage in the striatum was observed. (**B**) Enlarged image of the striatum. Most of the neurons were damaged, although there are intact neurons. (**C**) Cortical damage was frequently observed in the watershed area, which is the border between the anterior cerebral artery and the middle cerebral artery territory. (**D**) Enlarged image of the cortex. Many neurons were damaged. (**E**, **F**) The *arrowhead* indicates the boundary between the CA1 and CA3 subregions. No neuronal damage was seen in the CA3 subregion. Many neurons were damaged in the area between the *arrowhead* and the *arrow*. Almost all neurons were damaged in the medial part of the *arrow*. Scale bars = 400 μm (**A**, **E**); 100 μm (**C**); 50 μm (**F**); 25 μm (**B**, **D**)

4.3 Cell Death Detection

There are no gold standards for detecting cell death after tGCI. There are two approaches using biochemical or protein markers for cell death. One is to detect fragmentation of DNA, such as with terminal deoxynucleotidyl transferase-mediated deoxyuridine triphosphate-biotin nick end labeling (TUNEL),[19] in situ nick translation,[20] or immunohistochemistry of single-strand DNA.[21] The other approach uses immunohistochemistry of cleaved caspase-3,[22] cleaved caspase-9, cytochrome *c*, or cleaved poly(ADP-ribose) polymerase[23] to detect an enzymatic event or change in levels of protein expression, which are related to cell death. A combination of TUNEL and at least one other method is normally used to provide an index for cell death.

4.4 Other Evaluations

Behavioral analysis is used to detect neurological deficits, especially motor deficits. Screen test: briefly, the mice are placed on a 10 × 20 cm screen that can be rotated from 0° (horizontal) to 90° (vertical). Record the duration of time that the mouse can hold onto the screen after being rotated from 0° to 90° for a maximum of 15 s. Grade the time on a scale of 0–3, with 0 = falls off in 0–5 s; 1 = falls off in 6–10 s; 2 = falls off in 11–14 s; and 3 = stays on screen for full 15 s. Balance beam test: record the amount of time that the mouse can remain balanced on a horizontal wooden rod up to 30 s. Grade the time on a scale of 0–3, with 0 = unable to maintain grip or balance on wooden beam; 1 = remains on beam up to 10 s; 2 = remains on beam for 11–20 s; and 3 = remains on beam for 21–30 s. Prehensile-traction test: record the time that the mouse can cling to a horizontal rope for up to 5 s. Grade the time on a scale of 0–3, with 0 = hangs on for 0–2 s; 1 = hangs on for 3–4 s; 2 = hangs on for 5 s but cannot bring rear limb onto rope; and 3 = hangs on for 5 s and brings the rear limb on to rope. From these three tests, a maximal score of 9 and a minimal score of 0 can be obtained.[24] There are other ways to evaluate outcome, such as with another behavioral or physiological analysis.

5 Advantages, Complications, and Limitations

5.1 Advantages

1. A reliable and reproducible mouse model of tGCI is needed, since mice are the predominant species for transgene or gene knockout technology.
2. The two-vessel occlusion model is technically easy even with mice of relatively smaller size.
3. Continuous blood pressure monitoring is usually not necessary.
4. Evaluation of the plasticity of the PcomA can normalize individual differences in collateral flow through the Circle of Willis, and this model may be used with any mouse strain with different genetic backgrounds. For transgenic and mutant mice, it is recommended that wild-type mice with an identical or similar genetic background be used. This usually requires a backcross of about 8–10 generations to obtain the best genetic similarity between the wild-type and mutant animals.

5.2 Complications

Convulsion may occur, especially with longer ischemic duration. Mice that experience convulsion should be killed because they may have suffered additional brain damage.

5.3 Limitations

1. There may be inconsistencies in the results of the ischemic damage; therefore PcomA evaluation is necessary. An additional consideration is individual differences in collateral blood flow even in the same strain, possibly because of differences in the breeders or the presence of subspecies.[4] Moreover, a recent study demonstrated that mouse-strain differences in ischemic vulnerability were independent of rCBF and temperature.[12] There is a possibility that strain difference exists not only at the vasculature level but also at the cellular level.
2. A respirator is necessary because of respiratory disturbance.
3. Mortality is an important issue. Approximately 10% of mice die within 3 days after 5 min of two-vessel occlusion, and approximately 25% of mice die within 3 days after 10 min of two-vessel occlusion.

5.4 Conclusions

It is technically difficult to perform four-vessel occlusion or two-vessel occlusion with hypotension, which were originally developed in rats. These models are the gold standard for tGCI. In contrast, it has been demonstrated that C57/B6 is the most appropriate mouse strain for global ischemia because of the less-developed PcomA.[4] Therefore, the C57/B6 mouse strain and PcomA evaluation for tGCI are viewed as the current method of choice. However, further improvements are needed for reproducibility, reduced mortality, and clinical relevance for potential translational research.

Acknowledgments: This project was supported by National Institutes of Health grants P50 NS14543, R01 NS25372, R01 NS36147, and R01 NS38653. We thank Liza Reola, Bernard Calagui, and Trisha Crandall for technical assistance, and Cheryl Christensen for editorial assistance.

References

1. Kirino T. Delayed neuronal death in the gerbil hippocampus following ischemia. Brain Res 1982;239:57–69
2. Pulsinelli WA, Waldman S, Rawlinson D, Plum F. Moderate hyperglycemia augments ischemic brain damage: a neuropathologic study in the rat. Neurology 1982;32:1239–1246
3. Smith M-L, Auer RN, Siesjö BK. The density and distribution of ischemic brain injury in the rat following 2–10 min of forebrain ischemia. Acta Neuropathol (Berl) 1984;64:319–332
4. Yang G, Kitagawa K, Matsushita K, Mabuchi T, Yagita Y, Yanagihara T, Matsumoto M. C57BL/6 strain is most susceptible to cerebral ischemia following bilateral common carotid occlusion among seven mouse strains: selective neuronal death in the murine transient forebrain ischemia. Brain Res 1997;752:209–218
5. Fujii M, Hara H, Meng W, Vonsattel JP, Huang Z, Moskowitz MA. Strain-related differences in susceptibility to transient forebrain ischemia in SV-129 and C57Black/6 mice. Stroke 1997;28:1805–1810
6. Kitagawa K, Matsumoto M, Yang G, Mabuchi T, Yagita Y, Hori M, Yanagihara T. Cerebral ischemia after bilateral carotid artery occlusion and intraluminal suture occlusion in mice: evaluation of the patency of the posterior communicating artery. J Cereb Blood Flow Metab 1998;18:570–579
7. Murakami K, Kondo T, Kawase M, Chan PH. The development of a new mouse model of global ischemia: focus on the relationships between ischemia duration, anesthesia, cerebral vasculature, and neuronal injury following global ischemia in mice. Brain Res 1998;780:304–310
8. Wellons JC III, Sheng H, Laskowitz DT, Mackensen GB, Pearlstein RD, Warner DS. A comparison of strain-related susceptibility in two murine recovery models of global cerebral ischemia. Brain Res 2000;868:14–21
9. Olsson T, Wieloch T, Smith M-L. Brain damage in a mouse model of global cerebral ischemia. Effect of NMDA receptor blockade. Brain Res 2003;982:260–269
10. Sheng H, Laskowitz DT, Pearlstein RD, Warner DS. Characterization of a recovery global cerebral ischemia model in the mouse. J Neurosci Methods 1999;88:103–109
11. Panahian N, Yoshida T, Huang PL, Hedley-Whyte ET, Dalkara T, Fishman MC, Moskowitz MA. Attenuated hippocampal damage after global cerebral ischemia in mice mutant in neuronal nitric oxide synthase. Neuroscience 1996;72:343–354
12. Yonekura I, Kawahara N, Nakatomi H, Furuya K, Kirino T. A model of global cerebral ischemia in C57 BL/6 mice. J Cereb Blood Flow Metab 2004;24:151–158
13. Mizushima H, Zhou CJI, Dohi K, Horai R, Asano M, Iwakura Y, Hirabayashi T, Arata S, Nakajo S, Takaki A, Ohtaki H, Shioda S. Reduced postischemic apoptosis in the hippocampus of mice deficient in interleukin-1. J Comp Neurol 2002;448:203–216

14. Kofler J, Hattori K, Sawada M, DeVries AC, Martin LJ, Hurn PD, Traystman RJ. Histopathological and behavioral characterization of a novel model of cardiac arrest and cardiopulmonary resuscitation in mice. J Neurosci Methods 2004;136:33–44
15. Barone FC, Knudsen DJ, Nelson AH, Feuerstein GZ, Willette RN. Mouse strain differences in susceptibility to cerebral ischemia are related to cerebral vascular anatomy. J Cereb Blood Flow Metab 1993;13:683–692
16. Minamisawa H, Smith M-L, Siesjö BK. The effect of mild hyperthermia and hypothermia on brain damage following 5, 10, and 15 minutes of forebrain ischemia. Ann Neurol 1990;28:26–33
17. Kawase M, Murakami K, Fujimura M, Morita-Fujimura Y, Gasche Y, Kondo T, Scott RW, Chan PH. Exacerbation of delayed cell injury after transient global ischemia in mutant mice with CuZn superoxide dismutase deficiency. Stroke 1999;30:1962–1968
18. Pulsinelli WA, Brierley JB, Plum F. Temporal profile of neuronal damage in a model of transient forebrain ischemia. Ann Neurol 1982;11:491–498
19. Gavrieli Y, Sherman Y, Ben-Sasson SA. Identification of programmed cell death in situ via specific labeling of nuclear DNA fragmentation. J Cell Biol 1992;119:493–501
20. Wijsman JH, Jonker RR, Keijzer R, Van de Velde CJH, Cornelisse CJ, Van Dierendonck JH. A new method to detect apoptosis in paraffin sections: in situ end-labeling of fragmented DNA. J Histochem Cytochem 1993;41:7–12
21. Naruse I, Keino H, Kawarada Y. Antibody against single-stranded DNA detects both programmed cell death and drug-induced apoptosis. Histochemistry 1994;101:73–78
22. Gown AM, Willingham MC. Improved detection of apoptotic cells in archival paraffin sections: immunohistochemistry using antibodies to cleaved caspase 3. J Histochem Cytochem 2002;50:449–454
23. O'Brien MA, Moravec RA, Riss TL. Poly (ADP-ribose) polymerase cleavage monitored in situ in apoptotic cells. Biotechniques 2001;30:886–891
24. Sheng H, Laskowitz DT, Mackensen GB, Kudo M, Pearlstein RD, Warner DS. Apolipoprotein E deficiency worsens outcome from global cerebral ischemia in the mouse. Stroke 1999;30:1118–1124

Section II

Focal Cerebral Ischemic Models

Chapter 11

Transient Middle Cerebral Artery Occlusion Model in Rodents

Ying Mao, Wei Zhu, and Guo-Yuan Yang

Abstract Rodent transient middle cerebral artery occlusion (MCAO) is one of the widely used focal ischemia models. This model mimics human cerebral ischemic injury and provides a unique tool for the mechanistic studies of ischemia in brain tissues in vivo. Rodent ischemia and reperfusion models are commonly used because of a greater understanding of rodent genetics, the availability of specific antibodies to rodent epitopes and molecular probes in rodents, and the ability to study transgenic mouse strains. In this chapter, we discuss (1) the preparation of MCAO in rodents; (2) the intra-luminal MCAO techniques and their modifications; (3) the quantitative evaluation of model success; and (4) the advantages and limitations of MCAO models.

Keywords: Blood flow, Infarct volume, Middle cerebral artery occlusion, Rodent, Transient focal ischemia

1 Introduction

Ischemic stroke is the third most common disease-related killer in the United States. Developing reliable and reproducible transient focal cerebral ischemia models, which mimic features of human ischemic stroke syndromes, is necessary for the systematic study of pathophysiology and treatment of stroke. Rodent models of focal cerebral ischemia have been increasingly gaining acceptance in recent years because they are relevant to human clinical settings. The first commonly used technique of rat middle cerebral artery occlusion (MCAO) involves coagulation of the middle cerebral artery via a craniotomy.[1,2] The focal ischemia model has an advantage in comparing the difference in ischemic changes between ischemic and non-ischemic hemispheres. However, this permanent MCAO model is invasive and does not permit reperfusion. Post-ischemic blood restoration occurs frequently in human ischemic stroke, especially in cerebral embolism. To mimic this clinical condition, Koizumi and Zea-Longa developed a novel and relatively noninvasive transient MCAO model using an intraluminal suture.[3,4] The advantage of this model is its ability to induce focal reperfusion injury. Since then, several groups have modified this model to make it more feasible and reproducible. The greatest improvement

From: *Animal Models of Acute Neurological Injuries.*
Edited by: J. Chen (eds.) New York, Totowa, NJ

is the coating of the sutures using poly-L-lysine, a polycationic, polymerized amino acid.[5]

Yang et al. further developed a mouse transient MCAO model using the intraluminal suture technique.[6,7] The lesion induced by suture MCAO is involved both in the cortex and striatum. Working with rodents in ischemia or reperfusion research is advantageous because of our greater understanding of animal genetics, the availability of specific antibodies and molecular probes in mice, and our ability to study transgenic strains.[8] This transient focal ischemic model of mouse is widely used for the study of molecular mechanisms in ischemic brain injuries. To minimize the lesion, a modification of Tamura's model has been developed. This model induces less injury and localizes only in the cortex region.

Several other focal cerebral ischemia models in rodents have also been developed, such as the rat or mice embolic focal cerebral ischemia model[9–11] and the mini-stroke model.[12] Since these models are not as commonly used as the suture and distal MCAO models, we have not included them in this chapter.

2 Preparation of MCAO in Rodents

2.1 Anesthesia and Physiological Parameter Monitoring

2.1.1 Rats

Adult male Sprague-Dawley (or other strains) rats weighing 260–290 g are optimal, although the use of animals that are not within this range has also been reported. The most commonly used anesthesia for MCAO is isoflurane inhalation. Other inhalants include halothane (1–3% in a mixture of 70% nitrous oxide and 30% oxygen), ketamine (80–100 mg kg^{-1}) and xylazine (5–10 mg kg^{-1}). A PE-50 catheter is inserted into the femoral artery for continuous monitoring of arterial blood pressure and sampling for analysis of blood gases and blood pH. The temperature of both the temporal muscle and the rectum is measured by thermocouples (E type, OMEGA Engineering, Inc.). Brain temperature is regulated by using a heating lamp and a heating pad to maintain normothermia. Blood pressure is maintained at 100 mmHg or above, which is regulated by the intensity of the anesthetic. Blood gases are analyzed once during the experiment to maintain an arterial PaO2 level of 100 mmHg or higher (Fig. 1).

2.1.2 Mice

Adult male CD-1 (or other strains) mice weighing 30–35 g would be ideal for the procedure. The anesthesia for MCAO in mice is the same as for rats. As in the case of rats, a PE-10 catheter is inserted into the femoral artery for continuous monitoring of arterial blood pressure and sampling for the analysis of blood gases and blood pH. Measuring temporal muscle and rectum temperature and monitoring blood pressure and blood gas are the same as for rats.

2.2 Equipment

The optimal small animal surgery room needs to be kept quiet and pathogen-free, and the room temperature maintained at 25 ± 3°C. The following items are recommended for small animal surgery:

1. Operation microscope (OPMI 6-SH, Zeiss, Germany)
2. Bipolar coagulator (Malis precision-control bipolar coagulator, CMC-II-PC, Codman and Shurtleff Inc., Randolph, MA)

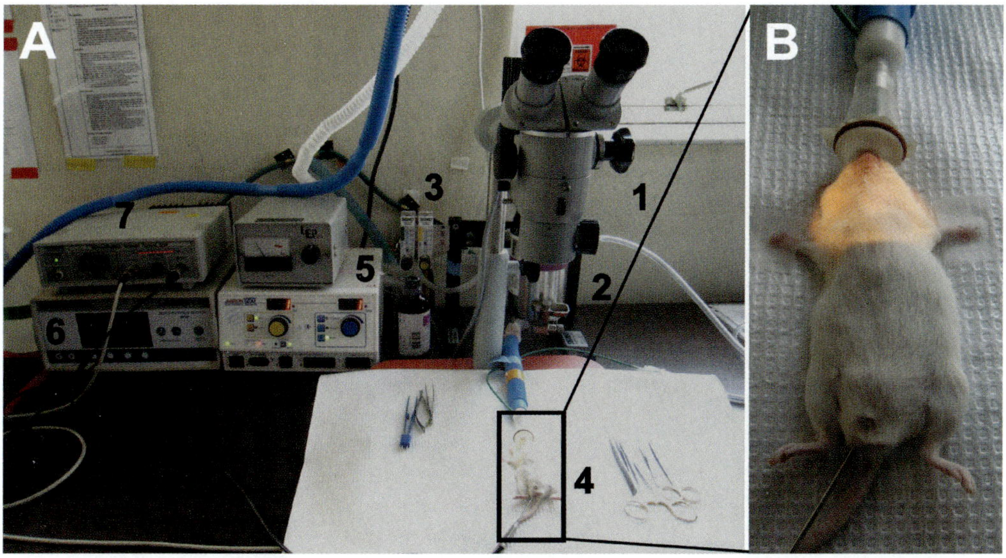

Fig. 1 (**A**) The animal surgical set for MCAO procedure. The basic set includes 1. Surgical microscope; 2. Vaporizer in anesthesia; 3. Flowmeter for air gas control; 4. Animal surgical table; 5. Bipolar coagulator; 6. Laser Doppler flowmeter; 7. Temperature controller. (**B**) Photograph showing a mouse under inherent anesthesia using isoflurane

3. Exercise and physiological system (AD Instruments, PowerLab/4SP, Castle Hill, Australia)
4. Temperature controller (Homeothermic Blanket Control Unit, Harvard Apparatus, Cambridge, MA)
5. Laser Doppler flowmeter (LASERFLO Blood perfusion monitor BPM^2, VASAMEDICS, St. Paul, MN)
6. Dry sterilizer (Simon Keller AG, Fine Science Tools, CH 3400, Foster City, CA)
7. Vaporizer (Summit Anesthesia Solutions, Dragerwerk AG Lubeck, Isofluorane Vapor 19.1, Bend, OR)
8. Stereotaxic frame (Model 900, DAVID KOPF INSTRUMENTS, Tujunga, CA)
9. pH/blood gas analyzer (Bayer, Radiolab 248, Tarrytown, NY)
10. High-speed micro drill (Fine Science Tools, Foster City, CA)
11. Surgical equipment (Fig. 2).

3 Transient MCAO Techniques and Their Modifications

3.1 Intraluminal Suture MCAO Model

3.1.1 Rat

Under the operating microscope, the left common carotid artery (CCA) is exposed through a midline incision. All other branches are isolated and coagulated along with the terminal lingual maxillary artery branches. The internal carotid artery (ICA) is then isolated and its extracranial branch, the pterygopalatine artery, is ligated close to its origin. Thus, the ICA, which is the only extracranial branch of the CCA, remains open. A 3-cm length of 3-0 nylon suture with a slightly enlarged and rounded tip is inserted into the transected

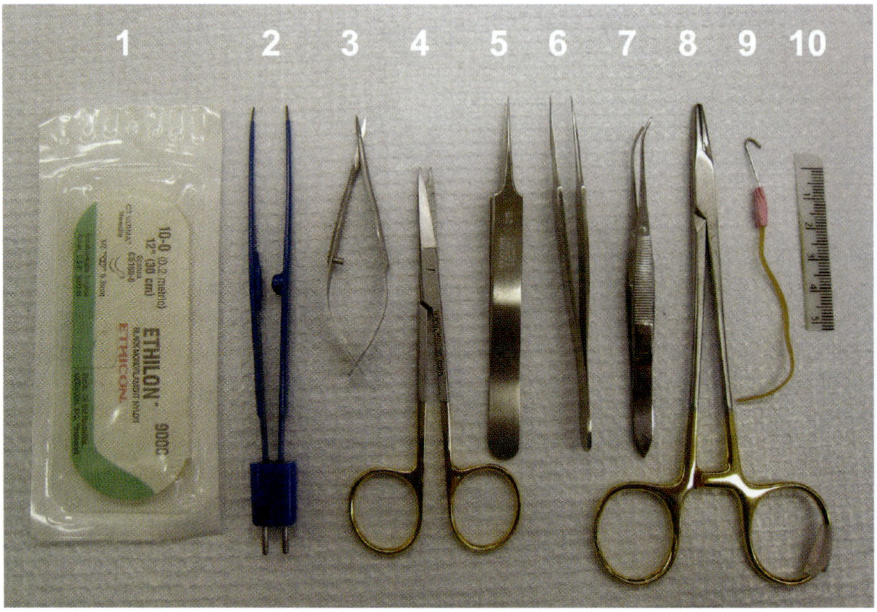

Fig. 2 Photograph showing a set of basic microsurgical equipment for the MCAO model. All items must be sterilized before use: 1. Nylon suture; 2. Bipolar forceps; 3. Microscissors for small vessels; 4. Surgical scissors for animal skin and tissue use; 5 and 6. Microforceps; 7. Tissue forceps; 8. Needle holder; 9. Skin hook for the exposure; 10. Ruler

lumen of the external carotid artery (ECA), and gently advanced from the ICA across to the opening of the MCA. The distance from the tip of the suture to the bifurcation of CCA is 17–19 mm. For reperfusion, the suture is withdrawn into the ECA to restore ICA–MCA blood flow. The skin is sutured and the animal is returned to the cage when it recovers from the anesthesia.[13]

3.1.2 Mouse

The method for mouse MCAO is similar to the rat model described above. The only difference is in the suture size (2 cm length of 5-0 nylon). The distance from the tip of the suture to the bifurcation of CCA is 10–11 mm (Fig. 3).[6,7]

3.2 Distal MCAO Mode[14]

3.2.1 Rat

The most commonly used anesthetic for the distal MCAO model is isoflurane inhalation. Under a surgical microscope, a 2-cm incision is made between the right orbit and the tragus. The temporal muscle is retracted laterally and a 3-mm diameter craniotomy is made just rostral to the foramen ovale. The dura is incised with a tiny hook and the MCA is exposed. The arachnoid is then opened and the MCA ligated with an 8-0 suture for the permanent MCAO. Interruption of the blood flow is confirmed visually under the microscope. For the transient distal MCAO, the MCA is temporarily occluded with a microclip or an 8-0 suture ligation. Immediately after, both CCAs are occluded with microclips and the rats undergo occlusion for 60 min. This model produces an infarct restricted to the cerebral cortex (Fig. 4). Reperfusion is achieved by removing the micro-clips or suture. Restoration of MCA blood flow can be verified visually. Blood

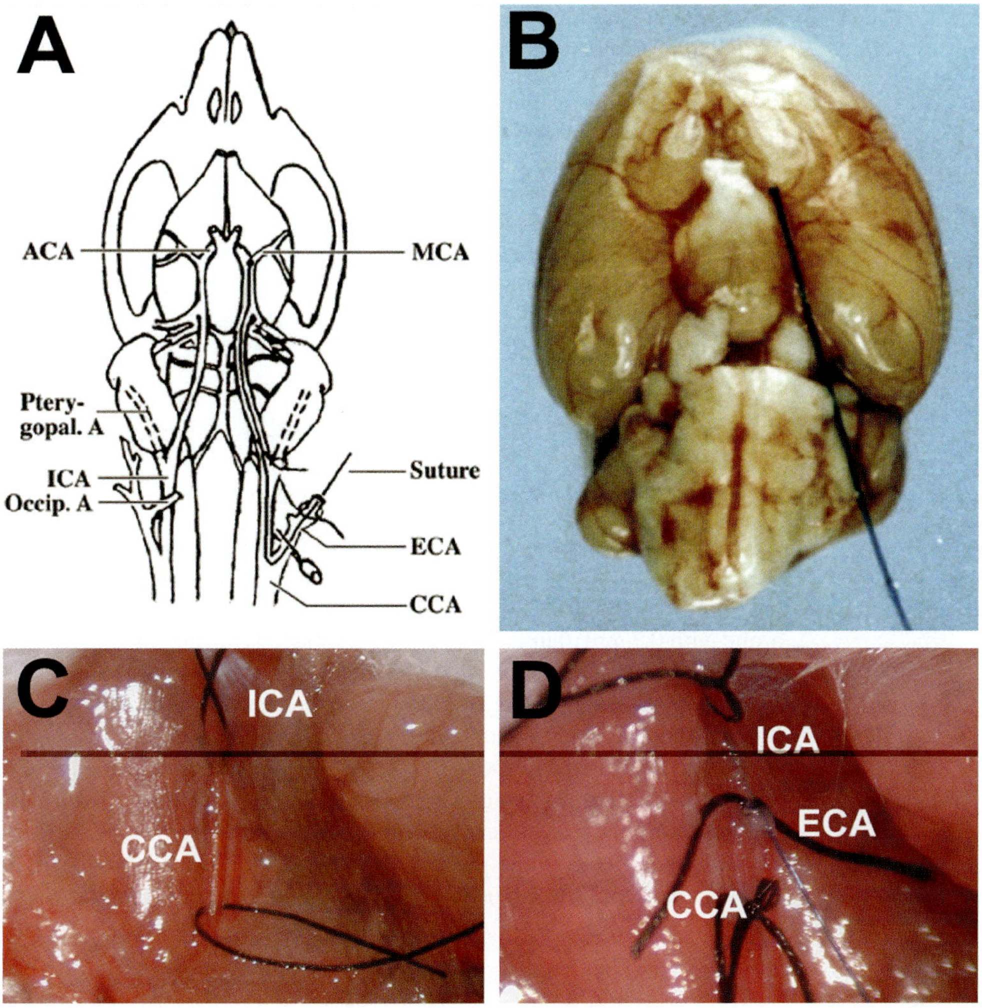

Fig. 3 (**A**) Diagram of cerebral anatomy in mice illustrating extracranial and intracranial vascular relations. Anterior communicate artery (ACA), middle carotid artery (MCA), internal carotid artery (ICA), external carotid artery (ECA), common carotid artery (CCA), and their branches are shown. (**B**) Photograph of mouse brain at autopsy showing intravascular suture, inserted through ICA, within the lumen of ECA. Suture occludes the origin of MCA from ICA. (**C**) Photograph showing the anatomy of CCA and ICA in the neck; ICA is looped by suture. (**D**) Photograph showing a right blue suture inserted into the opening of the ECA and advanced into the ICA

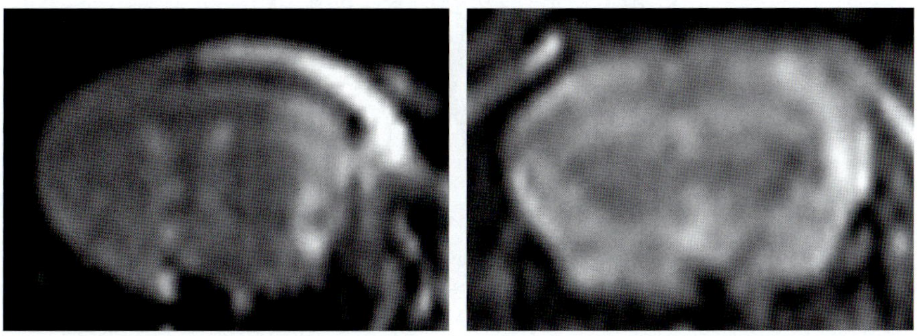

Fig. 4 MRI image showing the higher density signal in the right cortical region following 24 h of distal MCAO

flow is monitored during the MCAO using laser Doppler flowmetry (LDF). The temporal muscle is repositioned and the skin closed.

3.2.2 *Mouse*

The method for the mouse distal MCAO is similar to the rat distal MCAO model above. Under a surgical microscope, a 1-cm incision is made between the right orbit and tragus. The arachnoid is then opened and the MCA ligated with a 10-0 suture for the permanent MCAO. For the transient distal MCAO, the MCA is temporarily occluded with a microclip or a 10-0 suture ligation. Reperfusion is achieved by removing the microclips or the suture. Restoration of MCA blood flow can be verified visually. Blood flow is monitored during the MCAO using LDF. The temporal muscle is repositioned and the skin closed.

4 Quantitative Evaluation of Model Success

4.1 Focal Cerebral Blood Flow (CBF) Measurement

4.1.1 *Measurement of Local CBF Using Laser Doppler Flowmetry*

LDF can be used to verify that MCAO has occurred and to follow the flow over time. The BPM[2] laser Doppler flowmeter is equipped with a small-caliber probe with 0.7 mm diameter. The mice are placed in a stereotactic frame with a mouse holder, a midline incision is made, and a thermocouple is placed in the right temporal muscle and secured with cyanoacrylate glue. The pericranium over the coronal and sagittal sutures is exposed and kept moist with normal saline for LDF measurement.[15] The probe, held in a micromanipulator, is lowered until it comes into contact with the surface of the dura. Warmed saline is used during the experiment to slowly rinse around the probe and maintain a clear medium between the probe and the dura.[16] Stable baseline LDF reaings are obtained for 20 min prior to occlusion. Continuous digital display of

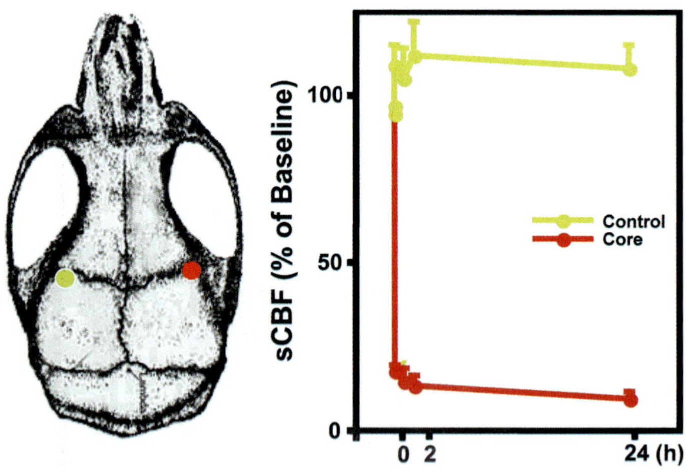

Fig. 5 CBF measurement. The *left side* shows a diagram of a mouse skull. *Two points* are marked to indicate the same area each time: the *yellow dot* is 3.5 mm lateral to the sagittal suture which indicates the control region; the *red dot* is 3.5 mm lateral to the sagittal suture on the *right side* which indicates the ischemic core region. Graph on the *right* shows the changes of regional CBF at 5 min, 1, 2 h, and 1 day after MCAO in the ischemic core and control point in the contralateral hemisphere

LDF values is averaged over 5-s intervals and recorded every 20 min during ischemia and early reperfusion. The CBF values are calculated and expressed as percentages of baseline values (Fig. 5).

4.2 Infarct Volume Measurement

4.2.1 Staining with 2,3,5-Triphenyltetrazolium Chloride (TTC)

This method is simple and fast, but not as accurate as the Cresyl violet staining method.[17,18] After MCAO, animal brains are removed immediately. From rats, six coronal slices of 1, 3, 5, 7, 9, and 11 mm distal from the frontal pole are dissected using a brain slicer, and from mice, four coronal slices of 1, 3, 5, and 7 mm. All brain slices are stained with 2% (wt/vol.) TTC in Dulbecco's phosphate buffer (pH 7.4) at 37°C for 20 min. The volume of infarction is calculated by multiplying the distance between sections (Fig. 6).[19]

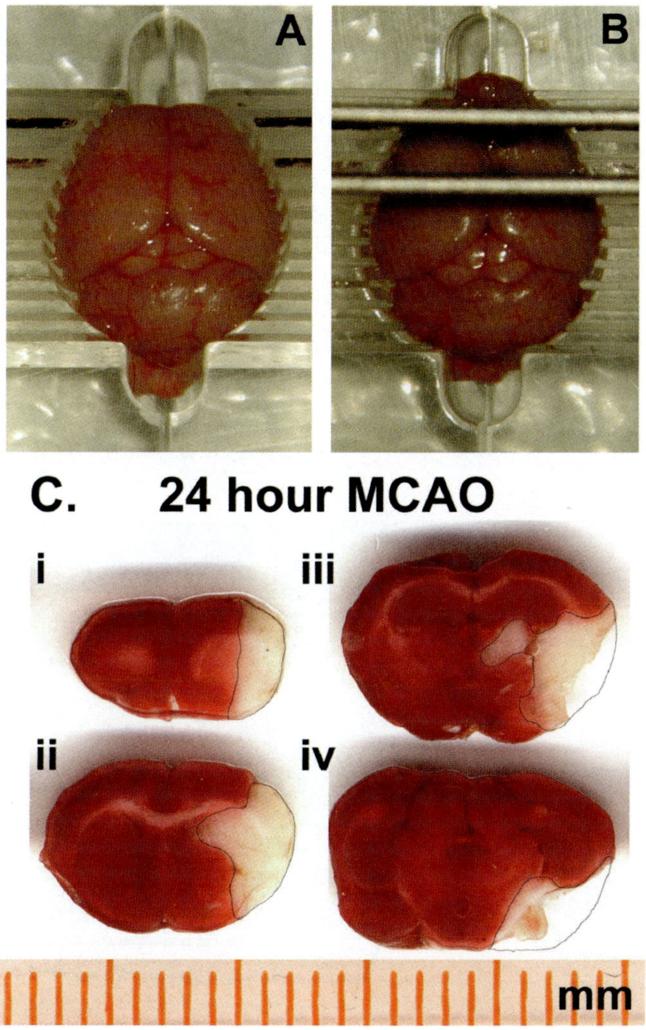

Fig. 6 Experimental sampling after ischemic injury. (**A**) Mouse brain matrix, which can be used for 1 mm coronal sectioning. (**B**) A mouse brain under sectioning. (**C**) A set of TTC-stained coronal sections, which is sectioned by using this mouse brain matrix

4.2.2 Cresyl Violet Staining Section

Following MCAO, the animals are killed and the brains are removed and frozen immediately in dry ice, in 2-methylbutane at −42°C for 5 min. Twenty coronal sections from the frontal pole, 20 μm in thickness at 200 μm intervals, are cut with a cryostat and mounted on slides. The sections are dried and then stained with Cresyl violet.[20] Non-Cresyl-violet staining is used to identify the infarct area. Using the NIH Image J software, the ischemic lesion area is calculated as the difference in the area of the non-ischemic hemisphere and the normal area in the ischemic hemisphere. The infarct volume is calculated by multiplying the infarct areas by the thickness of the sections (Fig. 7).

4.3 Neurobehavioral Evaluation

4.3.1 Body Asymmetry Test

This test is a simple and fast method to estimate the degree of ischemic brain injury. To measure motor asymmetry, animals are examined by the elevated body swing test as described.[21] Mice are examined for head swing while being suspended by their tails. The direction of the swing, either right or left, is recorded when the mouse turns its head sideways by an approximate 10° angle to the body's midline. The mouse is allowed to move freely for at least 30 s before the next swing is counted, and the trials are repeated 20 times for each animal. The frequency of head swings toward the non-injured side is counted and normalized as follows: (number of lateral swings in 20 tests − 10)/10 × 100%.

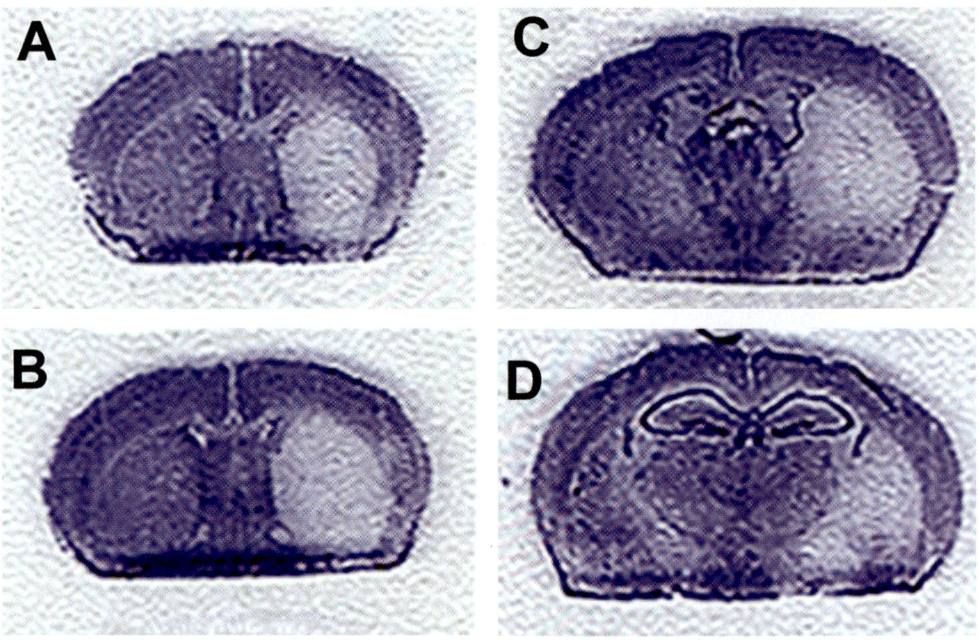

Fig. 7 A set of Cresyl-violet-stained mouse brain coronal sections. The mouse underwent 45 min of MCAO, followed by 24 h of reperfusion. The infarct areas are usually limited to the basal ganglia region

4.3.2 Water-Maze Learning

In the water maze test, the ability of the animal to locate a hidden submerged platform in a circular pool (142 cm in diameter, 50 cm deep) filled with warm, opaque water is assessed. Animals are trained first to locate a visible platform, and then to locate a hidden platform; in a subsequent trial the hidden platform is removed. Animals receive two training sessions per day for five consecutive days. Each session consists of three trials with a 10-min intertrial interval. The interval between the two daily sessions is 3 h. Once the animals locate the platform, they are allowed to remain on it for 10 s. The time to reach the platform (latency), path length, and swimming speed are recorded with a Noldus Instruments EthoVision video tracking system set to analyze two samples per second.

4.3.3 Rotor-Rod Test

The rotor-rod test is used to screen for motor deficits that might influence performance in the water maze testing. Rotor-rod balancing requires a variety of proprioceptive, vestibular, and fine-tuned motor abilities. The task requires the animal to balance on a rotating rod. Performance can be evaluated in trials of fixed speed; however, a slow acceleration during the test session limits interindividual variability in performance. After a 1-min adaptation period on the rod at rest, the rod is accelerated at 5 rpm every 15 s and the length of time for which the animal remains on the rod (fall latency) is recorded.

4.3.4 Novel Open Field Activity

To quantitate open field activity, animals are placed in brightly lit, individually automated infrared photocell activity cages interfaced with a computer. The open field is divided into nine zones, in which the center zones contain 6 × 6 photocells and the side zones 5 × 5 photocells. The data are converted into a zone map file, and the following parameters are calculated per interval: active time, distance moved, rearing times and events, corner entries, center entries, and percentage of time spent in the center. The open field activity is recorded for three consecutive 10-min intervals after a 1-min adaptation period. The number of fecal boli deposited during the 30-min period is recorded.

4.3.5 Novel Object Recognition Task

Animals are individually habituated to the open field for 3 days. During the training session, two novel objects are placed in the open field and each animal is allowed to explore for 15 min. During retention tests, the animals are placed back in the open field, in which one of the familiar objects is replaced by a novel object, and are allowed to explore for 5 min. The time spent exploring each object during the training and retention sessions is recorded. The ratio of the amount of time spent exploring any of the two objects (training session) or the novel one (retention session) to the total time spent exploring both objects is calculated and used to measure object recognition memory.

5 Advantages and limitations

5.1 Advantages

The MCAO model produces cerebral ischemia in the vascular territory of the MCA and gives rise to focal metabolic disturbance resulting in brain edema, selective neuronal necrosis, and infarction. The suture MCAO model is relatively simple to operate because no craniotomy needs to be performed; therefore, other possible effects, such as craniotomy, focal inflammation, and focal facial muscle damage, can be excluded. Suture MCAO producing ischemic brain injury involves both cortex and striatum; this injury further produces obvious sensorimotor dysfunction, which can be assessed by neurobehavioral evaluation.

The distal MCAO model produces ischemic brain injury that is restricted to the cortex of the ipsilateral hemisphere. The overall mortality rate remains low, although craniotomy is needed to expose and clip the MCA branch. Focal ischemia model producing injury is limited to the ipsilateral hemisphere and, therefore, the same region opposite to the injury in the contralateral hemisphere can be used for the control.

5.2 Limitations

5.2.1 Mortality

Approximately 10% of the rats die either during ischemia or shortly after reperfusion even in experienced hands. The main cause of the mortality is subarachnoid hemorrhage, which is caused by inappropriate insertion of the suture. Improving the top of the suture and practicing surgery more often can attenuate this mortality rate. Animal deaths also increase gradually following the time course of MCAO, especially after 5 days, caused by the animals' inability to eat and drink. Injecting saline and feeding them with milk chocolate may help reduce mortality.

5.2.2 Variation

Despite the reproducible outcomes, the transient suture MCAO model has a relatively higher variation of infarct volume. Zea-Longa has reported that the success rate in achieving infarction is only 56%.[4] Other studies have also reported that approximately 30% of the experimental animals have to be excluded. To reduce variability of the infarct volume, performing a pilot study to optimize the experiment (e.g., choosing appropriate suture to fit the diameter of animal ICA) is necessary. Furthermore, carefully controlling brain temperature, blood pressure, and cerebral blood flow during and immediately after MCAO, as well as the length of occlusion period, is extremely important. In addition, a suture coated with poly-L-lysine solution (0.1% (wt/vol), in deionized water; Sigma) may help reduce variability.[5]

References

1. Tamura A, Graham DI, McCulloch J, Teasdale GM. Focal cerebral ischaemia in the rat: 2. Regional cerebral blood flow determined by [14C]iodoantipyrine autoradiography following middle cerebral artery occlusion. J Cereb Blood Flow Metab 1981;1:61–9
2. Tamura A, Graham DI, McCulloch J, Teasdale GM. Focal cerebral ischaemia in the rat: 1. Description of technique and early neuropathological consequences following middle cerebral artery occlusion. J Cereb Blood Flow Metab 1981;1:53–60
3. Koizumi J, Yoshida Y, Nakazawa T, Ooneda G. Experimental studies of ischemic brain edema. I: a new experimental model of cerebral embolism in rats in which recirculation can be introduced in the ischemic area. Jpn J Stroke 1986;8:1–8
4. Zea-Longa E, Weinstein PR, Carlson S, Cummins R. Reversible middle cerebral artery occlusion without craniectomy in rats. Stroke 1989;20:84–91

5. Belayev L, Alonso OF, Busto R, Zhao W, Ginsberg MD. Middle cerebral artery occlusion in the rat by intraluminal suture. Neurological and pathological evaluation of an improved model. Stroke 1996;27:1616–22; discussion 1623
6. Yang GY, Chan PH, Chen J, et al. Human copper-zinc superoxide dismutase transgenic mice are highly resistant to reperfusion injury after focal cerebral ischemia. Stroke 1994;25:165–170
7. Yang GY, Liu XH, Kadoya C, Zhao YJ, Betz AL. Attenuation of ischemic inflammatory response in mouse brain using an adenoviral vector to induce overexpression of interleukin-1 receptor antagonist. J Cereb Blood Flow Metab 1998;18:840–847
8. Mao Y, Yang GY, Zhou LF, Stern JD, Betz AL. Focal cerebral ischemia in the mouse: description of a model and effects of permanent and temporary occlusion. Brain Res Mol Brain Res 1999;63:366–370
9. Zhang ZG, Zhang L, Ding G, et al. A model of mini-embolic stroke offers measurements of the neurovascular unit response in the living mouse. Stroke 2005;36:2701–2704
10. Zhang Z, Chopp M, Zhang RL, Goussev A. A mouse model of embolic focal cerebral ischemia. J Cereb Blood Flow Metab 1997;17:1081–1088
11. Zhang RL, Chopp M, Zhang ZG, Jiang Q, Ewing JR. A rat model of focal embolic cerebral ischemia. Brain Res 1997;766:83–92
12. Wei L, Rovainen CM, Woolsey TA. Ministrokes in rat barrel cortex. Stroke 1995;26:1459–1462
13. Zhao YJ, Yang GY, Domino EF. Zinc protoporphyrin, zinc ion, and protoporphyrin reduce focal cerebral ischemia. Stroke 1996;27:2299–2303
14. Won SJ, Xie L, Kim SH, et al. Influence of age on the response to fibroblast growth factor-2 treatment in a rat model of stroke. Brain Res 2006;1123:237–244
15. Dirnagl U, Pulsinelli W. Autoregulation of cerebral blood flow in experimental focal brain ischemia. J Cerebral Blood Flow Metab 1990;10(3):327–336
16. Dirnagl U, Kaplan B, Jacewicz M, Pulsinelli W. Continuous measurement of cerebral cortical blood flow by laser-Doppler flowmetry in a rat stroke model. J Cereb Blood Flow Metab 1989;9:589–596
17. Bederson JB, Pitts LH, Germano SM, Nishimura MC, Davis RL, Bartkowski HM. Evaluation of 2,3,5-Triphenylterazolium Chloride as a stain for detection and quantification of experimental cerebral infarction in rats. Stroke 1986;17(6):1304–1308
18. Lundy EF, Solik BS, Frank RS, et al. Morphometric evaluation of brain infarcts in rats and gerbils. J Pharmacol Meth 1986;16:201–214
19. Swanson RA, Morton MT, Tsao WG, Savalos RA, Davidson C, Sharp FR. A semiautomated method for measuring brain infarct volume. J Cereb Blood Flow Metab 1990;10:290–293
20. Yang GY, Pang L, Ge HL, et al. Attenuation of ischemia-induced mouse brain injury by SAG, a redox-inducible antioxidant protein. J Cereb Blood Flow Metab 2001;21:722–733
21. Borlongan CV, Saporta S, Poulos SG, Othberg A, Sanberg PR. Viability and survival of hNT neurons determine degree of functional recovery in grafted ischemic rats. Neuroreport 1998;9:2837–2842

Chapter 12

Three-Vessel Middle Cerebral Artery Occlusion Model

T.N. Lin, S.T. Chen, Y.Y. He, W.M. Cheung, and C.Y. Hsu

Abstract In the search for a more reproducible focal ischemic stroke model in the rat, we systematically interrupted blood flow to the right middle cerebral artery (MCA) territory by transient ligation of the right MCA, and both right and left common carotid arteries (CCAs) in succession. The regional CBF in this model was examined by laser Doppler flowmetry and [^{14}C] isopropyliodoamphetamine and [^{14}C] iodoantipyrine autoradiographic comparison. The right MCA cortex sustained severe ischemia, with a regional blood flow reduction by 88–92%, whereas only very mild ischemia was noted (reduction of blood flow by 10–20%) outside the right MCA cortex. A ligation-time-dependent focal infarct was consistently noted in the cerebral cortex supplied by the right MCA. The characteristic changes of ischemic necrosis were limited to the cortex, sparing the subcortical structures. The mortality rate was very low (less than 2%). Occlusion of the right MCA alone or together with the right CCA did not consistently cause gross infarction. Permanent ligation of the right MCA and both CCAs had a high mortality. Results from this study showed that a more consistent cortical infarction in the right MCA territory with low mortality can be achieved by interruption of collateral circulation following the right MCA ligation.

Keywords: Ischemic stroke, Cortical infarct, Anesthetics, Blood glucose, Edema, Gender, Angiogenesis, CBF, CBV, Transient Focal ischemia animal model

1 Introduction

The advantages of using rats for stroke research include the similarity of their intracranial circulation to that of humans,[1] the abundant neurochemical and genetics data derived from rat brain,[2–5] and the relatively low animal cost which is important for large-scale studies for statistical analysis. There are several models of cerebral ischemia developed in rats, which can be categorized by topography into focal or global and by chronology into reversible and

From: *Animal Models of Acute Neurological Injuries.*
Edited by: J. Chen (eds.) New York, Totowa, NJ

irreversible. Focal cerebral ischemia constitutes the bulk of stroke pathology in clinical practice, with infarct in the middle cerebral artery (MCA) region the major clinical presentation. Therefore, the methods for inducing focal cerebral ischemia are most commonly directed at the MCA and entail either extravascular ligation or intravascular occlusion. In this chapter, we will first review technical issues restricted to extravascular ligation of cerebral arteries to induce MCA ischemia and infarct. This is followed by surgical approaches we have taken to develop a 3-vessel MCA occlusion (MCAO) model which entails extravascular ligation of the distal MCA and both common carotid arteries (CCAs).

Unilateral occlusion of the CCA is a simple surgical procedure that results in focal cerebral ischemia in the territory irrigated by the CCA including the MCA region. However, unilateral or bilateral CCA occlusion has not been found to be useful in causing severe focal ischemia to result in cerebral infarction. This is because of extensive intracranial collateral circulation in rats provided by the Circle of Willis, leptomeningeal anastomoses, and dorsal collateral junctions. Unilateral CCA occlusion superimposed with systemic insult such as hypoxia or hypotension may increase the depth of focal cerebral ischemia in generating focal cerebral infarction. However, the added variables caused by a generalized metabolic derangement or systemic vascular dysfunction may confound the objectives of studying focal cerebral ischemia.

Occlusion of the MCA is a logical approach to induce focal cerebral ischemia in the MCA territory. Tamura et al.[6] attempted ligation of the proximal MCA through a subtemporal approach and reported consistent ischemic changes in the MCA territory, resulting in an MCA infarct. However, the surgical approaches to occlude a deep-seated proximal MCA truck are highly invasive, technically demanding, and associated with high mortality and other complications such as poor feeding because of the destruction of muscles and joint needed for chewing.

The distal MCA truck is superficially located on the cerebral cortex and can be ligated following a simple craniectomy. Coyle[7] applied this less invasive surgical approach to ligate the distal MCA truck above the rhinal fissure, but it did not produce consistent cerebral infarction in young Wistar rats.

We have undertaken to develop a reliable MCA infarction model, reasoning that the rich collateral circulation to the MCA territory is the major factor in preventing the development of significant focal cerebral ischemia to induce MCA infarction following distal MCA occlusion. Interruption of this collateral circulation to the distal MCA territory might be effective in inducing severe focal ischemia sufficient to cause a consistent MCA infarct. We compared regional blood flow affected the blood flow in the right MCA region with stepwise occlusion of the right MCA, the right MCA, and the right common carotid artery (CCA), and finally the right MCA, right and left CCAs and found that occlusion of the right MCA and both CCAs resulted in profound ischemia in the right MCA cortex to consistently cause a right MCA infarct. This 3-vessel MCA occlusion (MCAO) model is one of the most reproducible models of focal cerebral ischemia in rats, applying relatively noninvasive surgical procedures.

2 Materials

2.1 Surgical Tools

1. Venus tweezers 12 cm (No. 3, Regine, Switzerland)
2. Venus tweezers 11 cm (No. 5, Regine, Switzerland)
3. Needle holder 14 cm (Olsen-Hegar, Germany)
4. Iris forceps 10 cm (Klappenecker, Germany)
5. Hemostat 12.5 cm (13021-12, FST, Germany)
6. Retractor (Miltex 11-16, Germany)
7. Scissors 9 cm (Klappenecker, Germany)
8. Scalpel handle (Lawton 04-1050, Germany)
9. Semken forceps 12 cm, straight
10. Microclips, 8 mm

2.2 Surgical Materials

1. Injection needle (25, 26, 27 G; Terumo Co., Tokyo, Japan)
2. Syringe (1, 3, 5 ml; Terumo Co., Tokyo, Japan)
3. Surgical silk sutures 4-0 (Unik Surgical Sutures MFG. Co., Taiwan)
4. Surgical sutures, USP/EP with taper needle 10-0 (DAVIS/GECK Wayne, NJ, USA)
5. Bone wax (Ethicon, Ltd., UK)
6. Surgical blades No.10 (Feather Safety Razor Co., Ltd., Japan)
7. Taper point needle No.1
8. Rat brain matrix (Stoelting, USA)

2.3 Auxiliary Instruments

1. Homeothermic blanket system (Harvard Apparatus, South Natick, MA, USA)
2. Blood glucose analyzer (ACCU-CHEK Advantage II, Roche Diagnotics Co., Germany)
3. OPTI gas blood analyzer (OPTI Medical Systems, Inc.)
4. Multichannel recording system (Grass, Model 79D, Quincy, Ma)
5. Dental drill (Microvit, BEGO, Germany)
6. Operation microscope (Leica, Wild M650, Switzerland)
7. Fiber optic light source (Nikon, Japan)

3 Experimental Protocols

Male Long-Evans rats weighing 250–350 g (Charles River, Wilmington, Del.) were used. Housing and anesthesia concurred with guidelines established by the institutional animal welfare committee, in accordance with the Public Health Service Guide for the Care and Use of Laboratory Animals, U.S. Department of Agriculture regulations, and the American Veterinary Medicine Association Panel on Euthanasia guidelines.

The focal cerebral ischemia model has been detailed in previous publications.[8,9] The rats were allowed free access to water and rat chow (Wayne, Chicago, IL) until surgery. Surgery leading to focal cerebral ischemia was conducted under anesthesia with chloral hydrate (360 mg kg^{-1} body weight, IP). Rectal temperature was monitored and maintained at 37.0 ± 0.5°C with a homeothermic blanket system (Harvard Apparatus, South Natick, MA, USA).

3.1 Isolation of Common Carotid Arteries

The CCAs were isolated via a ventral midline cervical incision and both the right and left CCAs were isolated. Care was taken to separate the associated vagus nerve fiber without physical injury. Both CCAs were ligated following right MCA occlusion as detailed below.

3.2 Ligation of Right Middle Cerebral Artery

A 1.5-cm scalp incision was made at the midpoint between the right eye and the right ear. The temporalis muscle was separated in the plane of its fiber bundles and retracted in order to expose the zygoma and squamosal bone. A 2-mm burr hole was made with a dental drill 1 mm rostal to the anterior junction of the zygoma and the squamosal bone (Fig. 1). The drill pit was constantly cooled with saline at ambient temperature to avoid thermal or physical injury to the cortex during preparation of the burr hole. The dura mater was carefully pierced with a No. 11 scalpel blade and the MCA trunk was ligated immediately above the rhinal fissure with 10-0 suture. Complete interruption of blood flow was confirmed under an operating microscope. The temporalis muscle and overlying skin were allowed to fall back and were covered with a piece of gauze saturated with artificial cerebrospinal fluid (aCSF). Both CCAs were then occluded using nontraumatic aneurysm clips. After the predetermined duration of ischemia (15–120 min), the aneurysm clips on CCAs and

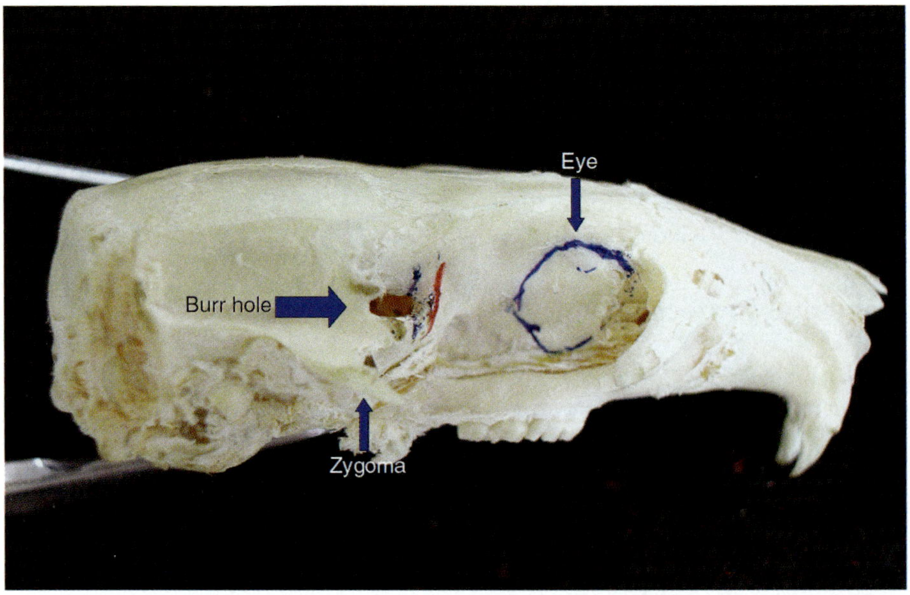

Fig. 1 The location of a 2-mm burr hole in the skull through which the distal MCA is ligated in the 3-vessel MCAO model

the MCA suture were removed and restoration of blood flow in all three arteries was observed directly under the microscope. The craniotomy was sealed with dental cement; the temporalis muscle and then the skin were sutured separately. The rats were allowed free access to food and water after recovery from anesthesia. All rats were kept in air-ventilated incubators at 24 ± 0.5°C until the end of the experiment.

3.3 Measurement of Metabolic and Physiological Parameters

Metabolic and physiological parameters were measured before, during, and after ischemia. These include the plasma glucose concentration (Advantage II, Roche Diagnostics Corporation), arterial blood gas, and pH (OPTI Gas Blood Analyzer, OPTI Medical Systems, Inc.). Arterial blood pressure and heart rate were monitored via an intra-arterial catheter in the right femoral artery, which was connected to a multichannel recording system (Grass, Model 79D, Quincy, MA), and maintained within normal ranges during surgery and ischemia as described previously.[10]

3.4 Triphenyl Tetrazolium Chloride (TTC) Staining

After a predefined reperfusion period (usually 24h), the animals were perfused intracardially with 200ml of cold 0.9% NaCl under deep pentobarbital anesthesia. The brain was removed carefully and cooled in ice-cold saline for 5min, and then was dissected into coronal 2-mm sections using a Jacobowitz brain slicer (Zivic-Miller Laboratories, Inc., Allison Park, PA). For the delineation of infarct area, the brain slices were incubated in isotonic phosphate-buffered saline (pH 7.4) containing 2% TTC (Sigma Chemical Co., St Louis, MO) at 37°C for 30min and then stored in 10% neutral-buffered formalin. For the purpose of photographic illustration, rats were perfused with 150ml normal saline and 50ml of TTC.

3.5 Morphometric Studies

For morphometric measurement of infarct volume, both direct and indirect methods were employed for the same brain samples. Morphometric determination of infarct volume by the direct method has been described previously.[8] Briefly, the cross-sectional area of infarction in the cerebral cortex of the right MCA territory of each brain slice was measured by a simple PCI Image Analysis System (Compix Inc., PA, USA). Total infarct volume for each brain was calculated by summation of the infarcted area of all brain slices (area of infarct in square millimeters × thickness [2mm]) from the same hemisphere. The indirect method[9] followed a formula similar to that proposed by Swanson et al.[11]: $RI = LT - RN$, where RI is the infarct volume in the right hemisphere measured by the indirect method, LT is the total cortex volume in the left hemisphere of the same brain, and RN is the non-infarcted cortex volume in the right hemisphere of the same brain. Total cortex volume in the right hemisphere of the same slice (RT) can be calculated as $RT = RN + RI$. The two formulas are valid only if $LT = RT$ in normal brain. This contention is supported by the measurement of LT and RT in four normal rat brains. The volumes for each category (non-infarcted cortex in the right hemisphere, total cortex in the left hemisphere, etc.) were derived from morphometric measurement of the respective category's area in each brain slice followed by summation of areas in the same category from all slices in a brain as described above for the direct measurement of the infarct volume. An automated measurement of the infarct size with scanned images of TTC-stained rat brains was also developed.[12]

4 Outcome Measures

In this stroke model, only the right MCA cortex sustained severe ischemia, with regional blood flow reduction by 88–92% based on laser Doppler flowmetry and $[^{14}C]$ isopropyliodoamphetamine $[^{14}C]$ and iodoantipyrine autoradiography. Only very mild ischemia was noted (reduction of blood flow by 10–20%) outside the right MCA cortex. This procedure consistently produces a finite cortical infarct in the right MCA territory. Different degrees of ischemia can be created in the cerebral cortex of the right MCA territory with increasing duration of ligation (Fig. 2).[8,13] The evolution of cerebral infarction was further examined by hematoxylin–eosin histochemical staining and in vivo monitoring by magnetic resonance imaging (MRI) (Fig. 3).[14] Post-ischemic vascular remodeling, edema, and necrosis/liquefaction were noted in this model (Fig. 4)[15–17] as seen in stroke patients. The characteristic changes were limited to the ipsilateral cortex with sparing of subcortical structures[18,19] and contralateral cortex.[20] The mortality rate was very low (less than 3%). Occlusion of the right MCA alone or together with the right CCA did not consistently cause gross infarction. Permanent ligation of the right MCA and both CCAs had a high mortality.[8]

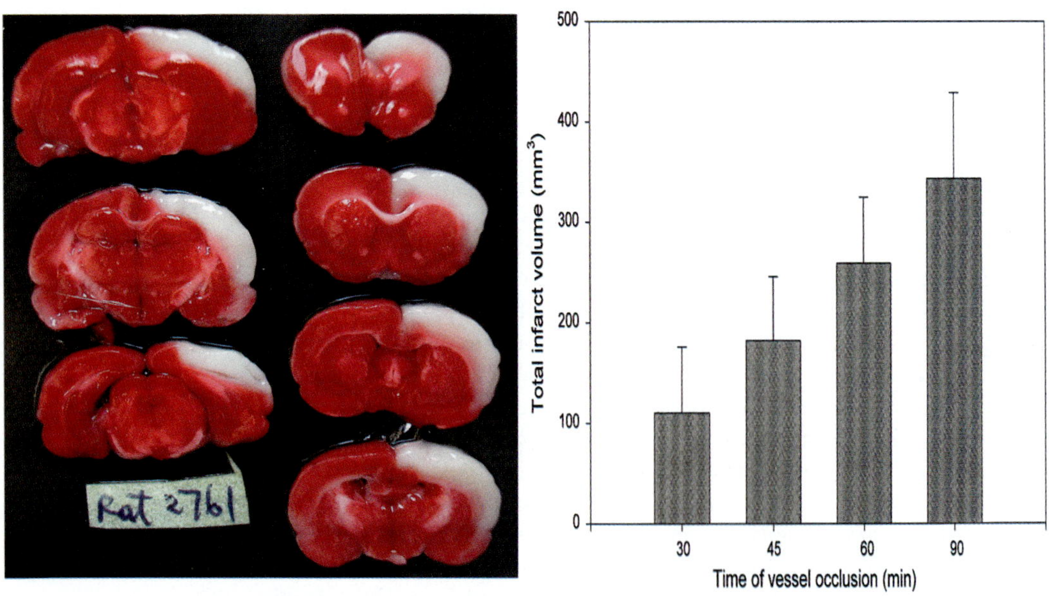

Fig. 2 Infarct area in the right MCA territory and the relationship between the duration of ischemic insult (occlusion time) and infarct volumes

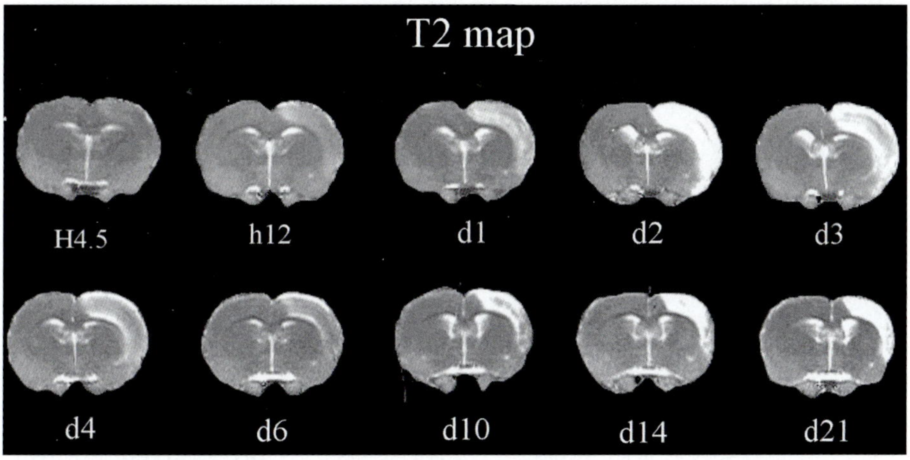

Fig. 3 The representative T2 map showing a right MCA cortical infarct in rats subjected to 60 min ischemia and various reperfusion periods

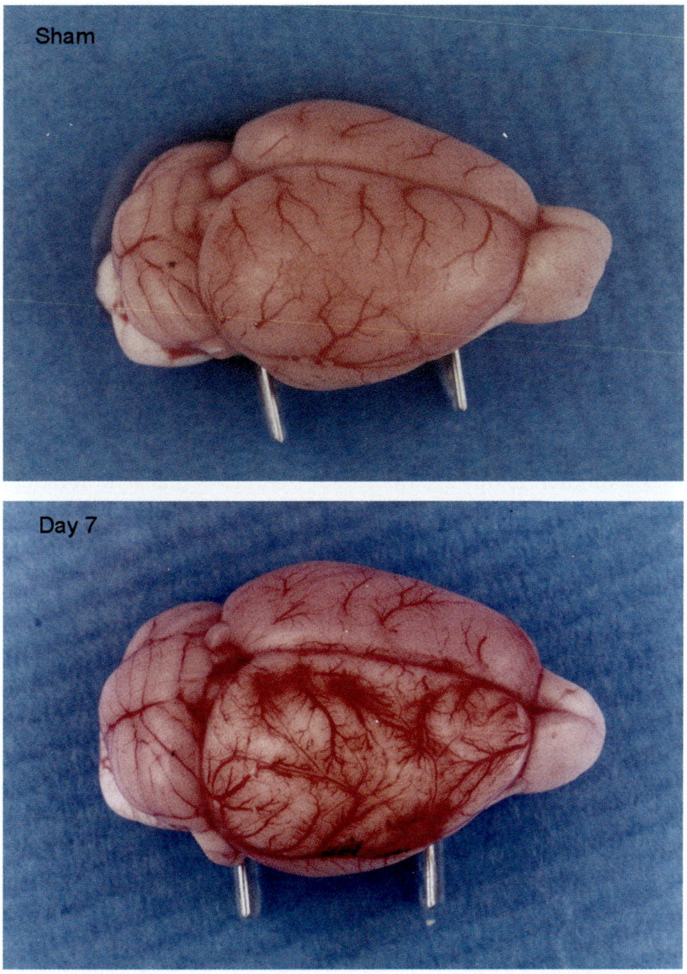

Fig. 4 Effect of focal cerebral ischemia reperfusion on the regional distribution of blood vessels in the cerebra cortex

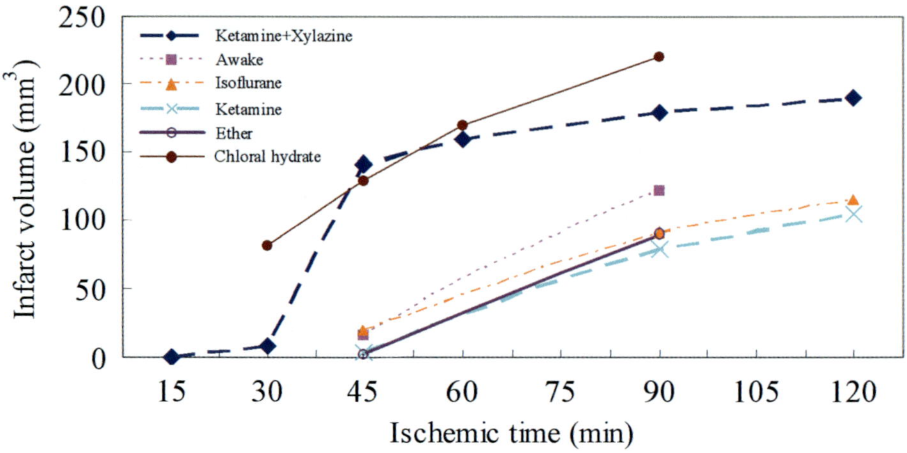

Fig. 5 Effect of anesthetic agents on the infarct volume. Rats were subjected to focal cerebral ischemia for 15, 30, 45, 60, 90, or 120 min and reperfused for 24 h

4.1 Effect of Anesthetic Agents on the Infarct Volume

There is considerable risk of cerebral ischemia during anesthesia and surgery. Anesthetic agents have been shown to have a profound effect on the pathophysiology of cerebral ischemia. Here, we compared the effect of various anesthetics on ischemic infarct volume (Fig. 5). We found chloral hydrate a desirable anesthetic for this stroke model. Chloral hydrate, a sedative, is used in the short-term treatment of insomnia, in relieving anxiety, in inducing sleep before surgery, or for treating alcohol withdrawal. It is also used after surgery for relieving pain. Better correlation between infarct volume and duration of ligation (30–90 min) was achieved with chloral hydrate than other anesthetic agnets such as ketamine, xylazine, ether, and isoflurane (Lin & Hsu., unpublished data). Another advantage of chloral hydrate is that it is not a controlled substance, reducing the regulatory requirements in purchasing and stocking.

4.2 Effect of Plasma Glucose on the Infarct Volume

It has been shown in clinical studies that the plasma glucose level is a major determinant of ischemic brain injury.[21] The 3-vessel MCAO model is characterized by the profound impact of plasma glucose levels on the ischemic outcome. Hyperglycemia resulted in significantly larger infarct volumes. Hyperglycemia is most detrimental during the period of ischemia. Lowing plasma glucose reduced the infarct volume. There is a therapeutic window for normalization of plasma glucose to be efficacious.[10,22] However, the molecular mechanisms underlying glycemic effects in ischemic brain are not fully understood. Fig. 6 shows that a 24-h pre-ischemic fasting reduces ischemic brain injury 24 h after ischemia[22]; this beneficial effect lasted up to 28 days of reperfusion. Although COX-1 gene therapy also reduced cerebral infarct at 24 h after ischemia,[23] no long-lasting protective effect was noted after 28 days of reperfusion. These results suggest the importance of plasma glucose on ischemic brain injury and indicate the need to long-term assess ischemic outcome.

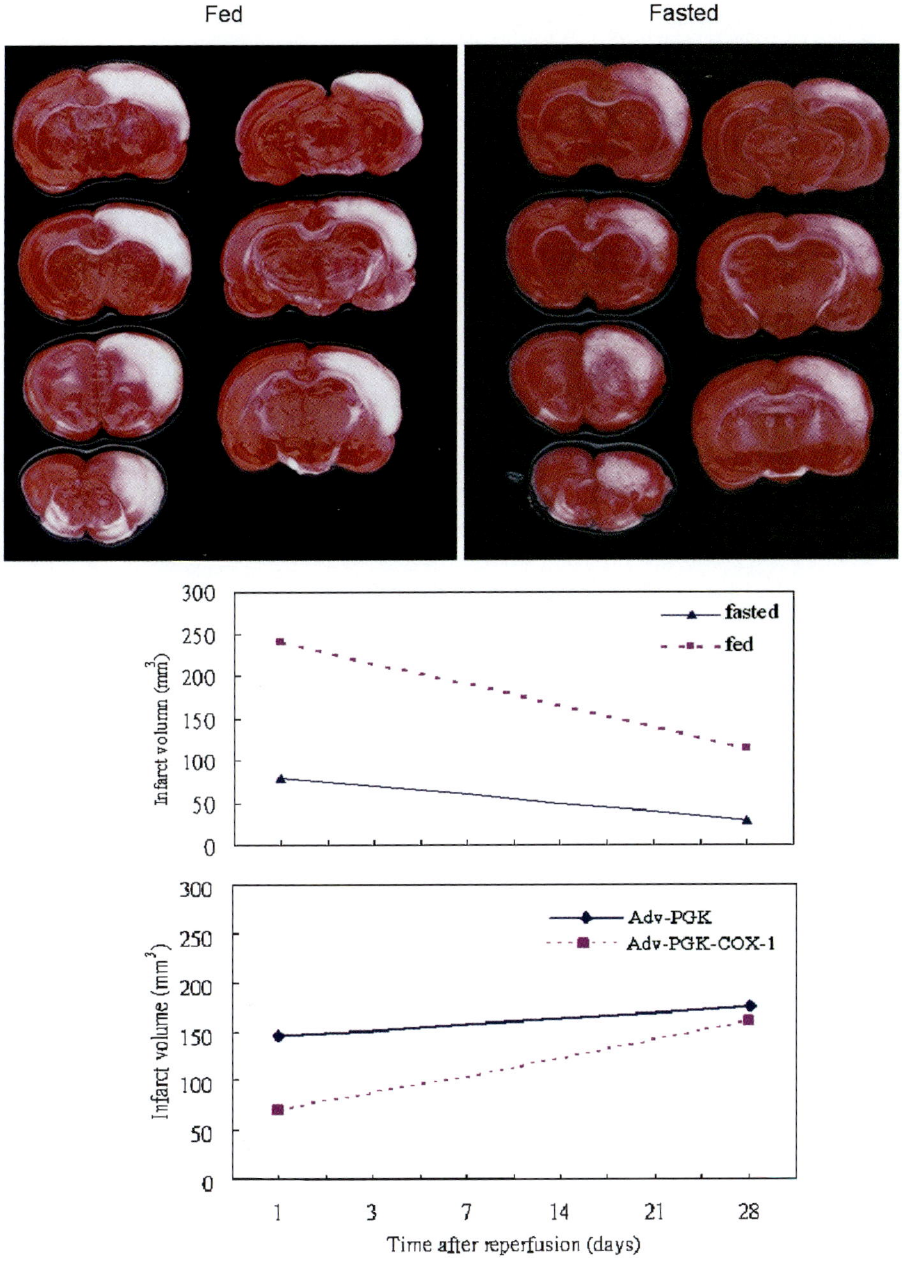

Fig. 6 Coronal brain sections showing the right MCA cortical infarct in fed (control) and fasted rats subjected to focal cerebral ischemia for 60 min and reperfusion for 24 h (*upper*). Evolution of the right MCA cortical infarct in rats subjected to preischemic fasting or COX-1 gene transfer (*lower*) prior to focal cerebral ischemic for 60 min

4.3 Effect of Brain Edema on the Infarct Volume

In animal models of focal cerebral ischemia, infarct volume is frequently determined by measuring the infarcted area in closely spaced sections of brain after special tissue staining to delineate the infarct in each brain slice. Morphometric analysis of infarct volume represents an objective and quantitative means to estimate the extent of ischemic brain injury and is a commonly employed measure to determine the efficacy of neuroprotective agents in preclinical trials. However, swelling of ischemic tissue may result in enlargement of the infarcted zone, leading to overestimation of infarct volume. Brain edema has been shown to artificially increase infarct volume by as much as 22%. Therefore, distortion of infarct volume by brain edema may compound the interpretation of therapeutic interventions. Severe brain edema evolving from 1 to 3 days after the ischemic insult in this model has been determined on the basis of cortical water content and MRI.[14,15] To minimize the error that may be introduced by brain edema, Swanson et al.[11] proposed a method to measure infarct volume on the basis of the volume of surviving normal gray matter, which was further confirmed by Lin et al.[9] It is clear that traditional direct measurement of infarct volume is associated with an overestimation during the development of brain edema in the first 3 days after ischemia. This artifact can be reduced with indirect measurement, which is based on non-infarcted cortex volume. (Refer to the "Morphometric Studies" section for the protocol to minimize distortion of infarct volume by brain edema.)

4.4 Effect of Gender on the Infarct Volume

Epidemiological studies have shown that the prevalence of cardiovascular disease (CVD) in women was lower than in men until the postmenopausal (over 55) age. When age is adjusted, the incidence of stroke is also lower in women than men. The percentage of individuals having a transient ischemic attack (TIA) or mini stroke is lower in women than men. Animal studies also have shown that there are gender differences in stroke outcome. In the 3-vessel

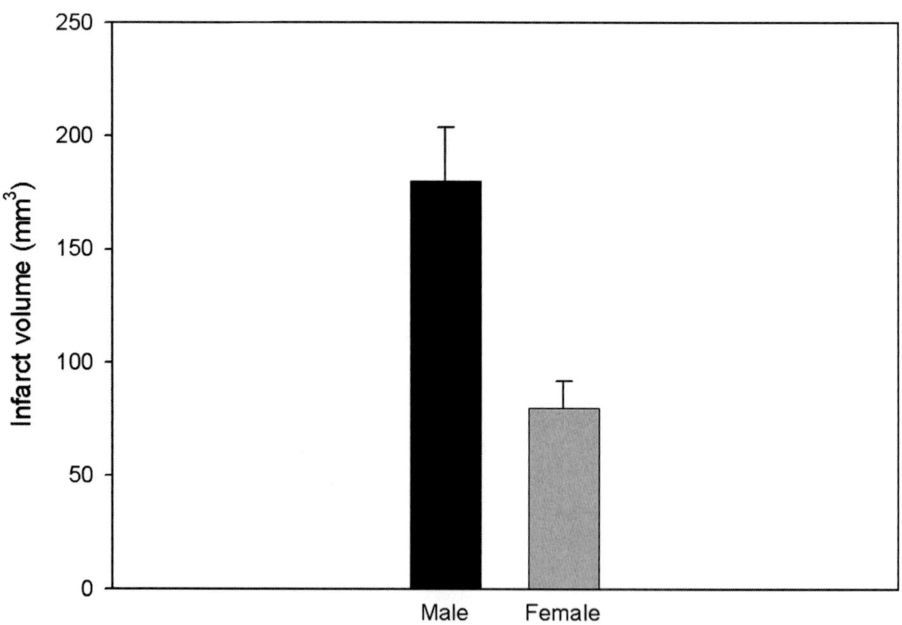

Fig. 7 Effect of gender difference on the infarct volume in rats subjected to focal cerebral ischemia for 60 min and reperfusion for 24 h

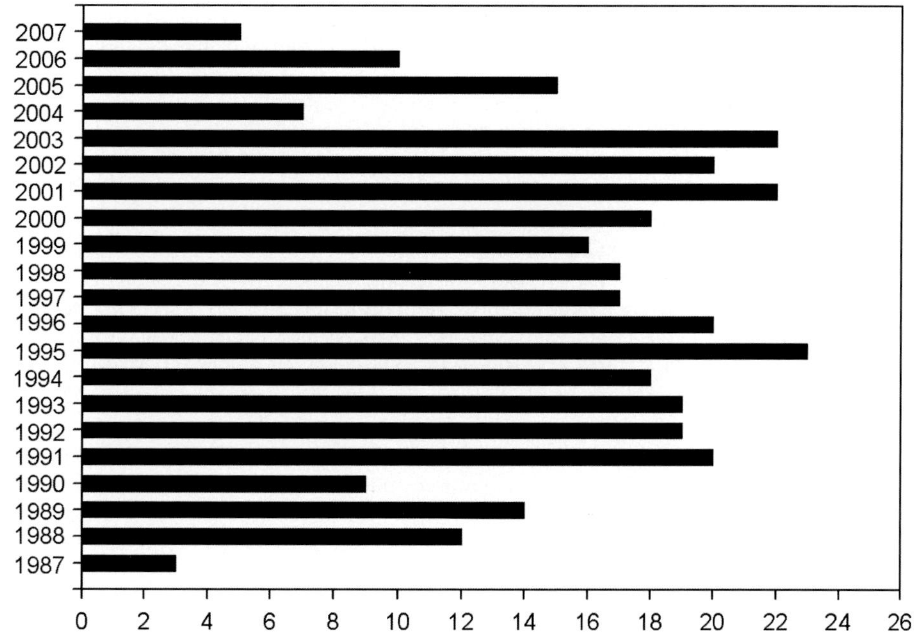

Fig. 8 Yearly citation reports on this MCAO model by Chen et al.[8] with "Web of Science"

MCAO model, males fare worse than female rats on the basis of infarct volume (Fig. 7). Gender difference should be considered when both male and female rats are mixed in studies applying the 3-vessel MCAO model.

5 Conclusion

Systematic interruption of regional cerebral blood flow to the right MCA as well as the right and left CCA results in one of the most reproducible models of focal cerebral ischemia in rats. In this 3-vessel MCAO model, the right MCA cortex sustains severe ischemia with reduction in regional cerebral blood flow by 88–92%. Only very mild ischemia is noted (reduction of blood flow by 10–20%) outside the right MCA cortex. A persistent and well-delineated infarct in the right MCA cortex develops in an occlusion-time-dependent manner. Post-ischemic edema, angiogenesis, apoptosis, and necrosis have been reported by applying this model. The characteristic ischemic changes are confined to the MCA cortex, sparing subcortical structures. This 3-vessel MCAO model with a consistent MCA cortex infarct and low mortality has been extensively used to study molecular mechanisms of ischemic brain injury and test new therapeutic agents (Fig. 8).

References

1. Yamori Y, Hone R, Handa H, Sato M, Fukase M (1976) Patbogenetic similarity of strokes in stroke-prone spontaneously hypertensive rats and humans. *Stroke* 7:46–53
2. Iversen IX, Iversen SD, Snyder SH (1978) Handbook of psychopharmacology, Vol. 9: Chemical pathways in the brain. London, Plenum Press

3. Chan PH, Fishman RA, Longar S, Chen S, Yu A (1985) Cellular and molecular effects of polyunsaturated fatty acids in brain ischemia and injury. *Prog Brain Res* 63:227–235
4. Wieloch T (1985) Neurochemical correlates to selective neuronal vulnerability. *Prog Brain Res* 63:69–85
5. Myhrer T (2003) Neurotransmitter systems involved in learning and memory in the rat: a meta-analysis based on studies of four behavioral tasks. *Brain Res Brain Res Rev* 41:268–287
6. Tamura A, Graham DI, McCulloch J, Teasdale GM (1981) Focal cerebral ischemia in the rat: I. Description of technique and early neuropathological consequences following middle cerebral artery occlusion. *J Cereb Blood Flow Metabol* 1:53–60
7. Coyle P (1982) Middle cerebral artery occlusion in the young rat. *Stroke* 13:855–859
8. Chen ST, Hsu CY, Hogan EL, Maricq H, Balentine JD (1986) A model of focal ischemic stroke in the rat: reproducible extensive cortical infarction. *Stroke* 17:738–743
9. Lin TN, He YY, Wu G, Khan M, Hsu CY (1993) Effect of brain edema on infarct volume in a focal cerebral ischemia model in rats. *Stroke* 24:117–121
10. Yip PK, He YY, Hsu CY, Garg N, Marangos P, Hogan EL (1991) Effect of plasma glucose on infarct size in focal cerebral ischemia-reperfusion. *Neurology* 41: 899–905
11. Swanson RA, Morton MT, Tsao-Wu G, Savalos RA, Davidson C, Sharp FR (1990) A semiautomated method for measuring brain infarct volume. *J Cereb Blood Flow Metab* 10:290–293
12. Goldlust EJ, Paczynski RP, He YY, Hsu CY, Goldberg MP (1996) Automated measurement of infarct size with scanned images of triphenyltetrazolium chloride-stained rat brains. *Stroke* 27:1657–1662
13. Liu TH, Beckman JS, Freeman BA, Hogan EL, Hsu CY (1989) Polyethylene glycol-conjugated superoxide dismutase and catalase reduce ischemic brain injury. *Am J Physiol* 256:H589–H593
14. Lin TN, Sun SW, Cheung WM, Li F, Chang C (2002) Dynamic changes in cerebral blood flow and angiogenesis after transient focal cerebral ischemia in rats. Evaluation with serial magnetic resonance imaging. *Stroke* 33:2985–2991
15. Hsu CY, Liu TH, Xu J, Hogan EL, Chao J, Sun G, Tai HH, Beckman JS, Freeman BA (1989) Arachidonic acid and its metabolites in cerebral ischemia. *Ann N Y Acad Sci* 559:282–295
16. Lin TN, Te J, Lee M, Sun GY, Hsu CY (1997) Induction of basic fibroblast growth factor (bFGF) expression following focal cerebral ischemia. *Brain Res Mol Brain Res* 49:255–265
17. Lin TN, Wang CK, Cheung WM, Hsu CY (2000) Induction of angiopoietin and Tie receptor mRNA expression after cerebral ischemia-reperfusion. *J Cereb Blood Flow Metab* 20:387–395
18. Dawson VL, Hsu CY, Liu TH, Dawson TM, Wamsley JK (1994) Receptor alterations in subcortical structures after bilateral middle cerebral artery infarction of the cerebral cortex. *Exp Neurol* 128:88–96
19. Chi SI, Wang CK, Chen JJ, Chau LY, Lin TN (2000) Differential regulation of H- and L-ferritin messenger RNA subunits, ferritin protein and iron following focal cerebral ischemia-reperfusion. *Neuroscience* 100:475–484
20. Cheung WM, Chen SF, Nian GM, Lin TN (2000) Induction of angiogenesis related genes in the contralateral cortex with a rat three-vessel occlusion model. *Chin J Physiol* 43:119–124
21. Capes SE, Hunt D, Malmberg K, Pathak P, Gerstein HC (2001) Stress hyperglycemia and prognosis of stroke in nondiabetic and diabetic patients: a systematic overview. *Stroke* 32:2426–2432

22. Lin TN, Te J, Huang HC, Chi SI, Hsu CY (1997) Prolongation and enhancement of postischemic c-fos expression after fasting. *Stroke* 28:412–418
23. Lin H, Lin TN, Cheung WM, Nian GM, Tseng PH, Chen SF, Chen JJ, Shyue SK, Liou JY, Wu CW, Wu KK (2002) Cyclooxygenase-1 and bicistronic cyclooxygenase-1/prostacyclin synthase gene transfer protect against ischemic cerebral infarction. *Circulation* 105:1962–1969

Chapter 13

A Thromboembolic Rat Model of Focal Cerebral Ischemia and Reperfusion with tPA

Yoshihiro Murata, Xiaoying Wang, and Eng H. Lo

Abstract In recent years, thromboembolic models of focal cerebral ischemia in rats have become a commonly used animal model to mimic human acute ischemic stroke, particularly for investigations into thrombolytic therapy with tissue type plasminogen activator (tPA). This chapter describes our approach with this model that is implemented using homologous blood clot injections into the middle cerebral artery (MCA) of rats. Reduction of regional cerebral blood flow (rCBF) and subsequent reperfusion with tPA can be monitored by laser Doppler flowmetry. A standard battery of outcomes is also described including infarct volume, hemorrhage volume (hemorrhagic transformation), mortality rate, and neurobehavioral testing.

Keywords: Spontaneously hypertensive rats (SHR), Wistar rat, Sprague-Dawley (SD) rat, Embolic stroke, Tissue type plasminogen activator, Reperfusion, Thrombolysis, Hemorrhage

1 Model Selection

Thromboembolic models of focal cerebral ischemia in rats are now increasingly used to mimic human thromboembolic stroke, particularly for investigation into thrombolytic therapies for acute focal cerebral ischemia. Recombinant human tissue plasminogen activator (tPA) is the only drug that has been approved by the U.S. Food and Drug Administration (FDA) for treating human acute ischemic stroke. Thrombolysis with tPA is an effective therapy for acute thromboembolic stroke.[1–3] However, hemorrhagic transformation, neurotoxicity, and short treatment time windows might comprise the major limitations for widespread implementation of tPA stroke therapy. In order to identify new strategies that might increase thrombolytic efficacy and reduce tPA-associated neurotoxicity and hemorrhagic transformation, a reproducible animal model is required. Here, we describe our approach to such an in vivo animal model of embolic stroke in rats. We emphasize that this rat model

From: *Animal Models of Acute Neurological Injuries.*
Edited by: J. Chen (eds.) New York, Totowa, NJ

is primarily based on the pioneering models originally developed by Zhang, Chopp, and colleagues.[4,5]

2 Materials

2.1 Animals

In the experimental stroke research literature, spontaneously hypertensive rats (SHR), Wistar rats, and Sprague-Dawley (SD) rats are among the most commonly used strains. For the present embolic clot model, there are several issues with regards to strain selection. From our experience, the different strains of rats may respond differently in terms of infarct volume, mortality, and rates of spontaneous recanalization (rate of failure). Infarct volume in SD rats has the highest variability, along with higher rates of spontaneous recanalization as well. This might be due to higher variability of collateral vessels in SD rats. Table 1 compares the rates of spontaneous recanalization after injection of 4 cm length of clot into the middle cerebral artery (MCA), as well as the ranges in infarct volume and mortality rate 24 h after ischemia. Compared with Wistar rats, SHR have higher mortality rates but larger infarct volumes and correspondingly lower rates of failure, i.e., spontaneous recanalization. Furthermore, aged rats also have higher mortality rates and larger infarction than younger ones. In our experience, rats in the 260–300 g range may provide reasonable results for this embolic clot model.

2.2 Anesthesia

Animals are anesthetized with 1–2% isoflurane under spontaneous respiration in a 30% oxygen/70% nitrous oxide mixture. Physiological parameters, including rectal temperature, mean arterial blood pressure, pH, pCO2, and pO2, should be monitored throughout the experiment.

2.3 Equipment

2.3.1 Anesthesia

- Isoflurane vaporizer (Smiths Medical Veterinary Division)
- Gas flow meter for oxygen and nitrous oxide (MDS Matrx)
- Temperature control system (FHC)

2.3.2 Surgery

- Surgical instruments (scissors, forceps, microvascular clip, etc.)
- Microscope (MZ8, LEICA)
- Drill (Fine Science Tool)

Table 1 Comparison of the thromboembolic model among the three rat strains.

	SD rats (n = 20)	Wistar rats (n = 100)	SHR (n = 15)
Infarct volume (mm^3 ± SD)	131 ± 162	280 ± 90	450 ± 48
Mortality rate (%)	10	11	31
Spontaneously recanalization rate (%)	78	40	27

SD Sprague-Dawley, *SHR* Spontaneously hypertensive rats

- Continuous laser Doppler flowmetry (Perimed AB)
- Polyethylene tubing (PE-50 (ID 0.58 mm, OD 0.965 mm), PE-10 (ID 0.28 mm, OD 0.61 mm) Becton Dickinson)
- Bipolar micro coagulator (Kirwan Surgical Product)
- Syringes (3 and 1 ml, Becton Dickinson)
- Needles (23 G, 30 G, Becton Dickinson)

2.3.3 Evaluation of Infarct Volume and Intracranial Hemorrhage

Quantitative Analysis of Infarct Volume

- 2,3,5-Triphenyltetrazolium chloride (TTC) (Sigma)
- Image J (NIH, http://rsb.info.nih.gov/nih-image/))

Spectrophotometric Hemoglobin Assay for Intracranial Hemorrhage

- Homogenizer (Power Gen 700, Fisher Scientific)
- Sonicator (Sonic Dismembrater, Fisher Scientific)
- Centrifuge (Centrifuge 5417R, Eppendorf)
- Drabkin's reagent (Sigma)
- Spectrophotometer (Spectronix 3000, Milton-Roy)

3 Procedures

The thromboembolic model used in our laboratory is closely based on the original models pioneered by Zhang, Chopp, and colleagues[4–8] with minor modification.

3.1 Preparation for Thromboembolic Model

3.1.1 Homologous Clot Prepared One Day Before the Experiment

1. Anesthetize a donor rat using the method mentioned above
2. Cannulate femoral artery and withdraw blood into 50 cm of PE-50 tubing (Fig. 1a)
3. Keep the clot in the tube for 2 h at room temperature, and subsequently retain for 22 h at 4°C

3.1.2 Modified PE-50 Catheter

Modified PE-50 catheter is made for injecting the clot:

1. Cut PE-50 tube into 40–50 cm
2. Hold both ends of the tube with forceps
3. Heat the middle of the tube with a heating lamp, and then pull and temper the heated tube (make the tip of the catheter of 0.3 mm outer diameter)
4. Cut the middle of the tube where the point was tempered to get two modified PE-50 catheter pieces (Fig. 1b)

3.2 Thromboembolic Model

3.2.1 Settlement and Measurement of rCBF

Measuring rCBF with laser Doppler flowmetry (LDF) is necessary to evaluate the reduction of CBF during MCA occlusion and also to document the degree of reperfusion after thrombolysis with tPA or other thrombolytic agents. Importantly, the clot may be autolyzed (spontaneous recanalization). Therefore, a prespecified cut-off of CBF reduction (quantitated as percentage

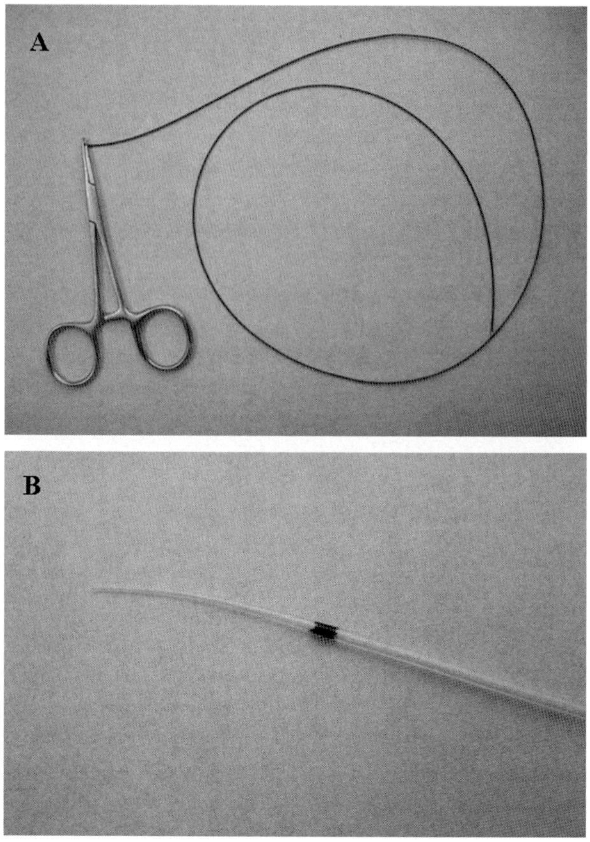

Fig. 1 (**A**) PE-50 tube containing the clot. (**B**) Modified PE-50 catheter (marked 15 mm from tip)

of pre-ischemic baseline levels) is necessary for stroke animal selection. In our experience, a cut-off of either 30% (immediately after clot injection) or 40% (right before thrombolytic treatment) appears to work well. However, these cut-off criteria should be independently assessed in each laboratory to achieve the degree of standardization as needed.

Anesthetize a donor rat using the method mentioned above:

1. Make an incision in the midline of the scalp
2. Dissect the temporal muscle from the bone with forceps, scissors, and dissector
   (At this time, bleeding from the muscle must be controlled with the bipolar coagulator, especially in the case of using tPA.)
3. Thin the bone at 2 mm posterior and 6 mm lateral to the bregma by drilling (Fig. 2a)
4. Place the LDF probe with glue and start monitoring rCBF (Fig. 2b)

3.2.2 Physiological Monitoring and Venous Line for Injection Routing

- Rectal temperatures are maintained at 37 ± 0.5°C with a thermostat-controlled heating pad

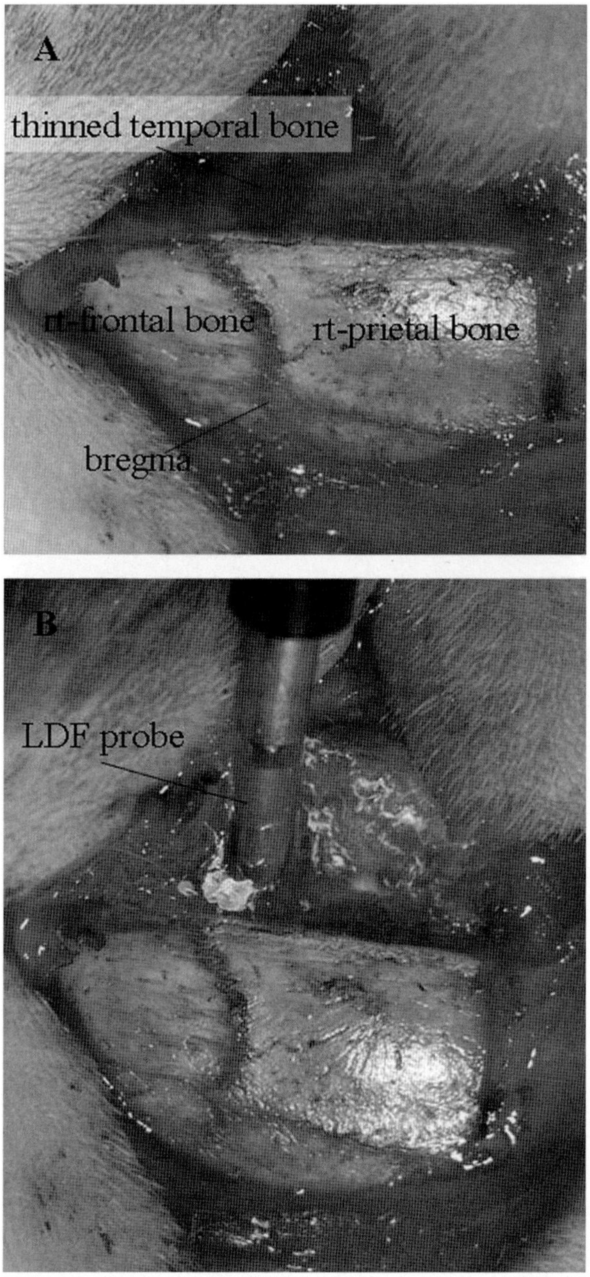

Fig. 2 (**A**) Dissection of the temporal muscle from temporal bone and thinning of the bone at 2 mm posterior and 6 mm lateral to bregma with a surgical drill. (**B**) Placement of the LDF probe with glue and monitoring of rCBF

- The femoral artery is cannulated and monitored for physiological parameters such as mean arterial blood pressure, pH, pCO2, and pO2 during the experiment
- The femoral vein is also cannulated for drug administration such as tPA

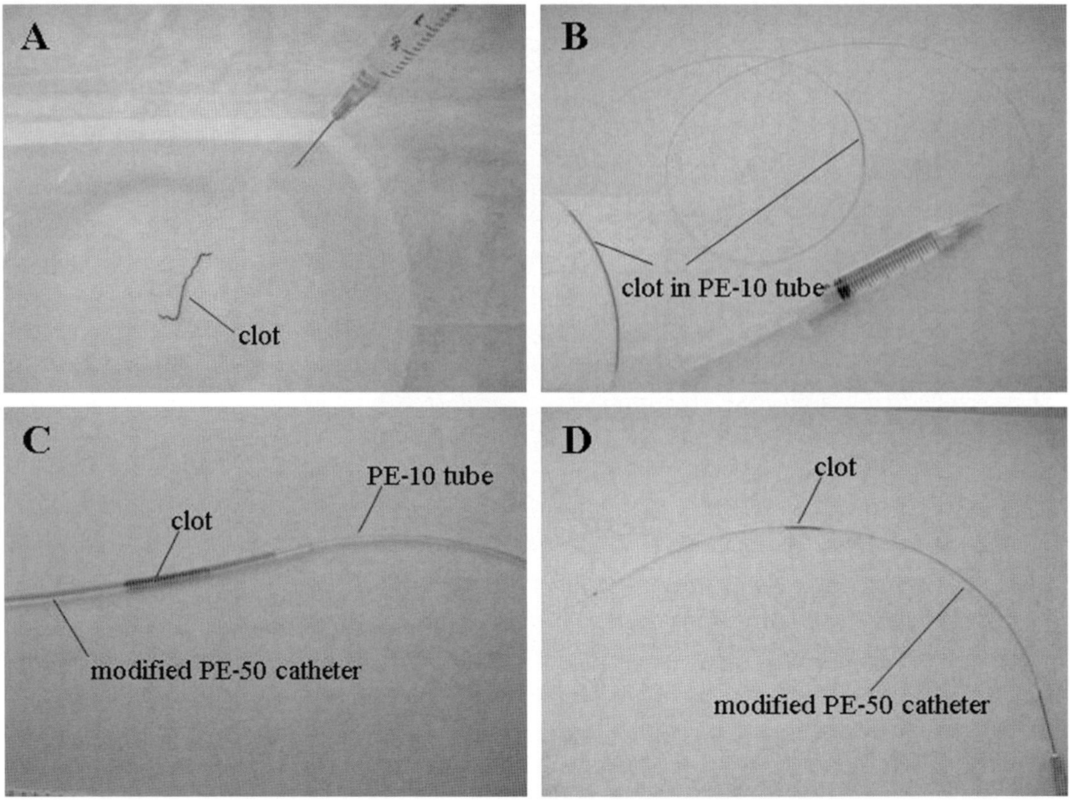

Fig. 3 (**A**) Transfer of the clot into a dish filled with saline from the PE-50 tube. (**B**) Aspiration of the clot into a 30-cm PE-10 tube and washing of the clot in the PE-10 tube. (**C**) Connecting the PE-10 tube to the bottom of a modified PE-50 catheter and shifting the clot into the modified PE-50 catheter. (**D**) The clot is contained within the modified PE-50 catheter

3.2.3 Preparation for Injecting the Clot

1. The PE-50 tubing containing the clot is cut to the optimal length (3–5 cm) and connected to a 3-ml syringe filled with saline with a 23 G needle
2. Transfer the clot into a dish filled with saline (Fig. 3a)
3. Aspirate the clot into a 30-cm PE-10 tube with a 3-ml syringe and 30 G needle and move the clot back and forth (wash) five times in the PE-10 tube (Fig. 3b)
4. Connect the PE-10 tube to the bottom of the modified PE-50 catheter and shift the clot into the modified PE-50 catheter (Fig. 3c)
5. Connect a 1-ml syringe with 23 G needle to the bottom of the modified PE-50 catheter (Fig. 3d)

3.2.4 Thromboembolic Model

1. Place the rat in a supine position, and make an incision in the midline of the neck
2. Expose the common carotid artery (CCA), the origin of internal carotid artery (ICA), the external carotid artery (ECA), the occipital artery (OA), and the superior thyroid artery (STA) under the microscope (Fig. 4a)

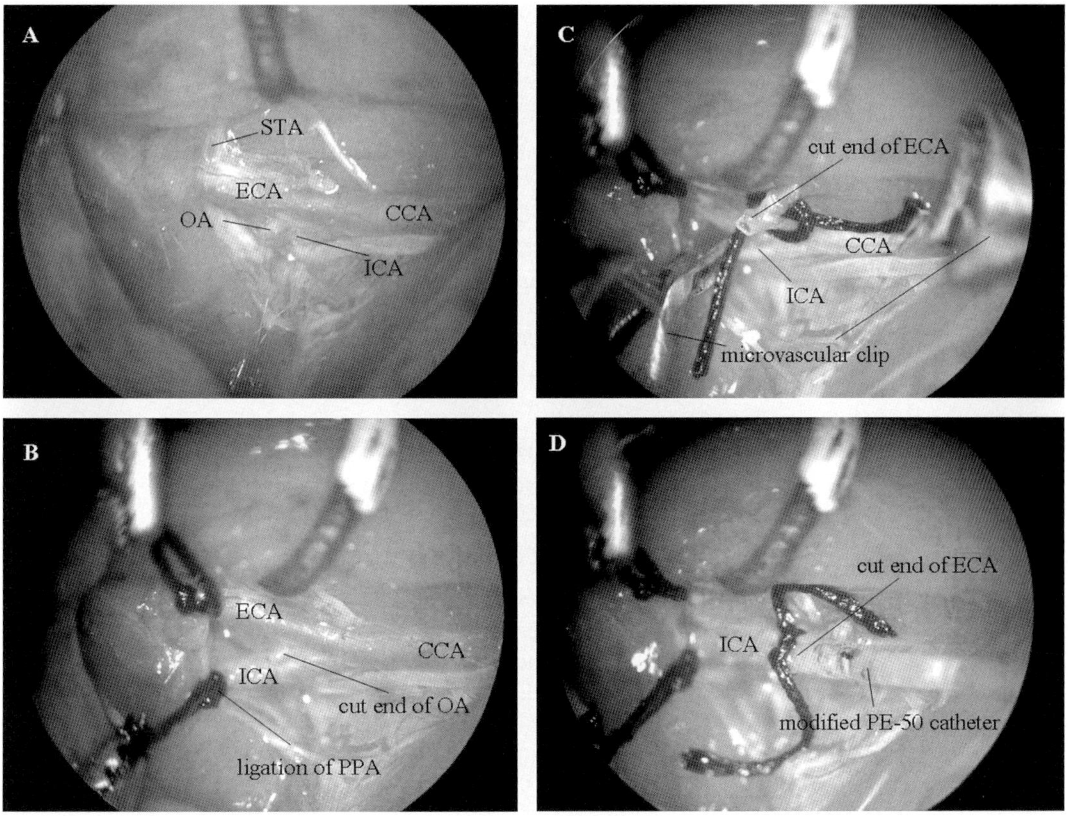

Fig. 4 Microscopic view of the arterial distribution during clot injection. CCA: common carotid artery, ICA: internal carotid artery, ECA: external carotid artery, OA: occipital artery, STA: superior thyroid artery, PPA: pterygopalatine artery

3. Cut the OA and the STA after coagulation
4. Expose the bifurcation of the ICA and the pterygopalatine artery (PPA) (Fig. 4b)
5. Ligate the origin of the PPA and the distal end of the ECA with 4-0 silk suture
6. Temporally clamp the proximal side of CCA and ICA with microvascular clip, and then cut the distal end of the ECA near the ligation
7. Tie the origin of ECA with a 4-0 silk suture (Fig. 4c). (This is important to control bleeding from cut end of the EC when the modified PE-50 catheter is inserted.)
8. Insert the modified PE-50 catheter about 15 mm long from the cut end of the ECA, then declamp the clip at the ICA and insert the modified PE-50 catheter into the ICA (Fig. 4d) (At this time, the tip of the catheter will be placed at 2–3 mm from the origin of the MCA)
9. Inject the clot gently with a small amount of saline (rCBF will be decrease to less than 30% of pre-ischemic baseline)
10. After 5 min, withdraw the catheter from the ECA
11. Close the wound with 4-0 silk suture

3.2.5 tPA Administration

tPA is administered intravenously at the indicated doses as a 10% bolus, and the remainder is infused continuously over 30 min using a syringe pump according to previously established protocols, which have been demonstrated to effectively induce clot lysis and cerebral reperfusion in rat embolic stroke.[9–12] A dose of 10 mg kg^{-1} of tPA is thought to be effective with perfusion improvement and beneficial outcomes within a 3-h time window in rodent brain. The relatively high dose of tPA is chosen on the basis of the approximately tenfold difference in fibrin-specific activity between human and rodent systems.[13]

However, the actual tPA dose should be independently assessed in each laboratory to ensure that adequate reperfusion is obtained for the purposes of each individual experiment.

4 Evaluation of Infarct Volume and Intracranial Hemorrhage (Hemorrhagic Transformation)

4.1 Measurement of Acute Infarct Volumes

4.1.1 At 24 h After Ischemia

1. Prepare seven coronal sections per brain (2 mm in thickness) (transcardial perfusion with saline is needed before making the sections if spectrophotometric assay of intracranial hemorrhage is also performed)
2. Put the sections into 2% TTC solution
3. Wait for about 15 min (living tissue will be stained red) and digitize pictures for image analysis (Fig. 5a)
4. Perform quantification of ischemic infarction volumes by a standard computer-assisted image analysis technique[7] using software such as Image J (NIH, http://rsb.info.nih.gov/nih-image/). Infarct volume is expressed in mm^3 or as percentage of the whole hemisphere

4.1.2 Measurements of Late Infarct Volumes

TTC staining for examining infarction volumes after 3 days of ischemic onset is inadequate. Nissle staining (Fig. 5b) or hematoxylin–eosin staining is commonly used.

4.2 Spectrophotometric Assay of Intracranial Hemorrhage

One of the major complications of tPA therapy is intracranial hemorrhage. Even without thrombolysis, hemorrhagic transformation may not also be uncommon after an ischemic stroke. The measurement of hemorrhagic transformation in the present rat embolic model is critical, especially to evaluate the safety concerns in thrombolytic therapy. The volume of intracerebral hemorrhage can be quantitated by a standard hemoglobin spectrophotometric assay.[6,14] Brain samples can be stored in a deep freezer for later analysis. We have previously found that after TTC staining, brain sections can be also used for hemoglobin spectrophotometric assay since TTC staining does not alter the reading of the assay.[6]

4.2.1 Making the Hemoglobin Standard Curve

A standard curve is obtained with the use of a "virtual" model of hemorrhage.

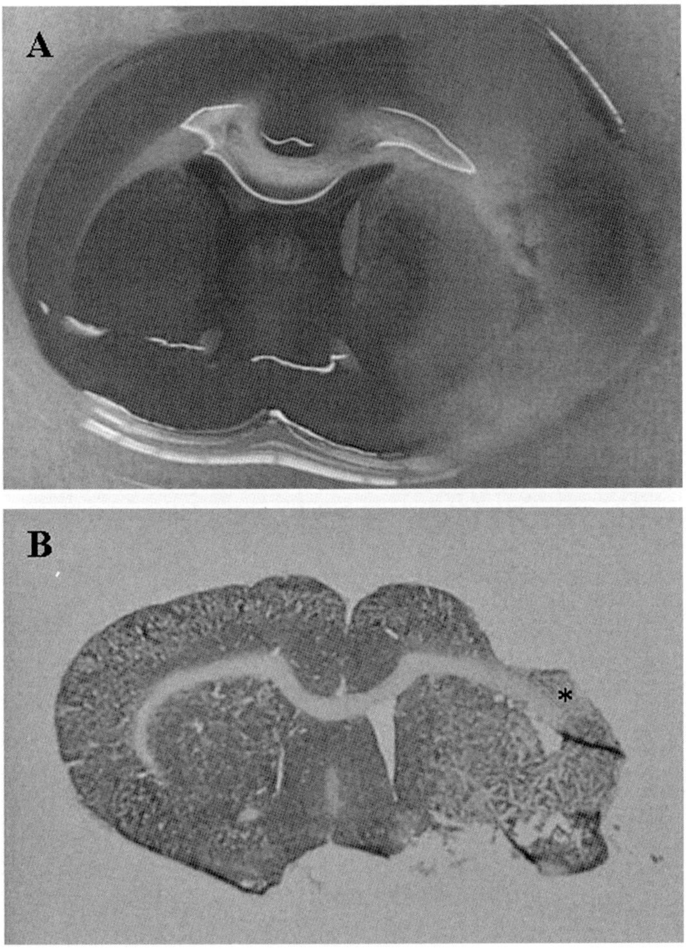

Fig. 5 (**A**) TTC staining: intact area turns red and infarct area presents white (*brown* in white area indicates hemorrhage). (**B**) Nissl staining: *Missing area and light blue area* (*asterisk*) in right hemisphere indicate infarct

1. Hemispheric brain tissue is obtained from normal rats after complete transcardial perfusion with saline in terms of pushing out intravascular blood.
2. Incremental volumes of homologous blood (0, 0.5, 1, 2, 4, 8, 16, 32, 50, 100, 200 µl) are added to each hemispheric brain sample and filled up with saline to reach a total volume of 3 ml.
3. Homogenization is carried out for 30 s at room temperature.
4. Sonication is carried out on ice for 1 min.
5. Centrifugation is done at 13,000 rpm for 30 min at room temperature.
6. Drabkin's reagent 1.6 ml (Sigma) is added into 0.4 ml sample supernatant (4/1 ratio), then shaken by vertexing.
7. A 15-min waiting period at room temperature follows. This reaction converts hemoglobin to cyanomethemoglobin, which has an absorbance peak at 540 nm, and whose concentration can then be assessed by reading at that wavelength.
8. The mixed samples are read at 540 nm with a spectrophotometer (Spectronix 3000, Milton-Roy, Rochester, NY, USA).

4.2.2 *Measurement for Stroke Brain Samples*

1. Complete transcardial perfusion with saline and take out the brains
2. Perform the procedures 3–7 mentioned in Sect. 4.2.1
3. Compare the reading with standard curve and calculate hemorrhage volumes with μl blood/g hemisphere

5 Limitations

The major limitation with the thromboembolic rat model is the degree of variability and/or failure caused by spontaneous recanalization. In our experience, the lowest rate of spontaneous recanalization is found in SHRs (see Table 1). Conversely, the mortality rates are the highest in SHRs compared with Wistar and SD rats (Table 1). In the end, the choice depends on the specific experiments and hypotheses involved. Furthermore, the reproducibility of this model is also heavily dependent on the experience of the operator and surgical nuances in each individual laboratory. Ultimately, however, this class of thromboembolic models may still be very useful for experimental stroke research in the context of testing antithrombosis and reperfusion therapies.

6 Discussion

6.1 Imaging of Thromboembolic Models

The pathophysiology of these embolic models can perhaps be best studied using in vivo imaging. Chopp's group[4,5] has most extensively studied changes in magnetic resonance imaging (MRI) associated with these models. In these studies, a reduction in CBF and a hyperintensity area were recognized by perfusion-weighted imaging and diffusion-weighted imaging respectively, after injecting the clot in the territory supplied by the MCA. rCBF in the ipsilateral parietal cortex was typically reported as dropping to approximately 30–40% of preischemic levels. The decrease of rCBF rapidly returned to preischemic level owing to the injection of tPA at 2 h after ischemia. These types of experiments have also been used to define MRI signatures associated with damage and recovery after an embolic stroke.

6.2 Hemorrhagic Transformation and Combination Therapy

As discussed previously, an advantage of the thromboembolic model is that we can investigate hypotheses related to tPA therapy. Thrombolysis with tPA can elevate the risks of intracranial hemorrhage after an ischemic stroke.[15,16] One of the major reasons for limiting the therapeutic time window (TTW) of tPA therapy is in fact related to this elevated risk of intracranial hemorrhage, even when tPA was administrated within 3 h.[1] Thus, the clinical challenge is how to extend the TTW for tPA, decrease or eliminate the risks of intracranial hemorrhage, and improve the efficacy of reperfusion therapy. The method for evaluating hemorrhagic transformation described in this chapter can be used for assessing potential combination therapies for ameliorating some of these complications in tPA therapy.

Using this type of thromboembolic model in rats, Zhang et al. reported that combination therapy with tPA plus a proteasome inhibitor PS-519[17] or bortezomib[18] reduced infarct volume and improved neurological outcome 6h after ischemia without increased incidence of hemorrhagic transformation. In another study, Zhang et al.[19] confirmed that combination with glycoprotein IIb/IIIa receptor inhibitor with full-dose and half-dose tPA reduced infarct volume and improved neurological outcome 4h after ischemia. Maeda et al. used a photochemically induced thrombotic focal ischemia model to demonstrate that combination therapy with tacrolimus (FK506)[20] plus tPA reduced infarct volume 3h after ischemia. Approaches such as these should prove very useful as we collectively search for effective ways to augment the use of tPA in ischemic stroke.

7 Other Thromboembolic and Thrombotic Models

There is a huge amount of literature in experimental stroke that have previously described a variety of embolic animal models. Listed below is a brief survey of the other models that have also been used. Each of them possesses its own features with advantages and disadvantages. More information on these different models can be obtained by looking at the specific papers, the details of which lie outside the scope of the present chapter.

(1) Multiple microembolic model in rats. This model was developed by Kudo et al.[21] and Meden et al.[22,23] via intracarotid injection of multiple homologous microembolic clots. The surigical approach is relatively simple. But there is a caveat in that the distribution of the emboli and arterial occlusions are, by definition, stochastic.
(2) Embolic stroke model in mice.[24,25] Many studies have extended the rat embolic stroke model to mice. The obvious advantage here is that one can then test more sophisticated biologic hypotheses using transgenic or knockout mice. A major difficulty here lies in the fact that these surgical procedures and corresponding survival outcome rates might be challenging in mice.
(3) Embolic stroke model in rabbits. This model was originally developed by Zivin et al.,[26] and it greatly contributed to the development of tPA thrombolytic treatments for stroke. The model is still widely used in the evaluation of preclinical thrombolytic agents.
(4) Embolic stroke model in monkeys.[27] Technically, this model may be difficult. There are only a few published papers. The use of in vivo (positron emission tomography (PET) or MRI) imaging is also an added strength of this approach. However, the costs may be correspondingly higher.
(5) Photochemical thrombotic model. This model has been used in rats[28] and mice,[29] wherein a combination of a photosensitive dye (Rose Bengal) and local photon illumination induces damage of vascular endothelial cells and consequently forms a clot in the targeted blood vessel. Branches of MCAs on the brain surface can be occluded, but the degree of clot formation or latency for thrombolysis-induced clot lysis may depend on the irradiation parameters selected. More recently, a two-photon laser approach has also been described, in which both the surface and the deeper cortical microvessels can be targeted.[30]

References

1. NINDS rt-PA Stroke Study Group. Tissue plasminogen activator for acute ischemic stroke. N Engl J Med 1995;333:1581–1587
2. Marler JR, Goldstein LB. Stroke: tPA and the clinic. Science 2003;301:1677
3. Reed SD, Cramer SC, Blough DK, Meyer K, Jarvik JG. Treatment with tissue plasminogen activator and inpatient mortality rates for patients with ischemic stroke treated in community hospitals. Stroke 2001;32:1832–1840
4. Zhang ZG, Zhang RL, Jiang Q, Raman SB, Cantwell L, Chopp M. A new rat model of thrombotic focal cerebral ischemia. J Cereb Blood Flow Metab 1997;17: 123–135
5. Zhang RL, Chopp M, Zhang ZG, Jiang Q, Ewing JR. A rat model of focal embolic cerebral ischemia. Brain Res 1997;766:83–92
6. Asahi M, Asahi K, Wang X, Lo EH. Reduction of tissue plasminogen activator induced hemorrhage and brain injury by free radical spin trapping after focal cerebral ischemia in rats. J Cereb Blood Flow Metab 2000;20:452–457
7. Sumii T, Singhal AB, Asahi M, Shimizu-Sasamata M, Suzuki M, Miyata K, Lo EH. Protective effect of pamiteplase, a modified tPA, in a rat model of embolic stroke. Neuroreport 2001;12:615–618
8. Sumii T, Lo EH. Involvement of matrix metalloproteinase in thrombolysis-associated hemorrhagic transformation after embolic focal ischemia in rats. Stroke 2002;33:831–836
9. Brinker G, Pillekamp F, Hossman KA. Brain hemorrhage after rtPA treatment of embolic stroke in spontaneously hypertensive rats. Neuroreport 1999;10:1943–1946
10. Chopp M, Zhang RL, Zhang ZG, Jiang Q. The clot thickens: thrombolysis and combination therapies. Acta Neurochir Suppl 1999;73:67–71
11. Kano T, Katayama Y, Tejima E, Lo EH. Hemorrhagic transformation after fibrinolytic therapy with tissue plasminogen activator in a rat thromboembolic model of stroke. Brain Res 2000;854:245–248
12. Meng W, Wang X, Asahi M, Kano T, Asahi K, Ackerman RH, Lo EH. Effects of tissue type plasminogen activator in embolic versus mechanical models of focal cerebral ischemia in rats. J Cereb Blood Flow Metab 1999;19:1316–1322
13. Korninger C, Collen D. Studies on specific fibrinolytic effect of human extrinsic plasminogen activator in human blood and in various animal species in vitro. Thromb Haemost 1981;46:561–565
14. Choudri TF, Hoh BL, Solomon RA, Connoly ES, Pinsky D. Use of a spectrophotometric assay to objectively quantify intracerebral hemorrhage in mice. Stroke 1997;28:2296–2302
15. Larrue V, von Kummer R, del Zoppo G, Bluhmki E. Hemorrhagic transformation in acute ischemic stroke: potential contributing factors in the ECASS study. Stroke 1999;28:957–960
16. NINDS rt-PA Stroke Study Group. Intracerebral hemorrhage after intravenous TPA therapy for ischemic stroke. Stroke 1997;28:2109–2118
17. Zhang L, Zhang ZG, Zhang RL, Lu M, Adams J, Elliott PJ, Chopp M. Postischemic (6-Hour) treatment with recombinant human tissue plasminogen activator and proteasome inhibitor PS-519 reduces infarction in a rat model of embolic focal cerebral ischemia. Stroke 2001;32:2926–2931
18. Zhang L, Zhang ZG, Liu X, Hozeska A, Stagliano N, Riordan W, Lu M, Chopp M. Treatment of embolic stroke in rats with bortezomib and recombinant human tissue plasminogen activator. Thromb Haemost 2006;95:166–173
19. Zhang L, Zhang ZG, Zhang R, Morris D, Lu M, Coller BS, Chopp M. Adjuvant treatment with a glycoprotein IIb/IIIa receptor inhibitor increases the therapeutic window for low-dose tissue plasminogen activator administration in a rat model of embolic stroke. Circulation 2003;107:2837–2843

20. Maeda M, Furuichi Y, Ueyama N, Moriguchi A, Satoh N, Matsuoka N, Goto T, Yanagihara T. A combined treatment with tacrolimus (FK506) and recombinant tissue plasminogen activator for thrombotic focal cerebral ischemia in rats: increased neuroprotective efficacy and extended therapeutic time window. J Cereb Blood Flow Metab 2002;22:1205–1211
21. Kudo M, Aoyama A, Ichimori S, Fukunaga N. An animal model of cerebral infarction. Homologous blood clot emboli in rats. Stroke 1982;13:505–508
22. Meden P, Overgaard K, Sereghy T, Boysen G. Enhancing the efficacy of thrombolysis by AMPA receptor blockade with NBQX in a rat embolic stroke model. J Neurol Sci 1993;119:209–216
23. Meden P, Overgaard K, Pedersen H, Boysen G. Effect of hypothermia and delayed thrombolysis in a rat embolic stroke model. Acta Neurol Scand 1994;90:91–98
24. Kilic E, Hermann DM, Hossmann KA. A reproducible model of thromboembolic stroke in mice. Neuroreport 1998;9:2967–2970
25. Atochin DN, Murciano JC, Gursoy-Ozdemir Y, Krasik T, Noda F, Ayata C, Dunn AK, Moskowitz MA, Huang PL, Muzykantov VR. Mouse model of microembolic stroke and reperfusion. Stroke 2004;35:2177–2182
26. Lapchak PA, Chapman DF, Zivin JA. Metalloproteinase inhibition reduces thrombolytic (tissue plasminogen activator)-induced hemorrhage after thromboembolic stroke. Stroke 2000;31:3034–3040
27. Kito G, Nishimura A, Susumu T, Nagata R, Kuge Y, Yokota C, Minematsu K. Experimental thromboembolic stroke in cynomolgus monkey. J Neurosci Methods 2001;105:45–53
28. Watson BD, Dietrich WD, Busto R, Wachtel MS, Ginsberg MD. Induction of reproducible brain infarction by photochemically initiated thrombosis. Ann Neurol 1985;17:497–504
29. Nagai N, Zhao BQ, Suzuki Y, Ihara H, Urano T, Umemura K. Tissue-type plasminogen activator has paradoxical roles in focal cerebral ischemic injury by thrombotic middle cerebral artery occlusion with mild or severe photochemical damage in mice. J Cereb Blood Flow Metab 2002;22:648–651
30. Nishimura N, Schaffer CB, Friedman B, Tsai PS, Lyden PD, Kleinfeld D. Targeted insult to subsurface cortical blood vessels using ultrashort laser pulses: three models of stroke. Nature Methods 2006;3:99–108

Chapter 14

Targeted Occlusion to Surface and Deep Vessels in Neocortex via Linear and Nonlinear Optical Absorption

David Kleinfeld, Beth Friedman, Patrick D. Lyden, and Andy Y. Shih

Abstract We discuss two complementary methods for the study of cerebral blood flow and brain function in response to the occlusion of individual, targeted blood vessels. These bear on the study of microstroke and vascular dysfunction in cortex. One method makes use of linear optical absorption by a photosensitizer, transiently circulated in the blood stream, to induce an occlusion in a surface or near-surface vessel. The second method makes use of nonlinear optical interactions, without the need to introduce an exogenous absorber, to induce an occlusion in a subsurface microvessel. A feature of both methods is that the dynamics of blood flow and functional aspects of the vasculature and underlying neurons in the neighborhood of the occluded vessel may be monitored before, during, and after the occlusion. We present details of both methods and associated surgical procedures, along with exemplary data from published studies.

Keywords: Ablation, Clot, Hypoxia, Ischemia, Microstroke, Photosensitizer, Plasma, Rodent, Two-photon

1 Introduction

There are several forms of vascular dysfunction that can contribute to the onset of blood flow disturbances and neurodegeneration.[1] We consider dysfunction that is caused by vessel occlusion, which leads to ischemia. Our focus is on techniques that can occlude individual, targeted single vessels. The importance of targeting is threefold. First, it provides a means to select a particular type of vessel, e.g., capillary versus surface artery. Second, It provides a means to select a vessel at a particular location, e.g., within a selected cortical column. Lastly, it provides a means to make both pre- and post measurements of hemodynamic and neuronal state variables in the region that neighbors the occluded vessel.

The diameter of vessels in neocortex ranges over an order of magnitude in size. Surface communicating arterioles, which form a supply network for the underlying parenchyma, span 20–50 μm in diameter; penetrating arterioles,

From: *Animal Models of Acute Neurological Injuries.*
Edited by: J. Chen (eds.) New York, Totowa, NJ

which transfer blood from the surface to the parenchyma, span 10–25 μm in diameter; and the microvessels and capillaries that supply the parenchyma span 5–10 μm in diameter.[2] A variety of techniques have been developed to induce occlusions in all these vessels (Table 1). One class of procedures involves the introduction of small particles into the blood stream, such as microspheres or small thrombo-emboli, that can be tailored to lodge in arterioles of a particular size and produce occlusions. However, the location of the occlusion cannot be targeted. A second class of procedures involves the local injection of a powerful vasoconstrictor, particularly endothelin-1, that can be targeted to specific arteries and arterioles but typically spreads within the extracellular space to impact many vessels rather than just the desired target. A final class makes use of physical blockages, including ligation by a fine suture or a fine cautery. This approach is appropriate for relatively large arteries and veins on the cortical surface. Our goal has been to exploit the physical localization of highly focused laser light to form occlusions.

We consider two complementary techniques to form occlusions, both of which require the concurrent use of in vivo confocal laser scanning microscopy or in vivo two-photon laser scanning microscopy (TPLSM) as a means to image the underlying vasculature and select a vessel for a targeted occlusion. In vivo TPLSM is particularly advantageous, as it allows fluorescently labeled structures to be visualized at depths of 500–1,000 μm below the cortical surface.[17] The first technique makes use of near-threshold optical activation of the photosensitizer Rose Bengal and is suitable for vessels at the surface of the brain.[18,19] These include the communicating arteries and the initial segment of the penetrating arterioles that branch off from the communicating arteries. The second makes use of plasma-mediated ablation through the nonlinear absorption of 100-fs, ultrashort laser pulses by blood plasma or the vascular wall.[20] This technique is suitable for subsurface vessels such as arterioles, capillaries,

Table 1 Methods for occlusion of individual small vessels in cortex.

Method	Vessel(s)	Targeted	Reperfusion	Representative references
Intravascular injection of microspheres (25–50 μm diameter)	Arterioles	No	No	3,4
Intravascular injection of macrospheres (300–400 μm diameter)	Arteries	No	No	5
Intravascular injection of thrombo emboli	Arteries, arterioles	No	Yes	6
Focal injection of vasoconstrictor (endothelin-1)	Arteries, arterioles	Partially	Yes	7–10
Suture ligation	Surface arteries	Yes	Yes	11–13
Microclip Thermo-coagulation Electro-coagulation	Surface arteries	Yes	No	14,15
Photosensitizer (Rose Bengal)	Surface arteries, penetrating arterioles	Yes	Incompletely tested	10,16,18,19
Plasma-mediated ablation (amplified 100 fs pulses)	Subsurface microvessels	Yes	No	19

and venules, as vessels above and below the target vessel are unaffected. However, this technique does not appear to be suitable for surface vessels, as the energy deposited can lead to rupture of the free surface.

2 Common Procedures

Our procedures are applicable to any laboratory animal, although most of our experiments to date involve the adult rat with an observation window placed over parietal cortex. The underlying vasculature is imaged with in vivo TPLSM,[21] with contrast generated by labeling of the blood plasma with high-molecular-weight fluorescein isothiocyanate-conjugated dextran[22–24] (fluorescein–dextran) (2 MDa, Sigma). Vessels throughout the cranial window are first mapped at low magnification, typically 4× (e.g., UB950, Olympus), to form a single map of an area of the exposed cortex. Target vessels for photodisruption are typically selected using the criterion that a portion of the vessel courses parallel to the cortical surface. We then map the local vascular architecture in the vicinity of the target vessel with a three-dimensional stack of images at high magnification, typically 40× (e.g., UM568, Olympus).

3 Photosensitizer-Mediated Occlusion of Surface Vessels

This technique involves the introduction of a photosensitizer, in our case Rose Bengal, into the blood stream and the subsequent irradiation of the target vessel with actinic light, green laser light in our case (Fig. 1a). The intensity of the

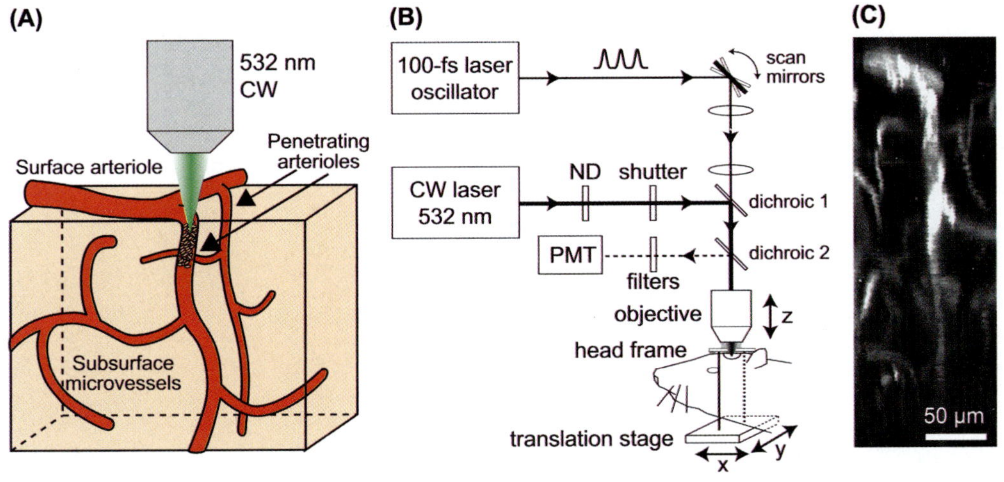

Fig. 1 Overview for forming intravascular occlusions in surface vessels using focal illumination of Rose Bengal. (**A**) Green laser light is tightly focused through an objective and directed into the lumen of a target vessel. The circulating Rose Bengal dye is photoactivated and initiates a clotting cascade. (**B**) Schematic of a two-photon laser scanning microscope system modified to include a continuous wave (CW) green laser for photothrombosis. The green laser is directed onto the optical axis of the near-infrared beam with dichroic mirror 1 (625 DRLP; Chroma). An approximately 3.5-mm hole was etched in the coating of dichroic mirror 2 (700 DCXRU; Chroma) to allow transmission of the green laser beam; this results in at most a 10% loss in collection efficiency. CW = continuous wave, ND = neutral density filter. (**C**) X–Z maximal projection of the TPLSM image stack showing a penetrating arteriole selectively occluded by photothrombosis. All panels adapted from Schaffer et al.[18]

light is adjusted to be at the threshold for photo-thrombosis in the plane of the target surface vessel. Light that propagates into the parenchyma will be lower in intensity and therefore below the threshold level to induce photothrombotic damage. This allows formation of the occlusion that is restricted to the surface vessel on the basis of three physical mechanisms: (1) absorption of the incident light by the target vessel; (2) divergence of the beam beyond the target vessel as a result of strong focusing; and (3) scattering of light by brain tissue. However, during the period of illumination small thrombi may be formed that travel downstream rather than stick to the site of the irradiated region.

The persistence of Rose Bengal within the blood plasma may be estimated from in vivo measurements of its fluorescence, which peaks near 580 nm (http://www.omlc.ogi.edu/spectra/PhotochemCAD/html/) in the plasma. In mouse, which has approximately a tenfold smaller blood volume than rat, Rose Bengal is cleared from circulation within 5 min[25] (Fig. 2).

3.1 Equipment

Formation of an occlusion with a photosensitizer is readily combined with TPLSM to enable the targeting and real-time monitoring of blood vessels. The typical setup makes use of a standard two-photon microscope, either homebuilt[26,27] or a commercial instrument (e.g., Ultima with optional high-speed two-photon optics set and an external detector; Prairie Technologies), in which a 20 mW or greater green ($\lambda = 532$ nm) laser light (e.g., GCL-100-M, CrystaLaser) is introduced into the beam pathway with a dichroic mirror (Fig. 1b). We used a fixed location of the beam, centered along the optical axis. The diameter of the beam is typically adjusted to pass through a hole in the emission dichroic (2 in Fig. 1b) and underfills the back aperture of the objective. The final spot size is approximately 3 μm in diameter and is brought to the same focus as the pulsed near-infrared beam for TPLSM using an appropriate telescope. In this case, the motorized/encoding X–Y stage can be used

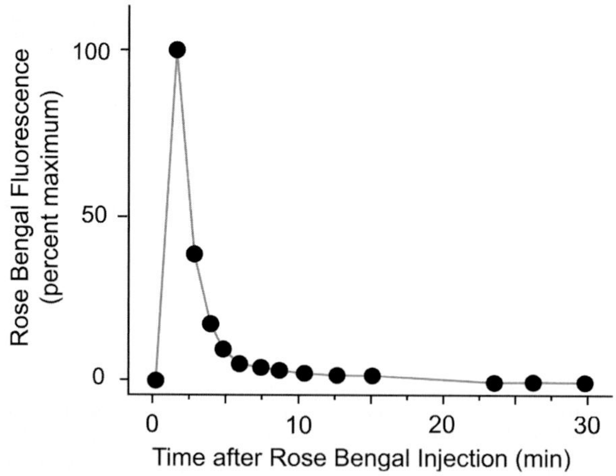

Fig. 2 Time scale for Rose Bengal clearance from circulation. Rose Bengal fluorescence in a cerebral arteriole was repeatedly measured after tail vein injection using TPLSM planar scans. Rose Bengal is largely cleared from circulation within 5 min. Data collected from mouse, courtesy of Zhang and Murphy[25]

to reposition the animal to change the location of photoactivation. Alternatively, the green actinic light may be introduced through a second set of scanners so that the position may be shifted electromechanically (e.g., Ultima with an optional second set of scanners). In all cases, the intensity of the laser light at the focus should initially be held at 0.1 mW using a tunable neutral density filter in the beam path, and increased as needed up to a maximum of 5 mW.

3.2 Procedure

(1) Prior to surgery

a. Choose a proper anesthetic

i. Isoflurane in a 30% O_2/70% N_2O gas mixture, i.e., 4% (v/v) induction and 1–2% (v/v) maintenance, provides adjustable anesthesia and the option of animal recovery

ii. Urethane provides long-lasting and stable anesthesia, i.e., 1,000 mg kg^{-1} initial dose with 100 mg kg^{-1} supplements as required, but is difficult to adjust and is inappropriate for recovery. When using urethane, a 30% O_2/70% N_2O gas mixture is delivered to maintain normal blood gas levels

iii. Other anesthetic agents are also suitable for this procedure[28]

b. Check for lack of toe pinch reflex to ensure an adequate level of anesthesia

c. Secure animal in a stereotaxic frame

d. Inject atropine, 0.05 mg kg^{-1}, intraperitoneal (IP) to reduce respiratory distress

e. Apply an opthalamic ointment to the eyes to retain moisture

f. Inject 2% (v/v) Lidocaine™ into the scalp subcutaneously before incision for local anesthesia

(2) Throughout surgery and recording (Fig. 3d)

a. Ensure heart and breathing rates are within a normal range, i.e., 300–400 and 60–120 events per minute, respectively, using a pulse oximeter (8,600 V; Nonin)

b. Maintain body temperature at 37°C using a feedback-regulated rectal probe and heat pad (Harvard)

c. Inject (IP) saline with 5% (w/v) dextrose, 3 mL kg^{-1} every 2 h to maintain body fluids and energy requirements

d. Install a femoral artery catheter to allow blood samples to be collected. Blood gas is monitored once every 2 h (Rapid Lab 248; Bayer). The normal range is 80–100 mmHg for pO_2, 35–45 mmHg for pCO_2, and 7.35–7.45 for pH

e. Monitor arterial blood pressure (XBP1000; Kent Scientific). The normal range is 80–130 mmHg for systolic pressure and 60–100 mm Hg for diastolic pressure. We use a tail-cuff device to measure blood pressure, but measurement through an intra-arterial catheter is also possible

Note that fluctuations in physiological parameters, often caused by variation in the depth of anesthesia, can greatly influence cerebral blood flow. It is important to maintain these variables in a normal range and to take note of abrupt changes during the experiment.

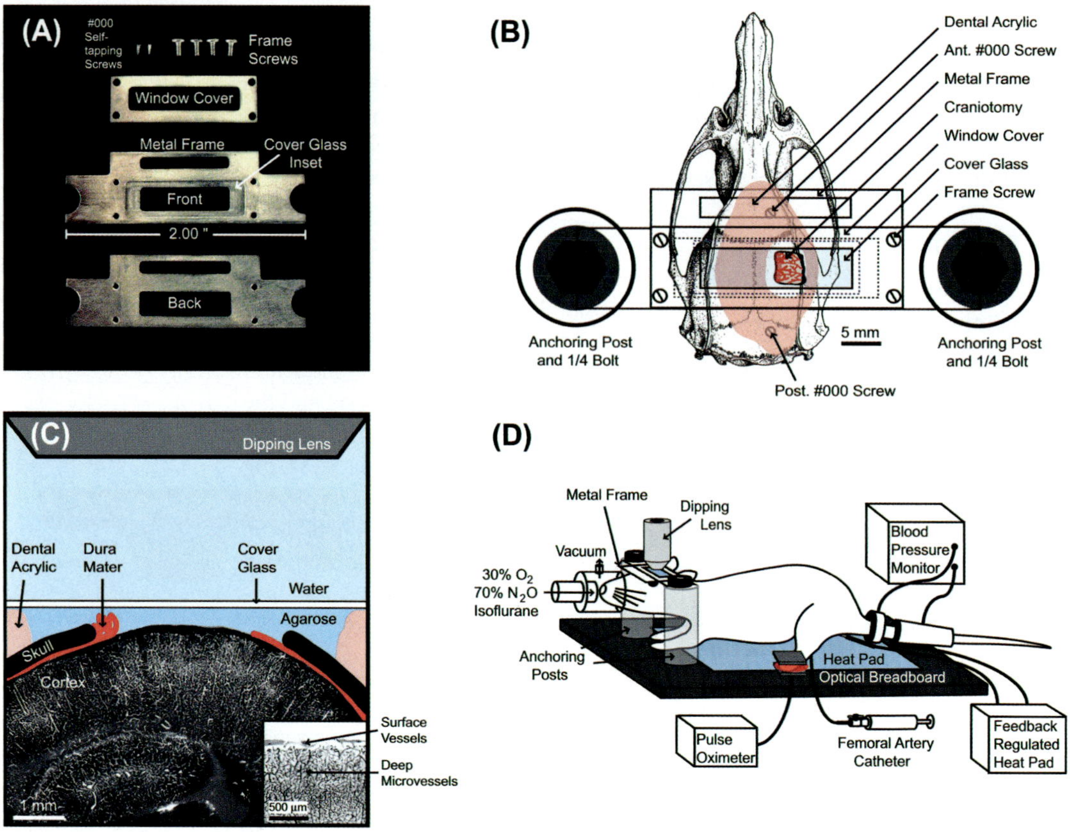

Fig. 3 Methods for in vivo imaging of blood flow through a cranial window. (**A**) To immobilize the head of the animal during imaging, we designed a metal frame that could be cemented to the skull, and then anchored to an optical setup. The frame is constructed of type 410 stainless steel with dimensions of 2.00 in. long, 0.61 in. wide, and 0.029 in. thick, and can be secured between two posts in a standard optical breadboard. An inset region, 0.015 in. deep, borders the frame window to hold a cover glass over the craniotomy. The window cover, 0.015 in. thick of type 301 stainless steel, is then secured to the frame with four screws, sandwiching the no. 1 cover glass in place. (**B**) Surgical procedure for acute imaging experiments. The animal is placed in a stereotaxic frame. Lidocaine™ is subcutaneously injected into the scalp for local anesthesia. An approximately 5-cm midline incision is made in the scalp, and the periostium is scraped to the sides with a bone curette. The scalp is retracted using hemostats. The right temporalis muscle is separated from the bone, without severing, and retracted. The skull surface is cleaned with cotton swabs, and bleeding from the skull is controlled with maintained pressure from the swab or superficial abrasion with a high-speed drill. A thin layer of VetBond™ is applied to the skull surface to aid attachment of the dental acrylic and frame. The location of the cranial window is marked in pencil. We typically place the window above the somatosensory cortex by centering a 4 × 4 mm craniotomy at –3.0 mm anterior–posterior and +4.5 mm medial–lateral. The edges of the marked cranial window are slowly thinned with a high-speed drill. The bone is flushed regularly with physiological saline to avoid overheating. Drilling is stopped just as the bone begins to crackle and cerebral spinal fluid begins to leak through. The resulting island of bone is gently lifted away horizontal to the skull surface using two forceps on opposite corners of the bone flap. The underlying dura is flushed with artificial cerebrospinal fluid (ACSF) and bleeding is controlled with small pieces of Gelfoam™ soaked in ACSF and KimWipes™, twisted to a fine point with the fingers. The cranial window is kept moist with a piece of moist Gelfoam™. Two burr holes are made with a 1/4 in. bit, anterior and posterior to the intended frame placement; self-tapping #000-3/32 screws are placed in the burr holes; and the threads are secured with a small dab of VetBond™. The metal frame is fixed tangentially to the clean, dry skull surface with dental acrylic. Extra dental acrylic is built up around the screws to provide further anchoring support. A small nick is made in the dura with a bent 27 G syringe needle. The dura is carefully teased apart with fine forceps (Dumont #5), starting at the nick and reflected to one side of the cranial window. The interior of the window chamber is filled with 37°C, 1.5% (w/v) low-melting-point agarose, dissolved in a modified version of ACSF (125 mM NaCl, 10 mM glucose, 10 mM HEPES, 3.1 mM $CaCl_2$, 1.3 mM $MgCl_2$, pH 7.4). The chamber is immediately sealed

(3) Surgery

a. Generate a craniotomy above the brain region of interest using a high-speed drill (Fig. 3b). We typically consider measurements over primary somatosensory cortex, which is the part of parietal cortex that nominally lies between –1 and –5 mm relative to the Bregma point and between 1 and 7 mm from the midline on the medial–lateral axis.[29]

b. Glue the metal frame to the skull with dental acrylic (Grip Cement; Dentsply). The frame rigidly holds the head of the animal to the optical apparatus (Fig. 3a, b). To achieve a reliable connection between acrylic and bone, the contact regions on the bone are cleaned of soft tissue, a thin layer of VetBond™ is applied, #000-3/32 self-tapping screws (Small Parts) are introduced to the anterior and posterior aspects of the skull, and screws, one of which passes through an opening in the frame, are then mechanically linked to the frame with dental acrylic (Fig. 3b).

c. Remove the dura after the dental acrylic has cured. It is crucial to avoid any bleeding, and therefore do not cut or tear across any large vessels in the dura. Removal of the dura greatly improves the optical field and is recommended.

d. Fill the interior of the chamber with 1.5% (w/v) low-melting-point agarose (A-9793; Sigma) dissolved in artificial cerebrospinal fluid (ACSF), which contains neither carbonate nor phosphate that cause precipitation when the solution is boiled to dissolve the agarose[30] (Fig. 3b, c). The temperature of the agarose must not exceed approximately 37°C before it is applied to the brain.

e. Immediately seal the chamber using a cover glass as a window (Fig. 3c). With the addition of a cover glass, the frame serves to provide a chamber that seals and protects the cortex. Re-sealing the craniotomy is crucial for the suppression of excessive motion which otherwise ensues as a result of cranial pressure fluctuations that are caused by heart beat and breathing.*

f. Stabilize the animals on a platform for imaging, using the frame as a head support (Fig. 3d). We prepare a separate plate that can be transported between surgical and imaging suites with the animal and all physiological monitoring devices assembled as one unit.

*The agarose between the cover glass and the cortex allows the cranial opening to be effectively resealed even if a narrow opening is left on the sides as a means to insert glass microelectrodes into cortex.[21]

Fig. 3 (continued) using a cover glass (no.1, cut to size), and the window cover is secured over the cover glass with four screws. The skin is sutured together around the frame, and agarose is trailed around the cover glass to hold water for the immersion lens. (**C**) A cross-sectional view of the cranial window. The inset diagram shows an inverted coronal view of surface vessels and deep microvessels that are targeted for occlusion. (**D**) A typical experimental setup used in our laboratory. The metal frame attached to the skull is immobilized between two anchoring posts inserted into an optical breadboard. Anesthesia is maintained with isoflurane in a 30% O_2/70% N_2O gas mixture. Blood pressure is measured with a tail-cuff device. The femoral artery catheter is used to collect blood samples for blood gas measurements, which is also the delivery route for fluorescein–dextran and Rose Bengal. Heart rate and blood O_2 saturation are continuously monitored using a pulse oximeter, which is useful in the assessing the depth of anesthesia. Rat skull diagram in panel B adapted from Paxinos[29]

g. Label the blood plasma with 2 MDa fluorescein–dextran through the femoral artery catheter as a means to generate contrast between the red blood cells (RBCs) and the plasma.† We inject a 1.0-mL kg^{-1} bolus of 5% (w/v) solution of fluorescein–dextran in physiological saline. This high-molecular-weight dye provides stable labeling, typically for 3 h, and further acts to stain the damage in occluded vessels.[18] The dye is supplemented as needed.

(4) Imaging and occlusion

a. Collect baseline vascular images at successive depths to construct a three-dimensional stack of images for maximal projection or volume-rendered views of the vasculature. Flow parameters, such as the speed of RBCs, are derived from line-scan data (Fig. 4a). Structural parameters, such as the cross-section of vessels, are derived from planar scans.

b. Select a vessel to target for occlusion.

c. Inject a 1.0 mL kg^{-1} bolus of 1% (w/v) solution of Rose Bengal in physiological saline through the catheter. Rose Bengal should be injected immediately before the formation of the occlusion.

d. To initiate photothrombosis, aim the focal point of the green laser light within the lumen of the target vessel but near the vessel wall. This will nucleate the clotting cascade and create an attachment point for the growing clot. The irradiation is periodically interrupted for brief epochs of imaging, typically 1 frame (0.2 s per frame) every 1 s for an 80% duty cycle, so that formation of the clot can be observed in real time. If a clot does not occur, close to the wall within 5–10 s of irradiation, and increase the power by factors of 2 until the clot begins to form.

e. Oscillate the relative positions of the animal and the beam, at 0.1–0.5 Hz, to traverse the diameter of the blood vessel. Occlusion requires 10–300 s of irradiation. Larger vessels with relatively fast RBC flow, i.e., surface vessels as opposed to penetrating arterioles, require more time to fully occlude; the exact time must be determined empirically.

f. A successful occlusion appears densely packed with immobile nonfluorescent RBCs often surrounded by a brighter region of stagnant fluorescent plasma (Figs. 1c, 4a, and 5a).

g. Collect postocclusion measurements, as desired.

h. The hypoxia marker pimonidazole hydrochloride (Hypoxyprobe™; http://www.hypoxyprobe.com) forms stable protein adducts in viable tissues with low oxygen content that can be detected by post hoc immunohistology. In particular, Hypoxyprobe™ can be used to locate relatively small zones of hypoxia caused by single-vessel occlusions (Fig. 5b). We typically inject Hypoxyprobe™ intravenously (IV) just after occlusion and 1 h before sacrifice.

(5) Histological examination

a. It is often desirable to examine the consequences of single-vessel occlusion using post hoc histology. The animal is transcardially perfused with

† The blood plasma may also be labeled through a tail-vein injection. Submerge the tail in warm water, *i.e.*, 37°C, for approximately 2 min and place a soft clamp at the base of the tail to dilate the veins. Starting as close to the tip of the tail as possible, insert a 24-gauge catheter and inject the bolus of fluorescein-dextran in physiological saline.

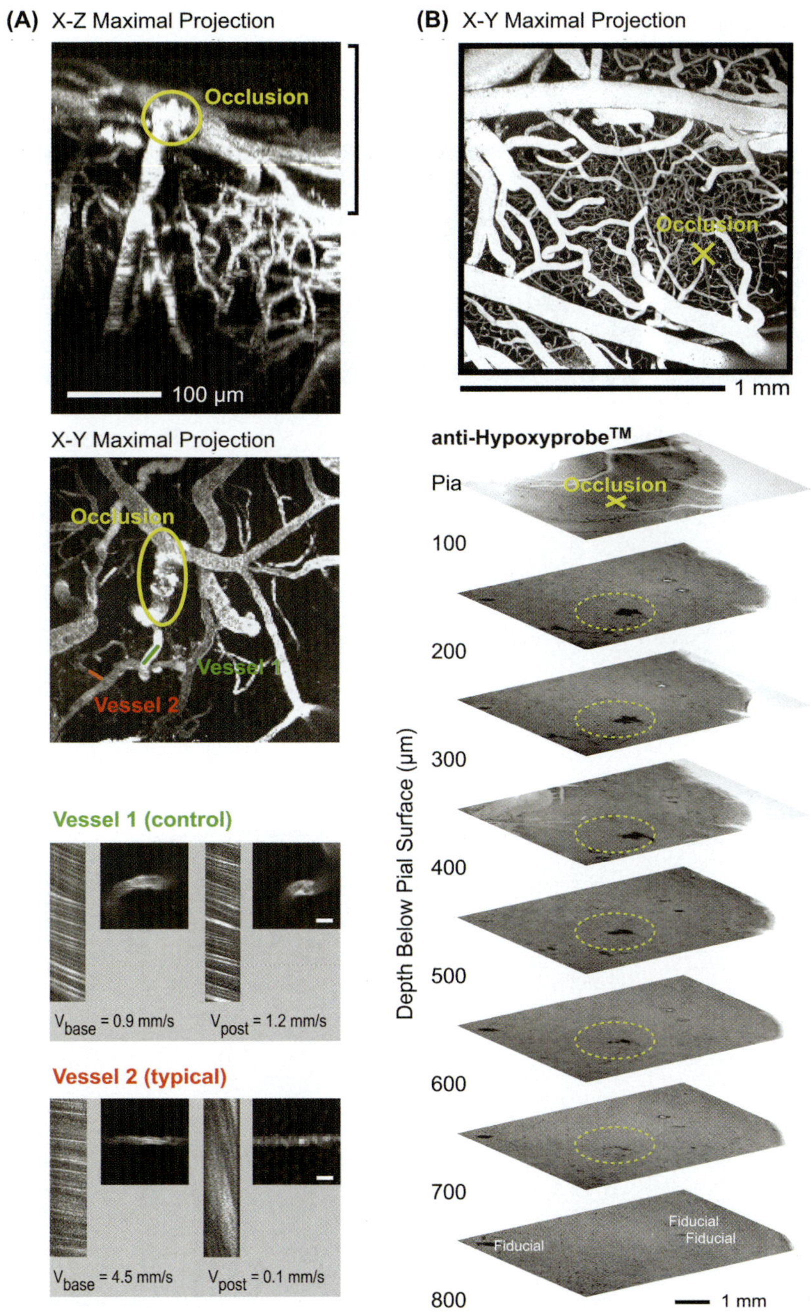

Fig. 4 Photothrombotic occlusion of penetrating arterioles using Rose Bengal. (**A**) Projected TPLSM image of a single penetrating arteriole occluded by photothrombosis (*yellow circle*) shown in the coronal plane (X–Z maximal projection), with surface arterioles near the top of the image. For the same field of interest, the surface vascular network is visible in the plane tangential to the cortical surface (X–Y maximal projection). The velocity of RBCs in two neighboring surface vessels (Vessel 1 and 2) was measured using repeated line scans. Vessel 1 shows no change in postocclusion RBC velocity (v_{post}) compared to baseline (v_{base}), whereas Vessel 2 shows essentially no flow after the occlusion. (**B**) Examination of the hypoxic volume generated after a single penetrating arteriole was occluded (*yellow cross*). Hypoxyprobe™ was injected (IV) after clot formation and 1 h before sacrifice. The animal was transcardially perfused with 100 mL of saline followed by 100 mL of 4% (w/v) paraformaldehyde. To locate the occluded vessel after brain extraction, fiducial markers were electrolytically placed near the target vessel before the brain was extracted from the skull. The entire ipsilateral cortex was removed and sandwiched between two glass slides separated by a distance of 2.5 mm. The tissue was postfixed for 24 h, equilibrated with 30% (w/v) sucrose, and sectioned on a freezing-sliding microtome into 100-μm sections tangential to the cortical surface. Immunostaining for Hypoxyprobe™ revealed a column of hypoxia, within the *dashed yellow circles*, that reached 700 μm below the pial surface. (**A**) adapted from Nishimura et al.[19]

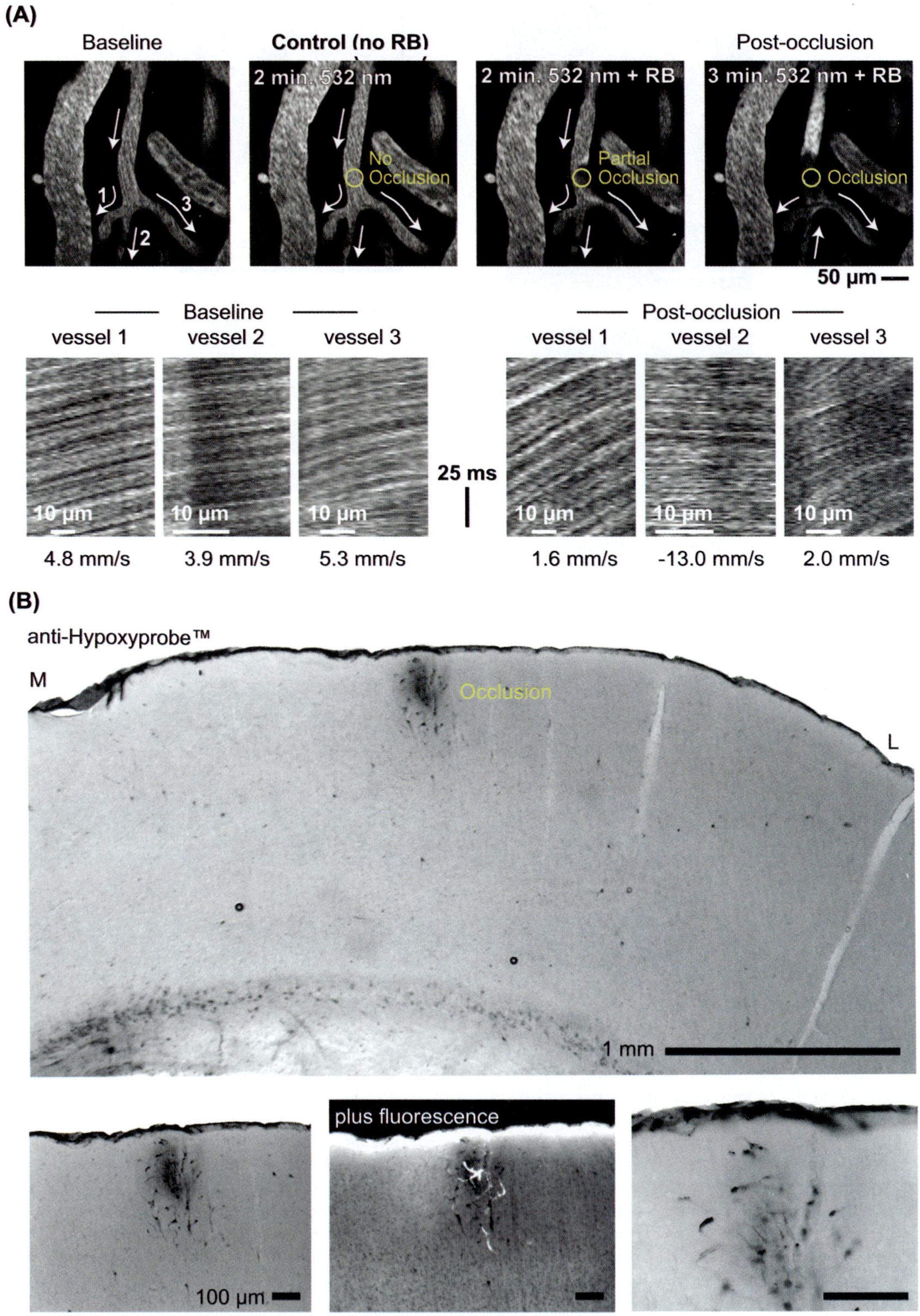

Fig. 5 Photothrombotic occlusion of surface arterioles using Rose Bengal. (**A**) Planar TPLSM images showing the time course for photothrombotic occlusion of a single surface arteriole. The first frame was taken at baseline, and *white arrows* indicate the directions of RBC flow as determined by line scans. Three vessels branching

saline followed by 4% (w/v) paraformaldehyde in phophate buffered saline, and the brain is removed, equilibrated with 30% (w/v) sucrose, and frozen-sectioned. Extravasated fluorescein–dextran is trapped within the vessel wall at or near the site of the occlusion (Figs. 4b and 5b).

3.3 Exemplary Data

The pial arterioles of the cortical surface arteries form a communicating network, from which penetrating arterioles branch and dive to perfuse underlying capillary beds. In the first experiment, we selectively occluded a penetrating arteriole (yellow circle in Fig. 4a). Baseline and postocclusion values of RBC velocities were measured in the target vessel, as well as neighboring microvessels, using continuous line scans[31] of a segment of the vessel in the plane of the cortical surface. The location of the occlusion appeared as a bright region of stagnant serum within the vessel lumen, and extravasated fluorescein-dextran indicated damage to the vessel wall. This occlusion was specific to the target vessel since the velocity of RBCs in a neighboring penetrating arteriole was unaffected (Vessel 1 in Fig. 4a). However, a downstream microvessel directly perfused by the target vessel experienced severe loss of blood flow (Vessel 2 in Fig. 4a). In a second animal, the hypoxic volume that resulted from the occlusion of a single penetrating arteriole was examined using post hoc histology (Fig. 4b). Immunostaining for Hypoxyprobe™ revealed a column of hypoxic tissue that extended 700 μm beneath the pial surface. This result is consistent with the idea that penetrating arterioles, which are largely devoid of collateral flow, act as bottlenecks in the supply of blood to the cortex.[19]

Occlusions placed within the redundant surface network[32] lead to a fundamentally different outcome. A typical example is shown in Fig. 5a, where a single vessel divides into downstream branches (first frame). Green laser light was focused in the lumen of the main arteriole (yellow circle). In the absence of circulating Rose Bengal, no occlusion was formed (second frame). However, immediately after injection of Rose Bengal, the irradiation initiates a clotting cascade against the vessel wall (third frame). Within 3 min of continuous irradiation, the vessel is completely occluded by a dense mass of clotted RBC with bright stagnant plasma directly upstream (fifth frame). As with the example given above, surface vessels that neighbor the target vessel are not affected by photothrombosis, which demonstrates the specificity of this technique.

Fig. 5 (continued) from the main arteriole, numbered 1–3, were examined for flow direction and RBC velocity before and after the occlusion. In the second frame, the *yellow circle* indicates the location of green laser irradiation. In the absence of circulating Rose Bengal, a 2-min period of irradiation did not form a clot. After the intravenous injection of Rose Bengal, irradiation initiated clot formation within 2 min and completely occluded the target vessel by 3 min. Surrounding vessels were unaffected by targeted photothrombosis. The occlusion appears as a dark mass of clotted RBCs with brightly fluorescent blood plasma directly upstream from the occlusion. Line scans show that the occlusion resulted in a rerouting of flow in downstream vessels. Notably, a flow reversal occurred in vessel 2, allowing vessels 1 and 3 to maintain near-baseline levels of RBC velocity. (**B**) Hypoxyprobe™ was injected (IV) 1 h after occlusion. The animal was transcardially perfused and the brain was prepared for histology (Fig. 4). The brain was cryosectioned along the coronal plane and immunostained for the Hypoxyprobe™. Hypoxia was restricted to a zone immediately beneath the site of occlusion. Lateral and medial directions are labeled by M and L, respectively. Below, a mixed brightfield fluorescent image, plus fluorescence, shows that the extravasated fluorescein–dextran overlaps with the hypoxic region. High-magnification images show specific Hypoxyprobe™ labeling of individual neurons. All panels adapted from Schaffer et al.[18]

One feature of disrupted flow in the surface arteriole network is the occurrence of robust flow reversals in downstream arterioles, which compensate for loss of flow caused by the occlusion. White arrows indicate baseline flow directions (Fig. 5a). Line scans show that vessel 2 reverses after the occlusion, which allows the flow in vessels 1 and 3 to be maintained at near-baseline levels. Post hoc Hypoxyprobe™ staining revealed a much smaller hypoxic region compared to penetrating arteriole occlusion (Fig. 5b), showing that re-routing of blood flow offers protection against the formation of a large damage zone. A small field of fluorescein–dextran extravasation is still visible in the vicinity of the occlusion and overlaps with the hypoxic region. High-magnification views of Hypoxyprobe-™stained tissue reveals the staining of individual neurons (Fig. 5b).

4 Plasma-Mediated Occlusion of Deep Vessels

This technique makes use of the dissociation of matter by high-fluence, 100–300-fs pulses of near infrared light. These pulses lead to ionization of materials such as blood plasma or vascular lumen within a femtoliter-sized focal volume of the incident laser pulse. The interaction of light towards the end of the pulse with the initial light-induced ionization results in a spatially limited release of mechanical energy in the form of a shock wave (Fig. 6a). This method allows occlusions to be induced below the surface of cortex, typically down to 500 μm

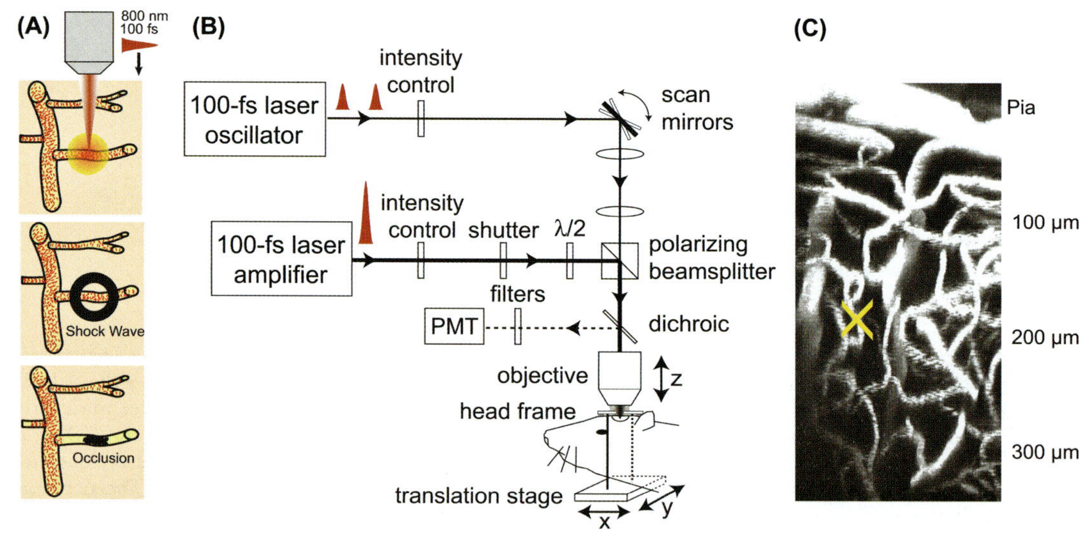

Fig. 6 Overview for forming intravascular occlusions in deep microvessels using amplified 100-fs laser pulses. (**A**) Laser pulses directed on a deep microvessel (*top image*) leads to a release of mechanical energy in the form of a shock wave (*middle image*). Repeated delivery of pulses, starting at low energies and gradually increasing, forms an occlusion within the target vessel (*bottom image*). (**B**) Schematic of a two-photon laser scanning microscope system modified for delivery of amplified 100-fs pulses. The dichroic is a 700 DCXRU (Chroma). (**C**) A projected X–Z TPLSM image stack showing deep microvessels (*yellow cross*) that are routinely targeted for selective occlusion using this technique. All panels adapted from Nishimura et al.[20]

but in principle down to 1,000 μm below the pia, without disruption of the neighboring tissue.

4.1 Equipment

Plasma-mediated occlusion is readily combined with TPLSM, as described above, to enable the targeting and real-time monitoring of blood vessels. The typical setup makes use of a standard two-photon microscope in which an amplified 100-fs light source is introduced in the beam path through a polarizing beam splitter (Fig. 6b). The imaging and photodisruption beams are focused to the same focal plane and the amplified beam is centered on the optical axis so that photodisruption occurs at the center of the imaged field. The energy per pulse of the photodisruption beam is varied with neutral density filters and the number of pulses is controlled by a mechanical shutter. There are multiple suppliers of "single-box" amplifiers, which consist of four devices in one housing, i.e., a 100-fs Ti:sapphire oscillator with associated pump to serve as the seed for the optical amplifier, together with a regenerative Ti:sapphire-based amplifier and its associated pump laser (e.g., Libra, Coherent Inc.).

The diameter of the amplified 100-fs beam is adjusted to overfill the back aperture of the objective using an appropriate telescope to form the smallest possible spot size. The energy per pulse of the amplified beam should initially be set at 0.03 μJ, which corresponds to a near-threshold fluence for damage of approximately 1 J cm^{-2} at the focus with a 40× immersion objective, consistent with the literature values for damage.

4.2 Procedure

The surgical and animal maintenance procedures are identical to those described for the photosensitizer-based occlusions. Similarly, histological examination of the tissue proceeds as described above. It is imperative to perform occlusions with concurrent imaging as a means to continuously monitor the state of the vasculature.

(6) Imaging and occlusion

a. Collect baseline vascular images at successive depths to construct a three-dimensional stack of images for maximal projection or volume-rendered views of the vasculature (Fig. 6c). Flow parameters, such as the speed of RBCs, are derived from line-scan data. Structural parameters, such as the cross-section of vessels, are derived from planar scans.
b. Select a subsurface microvessel to target for occlusion.
c. Aim the focal point of the amplified 100-fs beam near the center of the lumen and irradiate for approximately ten pulses at the minimum pulse energy. The first indication of photodisruption is visualized as the accumulation of fluorescein–dextran outside the vessel lumen; this corresponds to extravasation. If extravasation fails to occur, increase the energy by approximately a factor of 2 and irradiate again; extravasation typically occurs at 2–3 times the threshold energy and in the range of 0.1–0.5 μJ per pulse.
d. Once extravasation occurs, the irradiation is repeated with increased pulse energy, typically by less than a factor of 2, and increasing number of pulses, typically 10–100, until there is a complete cessation of RBC

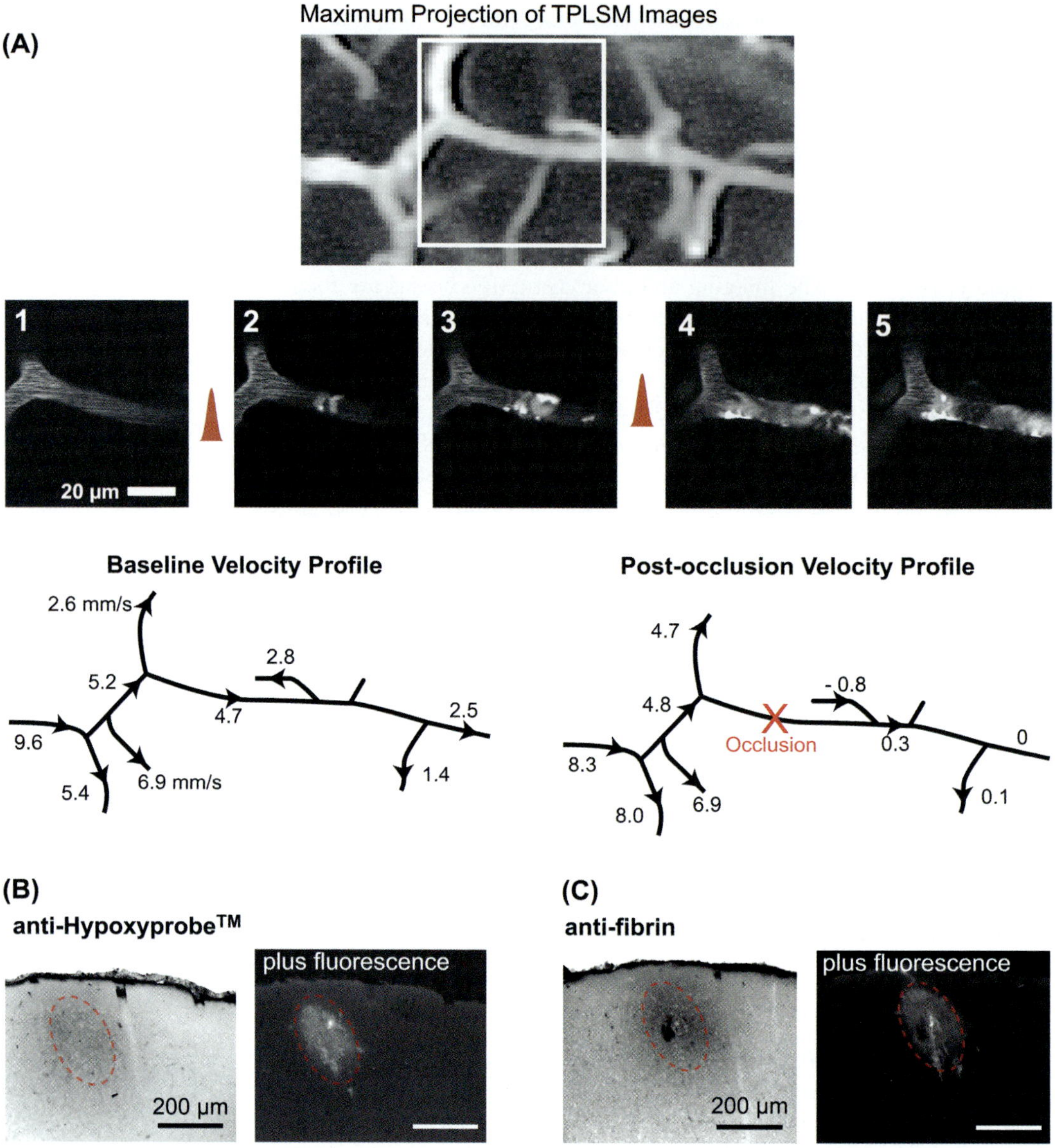

Fig. 7 Example of targeted deep microvessel occlusion using amplified 100-fs laser pulses. (**A**) Maximal projection of a TPLSM stack showing a tortuous network of subsurface microvessels. The second row shows planar images taken from a region of interest (*white rectangle*) depicting the time course for intravascular clot formation (frames 1–5). The *red pulses* indicate irradiation with multiple trains of 0.03-μJ pulses delivered at 1 kHz. After the first pulse, there was temporary cessation of RBC motion and swelling of the target vessel (frames 2 and 3). A second pulse led to limited extravasation and finally occlusion of the vessel lumen (frames 4 and 5). The third row shows vascular traces with baseline and postocclusion RBC velocity profiles, in millimeters per second, of the vascular network. *Arrowheads* denote the direction of RBC movement and the *red cross* marks the occluded microvessel. (**B**) Immunolabeling of tissue after deep microvessel occlusion. The *red ellipses* delineate a common area in the vicinity of the occlusion. Diffuse Hypoxyprobe™ immunostaining overlaps with affected region surrounding the occlusion, as marked by the territory of fluorescein–dextran extravasation. (**C**) Antifibrin immunostaining reveals localization of fibrin/fibrinogen clots, which appear as a mixed diffuse zone that surrounds a small dense region of intense staining. A mixed brightfield fluorescence image (plus fluorescence) shows that fibrin immunoreactivity is also confined to the immediate vicinity of the occlusion. All panels adapted from Nishimura et al.[20]

movement in the target vessel. Increasing the energy further may result in the rupture of the vessel and a hemorrhage.‡

e. A successful occlusion appears densely packed with immobile non-fluorescent RBCs, abutted by a brighter region of stagnant fluorescent plasma. The occlusion is further delimited by the uptake of fluorescein–dextran in the endothelial cells of the vessels.

4.3 Exemplary Data

We consider a microvessel, approximately 10 μm in diameter, that lie approximately 250 μm below the pia (Fig. 7a). Uniform flow is seen in the vessel prior to irradiation, as evidenced by streaks in the raster-scanned imaged (panel 1 in Fig. 7a). Labeling of the vessel wall with trapped fluorescein–dextran is observed, but flow is still present, subsequent to the first burst of pulses (panels 2 and 3 in Fig. 7a). A second burst of pulses leads to cessation of flow and a stagnant pool of fluorescent plasma behind the occlusion (panels 4 and 5 in Fig. 7a). Lastly, measurements of the speed and direction of RBC flow before and after formation of the occlusion indicate that the occlusion effectively halts flow in vessels that lie immediately downstream (Fig. 7a).

A histological examination of a plasma-mediated occlusion in a different preparation shows that the hypoxia marker Hypoxyprobe™ stains tissue in a diffuse manner near the site of the occlusion (Fig. 7b). This staining is highly overlapped with the extravasated fluorescein–dextran in this section. Clotting activity is observed in the targeted vessel through antifibrin immunoreactivity, a marker of fibrin/fibrinogen clots. The fibrin forms a dense core deposit with a more diffuse surround that is highly overlapped with the extravasated fluorescein–dextran in the section (Fig. 7c).

5 Synopsis

We have described two methods that, collectively, allow occlusions to be targeted to single vessels that lie anywhere within the upper 500–1000 μm of the cortex. These occlusions provide a means to address the consequences of vascular disease and microstrokes on a microscopic level. While past investigations have employed these relatively new methods to only acute studies, it is anticipated that the same occlusion and concurrent imaging techniques will be exploited for chronic investigations as well.

Acknowledgments: The authors are pleased to thank the Canadian Institutes of Health Research, the National Institutes of Health (NCRR, NIBIB and NINDS), the National Science Foundation, and the Veterans Medical Research Foundation for support.

‡ Preliminary findings suggest that irradiation of the wall of a penetrating arteriole with amplified 100-fs laser pulses, at a level deep to the surface, may be used to form an occlusion.[33]. This builds on work that used unamplified pulses with subthreshold energies to irritate the vasculature,[34] possibly by direct damage to RBCs through two-photon absorption of the incident 100-fs pulses.[35]

References

1. Iadecola C. Neurovascular regulation in the normal brain and in Alzheimer's disease. Nature Reviews of Neuroscience 2004;5:347–360
2. Scremin OU. Cerebral vascular system. In: Paxinos G, ed. The Rat Nervous System. 2 ed. San Diego: Academic Press; 1995
3. Miyake K, Takeo S, Kaijihara H. Sustained decrease in brain regional blood flow after microsphere embolism in rats. Stroke 1993;24:415–420
4. Sugi T, Schuier FJ, Hossmann KA, Zulch KJ. The effect of mild microembolic injury on the energy metabolism of the cat brain. Journal of Neurology 1980;223(4):285–292
5. Gerriets T, Li F, Silva MD, et al. The macrosphere model: evaluation of a new stroke model for permanent middle cerebral artery occlusion in rats. Journal of Neuroscience Methods 2003;122(2):201–211
6. Zhang Z, Zhang RL, Jiang Q, Raman SB, Cantwell L, Chopp M. A new rat model of thrombotic focal cerebral ischemia. Journal of Cerebral Blood Flow and Metabolism 1997;17(2):123–135
7. Fuxe K, Bjelke B, Andbjer B, Grahn H, Rimondini R, Agnati LF. Endothelin-1 induced lesions of the frontoparietal cortex of the rat. A possible model of focal cortical ischemia. Neuroreport 1997;8(11):2623–2629
8. Macrae IM, Robinson MJ, Graham DI, Reid JL, McCulloch J. Endothelin-1-induced reductions in cerebral blood flow: dose dependency, time course, and neuropathological consequences. Journal of Cerebral Blood Flow and Metabolism 1993;13(2):276–284
9. Sharkey J, Ritchie IM, Kelly PA. Perivascular microapplication of endothelin-1: a new model of focal cerebral ischaemia in the rat. Journal of Cerebral Blood Flow and Metabolism 1993;13(5):865–871
10. Zhang S, Boyd J, Delaney KR, Murphy TH. Rapid reversible changes in dendritic spine structure in vivo gated by the degree of ischemia. Journal of Neuroscience 2005;25:5333–5228
11. Wei L, Rovainen CM, Woolsey TA. Ministrokes in rat barrel cortex. Stroke 1995;26:1459–1462
12. Chen ST, Hsu CY, Hogan EL, Maricq H, Balentine JD. A model of focal ischemic stroke in the rat: Reproducible extensive cortical infarction. Stroke 1986;17(4): 738–743
13. Buchan AM, Xue D, Slivka A. A new model of temporary focal neocortical ischemia in the rat. Stroke 1992;23(2):273–279
14. Tamura A, Asano T, Sano K. Corrrelation between rCBF and histological changes following temporary middle cerebral artery occlusion. Stroke 1980;11:487–493
15. Katsman D, Zheng J, Spinelli K, Carmichael ST. Tissue microenvironments within functional cortical subdivisions adjacent to focal stroke. Journal of Cerebral Blood Flow and Metabolism 2003;23(9):997–1009
16. Markgraf CG, Kraydieh S, Prado R, Watson BD, Dietrich WD, Ginsberg MD. Comparative histopathologic consequences of photothrombotic occlusion of the distal middle cerebral artery in Sprague-Dawley and Wistar rats. Stroke 1993;24: 286–293
17. Helmchen F, Denk W. Deep tissue two-photon microscopy. Nature Methods 2005;2:932–940
18. Schaffer CB, Friedman B, Nishimura N, et al. Two-photon imaging of cortical surface microvessels reveals a robust redistribution in blood flow after vascular occlusion. Public Library of Science Biology 2006;4:258–270
19. Nishimura B, Schaffer CB, Friedman B, Lyden PD, Kleinfeld D. Penetrating arterioles are a bottleneck in the perfusion of neocortex. Proceedings of the National Academy of Sciences USA 2007;104:365–370

20. Nishimura N, Schaffer CB, Friedman B, Tsai PS, Lyden PD, Kleinfeld D. Targeted insult to individual subsurface cortical blood vessels using ultrashort laser pulses: Three models of stroke. Nature Methods 2006;3:99–108
21. Svoboda K, Denk W, Kleinfeld D, Tank DW. In vivo dendritic calcium dynamics in neocortical pyramidal neurons. Nature 1997;385:161–165
22. Kleinfeld D, Mitra PP, Helmchen F, Denk W. Fluctuations and stimulus-induced changes in blood flow observed in individual capillaries in layers 2 through 4 of rat neocortex. Proceedings of the National Academy of Sciences USA 1998;95:15741–15746
23. Kleinfeld D, Denk W. Two-photon imaging of neocortical microcirculation. In: Yuste R, Lanni F, Konnerth A, eds. Imaging Neurons: A Laboratory Manual. Cold Spring Harbor: Cold Spring Harbor Laboratory Press; 2000:23.1–23.15
24. Kleinfeld D, Denk W. Two-photon imaging of cortical microcirculation. In: Yuste R, Konnerth A, eds. Imaging in Neuroscience and Development. Cold Spring Harbor: Cold Spring Harbor Laboratory Press; 2005:701–705
25. Zhang S, Murphy TH. Imaging the impact of cortical microcirculation on synaptic structure and sensory-evoked hemodynamic responses in vivo. Public Library of Science Biology 2007;5:e119
26. Tsai PS, Nishimura N, Yoder EJ, Dolnick EM, White GA, Kleinfeld D. Principles, design, and construction of a two photon laser scanning microscope for in vitro and in vivo brain imaging. In: Frostig RD, ed. In Vivo Optical Imaging of Brain Function. Boca Raton: CRC Press; 2002:113–171
27. Majewska A, Yiu G, Yuste R. A custom-made two-photon microscope and deconvolution system. Pflugers Archives 2000;441:398–408
28. Short CE. Principles and Practice of Veterinary Anesthesia. Baltimore: Williams and Willisms; 1987
29. Paxinos G, Watson C. The Rat Brain in Stereotaxic Coordinates. San Diego: Academic Press; 1986
30. Kleinfeld D, Delaney KR. Distributed representation of vibrissa movement in the upper layers of somatosensory cortex revealed with voltage sensitive dyes. Journal of Comparative Neurology 1996;375:89–108
31. Dirnagl U, Villringer A, Einhaupl KM. In-vivo confocal scanning laser microscopy of the cerebral microcirculation. Journal of Microscopy 1992;165:147–157
32. Brozici M, van der Zwain A, Hillen B. Anatomy and functionality of leptomeningeal anastomoses: A review. Stroke 2003;34:2750–2762
33. Nishimura N, Rosidi NL, Mandell J, Iadecola C, Schaffer CB. Neighboring arterioles do not dilate to increase collateral flow after the occlusion of a cortical penetrating arteriole. In: Abstracts of the Society for Neuroscience Annual Meeting, Neuroscience 2007
34. Nimmerjahn A, Kirchhoff F, Helmchen F. Resting microglial cells are highly dynamic surveillants of brain parenchyma in vivo. Science 2005;308(5726): 1314–1318
35. Clay GO, Schaffer CB, Kleinfeld D. Large two-photon absorptivity of hemoglobin in the infrared range of 780–880 nm. Journal of Chemical Physics 2007;126:025102

Chapter 15

Middle Cerebral Artery Occlusion Suture Model for Mice

Feng Zhang, Steven H. Graham, and Jun Chen

Abstract This chapter introduces the procedure, assessment, application, limitations, and technical considerations of the popular focal brain ischemia model in mice, the middle cerebral artery occlusion (MCAO) suture model. Preparation of the suture, surgical procedures, examination of neurological function, and infarct volume assessment are described in detail.

Keywords: Mouse, Stroke, Brain ischemia, MCAO, Suture

1 Model Selection

Stroke is the leading cause of disability and the third leading cause of death in industrialized nations. Because there is no effective medication for neuroprotection against stroke, research in experimental cerebral ischemia is of great importance. There are three major animal models of cerebral ischemia: global ischemia, focal ischemia and hypoxia-ischemia. The middle cerebral artery occlusion (MCAO) suture model in rats is a popular tool for modeling focal cerebral ischemia. It was introduced by Zea Longa in 1989,[1] and later modified for mice by others.[2–4] The model effectively blocks blood flow into the MCA by insertion of a suture at the origin of the MCA. It is widely used for three main reasons: it mimics a thromboembolic artery occlusion in human stroke patients; it allows either permanent or transient ischemia; and, perhaps most importantly, it can be applied to transgenic or knock out mice.

2 Materials

2.1 Animal

Male adult C57BL/6 mice 8–10 weeks of age weighing 25 ± 3 g are most commonly used. Male mice of other strains such as CD1, 129J, and female mice are also used depending on the experiment.

From: *Animal Models of Acute Neurological Injuries.*
Edited by: J. Chen (eds.) New York, Totowa, NJ

2.2 Anesthesia

- Flowmeter (Kinet-o-meter)
- Isoflurane vaporizer (Fortec)
- Face and nose mask

2.3 Sutures (Ethilon)

- Black silk, 4-0
- Black silk, 10-0
- Black monofilament nylon, 8-0

2.4 Dentsulate Insulating Putty (Dentsply)

- One tube of base
- One tube of catalyst

2.5 Surgical Instruments

- Surgical microscope (Zeiss)
- Handheld cautery unit (Roboz)
- Shaver (Oster)
- Forceps (Roboz)
- Scissors (Roboz)
- Bipolar coagulator (Codman)
- Retractor (Roboz)
- Needle (Roboz)
- Needle holder (Roboz)

2.6 Measurement of Regional Cerebral Blood Flow

- PeriFlux System 5000 (Perimed AB, Stockholm, Sweden)
- Laser Doppler perfusion monitor (LDPM) probe (Perimed AB)

3 Procedures for MCAO

3.1 Preparation of Sutures

1. Cut 8-0 monofilament nylon sutures (straight) into 11 mm segments
2. Blunt one end of the suture segment with a cautery unit
3. Mix the Dentsulate insulating putty base and catalyst in a 1:1 ratio
4. Briefly insert the blunted end of the suture 4–5 mm deep into the Dentsulate mixture, then quickly withdraw
5. Remove extra silicone from the suture by scraping the suture against the forceps tip. The coat diameter should be around 180 μm
6. Allow the coat to dry before use, as shown in Fig. 1

3.2 Preparation of Mice

1. Give the mouse free access to water and food before the procedure
2. Place the mouse in a container, and flush with a gas mixture of O_2 and N_2O (30:70) with 5% isoflurane. Maintain anesthesia with 1.5% isoflurane through a face and nose mask under spontaneous respiration

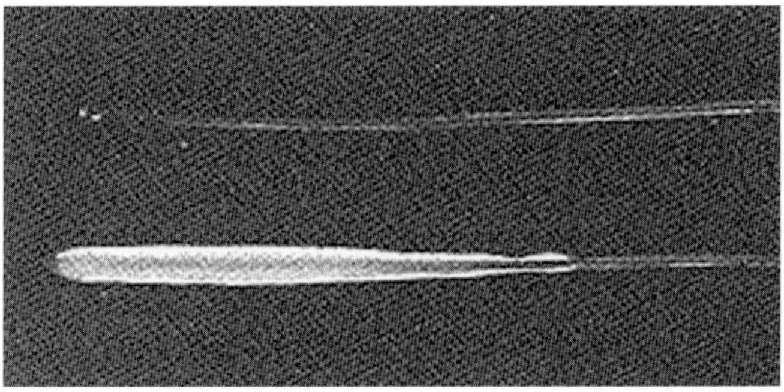

Fig. 1 Nylon sutures, 8-0. *Upper*, uncoated; *Lower*, coated with silicone

3. Shave the ventral aspect of the mouse's neck
4. Maintain the body temperature at 36–37°C throughout surgery

3.3 Step-by-Step Procedures on MCAO (Fig. 2)

3.3.1 Exposure of Carotid Arteries

1. Place the mouse in a supine position; spread and tape the limbs to the operating surface
2. Incise the midline of neck skin 1.5 cm, between the manubrium and the jaw
3. Divide the submandibular glands; dissect and retract digastric and sternocleidomastoid muscles to expose carotid artery triangle
4. Dissect and expose the left common carotid artery (CCA). Be careful with the trachea and vagus nerve
5. Isolate the occipital and superior thyroid arteries of the left external carotid artery (ECA), coagulate them with bipolar coagulator, and cut at their origins. Further, dissect ECA and expose the terminal lingual and maxillary arteries
6. Expose the proximal side of the left internal carotid artery (ICA)
7. Remove adipose tissue with the bifurcation of carotid arteries

3.3.2 Insertion and Introduction of Nylon Occluding Suture

1. Ligate permanently the distal end (lingual and maxillary arteries) of ECA with 10-0 silk suture
2. Ligate temporarily CCA with 8-0 silk sutures
3. Wrap 10-0 silk suture around the ECA, tie the suture loosely
4. Puncture or cut the distal end of the ECA or one of the terminal branches of ECA, and insert the coated end of the nylon occluding suture into ECA through the nick
5. Tighten the 10-0 silk suture around the ECA to avoid bleeding
6. Cut off the ECA proximal to the permanent ligations on it, and flip the ECA stump over caudally until it lines up with the CCA and ICA
7. Untie the temporary ligations on the CCA
8. Advance the occluding nylon suture 8.0 ± 0.5 mm deep into the ICA, measured from the arterial bifurcation. At this point, the intraluminal suture rides

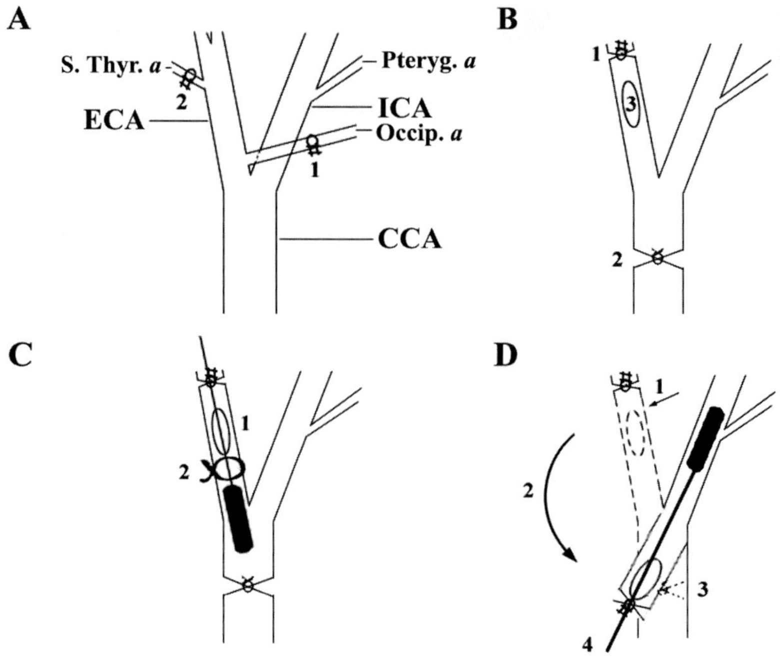

Fig. 2 Schematic diagrams of surgical procedures for middle cerebral artery occlusion (MCAO). (**A**) Anatomical structure of carotid arteries and the branches of ECA. 1. Coagulate and cut occipital artery; 2. Coagulate and cut superior thyroid artery. (**B**) Preparation of ECA. 1. Ligate the distal end of ECA permanently; 2. Ligate the CCA temporarily; 3. Nick ECA. (**C**). Inserting occluding suture into ECA. 1. Insert occluding suture into ECA through the nick; 2. Ligate ECA temporarily. (**D**) Positioning the occluding suture into the ICA. 1. Cut off ECA at distal end; 2. Flip over the ECA stump; 3. Untie the ligation around CCA; 4. Set the occluding suture into position (as seen in figure). ECA, external carotid artery; CCA, common carotid artery; ICA, internal carotid artery

over the origin of the MCA and blocks all sources of blood blow of the MCA, including the ICA and anterior carotid artery (ACA)

9. Further, tie the silk suture around the ECA to prevent the suture from sliding away
10. Close the incision, stop anesthesia. Let the mouse wake up and return it back to the cage

3.3.3 Restoration of Blood flow Into the MCA

If reperfusion is required, the occluding nylon suture can be withdrawn to restore the blood flow.

1. Re-anesthetize and position the mouse
2. Open the closed incision and expose the ECA stump
3. Holding the ECA stump with a pair of forceps, carefully pull the occluding nylon suture back from ICA to CCA
4. Close the incision and stop anesthesia. Let the mouse wake up and return it to cage

4 Measurement of Cerebral Blood Flow and Assessment of Ischemic Injury

The decease of regional cerebral blood flow (rCBF) in MCA territory after MCAO can be measured with laser Doppler flowmetry or ^{14}C-iodoantipyrine (14C-IAP) autoradiography. The results of focal cerebral ischemia and reperfusion are loss of function in neurons and glial cells and ultimately cell death. Therefore, ischemic injury to the brain can be measured by functional tests and histological examinations. Here we introduce two methods of rCBF measurement and three most commonly used tests.

4.1 Measurement of Regional Cerebral Blood Flow

4.1.1 Laser Doppler Flowmetry

rCBF is measured using a laser Doppler flowmeter (PeriFlux System 5000) with a LDPM probe. The probe is placed on the parietal bone over MCA region. rCBF is typically measured three times during the whole procedure. They are baseline rCBF before the occlusion of MCA, ischemic rCBF during the occlusion of MCA, and reperfusion rCBF after the withdrawal of the suture, as shown in Fig. 3.

4.1.2 ^{14}C-IAP Autoradiography

1. Prepare 15 µCi ^{14}C-IAP in 150 µL 0.9% sterile saline
2. Inject the ^{14}C-IAP intraperitoneally 30 min after the occlusion of MCA
3. Perform a heart stick at the left ventricle of heart exactly 2 min after the injection and withdraw as much blood as possible
4. Decapitate the mouse 2 min after the injection and quickly remove the brain, immerse the brain into cooled isopentane, and then store the brain at −80° in a freezer
5. Cut the brain into 20-µm thick sections and collect a series of sections throughout the length of the brain; let the sections dry
6. Expose the brain sections to X-ray films along with calibrated ^{14}C standards for 5–7 days. Measure optical densities of the interested areas using an MCID (Imaging Research Inc., St. Catharines, Ontario, Canada) image processing system (Fig. 4)

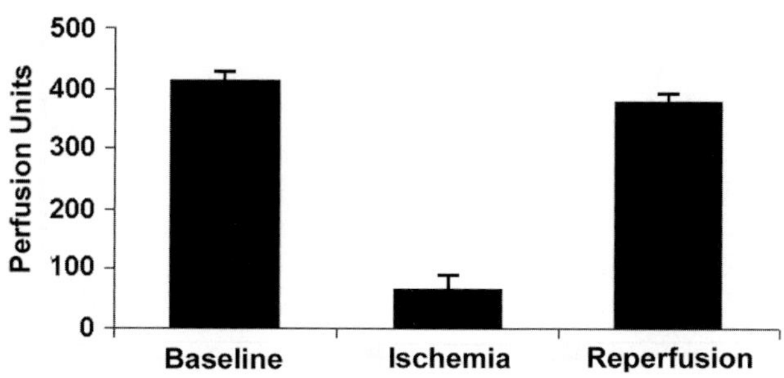

Fig. 3 Changes in regional cerebral blood flow (rCBF) before, during, and after MCAO evaluated in mice by laser Doppler flowmetry. MCAO in mice were induced with 8-0 coated suture for 60 min

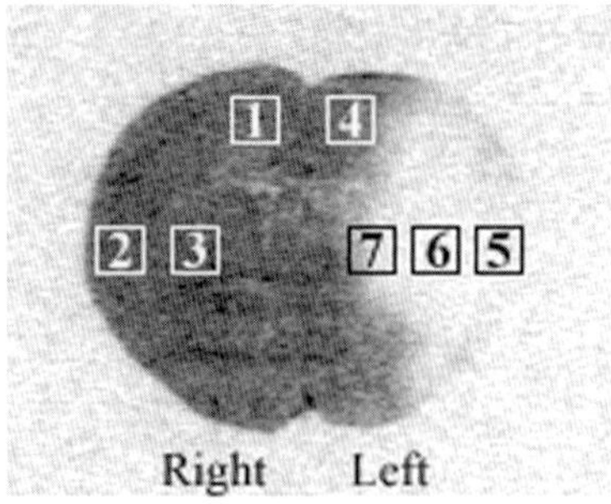

Area	CBF
1	48.40±5.97
2	65.50±7.40
3	53.10±5.82
4	31.04±4.25
5	17.79±4.57
6	20.56±4.50
7	35.04±5.68

Fig. 4 Representative ^{14}C-IAP autoradiograms of regional cerebral blood flow (rCBF) at the striatal level. MCAO in mouse was induced with 8-0 coated suture. The unit of rCBF is mL/100 g min^{-1}

7. Measure the amount of ^{14}C-IAP in 20 μL of blood by liquid scintillation counting
8. CBF is calculated from the blood and brain radio activities using the method introduced by Maeda, et al.[5]

4.2 Assessment of Ischemic Injury

4.2.1 Neurological Function Test

Neurological examination is a functional test and can be performed as soon as the mouse recovers from anesthesia. Neurological functions are scored on a five-point scale.[1]

- No neurological deficit = 0
- Failure to extend right forelimb fully when lifted by tail = 1
- Circling to the right = 2
- Falling to the right = 3
- No spontaneous walk or in a comatose state = 4

4.2.2 Measurement of Infarct Area Using TTC Staining

2,3,5-Triphenyltetrazolium chloride (TTC) is a substrate for mitochondrial enzymes. It stains viable tissues red, leaving dead tissues unstained as white (Fig. 5). A solution of 2% TTC in normal saline is used. The procedure for TTC staining is as follows:

1. Warm the TTC solution to 37°C
2. Euthanize the mouse 24–48 h after MCAO, and remove the brain
3. Fit the brain into microtome
4. Cut brain into 7–8 sections, 1 mm thick
5. Immerse sections into the prewarmed TTC solution, staining for 20 min at 37°C in dark
6. Wash the sections with saline
7. Fix the sections with 4% paraformaldehyde or formalin for 1–2 h

After staining and fixation, the brain sections can be photographed with a digital camera (Fig. 5). The infarct areas can then be analyzed using imaging software, such as MCID.

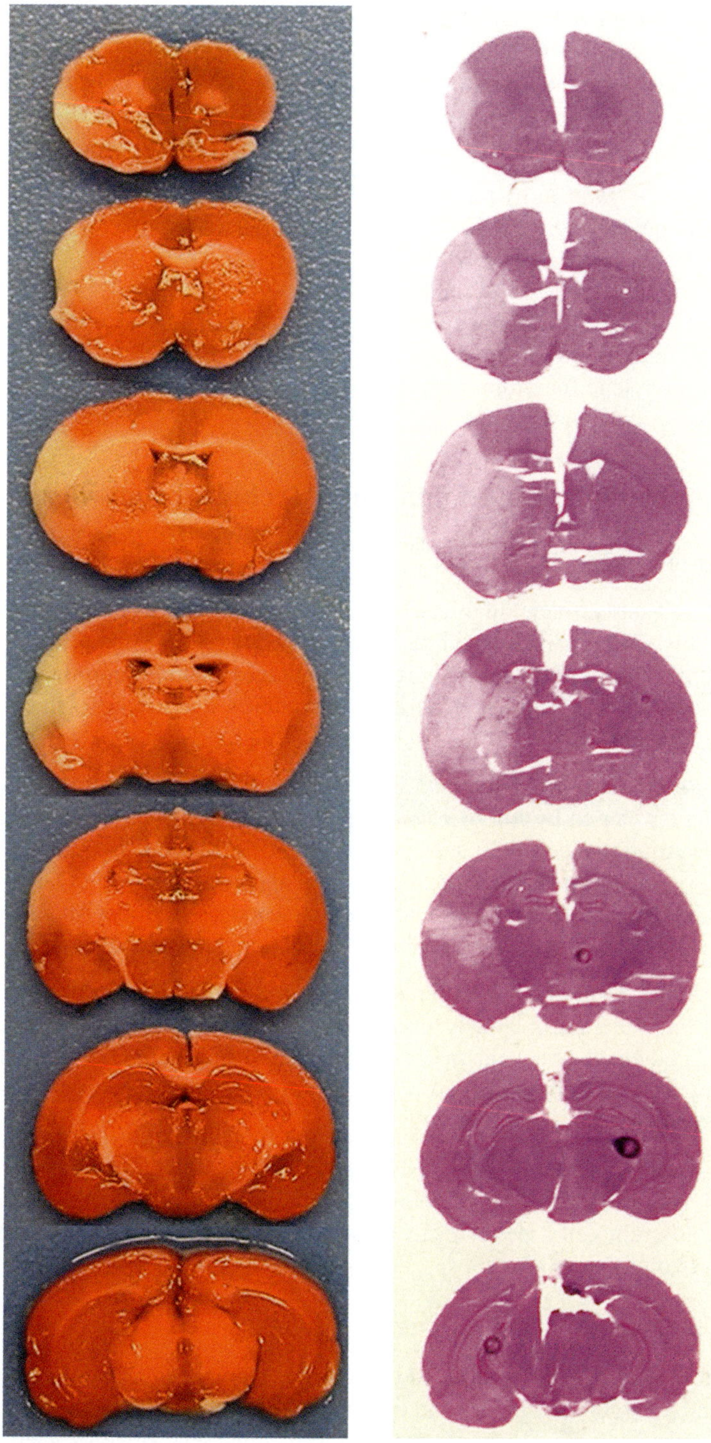

TTC **H&E**

Fig. 5 Triphenyltetrazolium chloride (TTC) and hematoxylin and eosin (H–E) stained brain sections in mice after 60-min MCAO and 48-h reperfusion

4.2.3 Histological Examination Using H–E Staining

Hematoxylin and eosin (H–E) are useful dyes for the histological study of the brain as well as for the assessment of infarct areas after brain ischemia (Fig. 5). Both fresh-frozen slides and formalin-fixed paraffin-embedded slides can be stained with H–E. The procedures for H–E staining are as follows:

1. Remove the paraffin by rehydrating the slides in water
2. Soak in hematoxylin for 10 min
3. Wash under tap water for 20 min
4. Soak in eosin for 15 s to 1 min
5. Dehydrate in 95% alcohol for 2 min, followed by 100% alcohol for 2 min. Repeat until excess eosin is removed
6. Clean in xylene for 2 min, repeat two times
7. Mount on coverslip

5 Applications, Limitations, and Technical Considerations

5.1 Applications

The MCAO suture model in mice is widely used to mimic focal brain ischemia in humans. The main advantage of this model is its compatibility with transgenic or knock out mice. Further advantages include reproducibility in ischemic injury, uncomplicated surgical procedures, and low costs in purchasing and housing.

5.1.1 Permanent and Transient Models of Ischemia

Brain injures caused by MCAO are composed of ischemic components such as adenosine triphosphate (ATP) depletion and necrosis. MCAO also induces reperfusional injuries such as free-radical injury and apoptosis. The MCAO model can produce either permanent or transient types of ischemia. Transient MCAO is also widely used as an ischemic preconditioning tool to induce ischemic tolerance.

5.1.2 Short-Term and Long-Term Ischemia

Short-term ischemia is usually used in pathological and pathophysiological studies, cell death mechanisms, neuroprotective mechanisms, and pharmacological studies. Long-term ischemia is very useful for neurogenesis and angiogenesis studies.

5.1.3 Systemic Reaction After Ischemia

This in vivo model of brain ischemia also allows studies on the interaction among different cell types with in brain, such as neurons, astrocytes, microglias, and oligodendrocytes.

5.2 Limitations

There are two major limitations to this model. The ischemic injury in mice is induced by surgery in healthy and pathogen-free animals. This may be different from the ischemic injury observed in human patients, since stroke patients usually have systemic disorders that may greatly affect the processes and outcome of ischemia. Therefore, the findings in mice cannot be completely applied to humans. The other limitation of this model is the difficulty in reproducing consistent infarct volumes.

5.3 Technical Considerations

In addition to the inconsistencies in infarct volume, bleeding and animal death can also be observed. The following are a few suggestions to overcome these limitations.

5.3.1 Suture Aspect

5.3.1.1 Size of the Suture
Unlike 5-0 and 6-0 monofilament nylon sutures, 8-0 sutures are very soft and flexible. The rates of subarachnoid hemorrhage and death following ischemia and reperfusion are extremely low with 8-0 sutures.

5.3.1.2 Length of the Silicone Coat
It is recommended to use a 4-mm coat for MCA occlusion. If the coat is too short, it will be very hard to advance it into the origin of the MCA; if it is too long, PCA or ACA occlusion may occur.

5.3.1.3 Diameter of the Silicone Coat
This is the key to successful MCAO. The optimal diameter for the silicone coat is around 180 µm. However, this can be adjusted in accordance with the body weight of the mouse.[4] If the coat is too thin, an infarct may appear only in the striatum or perhaps not at all. If the coat is too thick, it is difficult to advance the suture into the origin of MCA.

5.3.1.4 Depth of the Suture Insertion
The depth of suture insertion is determined by measuring the distance from the distal tip of the suture to the bifurcation of carotid arteries. If the suture is coated with 4–5 mm silicone coating, the tip should be inserted 8 ± 0.5 mm.

5.3.1.5 Preventing Loss of the Silicone Coat
To prevent the silicone coat from falling off the suture, blunt and enlarge the distal tip of the occluding suture before coating. This prevents the silicone coat from falling off when the suture is withdrawn to restore the blood flow.

5.3.2 Surgical Aspect

5.3.2.1 Body Temperature
Body temperature has obvious effects on the infarct volume following brain ischemia. Hyperthermia increases and hypothermia decreases the infarct volume. It is essential to maintain the body temperature during surgery and in the early stages of ischemia before the mouse fully recovers from anesthesia.

5.3.2.2 Vital Parameters
To prevent adverse effects on the mouse's vital parameters and a possible increase in infarct volume, efforts should be made to avoid stimulation or overstretching the trachea and vagus nerve during the surgery.

5.3.2.3 Pterygopalatine Artery
In the original MCAO model of rats, the pterygopalatine artery was ligated. However, the ligation of this artery is not necessary in mice. There exists the possibility that, without the ligation, the occluding suture sometime is advanced into the pterygopalatine artery. If this happens, the suture cannot be inserted more than 5 mm; and this disposition of suture can be easily corrected by redirecting the suture into ICA.

References

1. Longa EZ, Weinstein PR, Carlson S, Cummins R. Reversible middle cerebral artery occlusion without craniectomy in rats. Stroke 1989; 20:84–91
2. Yang G, Chan PH, Chen J, et al. Human copper-zinc superoxide dismutase transgenic mice are highly resistant to reperfusion injury after focal cerebral ischemia. Stroke 1994; 25:165–70
3. Connolly ES Jr., Winfree CJ, Stern DM, Solomon RA, Pinsky DJ. Procedural and strain-related variables significantly affect outcome in a murine model of focal cerebral ischemia. Neurosurgery 1996; 38:523–31; discussion 532
4. Tureyen K, Vemuganti R, Sailor KA, Dempsey RJ. Ideal suture diameter is critical for consistent middle cerebral artery occlusion in mice. Neurosurgery 2005; 56:196–200; discussion 196–200
5. Maeda K, Mies G, Olah L, Hossmann KA. Quantitative measurement of local cerebral blood flow in the anesthetized mouse using intraperitoneal [14C]iodoantipyrine injection and final arterial heart blood sampling. J Cereb Blood Flow Metab 2000; 20:10–4

Chapter 16

Intraluminal Middle Cerebral Artery Occlusion Model in Mice

Masaya Oda, Taku Sugawara, Kohei Kokubun, Kazuo Mizoi, and Hiroyuki Kinouchi

Abstract Intraluminal middle cerebral artery occlusion (MCAO) model in rodents has been widely used for cerebral ischemia studies. The procedure is relatively simple and noninvasive compared to the other focal cerebral ischemia models, and the ischemic insult is highly reproducible. The use of genetically modified mice has been attracting attention because it allows dissection of molecular and cellular mechanisms of various pathological conditions in the brain, and therefore the mouse intraluminal MCAO model has become more prevalent. We describe here our method of mouse intraluminal MCAO and characterize its effects on infarct areas, local cerebral blood flow, extracellular amino acids concentration, and physiological parameters.

Keywords: C57BL/6 mouse, Focal cerebral ischemia, Intraluminal occlusion, Cerebral infarct, Amino acids, Microdialysis, High-performance liquid chromatography

1 Model Selection

The rat intraluminal MCAO model was established more than two decades ago and has since been utilized to investigate the mechanisms of ischemia-induced brain damage. More recently, genetic modifications using transgenic techniques in mice have been widely used to elucidate molecular mechanisms of various pathological conditions and, therefore, mouse ischemia models have become prevalent. Compared to the other focal ischemia models, such as the direct middle cerebral artery (MCA) coagulation model,[1,2] the intraluminal MCAO model reported here is technically easier and less invasive because of the simple procedures without craniectomy.[3,4] This ischemia model has been found to be useful in various investigations on cerebral ischemia.

From: *Animal Models of Acute Neurological Injuries.*
Edited by: J. Chen (eds.) New York, Totowa, NJ

2 Materials

2.1 Animal

Our choice of the animal for the MCAO model is male C57BL/6 mice obtained from Charles River, Inc. (Japan). The C57BL/6 strain from this breeder is widely used as a background strain for genetic alternations, and shows little variation in intracranial collaterals and lacks the patent posterior communicating artery (PComA) on both sides.[5] For successful induction of the ischemia, the body weight of the animals has been shown to be important. Mice in the 25–30 g body weight range have been used in many ischemia studies and we also have found them to give more consistent outcomes than the ones outside this range. The mice are allowed to have free access to water and food before and after surgery.

2.2 Anesthesia

Anesthesia is induced with 2% halothane in a closed chamber and maintained with 1.0–1.5% halothane in 30% O_2 and 70% N_2O using a face mask. The rectal temperature is monitored and maintained at 37 ± 0.5°C with a thermal blanket throughout the surgical procedure.

2.3 Equipment

2.3.1 Anesthesia

Facemask
Halothane vaporizer
Flowmeter for O_2 and N_2O

2.3.2 Surgery

Surgical instruments (microscissors, forceps, knives, bipolar coagulator, etc.)
Microvascular clip
Stereotaxic device
Dissecting microscope
High-speed drill for implantation of vertical microdialysis probes
Temperature control unit (thermometer, temperature probe, and thermal blanket)
Silk suture, 6-0
Nylon monofilament suture, 5-0, with a round tip, blunted by heat (Dermaron 1756–41, Devis and Geck, Manati, Puetro Rico)

2.3.3 Recording of Regional Cerebral Blood Flow (rCBF)

Laser Doppler flowmeter (LDF, Laserflo BPM2, Vasamedics, St Paul, MN, USA)

2.3.4 In Vivo Microdialysis

Microdyalisis system (Eicom Corp., Tokyo, Japan)
Vertical microdialysis probes (0.22 mm outer diameter, 2 mm membrane length: Eicom Corp., Tokyo, Japan)
High-performance liquid chromatography apparatus (Eicom Corp., Tokyo, Japan)

2.3.5 Measurement of Infarct Size

Brain slicer (Mouse brain matrix, Harvard Apparatus, MA, USA)
Image scanner
NIH image analysis software

3 Procedures

Focal cerebral ischemia is induced by the occlusion of the MCA using the intraluminal filament technique.[6,7] (Figs. 1 and 2). Anesthesia is induced with 2% halothane in a closed chamber and maintained with 1.0–1.5% halothane in 30% O_2 and 70% N_2O delivered via a facemask. The animal is placed on the operating table in a supine position and the neck is extended with the use of a small pillow. The rectal temperature is monitored and maintained at 37 ± 0.5°C with a thermal blanket throughout the surgical procedure:

1. Through a ventral midline incision, the left external carotid artery (ECA), superior thyroid artery (STA), and occipital artery (OA) are exposed and the STA and OA are coagulated.
2. The left common carotid artery (CCA) is carefully freed from the adjacent vagus nerve, and the internal carotid artery (ICA) and pterygopalatine artery (PPA) are identified under the OA.

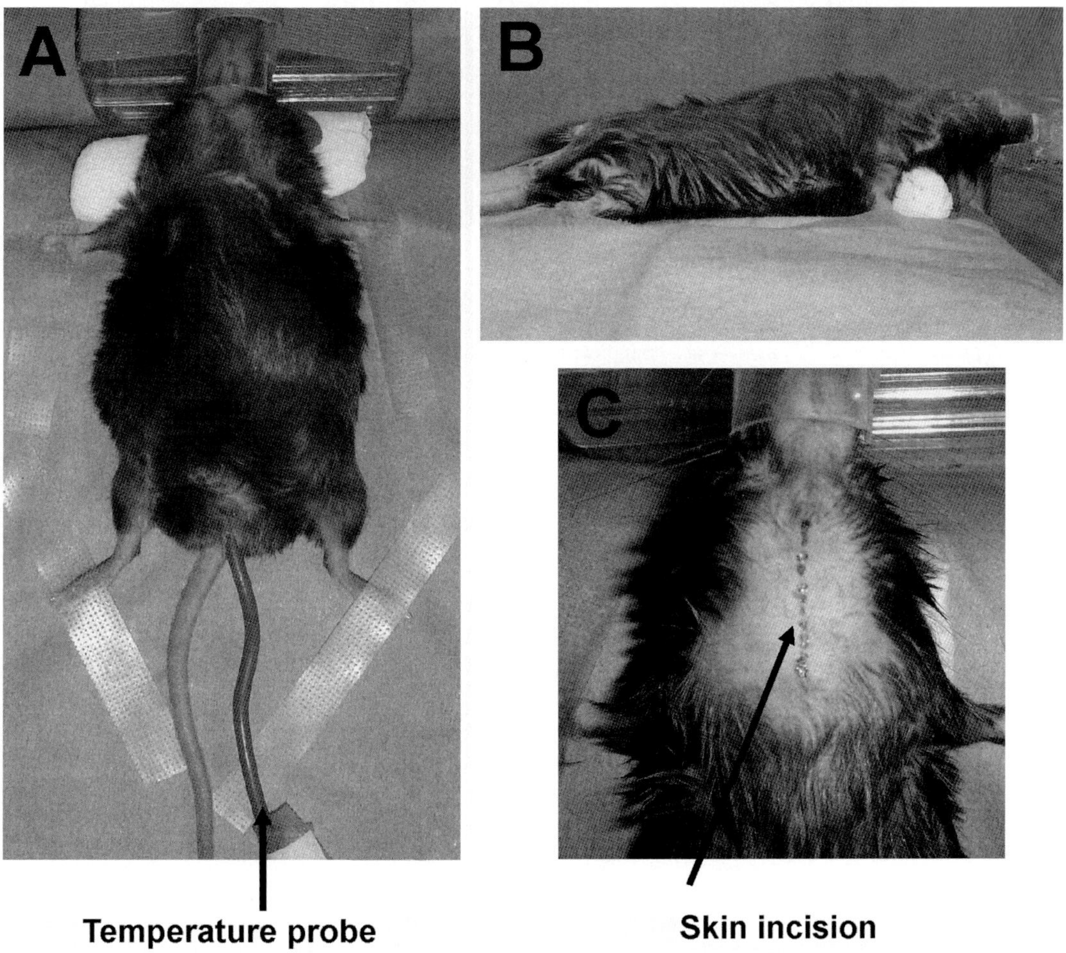

Fig. 1 Surgical setup. Mice are anesthetized with 2% halothane in a closed chamber and placed on an operating table in a supine position (**A**). The neck is extended with the use of a small pillow (**B**). Rectal temperature is monitored and maintained at 37 ± 0.5°C with a thermal blanket throughout the surgical procedure. Midline skin incision is to be performed as in (**C**)

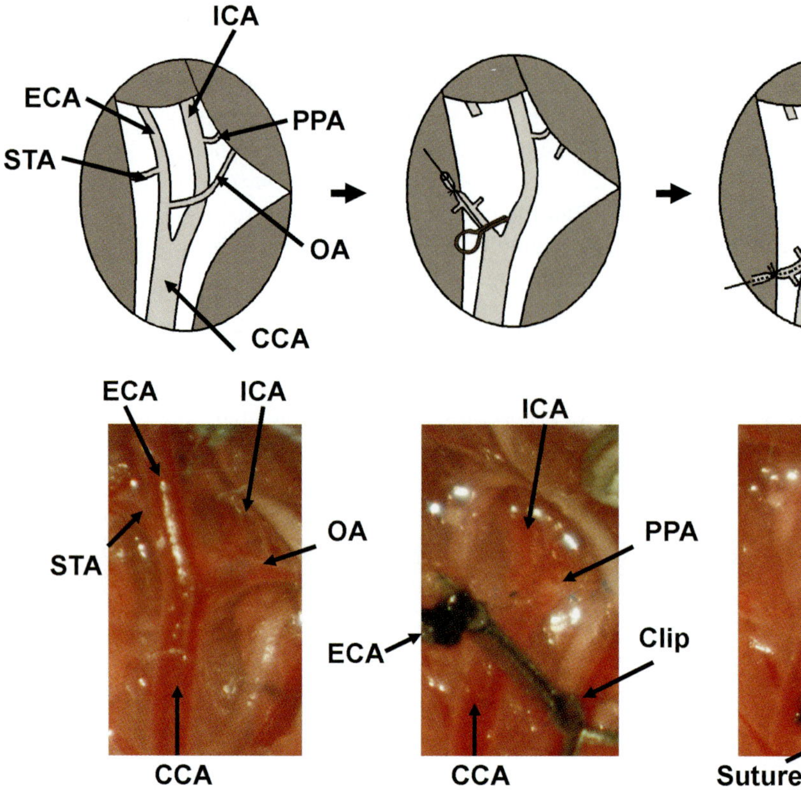

Fig. 2 Diagram of operative procedure (*above*) and corresponding views under a microscope (*below*). Through a ventral midline incision, the left external carotid artery (ECA), superior thyroid artery (STA), and occipital artery (OA) are exposed and the STA and OA are coagulated. The left common and internal carotid artery (CCA and ICA), and pterygopalatine artery (PPA) are identified. A microvascular clip is temporarily applied to the proximal ECA and a 5-0 monofilament suture with a round tip is inserted in ECA and advanced into the ICA

3. The distal ECA is ligated with a 6-0 silk suture just proximal to the lingual and maxillary artery bifurcation.
4. A microvascular clip is temporarily applied to the proximal ECA.
5. A 6-0 silk suture is loosely tied around the ECA between the clip and the distal ligature of the ECA.
6. An arteriotomy is carried out in the ECA between the distal and proximal ligatures and a 5-0 monofilament suture with a round tip is advanced into the ECA. After confirming that the suture tip is in the proximal ECA and adjacent to the clip, the proximal ligature is tightened enough to allow the suture to advance while preventing leakage of arterial blood from the ECA stump.
7. The clip is removed and the distal ECA is transected. The proximal ECA stump is retracted caudally.
8. A 5-0 nylon monofilament suture with a round tip is inserted into ICA 11 ± 0.5 mm from the bifurcation of the CCA until the LDF signal abruptly drops.

9. After 30 or 60 min of MCAO, the nylon suture is removed and the blood flow restoration is confirmed by the LDF signal. Mice in which the LDF signal during ischemia exceeds 10% of the pre-ischemic signal are excluded from this study.
10. The wound is sutured, and the animal is returned to its cage and allowed free access to water and food.

4 Outcome Evaluation

4.1 Infarct Size

Twenty-four hours after MCAO, the mice were deeply anesthetized and decapitated. The brain was removed and sectioned coronally into four 2-mm slices with a brain slicer. The slices were placed in 2% 2,3,5-triphenyltetrazolium chloride (TTC) solution at 37°C for 10 min, and fixed in a 10% buffered formalin solution. The sections were then placed on the image scanner and color images were obtained. The infarction area, appearing in white, was measured using the NIH image analysis software, and infarction volume was calculated by summing the infarct volume of sequential 2-mm-thick sections. Infarct areas were seen in the striatum and the dorsolateral cortex in the 60-min MCAO mice, whereas the 30-min MCAO mice demonstrated them only in the striatum, with a small portion of the cortex (Fig. 3a). The size of the infarct area in each slice is shown in Fig. 3b. The infarct volume in the 60-min MCAO mice was 54.5% larger than that of the 30-min animals (Fig. 3c).

4.2 Regional Cerebral Blood Flow Measurement

rCBF was measured by LDF. The flow probe (0.5 mm diameter) was placed on the cranial bone above the MCA territory (0.5 mm posterior and 4 mm lateral from the bregma) and away from large surface vessels. Steady-state baseline values were recorded before MCAO, and blood flow data were then expressed as percentages of the preocclusion baseline. rCBF immediately dropped after a suture insertion by approximately 90% of preocclusion value, and no significant difference was found in the percentage change during ischemia. After removal of the suture, rCBF returned to the preocclusion level (Fig. 4).

4.3 Profiles of Extracellular Amino Acid Concentrations in MCAO

Twenty-four hours before MCAO, vertical microdialysis probes were stereotaxically implanted in the left striatum of mice under anesthesia. The probe was coordinately implanted at 0.6 mm anterior and 2.0 mm lateral to the bregma and 2 mm ventral from the brain surface, according to the atlas of Franclin and Paxinos (1997). The external portion of the probes was fixed to the skull with dental cement. Throughout ischemia, dialysis probes were perfused with Ringer's solution (147 mmol L^{-1} NaCl, 2.3 mmol L^{-1} $CaCl_2$, 4.0 mmol L^{-1} KCl; pH 7.0) at a rate of 2 µL min^{-1}. The microdialysates (20 µL) were collected every 10 min. After a stabilization period of 60 min, the samples were collected from 60 min before MCAO to 90/120 min after MCAO in the 30/60-min MCAO models, respectively. Amino acid concentrations in the samples were measured by high-performance liquid chromatography.[8,9] The amino acid concentrations in all groups increased significantly at 20 min after

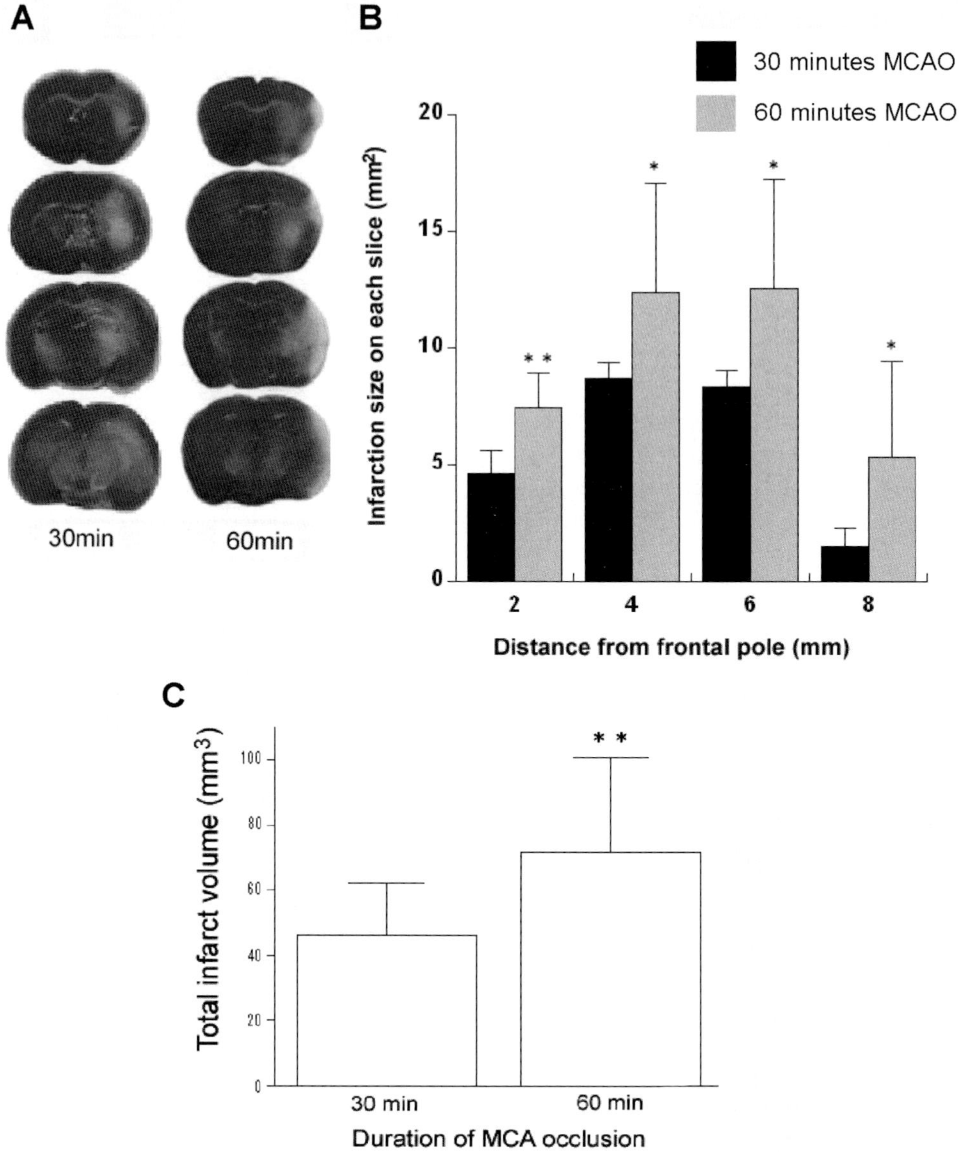

Fig. 3 Evaluation of brain injury after MCAO in the 30-min MCAO mice and the 60-min MCAO mice. A 2-mm slice of brain sample was stained with triphenyltetrazolium chloride for visualization of the infarct region (*white*). Representative photographs of each mouse group ($n = 10$) are shown (**A**). Infarct area (in mm^2) in each slice was measured (**B**). The infarct volume (in mm^3) in the 60-min MCAO mice was 54.5% larger than that of the 30-min MCAO mice (**C**). Values are mean ± SD. $^*P < 0.05$, $^{**}P < 0.01$

MCAO and reached their peak at 10 min after reperfusion. The peak concentrations of all amino acids were higher in the 60-min MCAO mice compared with in the 30-min MCAO mice (Fig. 5).

4.4 Physiological Parameters

The femoral arteries of the animals in this experiment were cannulated using a polyethylene catheter (PE-10, Clay Adams, NJ, USA) for blood pressure

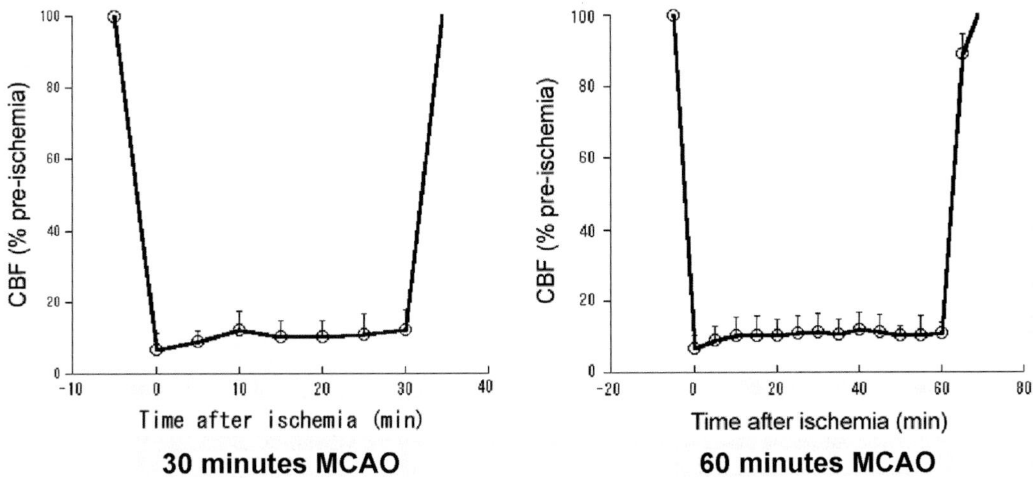

Fig. 4 rCBF measurement by LDF during MCAO in the 30-min MCAO mice and the 60-min MCAO mice. The blood flow declined reproducibly after thread insertion and returned to the control levels within 5 min after thread retraction. There was no significant change in rCBF during ischemia in both models. Values are mean ± SD (n = 10 in each group)

monitoring and serial blood gas sampling. Mean arterial blood pressure was monitored using a transducer amplifier (Nihon Koden, Tokyo, Japan) and pO_2, pCO_2, and pH are measured with a pH and blood gas analyzer. Rectal temperatures were measured and maintained at 37 ± 0.5°C by using temperature control unit during all surgical procedures and during recovery from anesthesia. These physiological variables were measured before and 60 min after MCAO in the 60-min MCAO model mice in this study and showed no significant differences (Table 1).

5 Advantages and Complications

5.1 Advantages

This mouse intraluminal MCAO model is less invasive than direct MCAO methods requiring craniectomy or electocoagulation of the brain surface. Also, MCAO of various durations can be obtained by restoring MCA blood flow by suture removal at appropriate time points without substantial technical difficulties. In our transient MCAO models, both 30- and 60-min MCAO induce consistent cerebral infarcts with a significant difference between the groups.

5.2 Complications

Hypothalamic damage following insertion of a monofilament suture into the ICA to occlude the MCA has been documented.[10,11] The hypothalamus is fed by the ACA, ICA, and the PComA. Because the suture head potentially occludes the ACA and ICA, rather than the MCA alone, the risk of hypothalamic injury should be considered. Also, subarachnoid hemorrhage is also a common complication after insertion of the monofilament suture into the ICA.

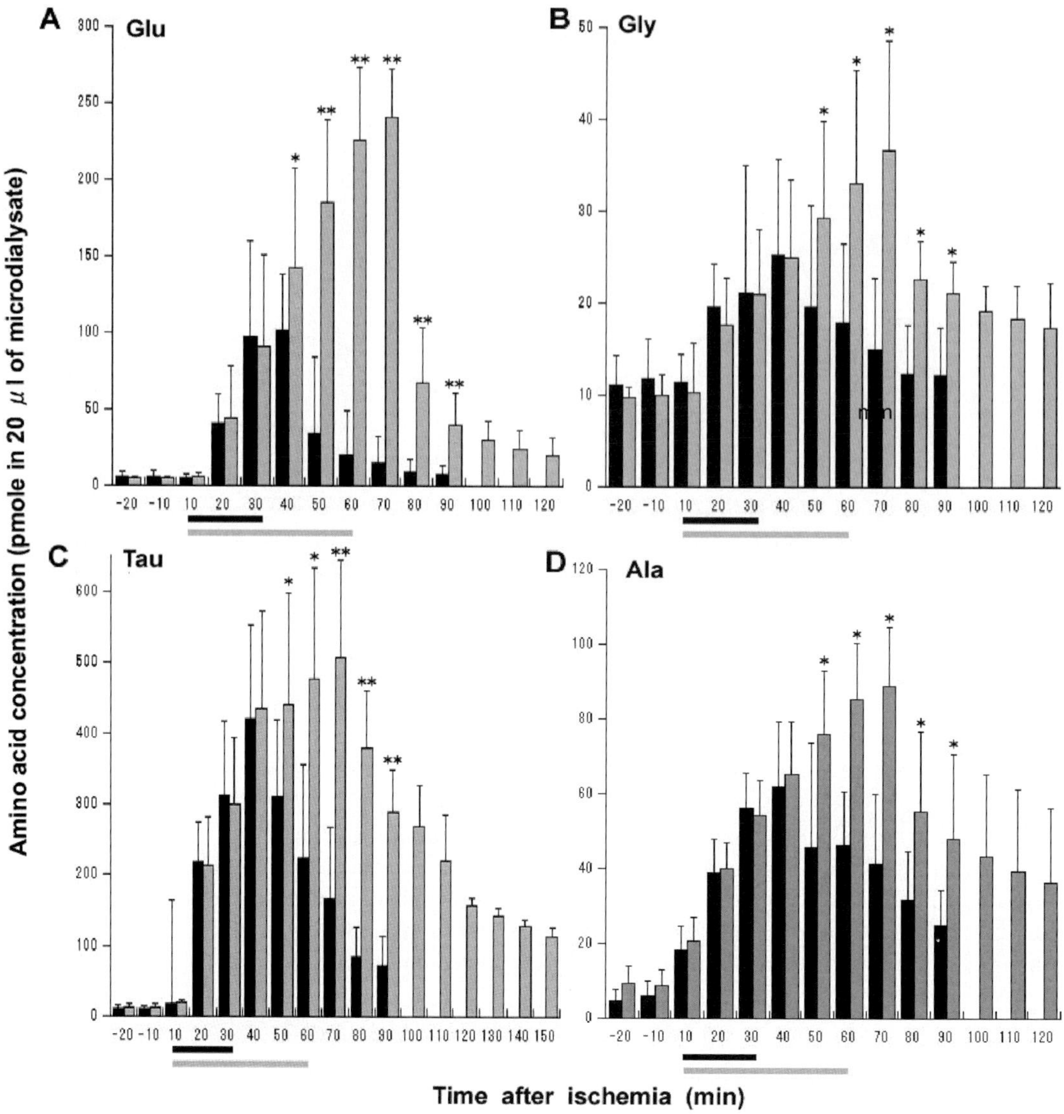

Fig. 5 Extracellular concentrations of amino acid neurotransmitters in focal ischemia. The horizontal bar stands for the MCAO period (30 and 60 min). Values are mean ± SD. $^{*}P < 0.05$, $^{**}P < 0.01$. The amino acid concentrations in all groups increased significantly at 20 min after MCAO and reached their peak at 10 min after reperfusion. The peak concentrations of all amino acids were higher in the 60-min MCAO mice compared with in the 30-min MCAO mice. Glu: glutamate, Gly: glycine, Tau: taurine, Ala: alanine

Table 1 Physiological parameters in the 60-min MCAO

	Pre-ischemia	Post-ischemia
MABP (mmHg)	102.1 ± 1.7	104.3 ± 3.5
pH	7.27 ± 0.05	7.28 ± 0.01
pO_2 (mmHg)	163.9 ± 12.7	169.2 ± 18.8
pCO_2 (mmHg)	42.6 ± 2.6	45.2 ± 2.3
Rectal temperature (°C)	37.1 ± 0.09	37.0 ± 0.06

Values are mean ± SD; $n = 10$. *MABP* mean arterial blood pressure

Approximately 10% of the mice die during or after ischemia. These complications occur often after excessive insertion of the monofilament suture into the ICA. Therefore, it is important to consider the suture being inserted into ICA once the LDF signal adruptly drops, but not until the resistance is felt.

References

1. Tamura A, Graham DI, McCulloch J, Teasdale GM. Focal cerebral ischaemia in the rat: 1. Description of technique and early neuropathological consequences following middle cerebral artery occlusion. J Cereb Blood Flow Metab 1981;1(1):53–60
2. Yamamoto M, Tamura A, Kirino T, Shimizu M, Sano K. Behavioral changes after focal cerebral ischemia by left middle cerebral artery occlusion in rats. Brain Res 1988;452(1–2):323–8
3. Kamii H, Kinouchi H, Sharp FR, Koistinaho J, Epstein CJ, Chan PH. Prolonged expression of hsp70 mRNA following transient focal cerebral ischemia in transgenic mice overexpressing CuZn-superoxide dismutase. J Cereb Blood Flow Metab 1994;14:478–86
4. Kinouchi H, Epstein CJ, Mizui T, Carlson E, Chen SF, Chan PH. Attenuation of focal cerebral ischemic injury in transgenic mice overexpressing CuZn superoxide dismutase. Proc Natl Acad Sci USA 1991;88:11158–62
5. Maeda K, Hata R, Hossmann KA. Differences in the cerebrovascular anatomy of C57black/6 and SV129 mice. Neuroreport 1998;9(7):1317–9
6. Longa EZ, Weinstein PR, Carlson S, Cummins R. Reversible middle cerebral artery occlusion without craniectomy in rats. Stroke 1989;20(1):84–91
7. Belayev L, Alonso OF, Busto R, Zhao W, Ginsberg MD. Middle cerebral artery occlusion in the rat by intraluminal suture. Neurological and pathological evaluation of an improved model. Stroke 1996;27(9):1616–22
8. Oda M, Kure S, Sugawara T, Yamaguchi S, Kojima K, Shinka T, Sato K, Narisawa A, Aoki Y, Matsubara Y, Omae T, Mizoi K, Kinouchi H. Direct correlation between ischemic injury and extracellular glycine concentration in mice with genetically altered activities of the glycine cleavage multienzyme system. Stroke 2007;38(7):2157–64
9. Shimizu-Sasamata M, Bosque-Hamilton P, Huang PL, Moskowitz MA, Lo EH. Attenuated neurotransmitter release and spreading depression-like depolarizations after focal ischemia in mutant mice with disrupted type I nitric oxide synthase gene. J Neurosci 1998;18:9564–71
10. Chen H, Chopp M, Zhang ZG, Garcia JH. The effect of hypothermia on transient middle cerebral artery occlusion in the rat. J Cereb Blood Flow Metab 1992;12(4):621–8
11. Garcia JH, Yoshida Y, Chen H, Li Y, Zhang ZG, Lian J, Chen S, Chopp M. Progression from ischemic injury to infarct following middle cerebral artery occlusion in the rat. Am J Pathol 1993;142(2):623–35

Chapter 17

Cerebral Hypoxia–Ischemia Model in Adult Mice

Faisal Adhami and Chia-Yi Kuan*

Abstract Cerebral hypoxia–ischemia in adult mice allows the study of the effects of hypoxia in a partially ischemic territory. This model may be clinically relevant in situations of general respiratory-circulatory failure (e.g., in shock) or ischemic tissue that displays a breakdown in oxygen metabolism during stroke progression. In this chapter, we outline this procedure in detail from ligation of the common carotid to administration of the hypoxic gas. We will also describe outcome measurements using behavioral deficits, Nissl stain infarct quantification, and fluorescein isothiocyanate-dextran (FITC-dextran) perfusion marking. Finally, we will briefly discuss the advantages, limitations, and complications with using cerebral hypoxia–ischemia in adult rodents.

Keywords: Ischemia, Hypoxia, No-reflow, Reperfusion, Thrombosis, Autophagy

1 Model Selection

Cerebral hypoxia–ischemia was originally developed for use in adult rats,[1] and later modified for use in neonates.[2] This procedure involves the occlusion of the right common carotid artery followed by systemic hypoxia. The application of this model in neonates has contributed much to the understanding of hypoxic–ischemic encephalopathy. Studies using hypoxia–ischemia in adults have also continued, using temperature control via a water bath.[3] We have also used this model, but found it difficult to attain consistent damage in regions other than the CA1 region of the hippocampus.[4] Hence, we modified this model to allow direct temperature control and other interventions under anesthesia during hypoxia.[5] This modification produced a large infarct across much of the lesion hemisphere in most animals. The lesion was largely caused by persistent thrombosis in the cerebral vasculature leading to a "no-reflow"-like phenomenon,[6,7] which is in contrast to the hyperemia seen in transient focal ischemia recovery.[8] The modified cerebral hypoxia–ischemia model is

*Division of Developmental Biology, Room 3464, Cincinnati Children's Hospital Research Foundation, 3333 Burnet Avenue, Cincinnati, OH 45229, USA, alex.kuan@chmcc.org

From: *Animal Models of Acute Neurological Injuries.*
Edited by: J. Chen (eds.) New York, Totowa, NJ

useful because of its simple and standardized surgical procedures, reliable pathologic outcome, and high throughput when used in conjunction with a multistation hypoxia system. The clinical relevance of this model centers on the fact that a perfusion-independent breakdown in oxygen metabolism marks tissue fated for imminent death in ischemic stroke,[9] thereby making this experimental paradigm useful for studying the mechanisms triggering infarction.

2 Materials

2.1 Animals

We have used C57Bl/6 inbred mice in most of our studies, which give consistent outcomes. We have also used CD-1 and BALB/c mice with reasonably consistent results.

2.2 Gas, Anesthesia System, and General Setup

- Disinfectant (Spore-Klenz) and sterile towels
- O_2 (100%)
- O_2 (7.5%) balanced with N_2
- Multistation anesthesia machine with isoflurane vaporizer (Harvard Apparatus 723024 and 723039)
- Anesthesia induction chamber
- Breathing circuit
- Heating pads (Homedics)
- Temperature controllers (Cole Parmer EW-89000-00) and rectal temperature probes (Cole Parmer NCI-00141PG)

2.3 Surgical Equipment

- Animal trimmer with fine comb (Harvard Apparatus 726110)
- Surgical bed (see Fig. 1b)
- Surgical instruments
- Two delicate fine forceps (e.g., Fine Science Tools 11373-12)
- One pair fine scissors (e.g., FST 14088-10)
  - One pair locking micro needle holder (e.g., FST 12060-01)
  - One pair large hemostat (e.g., FST 13008-12)
  - One pair large scissors (e.g., FST 14002-12)
  - One straight, large forceps (e.g., FST 11002-12)

Surgical supplies

- Cotton swabs
- Cotton gauze
- Isopropyl alcohol-soaked cotton swabs
- Betadine-soaked cotton swabs
- Alcohol-soaked wipes
- Medical tape
- Silk suture (5-0) spool for artery ligation (Harvard Apparatus 517680)
- Nylon monofilament suture (3-0) for wound closure
- Artificial tear gel (Fisher NC9860843)
- Water-soluble lubricant (Fisher NC9848567)

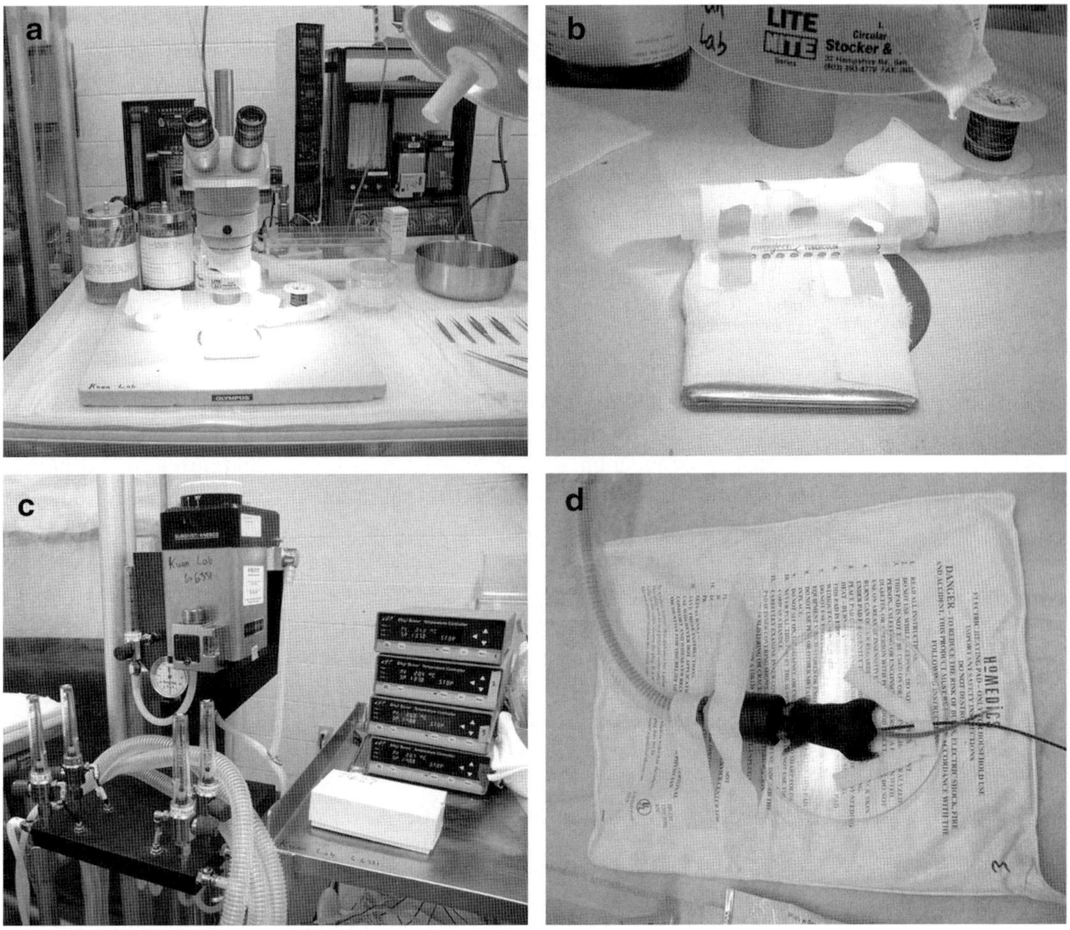

Fig. 1 *Setup of equipment for cerebral hypoxia–ischemia.* (**a**) Setup of the dissecting scope, with the breathing circuit running to the surgical bed. The scope has a light source attached to it, plus another overhead light source to provide better lighting and warmth. (**b**) Surgical bed and breathing mask for use during ischemia surgery. We made this with a cage card wrapped in gauze. Note the neck roll used here is the barrel of a 1-cc syringe. (**c**) Multistation hypoxia and temperature control units. This setup is for up to four animals at once. Close attention must be paid to the temperature and breathing rate, and appropriate adjustments made. (**d**) Hypoxia station setup. We used a standard retail heating pad for temperature control. The animal was kept on a Petri dish to dull and spread the heat, cut to allow flush seating of breathing mask and rectal temperature probe

3 Procedure

3.1 Setup

1. This procedure works best with two people, one performing the surgery (a 5–10 min procedure for each mouse) and the other administering hypoxia (adjusting anesthesia and temperature as needed).
2. Set up a sterile field using sterile towels soaked in disinfectant.
3. Prepare the anesthesia induction chamber and connect it to the breathing circuit. Feed this chamber with 2.5–3% isoflurane in 100% O_2, and enough flow to fill the chamber (our machine and breathing circuit work with 1–2 L min^{-1}; one should be able to smell isoflurane easily upon opening the chamber).

4. Place the dissecting microscope, surgical instruments, and other surgical supplies on the sterile field (see Fig. 1a). Set up the breathing circuit with 1.5–2% isoflurane in 100% O_2 and 1–2 L min^{-1} flow. Turn on the light sources to appropriately light and warm the surgical field. Cut two 1–2-cm (start longer until one feels comfortable with using shorter) lengths of 5-0 silk suture from spool to have them readily available for each animal.
5. Prepare the surgical bed (see Fig. 1b). We use a sterilized animal cage card for the base, but one could also use a sterile or wiped-down Petri dish or other small, hard plastic base. Use medical tape to secure a layer or two of sterile gauze to this base. Prepare a neck roll using either rolled sterile-gauze or the plunger of a sterile 1-mL syringe, secured with medical tape.
6. Set up heat-regulating system for hypoxia. Our customized system is set up as follows: wipe down with alcohol wipes and place heating pads at each station in the sterile field. Connect each heating pad to a temperature controller, connect the system to the power source, and turn it on. Set initial temperature to 37°C. Run rectal temperature probes to each station, and secure them temporarily (see Fig. 1c, d).
7. Connect the multistation anesthesia system with isoflurane vaporizer to the 7.5% O_2 source. We use the four-station model, so adjustments of main flow will need to be made for the two- or six-station model. Run breathing circuits to each station in the sterile field, and set up an initial main flow of 2 L min^{-1} (0.5 L min^{-1} $station^{-1}$) with 2% isoflurane and each individual station's diluting flow to 0.5 L min^{-1}. We precut holes into the membrane of the gas masks as a cross of about 2 cm length per cut (see Fig. 1c, d).

3.2 Surgery: Right Common Carotid Artery Occlusion

NOTE: All right/left references made are from the animal's perspective.

1. Anesthetize the mouse using isoflurane (3%) in the induction chamber for 2–3 min until the mouse is unresponsive to foot squeeze.
2. While holding mouse by the scruff of its neck lying belly-up in the palm, shave the ventral surface of the neck using a trimmer, getting rid of as much hair as possible. Apply artificial tear gel to the eyes. Re-anesthetize in the induction chamber for 1–2 min.
3. Place the mouse on the surgical bed, belly up, with the neck roll placed to push upwards the lower neck. Secure arms stretched out along neck roll at sides using small pieces of medical tape.
4. Prepare the surgical site for incision by cleaning, first with alcohol swabs, then with betadine swabs, and then with alcohol again, starting at the center of the incision site and moving outwards in a circular motion. The incision site should be about 2–4 mm to the right of the midline and about 2–4 mm superior to the clavicle.
5. Using a fine forceps and sharp fine scissors, make a 0.5-cm vertical incision, cutting just through the skin and the attached fascia to expose the underlying thyroid gland (see Fig. 2a).
6. Using the fine forceps, gently slide the thyroid (should move easily) so that the midline between the right and left lobes is visible. Carefully pull apart the fascia and tissue in midline to expose the underlying structures, and tuck the lobes of the thyroid to the sides (see Fig. 2a).
7. Locate the common carotid artery by tracing inferiorly the trachea medially and the sternocleidomastoid muscle laterally, pulling the sternocleidomastoid to the side as necessary. This is the inferior extreme of the anterior

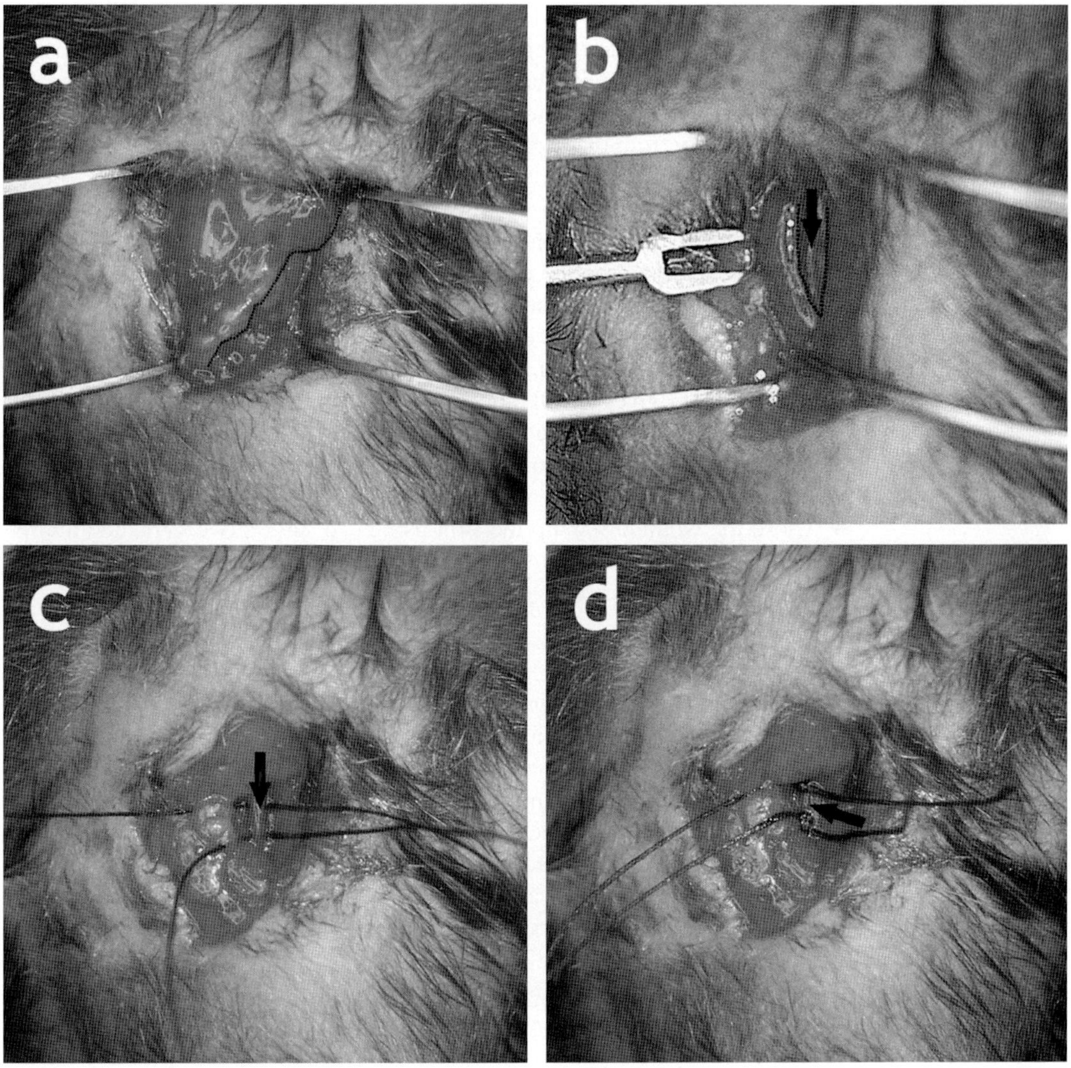

Fig. 2 *Surgical procedure for permanent right common carotid artery occlusion*. (**a**) After an incision 3–4 mm to the right of the midline, use delicate forceps to pull apart the fascia and expose the underlying thyroid tissue. Pull the thyroid gland to the right, until the edge is exposed (indicated *line*). (**b**) After securing the thyroid gland to the right, find the inferior part of the anterior triangle (shown by *line*), with the sternocleidomastoid muscle at the right (relative to the mouse) boundary and the trachea and overlying muscle at the left boundary. The carotid can be seen pulsing coming up from the apex of the anterior triangle (*arrow*). (**c**) Isolate the common carotid (*arrow*) with a delicate forceps, separating it from the vagus nerve and jugular vein without damaging these structures. (**d**) Using 5-0 silk sutures, tie two secure knots on the common carotid, and cut in between (*arrow*)

carotid triangle of the neck, well before the bifurcation of the internal and external carotids (see Fig. 2b). If the neck roll is positioned appropriately, the carotid should be apparent. Therefore, one may need to reposition the neck at this point. The carotid should be apparent as the large vessel visibly pulsing. It will be running in close proximity to the vagus nerve and the jugular vein laterally. Using a fine forceps, carefully pull a small opening on the fascia overlying the carotid, approaching initially from the medial side. Separating the fascia from the carotid without tearing the carotid is tricky, and care should be taken at this point. After fully exposing the carotid, insert a fine forceps below the carotid. Initially, it may be easier to pull up the carotid with the vagus nerve, separate them by pulling upwards with light tension, and inserting the other forceps in between to pull the carotid away from the vagus. The carotid artery is relatively tough, but one should avoid sudden movements or jerks while holding it up on the forceps (see Fig. 2c).

8. With the free hand, run the two precut lengths of 5-0 silk suture under the carotid. Next, release the carotid and tie sutures with double knots, one as inferior as possible and the other superiorly with enough gap to access with scissors. Make sure each knot is secure, but do not pull the suture through the artery by over-tightening. We feel that using the noted locking micro needle holder gives excellent control and grip to tighten the knot appropriately. There should be no visible pulse on the superior side of the inferior knot. Cut the excess suture and then cut the artery between the knots to ensure complete ligation of the common carotid (see Fig. 2d).
9. Close the wound. Sew up the skin using 3-0 suture, tweezers, and a large hemostat. One can use a single continuous suture along the length of the wound for the sake of speed.

3.3 Hypoxia

1. Make sure the bed is prewarmed on high for at least 10 min, and the breathing circuit is charged with anesthesia (7.5% O_2, 2 L min^{-1} 2% isoflurane on the main, 0.5 L min^{-1} of diluting flow for each station) for at least 5 min before transferring the animal.
2. Transfer the animal directly from the surgical bed to the hypoxia station, not allowing the animal to wake up. Put the mouse on the bed in the prone position, with the nose and mouth in the anesthesia mask, making sure that the mouth is not outside the gas mask. Using medical tape, secure the animal to the station (see Fig. 1d). Using plenty of water-soluble lubricant, insert the rectal temperature probe about 0.5–1 cm into animal's anus. Blood should not be seen on the probe when pulled out.
3. Mark starting time of hypoxia and starting temperature after 2–3 min to allow temperature readout to stabilize. For the best reproducibility, we have used a 40-min hypoxia. This gave an average infarct area of approximately 60% of ipsilateral hemisphere with only a limited number of mice with no infarct (~11%). The mortality produced with this was approximately 33% at 24 h. Reducing the amount of hypoxia increased survivability, but also increased variability.[5] Adjust hypoxia length accordingly.
4. During hypoxia, monitor respiratory rate (RR) and pattern, as well as temperature of the animal. Adjust the diluting gas flow to keep RR at 80–120 breaths per min (increase diluent flow if the RR drops too far, etc.). If the RR drops below 40, or if the animal starts gasping, one may need to allow the animal to recover in room air for a minute or two (this is rare if one closely monitor and control the RR). Turn off or on the heat output of the temperature controller to keep the animal's temperature between 36.5 and 37.5°C. Usually, the animal's temperature slowly rises and falls a couple times during 40 min of hypoxia, so it is important to monitor the animal body temperature and heat output carefully. If it is too hot, the mortality rate increases dramatically; if too cold, the injury is decreased.

5. After the hypoxia period is over, return animal to its cage until ready to process. After 15 min, note animal's behavior for reference (see outcome evaluation below).
6. Wipe down all surgical instruments, microscope, heating pads, rectal probes, gas masks, induction chamber, surgical beds, etc. with alcohol wipes. Re-sterilize surgical instruments.

4 Outcome Evaluation

We have used several methods to determine outcome in this model.[5] Here we describe the behavior score, Nissl stain for quantification of tissue damage, and fluorescein isothiocyanate-dextran (FITC-dextran) injection for assaying perfusion status. Owing to the high consistency in the spatial and temporal patterns of infarct progression in this model, one can also extract protein samples from specific regions of the brain to study biochemical events accompanying cell death.

4.1 Behavior

The neurological deficit score of the animal is usually predictive of the extent of injury in this procedure. This can be measured as early as 15–20 min post-hypoxia. This is noted on a six-point scale as reported before[5]: 0 – no noticeable deficit; 1 – ptosis of right eyelid and/or failure to extend forearm ipsilateral to injury when held just above bench; 2 – animal walks in large, loose circles consistently to the ipsilateral side; 3 – animal walks in small, tight circles consistently to the ipsilateral side; 4 – animal lies and does not walk and/or animal does barrel-rolls towards the ipsilateral side; 5 – animal dies in recovery from hypoxia.

4.2 Nissl Stain

To determine average infarct area in various experimental paradigms, we use Nissl stain of the frozen sections. This procedure allows us to note the extent and anatomy of injury in each animal in several sections throughout the forebrain. It provides consistently presentable sections and is relatively easy and quick to perform (see Fig. 2a, b). As an alternative, one could also do 2,3,5-triphenyltetrazolium chloride (TTC) stain on fresh sections,[5] which is quicker but, in our hands, yields less presentable data. The Nissl stain procedure is typically done at 24 h recovery as follows:

4.2.1 Solutions

Nissl Stain
  Cresyl fast violet (Chroma-Gesellschaft): 1 g
  Distilled water: 990 mL
  Acetic acid (10%): 10 mL
  Filter and keep at 37°C. Filter before each use
Avertin Anesthetic
  Stock is 1 g of 2,2,2-tribromoethanol in 1 mL *tert*-amyl alcohol
  Dissolve and store in dark at 4°C. For working solution, add 0.25 mL of stock to 10 mL of phospho-balanced saline (PBS), mix, and store in dark at 4°C

4.2.2 Preparation Procedure

(1) Inject approximately 0.25 mL Avertin intraperitoneally into the mouse, and allow it to fall asleep. An adequate depth of anesthesia is reached after a couple of minutes, when the mouse no longer responds to foot squeeze. Inject more Avertin if needed, 0.1 mL at a time.
(2) Once anesthetized, perform a transcardial perfusion first with saline, then with 4% paraformaldehyde. Leave in 4% paraformaldehyde (PF) overnight at 4°C.
(3) Drain PF, rinse with PBS, and soak brain at 4°C in 30% sucrose in PBS for 48 h, until brain has sunken.
(4) Freeze the brain in a tissue-freezing medium, and use a cryosectioner to obtain 50-μm sections. Mount on positively-charged slides and allow the slides to dry on a slide warmer for at least 30 min.

4.2.3 Nissl Procedure

(1) Put the slides in xylenes for approximately 3–10 min to remove fats that can interfere with the stain.
(2) Rehydrate slides by taking through an alcohol gradient: 100% EtOH for 3 min, 90% EtOH for 3 min, 70% EtOH for 3 min, and finally distilled water for 3 min.
(3) Incubate slides in prewarmed Nissl stain for 5–10 min at 50°C (varies depending on the age of the solution).
(4) Dip the slides in distilled water.
(5) Incubate slides in 70% EtOH for 2–5 min (observe the differentiation process; times will vary for this step depending on the thickness of the sections and strength of the stain. If stain is insufficient, dip slides in water and re-stain in Nissl solution. If the stain is too strong, expose briefly, with care, to 2% acetic acid in 70% EtOH before going to 90% EtOH. This differentiation can be rapid, and a bit uneven.)
(6) Dehydrate the slides again through the alcohol gradient: 90% EtOH for 3 min, and 100% EtOH for 3 min.
(7) Incubate the slides in xylene – two changes of approximately 5 min each.
(8) Remove the slides from xylene one at a time, remove excess liquid, and coverslip with Permount (Fisher, prod#: SP15-100). Place the slide on a bench top to let Permount solidify overnight.
(9) Quantification can be done by scanning sections into images, and tracing infarct areas using ImageJ or similar software.

4.3 FITC-Dextran Perfusion Status

To determine whether perfusion is impaired after hypoxia–ischemia, we found injection of FITC-dextran to be a useful perfusion marker, which produced reproducible and quantifiable images (see Fig. 3c, d).

4.3.1 Procedure

(1) The animals are exposed to hypoxia–ischemia and injected with anesthesia as described earlier at the recovery time desired.
(2) FITC-dextran (5 mg in 100 μL PBS of 2×10^6 MW; Sigma) is then injected with a small-bore needle into the left ventricle of the mouse.
(3) After 2 min, the animal is decapitated rapidly and the brain removed. It is post-fixed in 4% paraformaldehyde for 48 h.

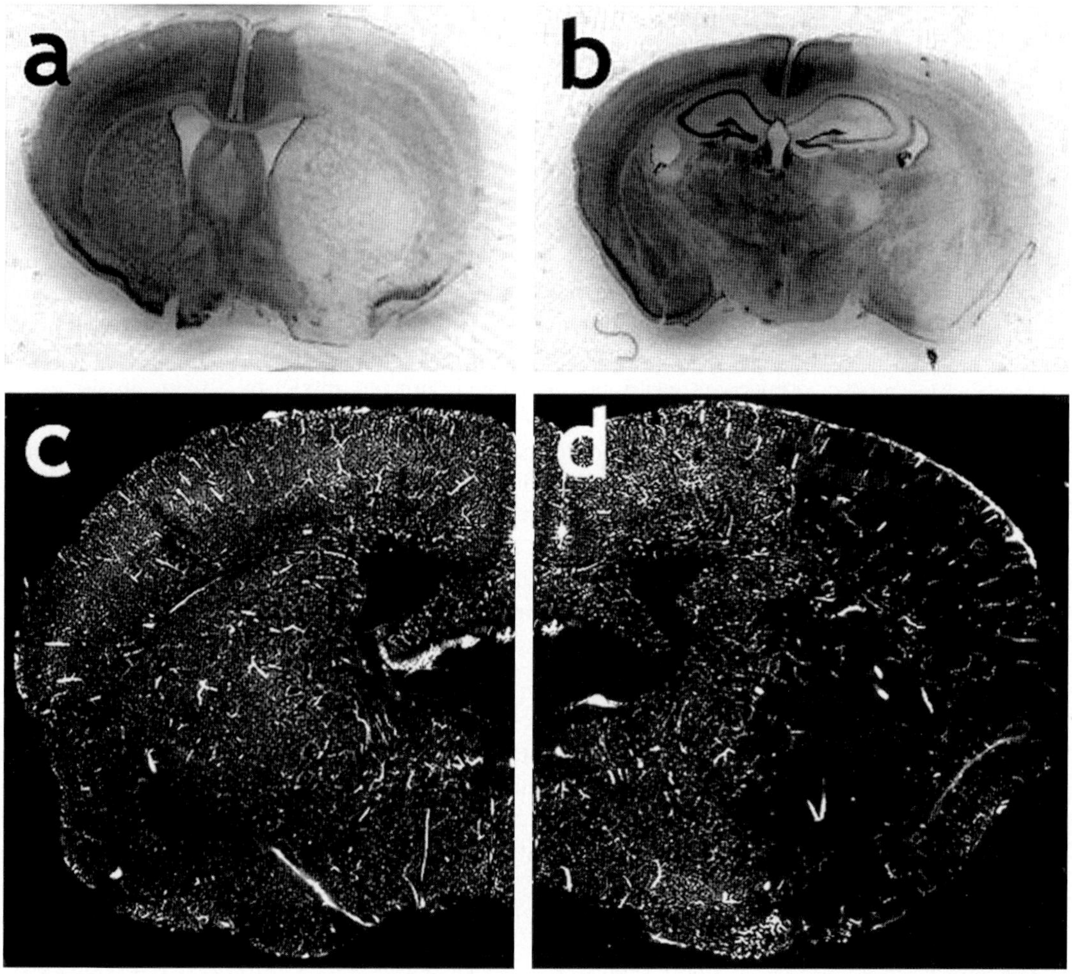

Fig. 3 *Examples of outcome measures after cerebral hypoxia–ischemia.* (**a**, **b**) Nissl stain at 24 h recovery at rostral (**a**) and caudal (**b**) levels of the brain. The damaged tissue is clearly indicated by the lack of Nissl stain. (**c**, **d**) FITC-dextran perfusion marks open vessels in all of the contralateral side of the brain (**c**), but shows a large territory of perfusion impairment on the lesion side (**d**) at 3 h of recovery

(4) Brain is then processed and sections are obtained as described earlier using the cryostat. Typically, we use 30-µm-thick sections.

(5) For quantification, images are obtained and analyzed using the ImageJ software. An index of perfusion can be measured by quantifying the number of FITC-positive pixels in relevant regions and dividing it by the total area of those relevant regions. This can then be normalized to the same measurement on the contralateral side to account for variability in animals or FITC-dextran administration.

5 Advantages, Complications, and Limitations

5.2 Advantages

(1) The surgical procedure of this model is simple and visible. Ligation of the common carotid artery can be quickly learned by experimenters with little prior surgical experience. With practice, the whole procedure can be completed in 5–10 min.
(2) Outcome of brain damage is predicable in more than 90% of all animals undergone this procedure, and the infarct is large and easily quantifiable.
(3) When combined with the multistation anesthesia system for hypoxia administration, this procedure is relatively high throughput. More than 20 mice can easily be challenged with this model by a team of two experimenters in 1 day.
(4) This procedure produces extensive thrombosis and perfusion impairment throughout the vasculature, and is therefore particularly useful for studies on the etiology and treatment of cerebral blood flow impairment in ischemic stroke.
(5) Autophagy is also seen induced in this model,[5] and therefore the model is relevant to studies aiming to understand the contribution of this process to stroke.

5.3 Complications

(1) During hypoxia, respiratory rate can vary greatly. Attention must be paid to this as well as the animal's body temperature, and adjustments to the anesthesia and temperature control mechanism need to be made promptly. Knowing when to adjust and what to expect in terms of animal response comes with experience.
(2) The contribution of isoflurane to the final pathology must be taken into account. Anesthesia is needed to keep the animals in place for temperature control during hypoxia, which we have found critical for making a reliable model. In comparative studies between two groups, if a change in response to isoflurane is noticed or expected in one group, another nonaffected anesthetic agent must be used for comparison of the outcomes.

5.4 Limitations

(1) In our experience, one serious limitation in this model is the one often found in injury models: Sv-129 mice seem to be mildly resistant. Therefore, genetically modified mice must be compared to littermates or extensively back-crossed to B6 lines before hypoxia–ischemia can provide reliable data.
(2) Despite the relative reliability of this model, there is still some variability from animal to animal. This is especially the case in lighter exposures to hypoxia, mixed genetic backgrounds, or other interfering treatments. Therefore, sufficient numbers must be used to get quantifiably significant data. The fact that this model is of high throughput helps to overcome this limitation.
(3) Mortality during recovery can be high, depending on the length of hypoxia used. With 40 min of hypoxia (7.5% Oxygen), about one-third of mice die

after 1 day. Therefore, less severe hypoxia needs to be used for long-term survival studies, and histological analysis is recommended at the end of a functional/behavioral study to document the extent of brain damage.

References

1. Levine, S., Anoxic–ischemic encephalopathy in rats. Am J Pathol, 1960. **36**:1–17
2. Rice, J.E. III, R.C. Vannucci, and J.B. Brierley, The influence of immaturity on hypoxic–ischemic brain damage in the rat. Ann Neurol, 1981. **9**(2):131–41
3. Vannucci, S.J., et al., Experimental stroke in the female diabetic, db/db, mouse. J Cereb Blood Flow Metab, 2001. **21**(1):52–60
4. Kuan, C.Y., et al., Hypoxia–ischemia induces DNA synthesis without cell proliferation in dying neurons in adult rodent brain. J Neurosci, 2004. **24**(47):10763–72
5. Adhami, F., et al., Cerebral hypoxia–ischemia induces intravascular coagulation and autophagy. Am J Pathol, 2006. **169**(2):566–83
6. del Zoppo, G.J. and T. Mabuchi, Cerebral microvessel responses to focal ischemia. J Cereb Blood Flow Metab, 2003. **23**(8):879–94
7. Majno G, A.A.I., Chiang J, Wright RL, No reflow after cerebral ischaemia. Lancet, 1967. **2**:569–70
8. Muller, T.B., O. Haraldseth, and G. Unsgard, Characterization of the microcirculation during ischemia and reperfusion in the penumbra of a rat model of temporary middle cerebral artery occlusion: a laser Doppler flowmetry study. Int J Microcirc Clin Exp, 1994. **14**(5):289–95
9. Sobesky, J., et al., Does the mismatch match the penumbra? Magnetic resonance imaging and positron emission tomography in early ischemic stroke. Stroke, 2005. **36**(5):980–5

Section III

Neonatal Hypoxia-Ischemia

Chapter 18

Cerebral Hypoxia–Ischemia in Neonatal Rats or Mice: A Model of Perinatal Brain Injury

Changlian Zhu*, Xiaoyang Wang, and Klas Blomgren

Abstract The most widespread animal model of perinatal brain injury is the neonatal rodent cerebral hypoxia–ischemia (HI) model. Ischemic brain injury is produced through unilateral common carotid artery ligation followed by moderate hypoxia. Carotid ligation alone does not induce any injury, nor does hypoxia alone, but the combination produces multifocal ischemic injury in the cortex, hippocampus, striatum, and thalamus in the hemisphere ipsilateral to the ligation, leaving the contralateral hemisphere uninjured. Blood flow is reduced in the ipsilateral hemisphere during hypoxia and returns to normal levels when normoxia is restored. Hence, this is a model of cerebral ischemia and reperfusion, despite permanent unilateral carotid artery ligation. The model is well characterized, easy to carry out, and allows long-term survival. The low cost allows inclusion of sufficient numbers of animals to study physiological and molecular mechanisms of perinatal HI brain injury, as well as testing potentially protective treatments and evaluating long-term functional outcome. A disadvantage is that the extent of injury may vary between animals, but this can be compensated for by using higher numbers of animals. Originally developed for rats, it has also been applied to the neonatal mouse, enabling use of genetically modified strains. The model can be used at different developmental stages, allowing comparison of, for example, the equivalent of preterm, term, juvenile, or adult brain injury.

Keywords: Neonatal, Hypoxia, Ischemia

1 Model Selection

Brain damage resulting from perinatal asphyxia remains an important cause of mortality and chronic disability in infants and children. Therefore, several different models of cerebral ischemia have been developed inducing either focal

*Center for Brain Repair and Rehabilitation, Institute of Neuroscience and Physiology, Göteborg University, SE 405 30 Göteborg, Sweden, changlian.zhu@neuro.gu.se

From: *Animal Models of Acute Neurological Injuries.*
Edited by: J. Chen (eds.) New York, Totowa, NJ

or global ischemia. Global ischemia occurs when all blood supply to the brain is interrupted or markedly reduced. This can be induced by intrauterine clamping of the carotid arteries or the umbilical cord. Focal ischemia affects one or more specific regions as defined by the vascular anatomy and is characterized by induction of infarctions within these regions. The model described here is induced by permanent occlusion of the common carotid artery followed by moderate hypoxia, resulting in multifocal infarctions in the hemisphere ipsilateral to the ligation, leaving the contralateral hemisphere uninjured. Because the model is based of the combination of vascular occlusion and hypoxia, the model is called a hypoxia–ischemia (HI) model. It is often called the Vannucci model in honor of Robert Vannucci who was instrumental in its development and characterization.[1,2] The HI model is technically easier to perform than, for example, transient occlusion of the median carotid artery (MCA). MCA occlusion is the classical rodent model to study stroke in adult animals, but it has also been adapted to 7-day-old mice.[3]

The model described in this chapter is by far the most commonly used animal model of perinatal brain injury. It was initially developed for 7-day-old rats, modified from the Levine preparation in the adult rat.[1,2] The 7-day-old postnatal rat was chosen because at this stage of development the animal's brain is microanatomically similar to that of a 32- to 34-week gestation human fetus or newborn preterm infant.[4,5] The cerebral cortical neuronal layering is complete, the germinal matrix is involuting, and white matter shows little myelination. The combination of unilateral common carotid artery ligation with 7.8% oxygen produces tissue injury in the cerebral cortex, hippocampus, striatum, and thalamus. Cerebral blood flow and metabolic correlates have been extensively characterized, and the model has been widely used to study pathophysiological mechanisms and neuroprotective strategies.[2] With the advent and increasing importance of genetically modified mouse strains, the model has also been adapted to the immature mouse,[6] and the degree of brain damage sustained by the animals depends on the mouse strain used.[7] Comparative studies using rats[8] and mice[9] of different ages have been performed, allowing assessment of the relative contribution of different mechanisms of brain injury at various stages of brain development.

2 Materials

2.1 Animals

The choice of animal for the HI model was originally 7-day-old Wistar or Sprague-Dawley rats with a body weight between 14 and 18 g (Fig. 1). The injury outcome is very similar between the two rat strains. In our experience, using inbred or outbred strains produce very similar results. The variability inherent to this model did not decrease when we used inbred rats. However, the numbers of pups per litter, the pups' body weight, and the exact age of the pups all have great influence on the extent and nature of the brain injury. To reduce variability, it is important to control the number of pups in each litter. Culling of litters to ten similarly sized rat pups (range: 8–12) is often used to ensure more uniform growth and subsequent brain injury. Careful monitoring of birth date is important since the tolerance to HI decreases substantially with brain development.[9,10]

Fig. 1 Seven-day-old rat pups. The pups have not opened their eyes and still lack a fur coat

Maturation of the mouse brain is similar to that of the rat in many respects. For practical reasons, 9-day-old mice, with a body weight ranging from 4 to 6 g, are often used for the neonatal mouse HI model. The brains of 5-day-old mice correspond to premature human brains of approximately gestational week 24–28. The brains of 21- and 60-day-old mice correspond to juvenile and young adult human brains, respectively.[9] Unlike the rat strains, different mouse strains display significant differences in their susceptibility to HI brain damage and mortality. CD1 mice are particularly susceptible to brain damage in this model, while 129Sv mice are resistant. The C57Bl/6 mice display a favorable profile with low mortality and reasonably consistent injury, and because it is a common genetic background for many genetically modified strains of mice, it is often used for HI studies.[7] The gender of the pups needs to be considered. Even though total brain injury was not different between male and female pups, the mechanisms of neuronal cell death after HI may be different.[11]

2.2 Anesthesia

Isoflurane or halothane is delivered from a vaporizer in a mixture of oxygen and nitrous oxide (1:1), 3.0% for induction and 1.0–1.5% for maintenance. The pups are anesthetized using a nose mask (Fig. 2). The duration of anesthesia should be less than 5 min.[12]

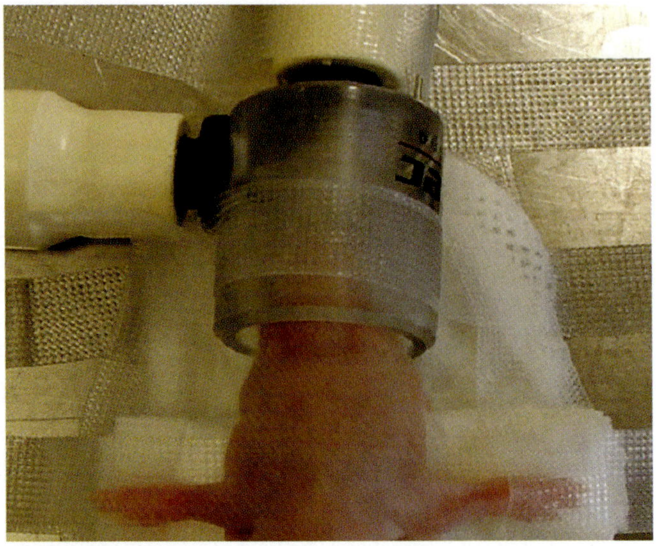

Fig. 2 Anesthesia using a nasal mask

2.3 Equipment

Anesthetic vaporizer
Surgical microscope
Nasal mask (Fig. 2)
Surgical tools, including scissors and finest tip forceps (Fig. 3)
Black monofilament nylon
Thermoregulated water bath
Incubator (Fig. 4)
Thermometer
Oxygen meter

3 Procedures

- Anesthetize the pups with isoflurane using a nasal mask (Fig. 2).
- Make a 0.5-cm cervical incision about 0.2–0.3 cm lateral from the midline to open the skin. We traditionally perform this procedure on the left common carotid artery, whereas other groups may use the right one.
- Separate the muscles covering the frontolateral neck with a fine-tip forceps to expose the left common carotid artery (Fig. 5). Be careful not to confuse the artery and the vein. The cervical vein is more superficial than the common carotid artery, which is located deeper under the muscles. The vein is bigger than the artery and the artery is more reddish and displays an obvious pulse. The artery is located medial to the vein (Fig. 5b).
- Dissect the common carotid artery free from the vagal nerve and other surrounding soft tissue.
- Perform ligation of the common carotid artery with two sutures (suture size: 6) and cut the artery between the two ligatures.
- Close the wound (suture size: 5) and infiltrate with a local anesthetic (Fig. 5c).

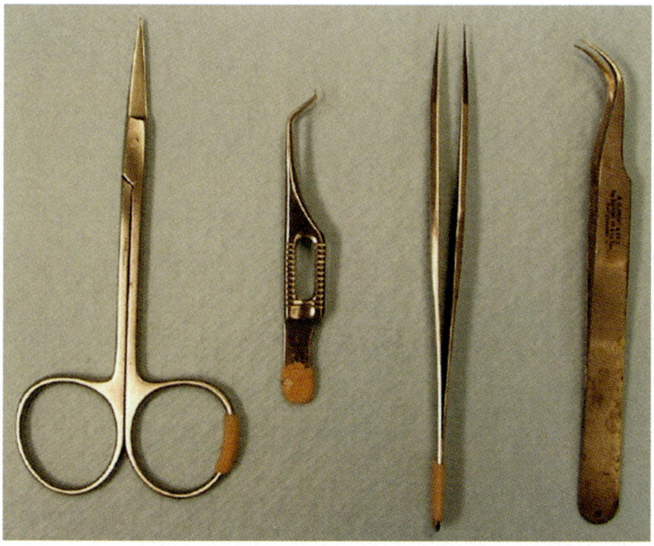

Fig. 3 Surgical tools. To the left is a pair of small scissors, followed by fine forceps

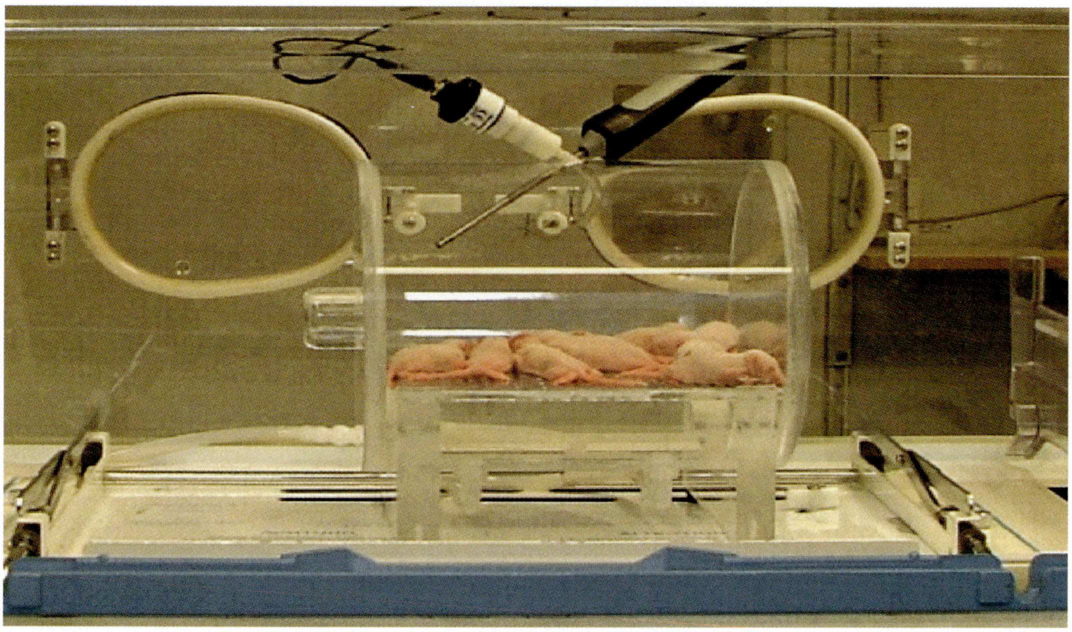

Fig. 4 Incubator for hypoxia with probes for monitoring of temperature and oxygen concentration. The larger incubator is the type used for babies in hospitals. It is used here to keep the ambient temperature constant around the smaller incubator containing the pups. Air or the hypoxic gas mixture is introduced through the inlets at the bottom of the smaller incubator and escapes at the top of the cylinder. The flow rate is kept constant at 3.0 L min^{-1} to ensure consistent partial oxygen pressure throughout the cylinder regardless of the number of pups present

- The pups should be kept at around 33°C ambient temperature before and after surgery. The body temperature of the pups will be close to the ambient temperature, since they have not yet developed their thermoregulation capacity. This is important to keep in mind, since lowering the body temperature just

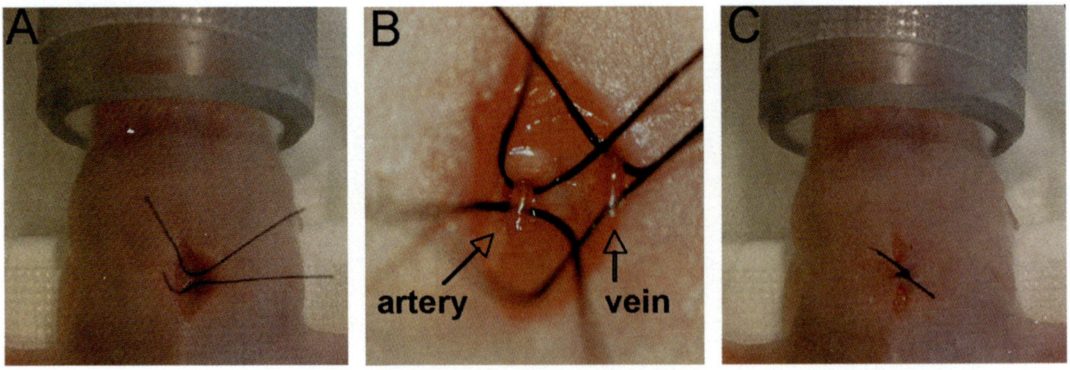

Fig. 5 *Surgical procedures*. (**a**) Cut the skin and dissect free the carotid artery; (**b**) At higher magnification, the difference between the common carotid artery (*left*) and the vein (*right*) is evident; (**c**) Close the wound with a suture size 5

1–2 degrees will yield smaller infarcts, and vice versa. The pups are not anesthetized during hypoxic exposure, but can nevertheless not maintain their body temperature.

- Put the pups back to the dam to recover for 1 h.
- After 1 h of recovery, put the pups in the incubator, keeping the temperature strictly to 36°C by flowing heated, moisturized air through the incubator for 10 min (Fig. 4). The gas flow is kept constant at 3.0 L min−1.
- Switch from air to 7.8% (for rats) or 10% (for mice) oxygen balanced with nitrogen for 50 min. Just like the air, the hypoxic gas mixture is heated and moisturized prior to flowing through the incubator and kept at a constant flow of 3.0 L min−1. The hypoxia time needs to be optimized in individual labs and may vary for different animal strains, temperature variations, different technical solutions for gas mixture and flow, or different ages of the animals.
- Switch from 7.8% (rats) or 10% (mice) oxygen to air for another 10 min.
- Put the pups back to the dam until sacrificed.

4 Outcome Evaluation

4.1 Time Points of Outcome Evaluation

Time points from immediately after to 24 h after HI are suitable to study cell-death-related mechanisms.[13,14] Brain injury evaluation, including infarct volume, atrophy, and tissue loss, is usually evaluated 3–7 days after HI. Evaluation of functional parameters, such as motor and cognitive function, is best performed later, from 3 weeks to 2 months after HI.[15] To study neurogenesis, animals should be sacrificed at least 4 weeks after BrdU labeling.[16]

4.2 Markers of Outcome Evaluation

The outcome of brain injury can be evaluated using gross morphology score, histological features, and/or behavior assessment.

Gross morphology scoring is the quickest way to evaluate brain injury immediately when the brain is dissected out after perfusion. The degree of injury is evaluated by inspection of the the brain surface. Grade 0: normal, equal size of the two hemispheres and no visible white lesion; grade 1: a small white lesion plaque; grade 2: hypotrophy and large cysts in the ipsilateral hemisphere; grade 3: only parasaggital viable tissue left along the entire midline; and grade 4: total loss of the ipsilateral hemisphere. Despite the rough nature of this method, the gross morphology scoring correlated well with histopathological scoring and tissue loss at 5 days after HI in 7-day-old rats.[12]

Brain injury in different regions can be evaluated using a semiquantitative neuropathological scoring system. Serial sections are stained with either hematoxylin–eosin (HE) or with an antibody against microtubule-associated protein 2 (MAP2) (Fig. 6). The cortical injury is graded from 0 to 4, 0 being no observable injury; 1 being one to three small isolated infarcts; 2 being one to several slightly larger, isolated infarcts; 3 being moderate confluent infarcts, and 4 being confluent infarction encompassing most of the hemisphere. The damages in the hippocampus, striatum, and thalamus are assessed both with respect to hypotrophy (shrinkage) (0–3) and observable cell injury/infarction (0–3), resulting in a neuropathological score for each brain region (0–6). The total score (0–22) is the sum for all four regions. The total neuropathological score is positively correlated with tissue loss.[12] To identify damaged or dying neurons, tissue sections can be stained with antibodies against, for example, active caspase-3, which is particularly prominent in the immature brain,[9] or markers of DNA damage, such as terminal deoxynucleotidyl transferase biotin-dUTP nick end labeling (TUNEL), or the so-called hairpin probe (Fig. 7).

Brain infarct volume is measured after MAP2 staining and image analysis of serial sections. The volumes are calculated according to the Cavalieri principle using the following formula: $V = \Sigma A \times p \times T$, where V is the

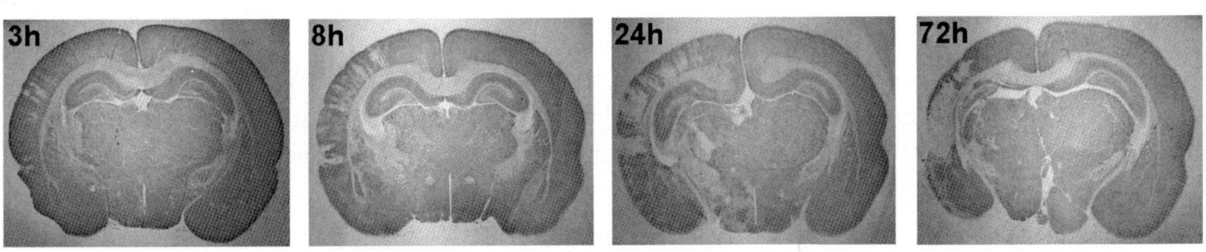

Fig. 6 *Development of injury with time*. The infarctions, as indicated by loss of MAP2 immunoreactivity, can be detected as early as 3 h post HI. The injured areas include the cortex, hippocampus, striatum, and thalamus. The injury in the cortex displays a columnar distribution, unlike the adult brain where a laminar pattern develops, and the injury has matured at 72 h after HI

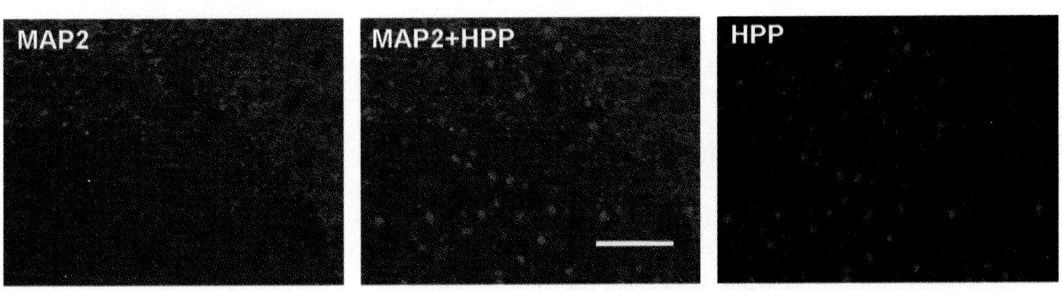

Fig. 7 *Double labeling of MAP2 and HPP*. Double labeling of MAP2 (*green*) and the DNA damage marker called hair-pin probe (HPP) (*red*), showing HPP-positive cells located in the MAP2-negative, injured area[13]

total volume, ΣA is the sum of the areas measured, p is the the inverse of the sections sampling fraction, and T is the section thickness. The MAP2-negative volumes reflect infarcts, whereas the MAP2-positive volume in the contralateral hemisphere minus the MAP2-positive volume in the ipsilateral hemisphere reflects tissue loss. The tissue loss reflects both infarct size and lack of growth that would otherwise have occurred.[12] Hence, infarct volume and tissue loss are essentially different entities, mainly because the brain is still growing when the injury occurs.

Behavior tests include sensorimotor function tests and learning or memory tests. Sensorimotor function can be tested using the grip–traction test for evaluation of muscle strength and paw function; the foot-fault test for detection asymmetries of paw misplacement; the postural reflex test and limb placing test for measuring fore- and hindlimb misplacements after different sensory stimuli.[15] Spatial/cognitive ability can be assessed using Morris' water maze.[17] Learning and memory function can be tested, for example, using the radial arm maze and object recognition test.

5 Advantages, Complications, and Limitations

5.1 Advantages

The progressive nature of HI bears many similarities to birth asphyxia with regard to blood flow changes and cellular metabolic derangements. The model is well characterized with respect to energy metabolism, neurotransmitters, hemodynamics, protein metabolism, and immunoinflammatory alterations.[1,18–20]

It is technically simple, and induces reproducible brain injury (>90% of rats exposed to HI have brain injury) and low mortality. The low cost allows inclusion of sufficient numbers of animals for neuroprotection studies.[1,21,22]

This is the only model that allows long-term survival with evaluation of neuropathology and neurological function. Blood flow is completely restored after HI, so in spite of permanent occlusion of one carotid artery, this is a reperfusion injury model.

This model is well characterized with respect to energy metabolism, neurotransmitters, hemodynamics, and other parameters. The pattern of "secondary energy failure" after HI corresponds to findings in human newborns after severe asphyxia.[18,23]

The model could readily be adapted to mice, allowing the use of genetically modified strains of mice.

In summary, the Vannucci immature rodent model has proven to be an excellent tool to study physiological and molecular mechanisms of perinatal HI brain damage as well as testing potentially protective treatments.

5.2 Limitations

The most important disadvantage is the considerable variation in the extent of brain injury. Therefore, larger numbers of pups are needed in each group in order to compensate for the variation.[1,24]

Because rodent pups are small, particularly mice, there are technical difficulties when it comes to cardiovascular monitoring and blood sampling.

There are also important differences when comparing newborn rodents with newborn infants, including lissencephaly and lack of multiorgan dysfunction. White matter lesions, which are often observed in infants, especially premature infants, are not easily produced in the model. The unilateral rather than global nature of the insult also differs from the clinical situation.

References

1. Rice JE III, Vannucci RC, Brierley JB (1981) The influence of immaturity on hypoxic–ischemic brain damage in the rat. Ann Neurol 9: 131–41
2. Vannucci RC, Vannucci SJ (2005) Perinatal hypoxic–ischemic brain damage: evolution of an animal model. Dev Neurosci 27: 81–86
3. Derugin N, Wendland M, Muramatsu K, Roberts TP, Gregory G, et al. (2000) Evolution of brain injury after transient middle cerebral artery occlusion in neonatal rats. Stroke 31: 1752–1761
4. Vannucci RC, Towfighi J (1999) Experimental models of hypothermic circulatory arrest. Semin Pediatr Neurol 6: 48–54
5. Vannucci RC, Connor JR, Mauger DT, Palmer C, Smith MB, et al. (1999) Rat model of perinatal hypoxic–ischemic brain damage. J Neurosci Res 55: 158–163
6. Ditelberg JS, Sheldon RA, Epstein CJ, Ferriero DM (1996) Brain injury after perinatal hypoxia–ischemia is exacerbated in copper/zinc superoxide dismutase transgenic mice. Pediatr Res 39: 204–208
7. Sheldon RA, Sedik C, Ferriero DM (1998) Strain-related brain injury in neonatal mice subjected to hypoxia–ischemia. Brain Res 810: 114–122
8. Hu BR, Liu CL, Ouyang Y, Blomgren K, Siesjo BK (2000) Involvement of caspase-3 in cell death after hypoxia–ischemia declines during brain maturation. J Cereb Blood Flow Metab 20: 1294–1300
9. Zhu C, Wang X, Xu F, Bahr BA, Shibata M, et al. (2005) The influence of age on apoptotic and other mechanisms of cell death after cerebral hypoxia–ischemia. Cell Death Differ 12: 162–176
10. Oakden E, Chiswick M, Rothwell N, Loddick S (2002) The influence of litter size on brain damage caused by hypoxic–ischemic injury in the neonatal rat. Pediatr Res 52: 692–696
11. Zhu C, Xu F, Wang X, Shibata M, Uchiyama Y, et al. (2006) Different apoptotic mechanisms are activated in male and female brains after neonatal hypoxia–ischaemia. J Neurochem 96: 1016–1027
12. Wang X, Zhu C, Wang X, Gerwien JG, Schrattenholz A, et al. (2004) The non-erythropoietic asialoerythropoietin protects against neonatal hypoxia–ischemia as potently as erythropoietin. J Neurochem 91: 900–910
13. Zhu C, Wang X, Hagberg H, Blomgren K (2000) Correlation between caspase-3 activation and three different markers of DNA damage in neonatal cerebral hypoxia–ischemia. J Neurochem 75: 819–829
14. Zhu C, Qiu L, Wang X, Hallin U, Cande C, et al. (2003) Involvement of apoptosis-inducing factor in neuronal death after hypoxia–ischemia in the neonatal rat brain. J Neurochem 86: 306–317
15. Bona E, Johansson BB, Hagberg H (1997) Sensorimotor function and neuropathology five to six weeks after hypoxia–ischemia in seven-day-old rats. Pediatr Res 42: 678–683
16. Qiu L, Zhu C, Wang X, Xu F, Eriksson PS, et al. (2007) Less neurogenesis and inflammation in the immature than in the juvenile brain after cerebral hypoxia–ischemia. J Cereb Blood Flow Metab 27: 785–794
17. Gustavsson M, Anderson MF, Mallard C, Hagberg H (2005) Hypoxic preconditioning confers long-term reduction of brain injury and improvement of neurological ability in immature rats. Pediatr Res 57: 305–309

18. Vannucci RC, Yager JY, Vannucci SJ (1994) Cerebral glucose and energy utilization during the evolution of hypoxic–ischemic brain damage in the immature rat. J Cereb Blood Flow Metab 14: 279–288
19. McRae A, Gilland E, Bona E, Hagberg H (1995) Microglia activation after neonatal hypoxic–ischemia. Brain Res Dev Brain Res 84: 245–252
20. Silverstein F, Johnston MV (1984) Effects of hypoxia–ischemia on monoamine metabolism in the immature brain. Ann Neurol 15: 342–347
21. Hagberg H, Ichord R, Palmer C, Yager JY, Vannucci SJ (2002) Animal models of developmental brain injury: relevance to human disease. A summary of the panel discussion from the Third Hershey Conference on Developmental Cerebral Blood Flow and Metabolism. Dev Neurosci 24: 364–366
22. Zhu C, Wang X, Huang Z, Qiu L, Xu F, et al. (2007) Apoptosis-inducing factor is a major contributor to neuronal loss induced by neonatal cerebral hypoxia–ischemia. Cell Death Differ 14: 775–784
23. Gunn AJ, Gunn TR (1998) The 'pharmacology' of neuronal rescue with cerebral hypothermia. Early Hum Dev 53: 19–35
24. Hagberg H, Bona E, Gilland E, Puka-Sundvall M (1997) Hypoxia–ischaemia model in the 7-day-old rat: possibilities and shortcomings. Acta Paediatr Suppl 422: 85–88

Chapter 19

Cerebral Palsy Model of Uterine Ischemia in Pregnant Rabbits

X. Ji, M. Derrick, and S. Tan*

Abstract Rabbits, like humans, are perinatal brain developers. The model of uterine ischemia mimics acute placental insufficiency states in humans such as placental abruption. A detailed description of the fetal model is presented. Following fetal hypoxia-ischemia, movement disorders are observed in postnatal rabbit kits akin to those observed in cerebral palsy in humans.

Keywords: Behavior, Brain, Dystonia, Fetus, Hypoxia, Newborn Infant, Premature Infant, Ischemia, Placental insufficiency, Uterus.

1 Introduction

In selecting an animal model for human perinatal disease, one must take into account the normal brain development of that animal. In mammalian perinatal models, rodents are most often used but rodents are postnatal brain developers. White matter and motor development occurs primarily after delivery. Other perinatal models use ungulates and nonhuman primates. Both ungulates and nonhuman primates are prenatal brain developers. At birth, most of the motor development has been completed. This probably is a protective strategy in the wild as the newborn needs to have some motor function in order to survive. Humans are unlike other primates in that for some reason, humans have become perinatal brain developers, just like rabbits.[1] Thus, for investigation of the etiology of motor deficits, the right model for perinatal diseases is probably not the closest animal relative to man, the chimpanzee or other great apes, but animals that are perinatal brain developers such as the rabbit.

The second important requirement for animal models is to mimic human disease conditions. One of the criticisms of the variations of the Levine model in rodents is that these are not reflective of any pathological conditions of humans in the perinatal period. For example, if one wants to establish a model

*Department of Pediatrics, 2650 Ridge Avenue, Evanston, IL 60201, USA, sidtan@northwestern.edu

From: *Animal Models of Acute Neurological Injuries.*
Edited by: J. Chen (eds.) New York, Totowa, NJ

Table 1

Advantages of the rabbit model	Comparison with other species
True fetal model	Postnatal models in rodents
Global hypoxia-ischemia	Unilateral, ischemic models in rodents
Workable size of brain from fetuses for MRI, biochemical and molecular biology analysis	Rodent fetuses and pups have very small brains
Rabbits, similar to humans, are perinatal brain developers especially for WM. Beginning myelination shortly after birth	Sheep and baboons are prenatal.
	Rodents are postnatal developers.
Limited neurobehavioral development at birth, similar to humans	Sheep and great apes almost fully developed at birth
Oxidant systems, such as xanthine oxidase are similar to humans	Rodents have very high levels of xanthine oxidase in circulation
Relatively inexpensive, litter size ~8, and easy to get enough *n* for experiments	Sheep and nonhuman primates are expensive

Hypoxia-ischemia results in immediate fetal bradycardia, logarithmic decrease in blood pH with increasing hypoxia-ischemia, parenchymal hemorrhage in preterm fetuses, and after preterm hypoxia-ischemia, cysts, hydrocephalus, and ventriculomegaly found at P1

of cerebral palsy, one has to look at the antecedents causing such a disease. Cerebral palsy is often believed to result from perinatal hypoxia-ischemia. One of the important conditions resulting in fetal hypoxia-ischemia is acute placental insufficiency that occurs in diseases such as placental abruption. Placental abruption has been shown to result in a higher incidence of cerebral palsy.[2] Modeling acute placental insufficiency states in animals causes a problem because this involves opening up the abdomen.[3] We were able to establish a model without entering the abdominal cavity that subjects the fetuses to hypoxia-ischemia.[4,5]

The major advantages of this model are listed in Table 1

2 Methods

2.1 Experimental Animal

- Species: Rabbit
- Breed/strain: New Zealand white (NZW)
- Sex: Female timed pregnant
- Weight range: 4.0–5.0 kg
- Vendors: Myrtle's Rabbits, Thompson Station, TN, or Covance Research Products, Denver, PA. Make sure the vendors have a constant time for visualized conception in order to ensure accurate timed pregnancies.
- Term gestation: 31.5 days
- Ages studied: 70% gestation (22 days, E22), 79% gestation (25 days, E25), and 92% gestation (29 days, E29)

2.2 Preoperative Evaluation

1. NZW rabbits need some time to recover from the stress of ground transport. Plan for added recovery time if air transport. It is best to wait 2–4 days after transport before experiments.

2. Conduct thorough physical examination before any procedure and make sure that there is not any condition that will compromise the planned experiment.
3. Monitor the rabbit dam's eating, drinking, urination, and defecation. Records must be kept and displayed on the cage. The normal values are as follows[6]:

   (a) Water intake: 50–100 mL kg^{-1} day^{-1}
   (b) Food intake: 50 g kg^{-1} day^{-1}
   (c) Fecal production: 15–60 g day^{-1}
   (d) Urine volume: 50–75 mL kg^{-1} day^{-1}
   (e) Rectal temperature: 38.0–40.0°C

4. When first starting it is advisable to draw blood for biochemical tests. This is also recommended if there is any doubt of the physical well-being of the animal. An ear vein can be used with a 23 G butterfly needle.

2.3 Preanesthesia Preparation

1. Fasting rabbits before anesthesia is unnecessary. Rabbits are thought to lack the vomiting reflex. In addition, eating certain amount of high-fiber diet (Teklad Global High Fiber Rabbit Diet-2031, Harlan Teklad, Madison, WI) before surgery can aid bowel movements after surgery, helping prevent postoperative constipation and possible ileus.
2. Weigh the rabbits before the procedure to accurately determine the dose of anesthetics and experimental compounds.
3. Check all records before operation to make sure that there are no preexisting contraindications to surgery, such as fever, refusal to drink or feed.
4. Preanesthetic sedation is not necessary. The investigator must be calm and soothing even when the rabbits are overactive. A calming strategy is to hold the rabbits in the arms for a couple of minutes, and cover the rabbits' eyes.
5. Restrain the rabbit in a Nalgene Rabbit Restrainer with pans (polypropylene, autoclavable, from Lab4Less) (Fig. 1). A struggling rabbit in an oversized restrainer can result in many complications including vertebral fracture, stripped nails and internal bleeding, and even death. Stuffing the rabbits into the restrainer should also be avoided.
6. Shave the hair of the left ear (usually left). It is suggested to use a shop vacuum to suck away the hair. Less powerful vacuum suction devices leave rabbit hair floating in the air.
7. Cannulation of margin vein and middle artery of the ear.
   (a) Preparation of cannulation tray:

   - One 22 GA (0.9 mm × 25 mm) 1.00 IN sterile shield I.V. catheter (BD, CE 0086)
   - One resealable male adapter (Hospira, Lot 35195 5H)
   - 70% alcohol prep, povidone iodine topical solution
   - Three 1.0 cm × 12.0 cm adhesive tapes
   - Sterile saline, 5-mL syringe, Heparin Lock Flush
   - Sterile gloves and a tray
   - One clean vessel Bulldog Clamp (Vascu-Statt 1001–574) or paperclip

   (b) Use povidone iodine and 70% alcohol preps to disinfect the distal part of the vein.

Fig. 1 The pregnant rabbit is restrained in a Nalgene Rabbit Restrainer with pans that are made up of polypropylene and are autoclavable (Lab4Less)

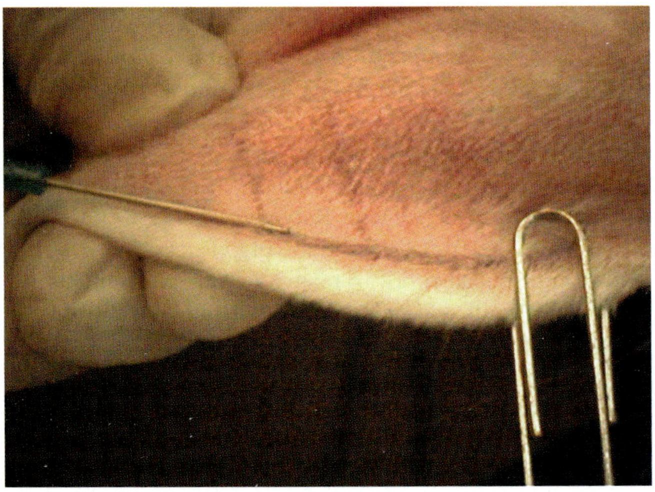

Fig. 2 A paperclip is used to block the lateral marginal vein of the left ear and the vein cannulated with a 22 GA (0.9 mm × 25 mm) sterile shield intravenous catheter (*left*)

(c) Choose the lateral marginal vein of either ear (usually the left), put the vessel clamp or paperclip at its proximal end to block the vein, and wait 1–2 min for the vein to dilate (Fig. 2). Use the left thumb and middle finger to pull the vein straight, use the index finger to push the middle part of the vein upward from the underneath to form a 30° angle. Keep the catheter parallel to the proximal part of the vein and punch the vein right at the angle with the tip bevel of the needle facing upward, slide the needle in 1.0 cm. Watch for blood

return and then slide the catheter forward over the introducer and withdraw the introducer, keeping the catheter in the vein.
(d) Connect the heparin-filled adaptor to the catheter tightly.
(e) Use the three strips of tape, one as an L-shape to secure the catheter to the ear.
(f) Flush the catheter with 2–3 mL of saline.
(g) The middle ear artery is accessed for an arterial line. To dilate the artery, tap the skin over the artery a few times. Cannulation is similar to the venous cannulation except that there is no need to clamp the vessel. The other difference is that once the catheter is in the artery, the operator must put pressure over the artery at the distal end of the catheter to prevent bleeding during the connection to the adaptor.

8. Vital signs: All vital signs should be monitored and recorded at a minimum of every 10 min. The normal values are as follows[6]: (1) respiratory rate: 32–60 min^{-1}, (2) heart rate: 200–300 min^{-1}, (3) arterial systolic pressure: 90–130 mmHg, (4) arterial diastolic pressure: 80–90 mmHg, (5) rectal temperature: 38.0–40.0°C.

2.4 General Anesthesia

1. Equipment

   (a) A OMNI-FLOW 4000 (Abbott Laboratories) multiple channel I.V. pump through which fluids and anesthetics can be infused together with separate speeds
   (b) A thermometer (Type K, Fisher Scientific) with accuracy of 0.1°C
   (c) A pulse oximeter for monitoring oxygen saturation of the arterial blood and pulse
   (d) Blood pressure monitor (Model 6000, Sensor Devices Inc.)
   (e) Bag and mask with oxygen supply. A premature baby mask offers the best fit for the rabbit's snout
   (f) Anesthetics: (1) fentanyl citrate, 1.0 mg per 20 mL per amp, (2) droperidol, 5 mg per 2 mL per amp
   (g) Solutions: (1) Lactated Ringer's injection, (2) 0.9% sodium chloride injection, (3) 5% sodium bicarbonate solution

2. Procedure of general anesthesia

   (a) Induction of general anesthesia:

      1. A dose of Fentanyl 0.075 mg kg^{-1} h^{-1}, and droperidol 3.75 mg kg^{-1} h^{-1} is given, mixed with normal saline or Ringer's (100 mL h^{-1}) IV through the ear vein.
      2. Induction lasts 5–10 min, during which the rabbit dam should be closely watched. There is a certain amount of variation between different rabbits in their response to anesthesia. Heart rate, respiration rate, the size of pupils and its reflex to light and touch, and the color of the skin should be observed and recorded to monitor the effect of anesthesia.
      3. Pupils will constrict and the respiration will become very shallow. This necessitates occasional use of bag and mask ventilation (initially ten breaths with 60% mask oxygen). A 50-mL bolus

injection of normal saline is routinely given in our laboratory as a precautionary measure against hypotension.

4. The dam will allow itself to be turned to its side by gentle handling at the end of induction. Attempts to abruptly move the animal or use too much force wake the animal up, and more anesthetic is eventually needed for induction.

(b) Maintenance of general anesthesia:

1. The rabbit is moved out of the restrainer to receive spinal anesthesia, and the maintenance period of the general anesthesia is started either before or after spinal anesthesia at the dose of Fentanyl 0.025–0.05 mg kg^{-1} h^{-1}, droperidol 1.25–2.5 mg kg^{-1} h^{-1} depending on the rabbit's condition.
2. The spinal anesthesia or epidural anesthesia is dealt in a later section.
3. After spinal or epidural anesthesia, the rabbit is laid supine on an operating board (custom made), and all four extremities are restrained using umbilical cord tape. The dose of anesthesia is dropped so that the levels of fentanyl and droperidol maintain the animal in deep sedation. The dam is allowed to breathe spontaneously. If any sign of pain is noted, the dose of anesthesia should be increased by 25%.

(a) The following signs usually indicate inadequate sedation:

1. Increased respiratory and heart rate
2. Vocalization or limb movements
3. Hyperreactivity to the pedal reflex, ear pinch, corneal reflex

(b) Deep general anesthesia leading to respiratory failure and cardiovascular collapse needs to be avoided. An anesthesia assistant is needed to closely watch the rabbits during the entire procedure and who can make timely adjustments to the dose of anesthetics and prevent possible cardiorespiratory collapse. The following signs suggest too much anesthetic dose:

1. Loss of muscle tone, including loss of resistance to opening of the mouth
2. Loss of reflexes, which include the corneal reflex, pedal reflex, and the ear pinch reflex
3. Dilatation of pupils
4. Slow heart rate, slow respiration rate, arrhythmias on ECG, and hypotension
5. Hypothermia
6. Lower SaO_2; abnormally low PaO_2, high $PaCO_2$; and low pH in arterial blood gas determinations

4. Arterial blood gas should be measured every 20–30 min. Oximeter readings are also very useful. The normal values for NZW rabbits for blood gases are as follows[6]:

(a) Arterial blood pH: 7.2–7.5 (our laboratory limits are 7.25–7.45)
(b) PaO_2: 85–102 mmHg
(c) $PaCO_2$: 20–46 Torr (our limits are 35–49)
(d) HCO_3^- 12–24 mmol L^{-1} (our limits are 20–24)
(e) SaO_2: 95–99%

5. Some precautionary measures need to be used during the maintenance period:
   (a) A temperature-controlled water pad is laid underneath the rabbit.
   (b) Occasional breaths with bag and mask and 40–60% FiO_2 may be needed.
   (c) Periodic suctioning or cleaning of the mouth secretions.
   (d) The head and chest should be slightly elevated if blood pressure is normal to prevent excessive pressure on the diaphragm by abdominal organs, which could affect the lung function
   (e) In addition to IV Ringer's infusion of 100 mL h^{-1}, an extra dose of 10 mL kg^{-1} may need to be given to maintain hydration and blood pressure in case of excessive blood loss. The rabbit will present with hypotension and fast weak pulses. The rabbit can also be placed in the Trendelenburg position in case of hypotension.
   (f) Heat lamp is often necessary to maintain temperature. This is usually necessary during aortic occlusion. Heating the IV line may be useful if hypothermia is a problem.

2.5 Spinal Anesthesia

1. Advantages: Spinal anesthesia allows the dam to breathe spontaneously and thus avoid the use of a ventilator. In addition, the amount of medications used for sedation or general anesthesia is considerably reduced. Thus, the confounding effect of drugs is reduced for any experiment.
2. Disadvantages: Occasionally the procedure results in postoperative paralysis from possible damage to nerve roots. In our experience, we have had 2 dams not moving their hindlimbs out of 50 survival surgeries, with a total incidence of 4%.
3. Contraindications to doing spinal anesthesia are as follows:

   (a) Local infections (skin diseases) at the injection site.
   (b) Cerebrospinal fluid (CSF) is mixed with blood, which does not clear.
   (c) Spinal deformity.
   (d) Operator is not familiar with the anatomy and procedure.

4. Preparation:
   (a) Shave the rabbit back over an area of 4 in. × 6 in. with electric clipper.
   (b) The rabbit is held by an assistant in a left lateral decubitus position when the operator is right handed and a right lateral decubitus position if left handed.
   (c) The rabbit's back should be positioned at and parallel to the edge of the operating table or bed so that the rabbit is within easy reach of the operator.
   (d) The assistant stands in front of the rabbit, stretches the front and back legs and crosses them in front of the rabbit and holds them firmly with the left hand, covering the dam's eyes with the right hand.
   (e) The spine is slightly bent to allow optimal expansion of the intervertebral spaces.
   (f) The skin is disinfected with povidone-iodine starting at the point of insertion and in ever-expanding circles. This is repeated twice. The povidone iodine is allowed to dry for 2–3 min, and then the skin is wiped with 70%

alcohol. The alcohol is then allowed to dry. Remember that all antiseptic solutions are neurotoxic, and care must be taken not to contaminate spinal needles or local anesthetics with the antiseptic solution.

(g) The rest of the rabbit skin is covered with sterile drapes leaving a very small area exposed.

5. Preparation of spinal tray (sterile):

   (a) 25 G spinal needle
   (b) Sterile gloves of different sizes
   (c) Sterile syringes of different sizes (1, 3, 5, 10, 20 mL)
   (d) Sterile gauze
   (e) Sterile drapes
   (f) Local anesthetic: 1% Lidocaine or 0.125% Bupivacaine
   (g) Spinal anesthetic: 2% Lidocaine or 0.25% Bupivacaine
   (h) Adhesive tapes

6. Anatomic landmarks: The injection of local anesthetic into subarachnoid space is usually carried out in the midline between L4-5 and L5-6 intervertebral spaces. A virtual line is drawn from one iliac crest to the other, which crosses the spinous process of L6. This is then used to determine the injection site.
7. Local anesthesia: 0.3–0.5 mL 1% Lidocaine is injected into the skin over the intervertebral space to produce a skin wheal. Another 2–3 mL is injected intradermally. This needle injection also functions as a *probing* needle to verify the intervertebral space between the two spines.
8. Procedure: Using the dominant hand to hold the spinal needle, pierce through the middle of the skin wheal, pointing cranially at a 10–30° angle. The operator can feel the resistance produced by supraspinous and interspinous ligaments when the needle is advanced through them. If the operator releases the needle at this moment, it maintains the proper direction. If it points laterally, it is likely that the needle is not in the interspinous space, and should be withdrawn. The bevel of the needle should face upward to avoid spinal nerve injury. A characteristic *pop* is felt when the subarachnoid space is entered. Needle advancement should stop as soon as the pop is felt. Advancing the needle further increases the risk of paresthesias and neuronal injury. Remove the stylet to allow CSF to flow freely. CSF appears within 3–5 s. Sometimes gentle aspiration is necessary to get some CSF. If there is still no CSF, rotate the needle very slowly in all the directions before careful advancing or withdrawing of the needle. Remember to put the stylet back before advancing the needle. Often CSF appears after slowly adjusting the needle's position. Attach the syringe containing the spinal anesthetic solution to the hub of the spinal needle while the other hand stabilized the spinal needle. The syringe should be free of air. The position of the needle in the subarachnoid space should be once more verified by gentle aspiration of CSF. Avoid careless movement of the needle during aspiration. Effort should be made to keep the syringe tightly attached to the spinal needle. For a surgical procedure less than 1.5 h, 0.3–0.5 mL of 2% Lidocaine or 0.25% Bupivacaine will produce satisfactory anesthesia; for a procedure between 1.5 and 2.5 h, we use 0.3–0.5 mL Marcaine Spinal (each milliliter contains 7.5-mg Bupivacaine HCl and 82.5-mg dextrose). Procedures running longer than that limit need

either leaving a catheter in subarachnoid space for epidural anesthesia, or requires additional general anesthesia.

9. Spinal anesthesia test: Normally after injection of local anesthetic into subarachnoid space, the motor blockage appears about 20 s later. In some cases it appears 1–2 min later. The hindlimbs should be completely relaxed. Anesthesia can be evaluated by using skin pinch stimulus with a hemostat moving from lower abdomen to cephalad positions.

2.6 Epidural Anesthesia

Epidural anesthesia is an alternative method used in our lab for some lengthy procedures. Its advantages are that it can provide prolonged anesthesia, it is safer for various procedures, and produces less severe complications than spinal anesthesia. The disadvantages are that technically it is more difficult to master, it can cause total spinal anesthesia, or incomplete anesthesia when it fails. The approaches to the lumbar epidural space are similar to those described for spinal anesthesia.

1. Preparation of the spinal tray (sterile):

   (a) 9.0-cm 18-gauge epidural needle (standard Touhy needle)
   (b) 20-gauge epidural catheter with connector
   (c) Sterile gloves of different sizes
   (d) Sterile syringes of different sizes (1, 3, 5, 10, 20 mL)
   (e) Alcohol wipes
   (f) Sterile gauze
   (g) Sterile drape, adhesive tapes
   (h) Local anesthetic for the skin infiltration (1% Lidocaine or 0.125% Bupivacaine) and epidural space injection (2% Lidocaine or 0.25% Bupivacaine), all of which contain epinephrine 1:200,000
   (i) A 5-mL syringe filled with anesthetic

2. Procedure: There are two methods used to identify entrance of the epidural needle into the epidural space: Loss-of-resistance (LOR) and hanging drop. The LOR technique is the most popular, and easier to learn.

   (a) Skin incision: Using a large-diameter needle or a no. 11 sharp surgical blade, make a tiny incision to facilitate the skin passage of a blunt epidural needle.
   (b) Passage through the ligaments: Spread the index and middle fingers of the left hand to fix the injection site in between; a 9-cm epidural needle is held between the thumb, index, and middle fingers of the right hand and is advanced through the skin incision. Its bevel tip should face cephalad when bilateral blockage is desired. When the needle passes the supraspinous ligament and intervertebral ligament, the operator feels consistent resistance. The distance from skin to the ligamentum flavum is ~1.5 cm. Stop and check if the needle stays in the midline before reaching epidural space. If the needle is deviated or can be moved freely, it is not in the midline and needs to be replaced.
   (c) Take out the stylet. Then one has choice of two techniques.
   (d) Loss-of-resistance technique: The LOR syringe is filled with either saline or air in a 2–4-mL syringe. Apply certain pressure on the syringe plunger. The saline or air should not move since the needle tip is in the

tough ligament tissue. Keep continuous pressure on the plunger of the syringe while the needle is advanced very slowly (one millimeter each time). The needle shaft is controlled near the skin by left thumb and index while it is advanced by the right hand. Stop after every advance of 2 mm and check whether the epidural space has been reached. The operator can feel as the needle tip advances into the ligamentum flavum because it is tougher than interspinous ligament. When the needle tip reaches epidural space, the resistance against plunger suddenly disappears (thus LOR), and saline or air partially disappears into the epidural space. The advancement of the needle should be stopped immediately to prevent the needle from punching through the dura mater to injure the spinal cord. One should immediately stop injecting any more saline or air as it may lead to partial anesthesia. The depth of the epidural space from skin in adult NZW White rabbit with 20° caudal-side angle is about 2.0 cm.

(e) *Hanging drop* technique: After the ligamentum flavum has been reached, take the stylet out and put a drop of saline into the hub of the epidural needle. When the ligamentum flavum is passed, and the epidural space is reached, the drop will suddenly disappear.

(f) Aspiration test and injection of testing dose through the needle: Gentle aspiration is carried out in several directions to ensure that there is no CSF coming out. The needle is secured by the thumb and index finger of the left hand. After a negative aspiration test, a test dose of anesthetic, 0.5–0.75-mL 0.25% Bupivacaine or 2% Lidocaine with epinephrine, can be injected. It takes 5 min for test dose to take effect: loss of muscle tone of lower part of the rabbit's body and back legs and absent skin response to painful stimuli. You can also choose to give test dose through the catheter later.

(g) Insertion of the epidural catheter: After the epidural space is identified, an epidural catheter is threaded through the epidural needle. The catheter is threaded in short increments by grasping the catheter 1 cm from the hub of the needle and advancing it through the needle. Once at the 12-cm mark at the needle hub, the catheter has reached the tip of the epidural needle and is about to enter the epidural space. The catheter is usually advanced into the space 1.5–2.0 cm, which means that the catheter should reach the 14–15-cm mark at the needle hub. The epidural needle can then be removed slowly with one hand while the other hand continues to thread the catheter through the needle. The length of the catheter inside the space should be rechecked and may be adjusted. Never attempt to pull back the catheter when the needle has not been withdrawn first. It will cut off the catheter tip.

(h) Securing the epidural catheter: Connect the catheter to an epidural connector, take the cap off, and lower it to a position below the epidural space. If the catheter tip has been misplaced in the subarachnoid space, or in veins, CSF or blood would appear by gravity. This procedure should be used to discover any misplaced catheter before anesthetic administration. The catheter is secured at the site of puncture by sterile dressing and adhesive tapes.

(i) Epidural test dose: This tests whether the epidural catheter tip is misplaced in the subarachnoid space or a blood vessel. A negative test result does not guarantee the correct placement of the epidural catheter in the epidural space, but decreases the likelihood that the catheter tip is in a blood vessel or the subarachnoid space. 0.5-mL 2% Lidocaine is injected through the catheter. Wait 5 min for response.

(j) Injection of epidural anesthetic solution: Drugs should be double-checked before each injection. 2-mL 2% Lidocaine or 0.25% Bupivacaine both with 1:200,000 epinephrine is injected slowly and effect lasts 60–90 min.

(k) Removal of the catheter: Removal of the epidural catheter is a simple process.

2.7 Femoral Arterial Catheterization

In our rabbit model, an aortic balloon catheter is introduced into the distal aorta. The balloon is inflated and results in uterine ischemia. The uterine ischemia results in fetal hypoxia-ischemia. A single-lumen balloon catheter can be used. If a double-lumen catheter is employed, blood gas sampling, blood pressure monitoring, and drug administration can be carried out simultaneously during the same hypoxia period.

1. Equipment:

   (a) A sterile arterial catheterization package (gauze, hemostat, needle holder, microneedle holder, scalpel, drapes, and 2-0, 7-0 sutures)

   (b) A sterile instrument tray containing single-use micro blood vessel clips, micro vascular introducer, 1.0, 5, 10-mL syringes one of each, one three-way stopcock, and 4 French arterial embolectomy catheter (usable length 40 cm – single lumen or double).

2. Preparation of the surgical area: The left femoral artery has been traditionally chosen as the site of arterial catheterization by our laboratory. The left groin area is shaved by an electric clipper 2–3 in. around the incision site. Care should be taken not to injure the local skin. It may later become the source of infection, which sometimes leads to postoperative laceration of the groin wound. Use povidone iodine and 70% alcohol to disinfect the area. The disinfected area is covered with sterile drapes, exposing only the area overlying left femoral artery in the center.
3. 3–5-mL 1% Lidocaine or 0.125% Bupivacaine is injected subcutaneously along the planned incision site.
4. Exposure of femoral artery: In most instances a 3–5-cm longitudinal incision along the course of left femoral artery provides adequate exposure to femoral artery. Incision is made smoothly by a surgical scalpel (no. 10). Some subcutaneous bleeding spots will be seen and can be cauterized by an electrocautery. When the deep fascia is reached, use two fingers to separate subcutaneous fat tissue, and a femoral neurovascular bundle will be identified right underneath the deep fascia. The femoral artery is located right below the inner part of the inguinal ligament with its course running downward in a longitudinal shallow ditch between two muscle groups. Palpation of the artery helps determine the position of the incision. At the lateral middle point of the neurovascular bundle, make a small (3–5 mm) incision on the fascia over the bundle with a sharp scalpel, and then pick the fascia up with a tissue forceps, extend the fascia incision up

and down with tissue scissors along the fiber direction of the bundle to about the length of the incision. This will usually give enough exposure to the neurovascular bundle. Pick up both edges of the incision site and use the handle of the scalpel to separate the fascia from the underneath structures. The femoral neurovascular bundle is sometimes covered with a thin layer of muscles, which can be easily removed by careful dissection with fine scissors. The three structures of the bundle are as follows: (1) femoral vein – front-median with larger diameter and dark blue color, (2) femoral artery retrolateral to the femoral vein with characteristic pulsations, fresh red color, and smaller diameter, and (3) femoral nerve in the lateral groove, a thin silver-color bundle. The mnemonic is VAN for vein, artery, and nerve. Using fine tweezers pick up the outer membrane of the artery and cut open with microscissors, and then a hemostat is inserted to separate the membrane from femoral artery wall along its course. During this process some small branches of the artery may be met and should be tied with 4-0 silk sutures before cutting. Do not cut these branches without tying them because then the bloody field will make visualization difficult. The femoral artery should be exposed at least 2.0 cm to allow enough exposure for catheterization (one can use hemostat only to do this dissection step if one is leery of using scissors at this juncture). A poorly exposed artery is the most common reason for failed catheterization. The isolated artery is held by a microclip at its proximal end and looped with two 2-0 silk sutures both proximally and distally. The two loop sutures then will be secured to control unexpected hemorrhage and help facilitate the incision of the artery and introduction of the catheter into the femoral artery.

5. Catheterization: With left hand holding up the distal loop to raise the middle part of the artery, a 1–2-mm incision is made in the longitudinal axis of the artery. The incision is only carefully made in the front wall of the artery. Usually, a little bit blood can be seen where the incision is made to help identify the incision. Releasing the proximal clip shortly and letting the blood come through the incision can also help identify its location (Watch out! Blood can spatter into your eyes and face! Did you wear goggles?). Hanging up the proximal loop with the left hand to stop the blood flow and releasing the clip near it, an assistant could help open the incision with a tiny L-shaped retractor. The operator can then insert a 4 French Fogarty embolectomy balloon catheter at the distal opening. The catheter is inserted 10 cm from the incision. This distance will place the balloon in the descending aorta just above the uterine arteries and below the renal arteries. The balloon is inflated with 0.2–0.3 mL of saline. The looped sutures are used to secure the catheter to the artery. This is important to prevent accidental removal of the catheter with its hemorrhaging consequences.
6. Closure of vessel and incision: After the end of uterine ischemia the balloon is deflated and the catheter withdrawn. Just before final removal a clip is placed proximal to the incision in the artery. Then the catheter is completely withdrawn. There are two more precautionary measures to implement at this point. The incision site is flooded with heparin lock flush solution and normal saline. Then, the clip is released for an instant to allow for the arterial blood to flush away a possible blood clot formed during procedure. The bloody fluid in the site is then cleared and the artery visual-

ized again. 1.0-mL 1% Lidocaine is dropped around the vessel to vasodilate the femoral artery. The opening in the femoral artery is then closed with 4–6 interrupted over-and-over stitches using 7-0 silk sutures. This suturing is made easier by visualizing with magnifying loupes. Attention should be taken not to narrow the vessel lumen and prevent blood flow. After suturing the vessel, pressure should be applied on the suturing site with warm-saline-wet gauze, the clips removed to let blood flow through, and to check for obvious leakage through the sutures. Minor leakage often can be stopped by application of pressure for 3–5 min. Interrupted over-and-over suture with 4-0 silk sutures are used to close fascias and soft tissue. The skin incision site is prepped with 70% alcohol before closing. An interrupted vertical mattress suture is used to close the skin, which gives more exact apposition, approximation, and more strength to the sutures and reduces the chance of postoperative reopening.

7. Trouble shooting:
   (a) Difficulty in distinguishing between femoral artery and vein. 2% Lidocaine can be used to dilate the artery and visualize the pulsations.
   (b) The femoral artery has an outer membrane that needs to be stripped gently. This will help in the insertion of the catheter. Care must be taken not to tear the artery in this process.
   (c) Arterial spasms can prevent insertion of the catheter. One must be careful not to use force because the artery can tear or the catheter can break through the arterial wall. Dripping 1.0-mL 1% Lidocaine or warm saline can help dilate the artery and make the insertion easier.
   (d) The catheter egresses the artery through another hole at the back side. Withdraw the catheter. Reintroduce the L-shaped introducer and separate the walls of the artery at a more proximal location. Then insert the catheter again.

2.8 Laparotomy

This procedure is provided for the investigator if there is a need for direct manipulation of the fetus or direct administration of drugs to the fetus.

1. Equipment:
   (a) A sterile laparotomy package (gauzes, hemostats, needle holder, micro-needle holder, scalpel, drapes, retractors, steel bowls)
   (b) Different size sutures (2-0, 4-0, 7-0)
   (c) Different size syringes (1.0, 5, 10 mL, one of each)
   (d) Sterile gloves, tray, 2% Lidocaine, or 0.25% Bupivacaine and normal saline
2. Preparation of the surgical area: The hair on the abdomen is shaved carefully using an electric clipper from the xiphoid process to the pubic symphysis, including two groin areas, and extending to the flank (Fig. 3). Use 70% alcohol to clean the area three times before povidone-iodine-soaked gauzes. The operator should start wiping the area from upper middle downward and from the midline to the periphery. The pubic area should be wiped and separate gauzes should be applied. Allow 2–3 min to dry, and then apply alcohol with the same method to wipe off the povidone iodine. The disinfected area is then covered with sterile drapes, leaving the midline incision area in the

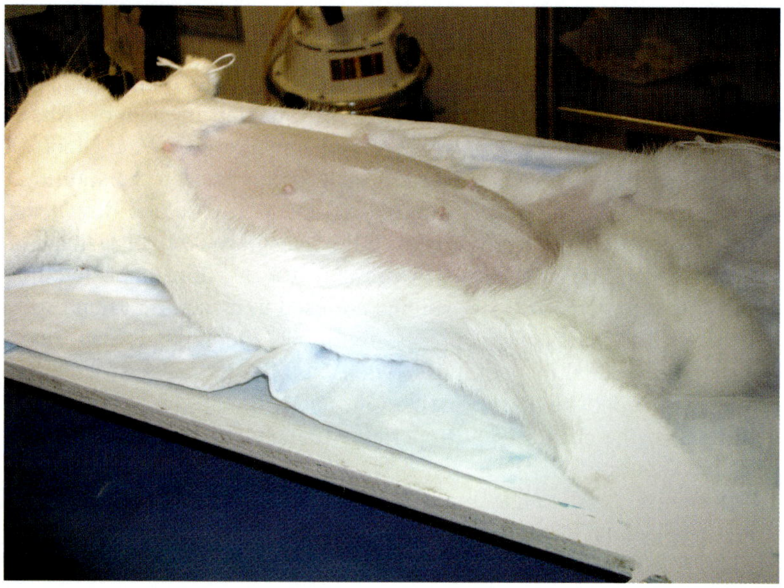

Fig. 3 After spinal anesthesia, the sedated pregnant rabbit is laid on her back. The left groin area and abdomen are shaved of all fur using clippers and the fur is removed by a shop vacuum (background) in preparation for catheterization of the left femoral artery and laparotomy

center exposed. If local anesthetic infiltration is needed, 1% Lidocaine or 0.125% Bupivacaine can be given along the incision.

3. Incision and exploration: An incision is made in the lower midline area, starting 2–3 cm above the umbilicus and extending in a cephalocaudal direction downward for 10–15 cm. The mammary tissue full with milk can be seen after the skin is incised. Attention should be paid not to damage the mammary tissue. Some small veins cross the midline. They should be cauterized or tied and cut with tissue scissors. Then the midline of the abdominal fascia is exposed. Together with an assistant, two toothed forceps are used to pick up the abdominal muscle and fascia layer making a small tent and ensuring no abdominal contents are caught in the tent. A small longitudinal incision is made in the middle with a scalpel; fine tissue scissors is then used to extend the incision and cut the peritoneum. As the incision progresses downward a fat-covered bladder will appear underneath, which can be mistaken for fat tissue and accidentally incised by an inexperienced operator. The bladder can be pushed gently downward with gauze over a right index finger. The peritoneum can then be picked up to avoid the possible intestinal damage and incised with tissue scissors. A few bleeding points are often met in the midline and must be clamped, tied, or controlled by electrocoagulation. While protecting both edges of the incision with saline-soaked gauzes, hold the edges with retractors to expose the abdominal cavity. Exploration should be very gentle. The intestines are pushed aside carefully and packed away with warm, moist

gauze pads, or placed in a plastic bag and moistened with warm saline to minimize heat and evaporative water loss. The uterus filled with fetuses is usually inspected inside abdominal cavity, and can only be taken outside for a short period. It is easy to cause intrauterine hemorrhage and placental separation.

4. Injection of experimental drugs: After the number and position of the fetuses are determined, an experimental drug solution can be injected with a 1-mL TB syringe into the fetal peritoneal cavity, subarachnoid space, or nasal cavity through the wall of the uterus. Sometimes an ultrasound is used to guide the injection. Be sure to avoid the placenta and inject through other side of the placenta.

2.9 Hysterotomy

1. Following laparotomy the uterus is isolated. The incision is made in the avascular part of the uterus in a longitudinal plane. Fetuses can be safely brought out of uterus, the umbilical cord ligated, and the uterus can be sutured up with 3-0 sutures.
2. Closure of abdominal incision: After finishing all the steps of operation, use warm saline to wash the area. The peritoneum is closed by a running over-and-over suture with 2-0 silk in a cephalocaudal direction while the intestines and uterus are protected by a saline-wet gauze pad. Be sure not to suture the internal organs with peritoneum, which could lead to postoperative ileus. Check the leakage and use interrupted over-and-over suture with 2-0 silk to close fascia. Prep the skin with alcohol and close subcutaneous soft tissue with 4–0 silk interrupted over-and-over suture. Be sure to check and have all bleeding points tied or cauterized to prevent subcutaneous hematoma that could be a reason for postoperative wound infection or laceration. An interrupted vertical mattress suture is used to close the skin.

3 Neurobehavioral Abnormalities at Postnatal Day 1 (P1) and 11 (P11)

A range of neurobehavioral deficits is observed in the P1 newborn kits following fetal hypoxia-ischemia at E22.[4] There is increasing stillbirths with increasing hypoxia-ischemia. There are increasing neurobehavioral deficits following greater than 30 min of sustained fetal hypoxia-ischemia at E22. What is striking is the observation of hypertonia and resulting postural abnormalities in ~75% of the survivors. Some of the stillbirths also have postural abnormalities. We are able to keep alive some of the severely affected kits, born following hypoxia-ischemia at E22, by careful nursing and orogastric feeding of rabbit milk. In all cases with hypertonia and postural deficits, the hypertonia persists and the motor deficits at P11 are a stark contrast to the motor capabilities of normal controls.[5]

Acknowledgments: Supported by grants from NIH, HD01138, NS41476, NS43285, and NS051402 (ST).

References

1. Harel S, Shapira Y, Hartzler J, Teng EL, Quiligan E, Van Der Meulen JP. Neuromotor development in relation to birth weight in rabbits. Biol Neonate 1978; 33(1–2):1–7.
2. Matsuda Y, Maeda T, Kouno S. Comparison of neonatal outcome including cerebral palsy between abruptio placentae and placenta previa. Eur J Obstet Gynecol Reprod Biol 2003; 106(2):125–129.
3. Iwasa H, Aono T, Fukuzawa K. Protective effect of vitamin E on fetal distress induced by ischemia of the uteroplacental system in pregnant rats. Free Radic Biol Med 1990; 8(4):393–400.
4. Derrick M, Luo NL, Bregman JC, et al. Preterm fetal hypoxia-ischemia causes hypertonia and motor deficits in the neonatal rabbit: a model for human cerebral palsy? J Neurosci 2004; 24:24–34.
5. Tan S, Drobyshevsky A, Jilling T, et al. Model of cerebral palsy in the perinatal rabbit. J Child Neurol 2005; 20(12):972–979.
6. Suckow MA, Douglas FA. The Laboratory Rabbit. CRC Press, Boca Raton, FL, 1997.

Chapter 20

Model of Neonatal Focal Cerebral Ischemia-Reperfusion

Zinaida S. Vexler* and Donna M. Ferriero

Abstract It has become clear that susceptibility to hypoxic-ischemic, excitotoxic, and oxidative insults is increased at specific stages of postnatal brain maturation, affecting both the acute and lasting injury patterns seen after the insult. For decades, a model of hypoxia-ischemia in postnatal days 7–9 (P7-P9) was the only rodent model relevant to ischemic brain injury in the neonate. This model is more likely to mimic global hypoxic-ischemic encephalopathy rather that focal transient ischemia of a single artery, such as the middle cerebral artery (MCA). To study the mechanisms of arterial pediatric stroke, a transient MCA occlusion model was first developed in juvenile rats. Since focal occlusion of the MCA is more commonly seen in term human babies than children and brain maturation affects the response to cerebral ischemia, we developed a transient MCA occlusion in P7 rats to satisfy the need for age-appropriate stroke models. We produced a series of models of different stroke severity by varying the duration of MCA occlusion and the age of the animals. In this chapter, we will describe in detail surgical procedures to induce MCA occlusion, ways to ascertain that the procedure is successful, and discuss factors and limitations that can affect short and longer-term injury outcomes.

Keywords: Arterial stroke, Middle cerebral artery occlusion, Neonate, Neonatal, Age, Reperfusion

1 Introduction

The incidence of arterial stroke in newborns – at least 1 in 4,000 live term babies[1] – is similar to that in the elderly. The traditional view that neonates have greater resistance to CNS injury than adults has not been confirmed. Rather, it has become clear that at specific early stages of brain maturation

*Department of Neurology, Box 0663, San Francisco, CA 94143-0663, Zena.Vexler@ucsf.edu

From: *Animal Models of Acute Neurological Injuries.*
Edited by: J. Chen (eds.) New York, Totowa, NJ

susceptibility to hypoxic-ischemic, excitotoxic and oxidative insults may be increased.[2–4] The exact reasons for differences in response are yet to be understood but few known major contributing factors include differences in energy metabolism,[5,6] the status of excitatory and inhibitory receptors,[2,7–9] oxidative defenses,[4,10,11] inflammatory responses,[12] and a substantially more prominent role of caspase-3-mediated apoptosis in the immature posthypoxic-ischemic brain than in that of the adult.[13–15] These data have been obtained in a model of hypoxia-ischemia (HI), a model that consists of a combined ligation of the common carotid artery of one hemisphere and systemic hypoxia for various periods of time.[16] Because of the presence of a systemic hypoxic exposure, HI model, which was essentially the only available model related to cerebral ischemia in neonatal rodents for many years, mimics hypoxic ischemic encephalopathy (HIE) more closely than it does focal stroke.

To better understand mechanisms of pediatric arterial stroke, age-appropriate animal stroke models have been developed by modifying the monofilament MCA occlusion model established by Longa et al. in 1989[17] for studying ischemic brain injury in adult rats. Ashwal et al.[18] first developed a model of transient middle cerebral artery (MCA) occlusion in juvenile, postnatal day 14–18 (P14-P18) spontaneously hypertensive rats. Considering that P14 is the age that corresponds to an age of a toddler and, based on data in adult rats, reperfusion is poorer and infarct is more extensive in spontaneously hypertensive when compared with other rat strains, we developed a monofilament MCA occlusion model in normotensive P7 rat pups.[19] The MCA occlusion model developed by our group produces severe disruption of cerebral blood flow in the region of the occluded MCA and confirmed partial reperfusion upon suture removal, as is evident from contrast-enhanced MRI studies.[20] Developing a model of transient suture occlusion is important because reperfusion is a common situation seen in human neonates with arterial stroke.[1,21] Another age-appropriate rat stroke model was developed by Renolleau et al.[22] with the use of a combined permanent MCA occlusion and transient occlusion of the common carotid artery in P7 rats. The transient monofilament MCA occlusion model in P7 rat proved useful for short-term stroke studies.[19,20,23] This surgical procedure was, however, associated with limited ability of neonates to thrive over longer postreperfusion periods.[24] To overcome problems in gaining weight by injured pups over time and therefore make the model suitable for long-term studies, we have recently modified the surgical procedure to produce a less invasive surgical procedure,[25] a procedure that does not require the use of vessel electocoagulation and use of aneurysm clips.

In this chapter, we will describe in detail the modified surgical procedure to induce MCA occlusion, ways to ascertain that the procedure is successful, and discuss factors and limitations that can affect short and longer-term injury outcomes. We will also discuss the effects of different duration of MCA occlusion in P7 rats, including permanent occlusion, on injury and the development of transient MCA occlusion in P10 rats.

2 Detailed Methods

2.1 Animals

All animal research should be approved by institutional Committee of Animal Research. The described experimental procedures were performed in accordance with NIH guidelines for humane handling of animals with prior approval from the Committee of Animal Research at University of California San Francisco. Female Sprague Dawley rats with a 6-day-old litter (10–11 pups per litter) were obtained from Simonson Labs (Gillroy, CA). The mother and her litter were given food and water ad libitum and housed in a temperature/light controlled animal care facility until the pups are 7-days old.

2.2 Suture Preparation

Proper suture coating is critical to achieve complete blockage of cerebral blood flow to the MCA and to prevent intracerebral hemorrhage. The coating is accomplished as follows:

A 6–0 dermalon monofilament suture (United States Surgical, Norwalk, CT) is cut into 12 mm segments. A 2 mm end of each suture is lightly scratched on all sides with fine, 400 silicone carbide paper, and the tip rounded by heat.[19] To prepare a silicone-based polysiloxane impression material (CutterSil, Miles Inc. Dental Products, South Bend, IN), 0.6 ml of the base is mixed for 20–30 s with one drop of the liquid activator on a clean nonporous surface. Under the microscope, the 2 mm scratched end of the suture is dipped in the silicone, lightly wiped against paper to remove excess of silicone, and immediately taped upside down. When executed properly, this results in a finely coated, smooth, tapered suture, resembling a baseball bat. The adhesion and hardening process is quick, and the sutures may be used within 10 min of preparation. Each suture is examined under a microscope prior to use, and the diameter measured with a micrometer. The ideal dimensions are a distal width of 0.18 ± 0.02 mm. Sutures that do not have a smooth continuous taper or those with a distal tip width of >0.2 mm or <0.14 mm are rejected. The silicone is stripped and the suture is recoated later.

2.3 Animal surgery

2.3.1 The P7 pups are anesthetized utilizing 3% isoflurane in a mixture of 70%N_2O/30%O_2, and then the isoflurane is reduced to 1–1.5%, as needed. Throughout the 15–20 min procedure, the animal temperature is maintained at 38°C by a heating pad and an overhead lamp.

2.3.2 The cervical incision is made directly over the right common carotid (CCA), external carotid (EC), and internal carotid (IC) bifurcation. Beginning at its origin, a 1 mm segment of IC is carefully dissected and tied off with a single strand of silk at its origin.[25] A second silk suture is looped around the IC, just above the ptarygopalatine artery (PPA), and gently pulled laterally to prevent retrograde blood-flow. A small incision is made in the proximal isolated IC vessel segment.

2.3.3 The coated 6–0 occluding suture is inserted and advanced 7.5–9.0 mm, depending on the pup weight. In smaller pups, 12–14 gm, the suture is advanced 7.5–8.5 mm; pups over 15 gm have the suture filament advanced 8.5–9 mm (Fig. 1a).

2.3.4 The lower knot is gently tightened, securing the suture, and the skin incision is closed with three interrupted 7–0 silk sutures.

2.3.5 The pups are transferred in a container placed on a heating pad under the lamp until the animals are recovered from anesthesia. Hyperthermia must be avoided.

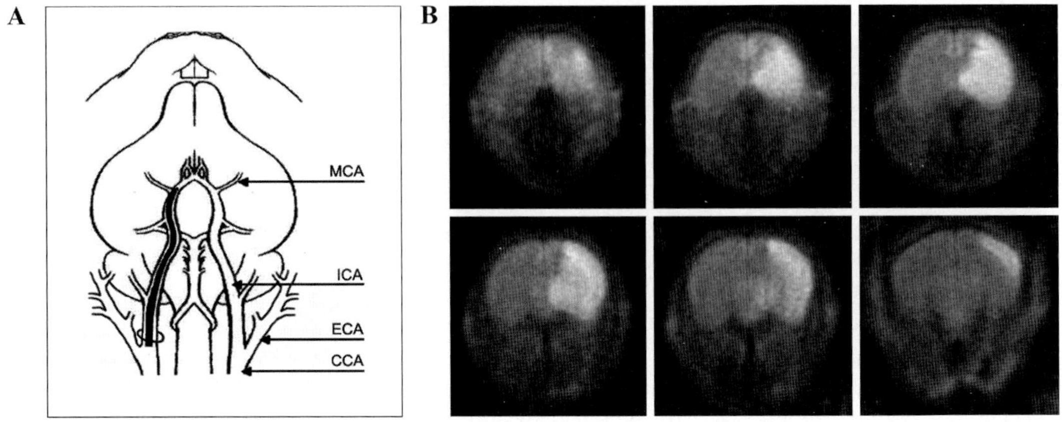

Fig. 1 (**a**) A schematic diagram of the suture insertion and advancement to occlude the MCA. CCA – common carotid artery; ECA – external carotid artery; ICA – internal carotid artery. B. An example of anterior-to-posterior injury distribution pattern on DW-MRI 2 hr post-MCA occlusion. DW hyperintensity is observed in brain regions with slowed water diffusion

2.3.6 The pups are returned to dams after they are fully recovered from anesthesia.

2.3.7 For reperfusion, following anesthesia (described in 2.3.1), the middle skin suture is removed, both knots are removed, and the suture filament together with the coating is gently pulled out and a small piece of surgical with gentle pressure is applied to the arteriotomy. Typically, there is no blood loss during either occlusion or reperfusion.

2.3.8 The pups are transferred in a container placed on a heating pad under the lamp until animals are recovered from anesthesia, as described in 2.3.5.

2.3.9 Pups are returned to dams after they are fully recovered from anesthesia.

2.3.10 Animal weights are recorded daily. If no weight gain is seen within 2 days after surgical procedure upon the development of surgical techniques, animals are to be gavage-fed with 2% condensed milk until signs of weight gain are evident.[24]

3 Noninvasive Delineation of Injury Patterns

Noninvasive injury delineation by MRI is not required but is helpful during the development of the procedure to confirm that the desired injury pattern has been achieved in individual animals. Upon development of the surgical procedure, MRI can also be used as a criterion for excluding uninjured animals in drug-studies.

3.1 Diffusion-weighted magnetic resonance imaging (DW-MRI) is a noninvasive method, which is sensitive to reduced water diffusivity in the tissue following occlusion-induced disruption of CBF. DW-MRI provides information on the presence of injury in minutes; animals need to be anesthetized for imaging. Therefore, the technique can be used for delineation of the spatial pattern of early injury during the occlusion.[20,25] An example of the anterior-to-posterior injury distribution on DW-MRI 2h after MCA occlusion is shown in Fig. 1b.

3.2 Contrast-enhanced MRI can ascertain the completeness of CBF reduction in the MCA vascular territory during the occlusion and restoration of CBF upon retraction of the suture.[20] This method requires intrajugular injection of the magnetic susceptibility contrast.

4 Caveats, limitations, and concerns

4.1 Inadequate Occlusion of the MCA

As described in Sect. 2.2, a combination of factors can affect the successful occlusion of the MCA. The weight of animals can vary substantially even within the same litter thus affecting the length of the suture advancement and the thickness of the suture needed for proper placement of the tip of the suture in proximity to MCA. The use of a shorter than necessary suture will not result in blockage of the MCA. In contrary, the use of a longer than needed suture may produce partial rather than nearly complete disruption of cerebral blood flow through the MCA as the thickest portion of the suture, its coated tip, will be far from the MCA. Unusually, high branching of the MCA, which is occasionally seen, can provide sufficient residual blood flow, thus precluding from complete occlusion.

4.2 Insufficient Suture Coating

An uncoated suture may perforate the vessel. Thick or uneven coating will make the suture more rigid, making it more difficult to advance or retract. Excessive coating may affect the ability to reperfuse because of vasospasm and affected responsiveness of the vessel. Reuse of the coated suture may affect consistency of results, as the suture will become more rigid and/or the coating may slide off.

4.3 Excessive Anesthesia

Multiple lengthy anesthesia episodes may affect the ability of animals to thrive, and therefore the duration of anesthesia should be kept to a necessary minimum, with frequent checking of pinch and other reflexes. Because of the small size of P7 rats, physiological monitoring is limited as MABP monitoring or blood gas analysis are currently not available. The type of anesthetic may affect time necessary for pups to recover form anesthesia. Isoflurane is preferable as animals recover from this agent rapidly. Animals should be put back to their dam for nursing as soon as they are fully awake. If surgery and recovery from anesthesia take longer than what we described in Sect. 2, animals may become dehydrated and saline is to be injected (i.p., 0.1–0.2 ml).

4.4 Lack/Delay in Weight Gain

4.4.1 Like with stroke models in adults and HI model in neonates, essentially all animals lose weight over the first 24-h period after the insult. However, weight loss can persist over time, as we described in our previous publications,[24,25] especially following the originally described procedure.[19] The exact reasons for lack of weight gain after MCAO are not known but the coagulation and/or ligation of several arteries, including the external carotid, occipital, superior thyroid, lingual, and maxillary arteries may reduce or completely cut off blood flow to the ipsilateral neck muscles, tongue, salivary glands. This will and the face is the likely possibilities. Compared with original procedure, the modified procedure described in this chapter is less invasive because of elimination of bipolar coagulation, electrocautery, and use of aneurysm clips from the surgical procedure.

4.4.2 Reduced nursing can occur in case of frequent moving cages from facility to facility. Normal feeding can be reduced because of frequent removal of pups from the cage.

4.4.3 Gavage feeding with milk may be necessary if weight gain of injured pups is diminished.[24,25]

4.5 The Incidence of Injury

Intralitter variations in the incidence may occur.

4.6 The Route of Suture Insertion

Insertion of suture through the ECA is the common way of occlusion of the MCA, which produces consistent injury acutely. The limitation is that this surgical procedure interferes substantially with animal grow and long-term survival. The surgical procedure described in this chapter that involves suture insertion through the ICA results in an overall similar injury pattern as the later surgical procedure.[25]

4.7 Variable Extent of Recovery Over Time

With the use of combined multiple MRI sessions during MCA occlusion, 24 h and 7 days after reperfusion following a 3 h MCA occlusion, we showed that individual pups with nearly identical spatial injury pattern within 24 h may recover to a different extent by 7 days post-MCA occlusion,[25] from no to a nearly complete recovery, based on histology.

4.8 The Effect of Gender

Gender may affect injury evolution and outcome, as has been shown for HI.[26,27]

5 Advantages and Unknowns

5.1 Varying the time of MCA occlusion in animals of the same age results in lesions of different size. The use of a 3 h MCA occlusion[24,25] produced a larger lesion than a 2 h[28] MCA occlusion in P7 rats. Lack of reperfusion is associated with even larger injury.[23]

5.2 Permanent occlusion of the MCA favors a necrotic type of cell death whereas occlusion-reperfusion induces both apoptosis and necrosis.[23] Following transient MCA occlusion, apoptotic neuronal death is substantial both in the ischemic core and in penumbra[23]; no data on the presence of continum cell death, which has been reported after HI in P7 rats,[29] are available.

5.3 Age of neonatal rats and injury outcome. MCA occlusion in neonatal rats of different ages may require different duration of occlusion to achieve similar injury severity histologically but no systematic comparative studies have been performed. In addition to focal stroke model in P7 rats, we have developed a similar model in P10 rats.[30] The latter model produces consistent injury pattern subchronically[31] and animals survive for months.[32]

6 Summary and Future Directions

Several focal stroke models in neonatal rats have become available over the last decade, making it possible to study mechanisms of neonatal stroke experimentally. These models are based on occlusion of MCA in P7[19,20,22–25,28,33] and P10[30–32,34,35] rats. Reperfusion following monofilament occlusion is induced in some but not all models. The completeness of reperfusion may vary depending on the way MCA occlusion is induced (transient placement of a monofilament to block the MCA compared with transient CCA ligation combined with permanent MCA ligation) and duration of the occlusion (transient compared with permanent). Another important aspect of availability of models in rats of different ages is the ability to determine maturational differences in the response to stroke within the neonatal period (stroke in P7[19,22,25] compared with P10 rats[30,35]) as well as comparison to stroke effects in juvenile rats (P14-P17[18]). Although much is to be learned about the role of gender and strain on short and long-term outcomes, availability of several models of focal cerebral ischemia provides ways to advance the field of neonatal stroke.

Other common conditions in human newborns are venous thrombosis and intracerebral hemorrhage. So far, there have been no age-appropriate models developed for those diseases.

Acknowledgments: This study was funded by NIH grants NIH NS44025 to ZSV and NS35902 to DMF and ZSV and AHA GIA grant 0655236Y to ZSV. The authors thank Nikita Derugin for critical reading of the manuscript and Joel Faustino for technical assistance.

References

1. deVeber G, Roach ES, Riela AR, Wiznitzer M. Stroke in children: recognition, treatment, and future directions. Semin Pediatr Neurol 2000; 7:309–17.
2. Ikonomidou C, Bosch F, Miksa M, et al. Blockade of NMDA receptors and apoptotic neurodegeneration in the developing brain. Science 1999; 283:70–4.
3. Olney JW, Wozniak DF, Jevtovic-Todorovic V, Farber NB, Bittigau P, Ikonomidou C. Drug-induced apoptotic neurodegeneration in the developing brain. Brain Pathol 2002; 12:488–98.
4. Sheldon RA, Jiang X, Francisco C, et al. Manipulation of antioxidant pathways in neonatal murine brain. Pediatr Res 2004; 56:656–62.
5. Vannucci RC, Vannucci SJ. Glucose metabolism in the developing brain. Semin Perinatol 2000; 24:107–15.
6. Vannucci SJ, Simpson IA. Developmental switch in brain nutrient transporter expression in the rat. Am J Physiol Endocrinol Metab 2003; 285:E1127–34.

7. Ikonomidou C, Mosinger JL, Salles KS, Labruyere J, Olney JW. Sensitivity of the developing rat brain to hypobaric/ischemic damage parallels sensitivity to N-methyl-aspartate neurotoxicity. J Neurosci 1989; 9:2809–18.
8. McDonald JW, Johnston MV. Physiological and pathophysiological roles of excitatory amino acids during central nervous system development. Brain Research. Brain Research Reviews 1990; 15:41–70.
9. Bickler PE, Fahlman CS, Ferriero DM. Hypoxia increases calcium flux through cortical neuron glutamate receptors via protein kinase C. J Neurochem 2004; 88:878–84.
10. Aspberg A, Tottmar O. Development of antioxidant enzymes in rat brain and in reaggregation culture of fetal brain cells. Brain Res Dev Brain Res 1992; 66:55–8.
11. Fullerton HJ, Ditelberg JS, Chen SF, et al. Copper/zinc superoxide dismutase transgenic brain accumulates hydrogen peroxide after perinatal hypoxia ischemia. Ann Neurol 1998; 44:357–64.
12. Hedtjarn M, Mallard C, Hagberg H. Inflammatory gene profiling in the developing mouse brain after hypoxia-ischemia. J Cereb Blood Flow Metab 2004; 24:1333–51.
13. Cheng Y, Deshmukh M, D'Costa A, et al. Caspase inhibitor affords neuroprotection with delayed administration in a rat model of neonatal hypoxic-ischemic brain injury [see comments]. J Clin Invest 1998; 101:1992–1999.
14. Hu BR, Liu CL, Ouyang Y, Blomgren K, Siesjo BK. Involvement of caspase-3 in cell death after hypoxia-ischemia declines during brain maturation. J Cereb Blood Flow Metab 2000; 20:1294–1300.
15. Zhu C, Wang X, Xu F, et al. The influence of age on apoptotic and other mechanisms of cell death after cerebral hypoxia-ischemia. Cell Death Differ 2005; 12:162–76.
16. Rice JEd, Vannucci RC, Brierley JB. The influence of immaturity on hypoxic-ischemic brain damage in the rat. Ann Neurol 1981; 9:131–141.
17. Longa EZ, Weinstein PR, Carlson S, Cummins R. Reversible middle cerebral artery occlusion without craniectomy in rats. Stroke 1989; 20:84–91.
18. Ashwal S, Cole DJ, Osborne S, Osborne TN, Pearce WJ. A new model of neonatal stroke: reversible middle cerebral artery occlusion in the rat pup. Pediatr Neurol 1995; 12:191–6.
19. Derugin N, Ferriero DM, Vexler ZS. Neonatal reversible focal cerebral ischemia: a new model. Neurosci Res 1998; 32:349–353.
20. Derugin N, Wendland M, Muramatsu K, et al. Evolution of brain injury after transient middle cerebral artery occlusion in neonatal rat. Stroke 2000; 31:1752–1761.
21. Ferriero DM. Neonatal brain injury. N Engl J Med 2004; 351:1985–95.
22. Renolleau S, Aggoun-Zouaoui D, Ben-Ari Y, Charriaut-Marlangue C. A model of transient unilateral focal ischemia with reperfusion in the P7 neonatal rat: morphological changes indicative of apoptosis. Stroke 1998; 29:1454–60; discussion 1461.
23. Manabat C, Han BH, Wendland M, et al. Reperfusion differentially induces caspase-3 activation in ischemic core and penumbra after stroke in immature brain. Stroke 2003; 34:207–13.
24. Fox C, Dingman A, Derugin N, et al. Minocycline confers early but transient protection in the immature brain following focal cerebral ischemia-reperfusion. J Cereb Blood Flow Metab 2005; 25:1138–49.
25. Derugin N, Dingman A, Wendland M, Fox C, Vexler ZS. Magnetic Resonance Imaging as a Surrogate Measure for Histological Sub-Chronic Endpoint in a Neonatal Rat Stroke Model. Brain Res 2005; 1066:49–56.
26. Hagberg H, Wilson MA, Matsushita H, et al. PARP-1 gene disruption in mice preferentially protects males from perinatal brain injury. J Neurochem 2004; 90:1068–75.
27. van den Tweel ER, van Bel F, Kavelaars A, et al. Long-term neuroprotection with 2-iminobiotin, an inhibitor of neuronal and inducible nitric oxide synthase, after

cerebral hypoxia-ischemia in neonatal rats. J Cereb Blood Flow Metab 2005; 25:67–74.

28. Dingman A, Lee SY, Derugin N, Wendland MF, Vexler ZS. Aminoguanidine inhibits caspase-3 and calpain activation without affecting microglial activation following neonatal transient ischemia. Journal of Neurochemistry 2006; 96:1467–79.
29. Northington FJ, Graham EM, Martin LJ. Apoptosis in perinatal hypoxic-ischemic brain injury: How important is it and should it be inhibited? Brain Res Brain Res Rev 2005; 50:244–57.
30. Mu D, Jiang X, Sheldon RA, et al. Regulation of hypoxia-inducible factor 1alpha and induction of vascular endothelial growth factor in a rat neonatal stroke model. Neurobiol Dis 2003; 14:524–34.
31. Mu D, Chang YS, Vexler ZS, Ferriero DM. Hypoxia-inducible factor 1alpha and erythropoietin upregulation with deferoxamine salvage after neonatal stroke. Exp Neurol 2005.
32. Chang YS, Mu D, Wendland M, et al. Erythropoietin improves functional and histological outcome in neonatal stroke. Pediatr Res 2005; 58:106–11.
33. Denker S, Ji S, Lee SY, et al. Macrophages are Comprised of Resident Brain Microglia not Infiltrating Peripheral Monocytes Acutely after Neonatal Stroke. J Neurochem 2007; 100:893–904.
34. Ashwal S, Tone B, Tian HR, Chong S, Obenaus A. Comparison of two neonatal ischemic injury models using magnetic resonance imaging. Pediatr Res 2007; 61:9–14.
35. Ashwal S, Tone B, Tian HR, Chong S, Obenaus A. Serial magnetic resonance imaging in a rat pup filament stroke model. Exp Neurol 2006; 202:294–301.

Part III

Cerebral Hemorrhage Animal Models

Section I

Subarachnoid Hemorrhage

Chapter 21

Monofilament Perforation Subarachnoid Hemorrhage Rat Model

Julian Cahill and John H. Zhang*

Abstract The monofilament puncture model for the induction of a subarachnoid hemorrhage in the rat is a difficult procedure to master. It is, however, important to do so, not only for subarachnoid hemorrhage research but also for stroke models, as the procedure is very similar. The monofilament model is by far the best model for investigating subarachnoid hemorrhage. This model, although technically more difficult compared with other techniques, has become more popular of late as the injury created is more authentic. The following description will explain step by step how to master this procedure with ease.

Keywords: Subarachnoid hemorrhage, Monofilament puncture model, Vasospasm

1 Model Selection

There are a number of methods by which a subarachnoid hemorrhage (SAH) can be induced in the rodent model. The most popular methods are the monofilament puncture method and the single or double hemorrhage injection models.[1–4] The injection model is often used for large animals such as the canine model, as the puncture model is difficult to perform in larger animals. Additionally, the puncture model is associated with a much higher mortality rate compared with the injection model, which can be very expensive, especially when you are using canine or primate subjects.

In the injection model, autologous blood is injected through the cisterna magna or the prechiasmatic cistern.[5] The best method by which a SAH is induced in the rodent model is currently unclear. Prunell et al. compared the three aforementioned methods and found in favor of the prechiasmatic injection model.[5] However, the injection model does not damage the vessel, and in addition it is well recognized that the injury and subsequent degree of

*Division of Neurosurgery, Loma Linda University Medical Centre, California 92354, USA, johnzhang3910@yahoo.com

From: *Animal Models of Acute Neurological Injuries.*
Edited by: J. Chen (eds.) New York, Totowa, NJ

vasospasm is not as severe in the injection model compared with the monofilament model,[6] which matches the human condition more closely. This is an extremely important fact. The etiology of vasospasm is currently unknown and may lie within the vessel wall. This would clearly be missed in any model, using the injection method. Major advantages of the monofilament model include the following:

1. Damage to the cerebral vessels, thereby reflecting the clinical condition more closely.
2. Rapid increase in the intracranial cerebral pressure (ICP) and therefore the subsequent global ischemia is more acute.
3. This method does not require a craniotomy and avoids potential damage to the surrounding structures.
4. With practice this method can be reproduced easily and rapidly.
5. The monofilament model can easily be adapted to stroke research.
6. Results in a more accurate degree of vasospasm, which can then be studied easily.
7. The monofilament model is a single procedure and does not therefore require additional anesthetics and procedures.
8. Allows for the easy confirmation of SAH, i.e., by the presence of blood whereas other procedures require histological confirmation of SAH.
9. Overall the main strength of this model is that it closely matches the human model in all parameters including morbidity, mortality, and severity.

2 Materials

Adult Sprague-Dawley (SD) rats weighing between 280–350 g are used for this procedure. The following equipment is needed to complete the surgery:

1. Anesthetic equipment, including intubation equipment. Alternatives include a tracheostomy or simply supplemental oxygen but this is not recommended
2. Hair clippers (with #40 blade) and skin disinfectant, e.g., iodophors (Betadine) or chlorhexidine (Novasan)
3. Sterile surgical drapes, gown, and gloves
4. Operating surface (nonporous, reusable, and cleanable), well ventilated and well lit working area
5. Operating microscope (Kent Scientific)
6. MV-9 electro surgical unit (Kent Scientific)
7. Standard operating surgical instruments and micro instruments if available. This should include a scalpel (disposable), fine forceps, vascular clamps, needle holder, retractors, and microscissors
8. Suction if available or Q-tips (preferred)
9. Sterile sutures, 4/0 nylon monofilament suture, 10/0 silk ties, and 3/0 silk suture

3 Procedures

1. As with any surgery, it is important to maintain a sterile environment and to use sterile equipment. All instruments should be autoclaved after

each procedure. The surgeon should wear an appropriate sterile gown and gloves. The animal should be prepared in advance and in a different area to where the surgery is going to take place. This includes preparation of the skin, shaving where appropriate, and the liberal use of a suitable skin disinfectant.

2. For the monofilament procedure, the animal is placed supine with the neck gently extended. Care must be taken to secure the endotracheal tube if this method of ventilation has been chosen. Once the animal is secured, additional sterile drapes can be applied as appropriate to maintain the sterile environment.
3. A vertical incision should be made 5 mm lateral and parallel to the midline on the animal's neck extending approximately 3 cm from the mandible caudally. This procedure is usually performed on the animal's left-hand side, in order for the SAH to be on the dominant side of the brain.
4. Once accomplished, the subcutaneous tissue, salivary glands, and lymphoid tissue are all reflected laterally. This allows for a good exposure showing the trachea in the midline, which is covered by the sternohyoid muscle. Laterally and superiorly, the sternomastoid muscle can be seen curving around the superior end of the sternohyoid and inferolaterally the omohyoid makes up the third side of the triangle.
5. Using blunt dissection, the muscles aforementioned are separated as follows. The omohyoid muscle is retracted laterally, as is the sternomastoid muscle. The inferior portion of the sternohyoid is reflected medially over the trachea.
6. Once this has been achieved, the common carotid artery (CCA) can be seen lateral to the trachea extending superiorly under the lateral edge of the hyoid cartilage, which lies under the sternomastoid, which is reflected. The hyoid cartilage is then gently reflected superiorly. This allows for complete exposure of the common, external (ECA) and internal carotid (ICA) arteries, and their associated branches.
7. The carotid vessels are also surrounded by a perivascular sheath which is incised. These vessels have a number of branches in the neck, and care must be taken to avoid cutting them. Some bleeding may occur during the dissection but this usually stops with pressure (Fig. 1). Additionally, diathermy can be used for persistent bleeding.
8. In the region of the division of the CCA into the ECA and the ICA lies the occipital artery, a branch of the ECA that travels superiolaterally, overlying the ICA for 2–3 cm. This is ligated, with a 10/0 silk tie and divided close to the ECA. A curved forceps is useful here to get your suture around the vessel. The occipital artery adheres tightly to the underlying ICA. It can be difficult to separate the two.
9. Another branch of the ECA can be seen more distally which travels medially; this is the superior thyroid artery. This vessel is also ligated and divided. It is not necessary to dissect out the superior thyroid artery completely. Simply ligate the vessel as it emerges from the ECA, as you will not encounter the vessel again in this procedure.
10. Finally, the ECA gives a third branch, the lingual artery. Just proximal to this division, the ECA is ligated and divided. The stump of the ECA is then reflected inferiorly (Fig. 2). The tie on the ECA is usually left long to easily pick it up later on.

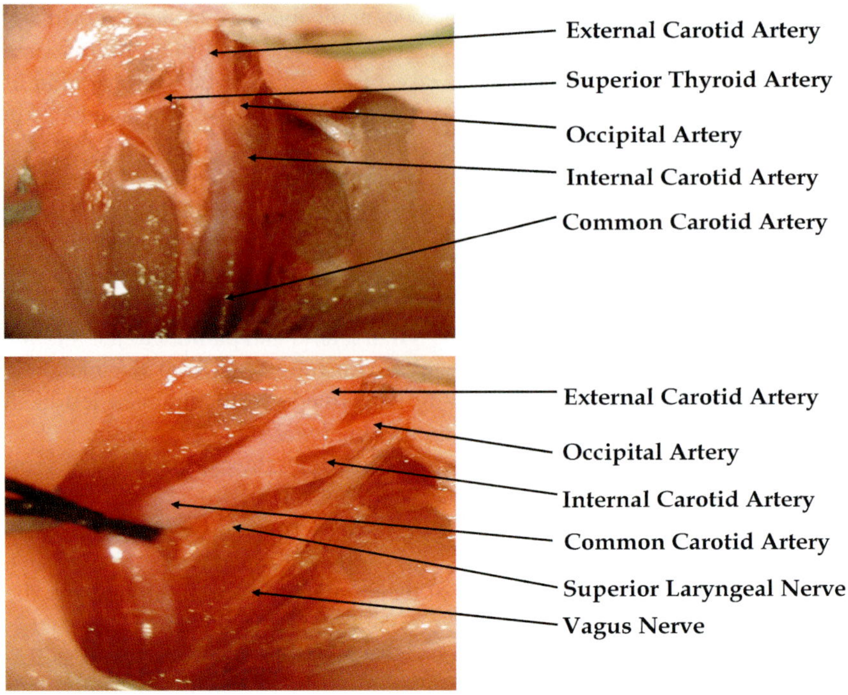

Fig. 1 Dissection of the carotid artery. After displacing the strap muscles of the neck laterally, the carotid sheath comes into view, which is dissected out of its sheath. Care must be taken to leave the vagus nerve undisturbed

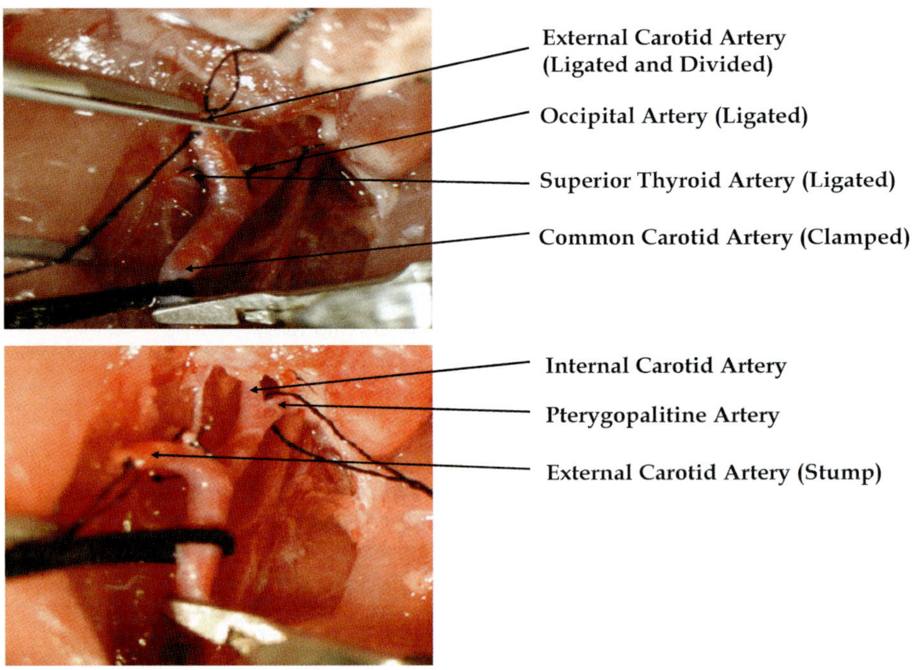

Fig. 2 External carotid artery ligation. The ECA is ligated to form a stump. The remaining branches of the vessel are also ligated to form a path for the insertion of the monofilament suture

11. Once this is accomplished, the ICA is dissected clear of the surrounding connective tissue. At its distal end, the ICA gives a branch, which travels laterally, this is the pterygopalatine artery. Care must be taken to completely free the CCA, the ICA and its branch, the pterygopalatine artery from the base of skull to as far caudally as is comfortable to retract.
12. Using a 10/0 silk tie, the base of the stump of the ECA is tied loosely. Furthermore, an 8 cm straight 4/0 nylon suture is fashioned. Using indelible ink, the tip is marked. Further marks are placed at 10, 20, and 30 mm. This allows you to gauge how much of the suture has passed through the vessel later on. The tip of the suture is then cut obliquely so as to sharpen it.
13. A vascular clamp is then placed across the base of the CCA as far proximally as possible. Another clamp is placed over the pterygopalatine artery at its base, parallel to the ICA. It is important to ensure that this clamp lies flush with the ICA, otherwise the suture can get stuck here when cannulating the ICA. Alternatively, the pterygopalatine artery can be ligated flush with the ICA. A final clamp is placed across the ICA as it penetrates the base of the skull.
14. Using microscissors the stump of the ECA is incised but not divided. This incision should be as small as possible to avoid tearing the vessel as it is manipulated. The 4/0 prepared nylon suture is then cannulated into the stump of the ECA, which has been deflected proximally producing a straight passage from the ECA to the ICA (Fig. 3).

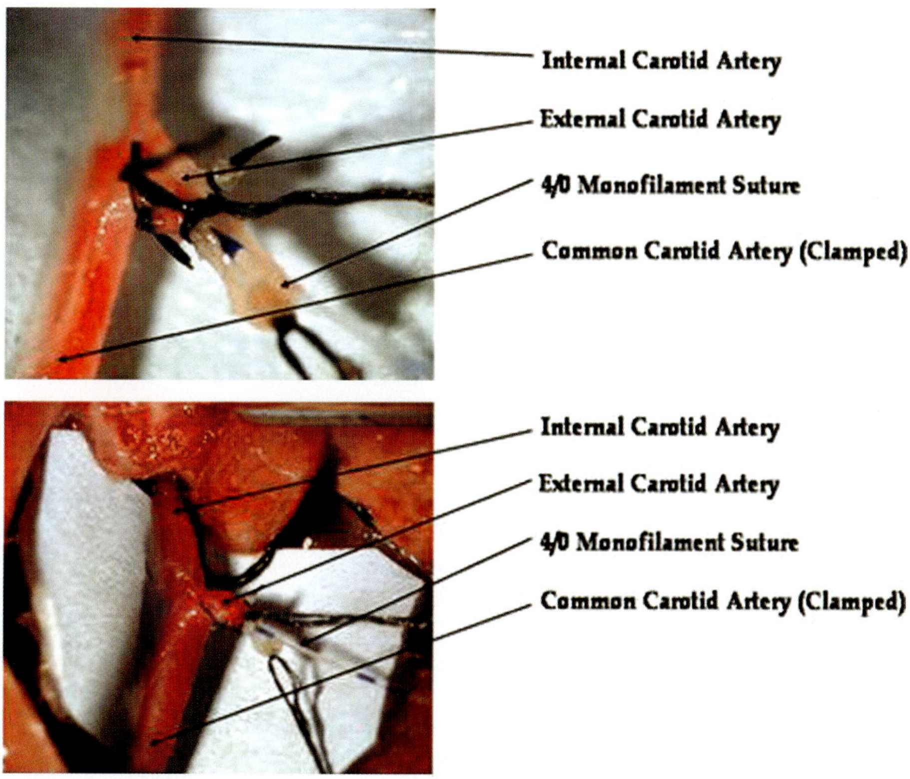

Fig. 3 Monofilament suture insertion. After clamping the ICA and the CCA, the ECA is cannulated using a 4/0 monofilament suture. Once the suture has advanced beyond the pterygopalatine artery, the clamp on the ICA is removed and the suture can be advanced further

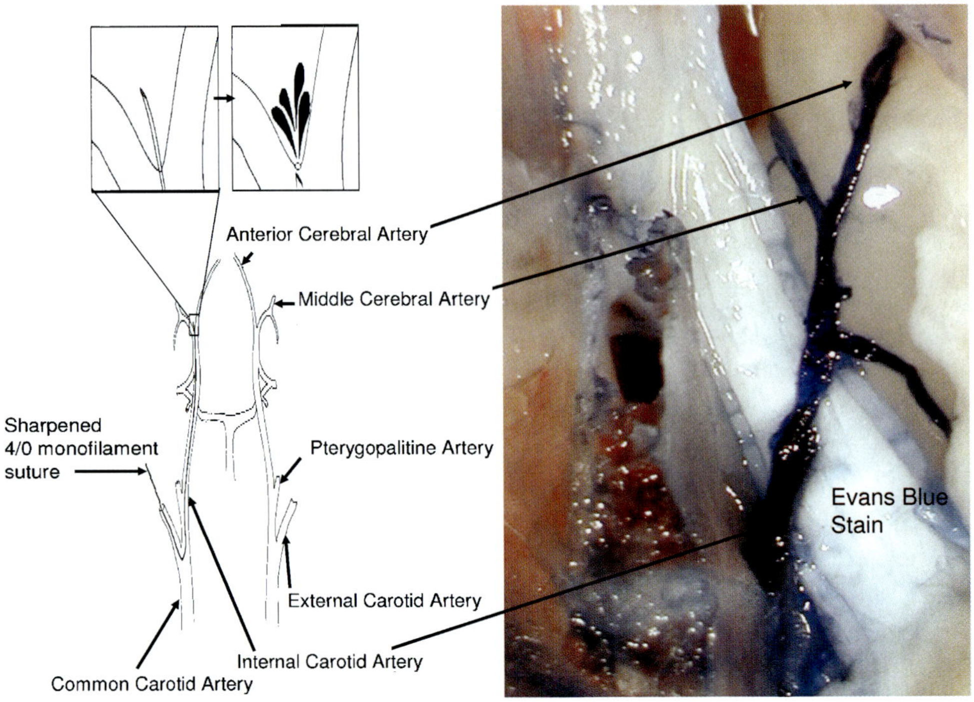

Fig. 4 Cartoon showing the anatomy of the circle of Willis. The monofilament suture is inserted through the ECA stump, into the ICA to eventually emerge between the ACA and the MCA. Cartoon adapted from Ref. 7

15. The 4/0 nylon suture is then advanced to the clamp across the ICA, bypassing the pterygopalatine artery. Once this is done, the silk tie at the base of the ECA is tied, but not completely.
16. The clamp across the ICA is then removed and the 4/0 nylon suture is advanced superiorly. It is approximately 27 mm from the junction of the ECA and the ICA to the junction of the MCA and the anterior cerebral artery (ACA) (Fig. 4). No resistance should be felt up to this point. If resistance is felt, the tip of the suture has hit the vessel wall. If this occurs simply retract the suture slightly, gently rotate it, and try advancing the suture again.
17. Initially the suture will travel through the ICA with ease. At approximately 25–30 mm resistance is felt. By advancing the suture forward slowly, it is possible to puncture the junction of the ACA and the MCA (Fig. 5).
18. Once a puncture has been successfully performed, the suture is withdrawn and the stump of the ECA is ligated. Both of the clamps across the CCA and the pterygopalatine artery are then removed, and the retractors are also removed.
19. After meticulous hemostasis has been achieved, the incision is closed using a 3/0 silk suture.

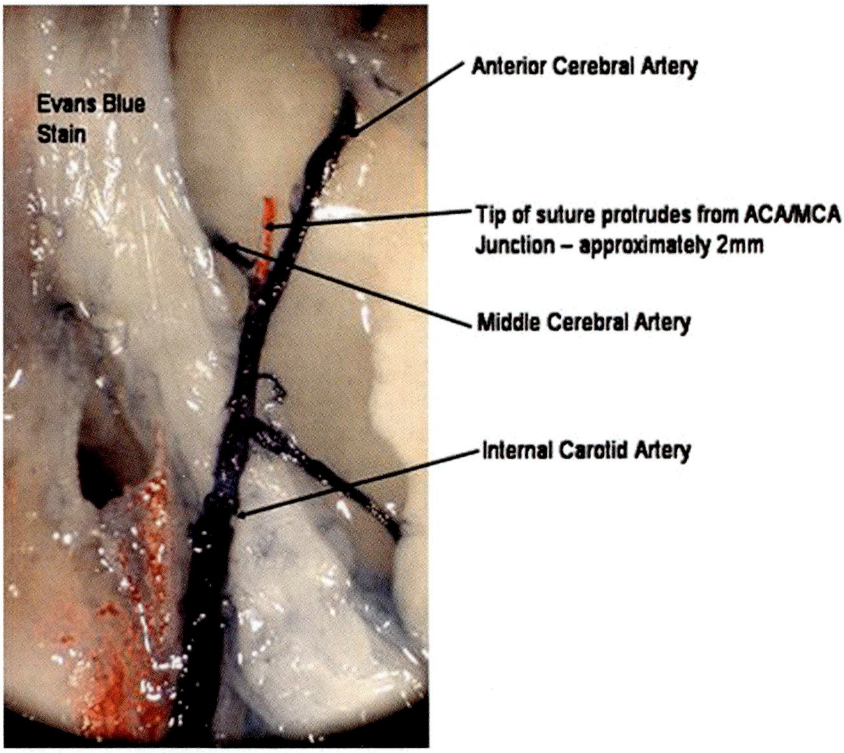

Fig. 5 The emergence of the monofilament from the junction of the ACA and the MCA. The monofilament suture is used to penetrate the junction of the ACA and MCA arteries to a depth of approximately 2 mm. This allows for a SAH to occur without an excessive mortality

4 Outcome Evaluations

1. Neurological outcome: Based on a neurological outcome scale such as that devised by Garcia et al, 1995.[8]
2. Brain edema: Brain water content can be measured using the formula (wet weight–dry weight/wet weight) × 100%.
3. Histology and immunohistochemistry: To confirm the presence of vasospasm and subsequent cellular apoptosis.

5 Applications, Complications, and Limitations

1. This model is useful for experiments concerning not only SAH but stroke as well. This model is not appropriate for use in large animal studies, largely because of the inhibitive cost associated with the high mortality.
2. Avoid using large amounts of alcohol-based solutions when preparing the animal for surgery, as this can lead to hypothermia.
3. The rat's skin is quite tough, particularly compared with human skin. A surprising degree of pressure will be required to incise the skin in the rat.

4. Care must be taken, to avoid excessive retraction on the trachea throughout the procedure as this leads to severe postoperative respiratory difficulties.
5. The hyoid cartilage should be gently reflected superiorly. This causes less trauma and fewer postoperative complications rather than excising the cartilage, which some authors recommend.
6. Do not be tempted to simply ligate the occipital artery without first freeing its attachments to the ICA. The ICA will need to be dissected out later in the procedure and if the occipital artery is still adherent to it, the risk of perforating the vessel increases.
7. Do not be tempted to use the tie on the ECA to manipulate the vessel as this can easily lead to tearing of the vessel resulting in catastrophic hemorrhage.
8. It is very important to use a straight length of the suture material. If the length of suture is curved, it can easily penetrate the wall of the ICA when the vessel is being cannulated. The tip of the suture should be carefully examined under the microscope to ensure that it is as sharp as possible. This will allow the suture to easily pass through the vessel wall at the junction of the ACA and the MCA. Some authors recommend different sized suture material, with which to make a puncture. In my experience, the 3/0 nylon suture results in a very high mortality rate while a 5/0 suture can be difficult to puncture the vessel wall with.
9. The purpose of the tie on the ECA is to prevent bleeding, once the clamp on the ICA is released. However, it is important to keep in mind that the suture must travel through the tie so it cannot be completely occluded. It is a good idea to release the clamp on the ICA and check the degree of bleeding from the ECA before removing the clamp completely. If the degree of hemorrhage is excessive, the clamp can be easily replaced and the tie on the ECA secured more firmly.
10. It is important not to advance the monofilament suture too far as this will result in the death of the animal, as the cortex will be penetrated by the suture. By not advancing far enough, the animal will not have a SAH. If the suture is advanced and retracted and advanced again you may end up with multiple holes in the vessel. This will result in a large SAH, which usually leads to the death of the animal. While advancing the suture, once resistance is felt, it is possible to feel a give as the suture penetrates the vessel wall. This is the most difficult part of the procedure as the "give" that can be felt as the suture penetrates the vessel is quite subtle. Furthermore, the procedure at this point is completely blind. There is no easy description of this aspect of the procedure. If you are using additional monitoring equipment such as cerebral blood flow monitoring or even blood pressure measurements, changes in either of these may indicate that a SAH has occurred. Please note that this is not a guarantee. Remember that the full extent of the injury will not be obvious until the clamps have been removed. With practice, however, the point of puncture will become easier to determine.
11. The suture used to close the skin does not really matter, unless the animals are expected to survive in the long term. If this is the case, a 3/0 viacryl rapide subcutaneous suture would be ideal as it would avoid the need to remove the sutures. At the point of removing the clamps, the full effect of the SAH will become evident. To successfully mimic a true SAH, the

vascular clamps should be removed as rapidly as possible after the puncture. This is to avoid a trickle of blood entering the subarachnoid space and slowly increasing the intracranial pressure prior to removing the clamps. Furthermore, if the removal of the clamps is delayed the puncture site may close secondary to local vasospasm or clot formation.

12. For a sham-operated group of animals, the operational procedure is essentially the same. The suture is made as before, except that the tip of the suture is not sharpened. The suture is placed as before into the ICA via the stump of the ECA and advanced. On this occasion, however, the suture is not advanced to more than 25 mm or until resistance is felt. The reasoning behind this is derived from the observation that the placement of a catheter within a vessel in humans can lead to spasm. This is often observed at angiography.[9] By placing the suture within the vessel in the rat, it is possible to exclude the possibility of the suture being responsible for vasospasm in the experiment as opposed to the SAH itself. Furthermore, it is possible to show that that the differences observed between sham-operated animals and those having a SAH are due to the hemorrhage and subsequent vasospasm as opposed to any vasospasm caused by the presence of the suture within the vessel.[10]
13. Throughout this procedure, it is recommended that continuous assessments of the animal's vital parameters are performed. This should include regular arterial blood gas analysis (ABG's) and subsequent adjustment of the anesthetic machine as appropriate. Either invasive or noninvasive measurements of blood pressure and heart rate are also recommended. Glucose and hematocrit levels pre and postoperatively may also be measured. Because of their large surface to body weight ratio rodents tend to loose heat rapidly and should always be kept warm. Throughout the procedure, a thermal blanket and rectal temperature {RET-2 rectal probe, (Kent Scientific)} measurement should be employed to prevent hypothermia, which can be protective in SAH. After the procedure is performed, it can take up to 2 h for the animal to regain consciousness and be stable enough to breathe on its own. Cerebral blood flow measurements can be extremely useful during this time, as well as the other parameters mentioned to gauge when the anesthetic should be lightened and eventually removed. Once the animal is self ventilating, extubation can be performed.

Acknowledgments: The authors wish to acknowledge Mr. Gen Kusaka for his help with the intraoperative photographs.

References

1. Cahill, J., Calvert, J. W., Solaroglu, I., & Zhang, J. H. (2006). Vasospasm and p53-induced apoptosis in an experimental model of subarachnoid hemorrhage. *Stroke* **37**, 1868–1874.
2. Yamaguchi, M., Zhou, C., Nanda, A., & Zhang, J. H. (2004). Ras protein contributes to cerebral vasospasm in a canine double-hemorrhage model. *Stroke* **35**, 1750–1755.
3. Zhou, C., Yamaguchi, M., Kusaka, G., Schonholz, C., Nanda, A., & Zhang, J. H. (2004). Caspase inhibitors prevent endothelial apoptosis and cerebral vasospasm in dog model of experimental subarachnoid hemorrhage. *J. Cereb. Blood Flow Metab.* **24**, 419–431.

4. Zubkov, A. Y., Aoki, K., Parent, A. D., & Zhang, J. H. (2002). Preliminary study of the effects of caspase inhibitors on vasospasm in dog penetrating arteries. *Life Sci.* **70**, 3007–3018.
5. Prunell, G. F., Mathiesen, T., Diemer, N. H., & Svendgaard, N. A. (2003). Experimental subarachnoid hemorrhage: subarachnoid blood volume, mortality rate, neuronal death, cerebral blood flow, and perfusion pressure in three different rat models. *Neurosurgery* **52**, 165–175.
6. Gules, I., Satoh, M., Clower, B. R., Nanda, A., & Zhang, J. H. (2002). Comparison of three rat models of cerebral vasospasm. *Am. J. Physiol Heart Circ. Physiol.* **283**, H2551–H2559.
7. Bederson, J. B., Germano, I. M., & Guarino, L. (1995). Cortical blood flow and cerebral perfusion pressure in a new noncraniotomy model of subarachnoid hemorrhage in the rat. *Stroke* **26**, 1086–1091.
8. Garcia, J. H., Wagner, S., Liu, K. F., & Hu, X. J. (1995). Neurological deficit and extent of neuronal necrosis attributable to middle cerebral artery occlusion in rats. Statistical validation. *Stroke* **26**, 627–634.
9. Chang, K. S., Wang, K. Y., Yao, Y. W., Huang, J. L., Lee, W. L., Ho, H. Y., Hsueh, C. W., Huang, D. S., Chen, Y. T., & Ting, C. T. (2000). Catheter-induced coronary spasm–a view of mechanical factors and experience with selective left coronary arteriography. *Zhonghua Yi. Xue. Za Zhi. (Taipei)* **63**, 107–112.
10. Kusaka, G., Ishikawa, M., Nanda, A., Granger, D. N., & Zhang, J. H. (2004). Signaling pathways for early brain injury after subarachnoid hemorrhage. *J. Cereb. Blood Flow Metab.* **24**, 916–925.

Chapter 22

Blood Injection Subarachnoid Hemorrhage Rat Model

Hartmut Vatter*, Erdem Gueresir, and Volker Seifert

Abstract The pathophysiological mechanisms after subarachnoid hemorrhage (SAH) may be divided in an acute phase after the bleeding, which includes increased intracranial pressure, reduced cerebral blood flow (CBF), and delayed consequences of the subarachnoid blood clot. The subarachnoid blood infusion in rat injection models could be classified in to volume and pressure-dependent and furthermore in to single or double hemorrhage. Investigating the acute phase pressure-dependent models seems to mimic the pathological sequel superior to volume dependent. However, the development of delayed mechanisms, like a reduced CBF due to cerebral vasospasm or other pathological mechanisms seems mainly to be dependent on the amount and the duration of the subarachnoid blood. Accordingly, the double cisterna magna blood injection model will be presented in this chapter, in which a sufficient subarachnoidal blood distribution and contact time could be achieved with a reduced acute phase effect. Therefore, this model seems to be predisposed for investigation on delayed consequences of SAH.

Keywords: Rat, Subarachnoid hemorrhage, Vasospasm, Blood injection cisterna magna

1 Model Selection

The pathophysiological mechanisms after subarachnoid hemorrhage (SAH) mainly consist of an initial acute phase after the bleeding and delayed consequences of the subarachnoid blood clot. The main characteristics of the acute phase are elevation of intracranial pressure (ICP), reduced cerebral perfusion, and cerebral blood flow (CBF).This may lead to an acute injury cascade resulting in cerebral injury during the aneurysm rupture.[1]

However, delayed neurological deterioration occurs in about 30% of patients suffering from SAH,[2] and seems to be more capable for potential treatment or

*Department of Neurosurgery, Johann Wolfgang Goethe-University, Frankfurt/Main, Germany, H.Vatter@em.uni-frankfurt.de

From: *Animal Models of Acute Neurological Injuries.*
Edited by: J. Chen (eds.) New York, Totowa, NJ

prophylaxis than the initial insult. Cerebral vasospasm (CVS) leading to cerebral ischemia and infarction is the classic cause of delayed neurological deterioration. Recent clinical investigations, however, demonstrated prevention of vasospasm by an endothelin receptor antagonist without improved outcome. Therefore, the causes of the delayed neurological deterioration continue to be poorly understood,[3] supporting the need of further experimental investigations. In parallel with the need for experimental data, ethical concerns and cost issues have forced the use of small affordable animals, which made rats becoming increasingly popular.[4]

Induction of experimental SAH in the rat was developed by vessel disruption and blood injection[5] models. The blood injection models may further be classified in volume and pressure-dependent blood infusion methods.[4] In pressure-dependent infusion models mainly the middle arterial pressure is used leading to a pronounced acute injury mimicking the initial bleeding[1] before delayed mechanisms can be observed.

The development of delayed mechanisms, like a reduced CBF due to CVS or other pathological mechanisms, however, seems mainly to be dependent on the amount and the distribution of the subarachnoid blood. Additionally, the duration of the blood degradation seems to be a crucial factor for these delayed effects.[2] Therefore, volume-dependent injection models are focused on these effects. However, even slow and careful blood application in the subarachnoid space results in an acute effect consisting of increased ICP and reduced CBF, which depends on the amount of injected blood.[1,6] Accordingly, the amount of the singular blood application seems to be limited to 300 ∝l or less in rats if the focus of the investigation should be on delayed mechanisms.

In single hemorrhage models, a biphasic development of CVS was demonstrated in several investigations with the first peak less than 1 h after the injection and a second peak on day one or two. Maximum reduction of the angiographic diameter of cerebrovasculature was roughly 20–50% of control subjects.[6] The relatively short time course of the development of delayed deficits may be caused by the increased blood degradation in rodents compared with humans.[5] This could intersperse initial and delayed pathological mechanisms. Therefore, double hemorrhage models were developed,[7] in which the time course of delayed deficits is extended to reflect the clinical situation in a better way. In these models, maximum constriction of cerebral arteries occurs between day 3 and 5 after the initial bleeding. Furthermore, the clinical and the angiographic time course and the reduced CBF was characterized recently, and revealed a maximum effect on day 5.[6,8,9] Accordingly, the rat double hemorrhage model, which is described in the following, seems to be an appropriate and assessable small animal model for investigations on delayed pathological effects of SAH.

2 Materials and Methods

Adult male Sprague-Dawley or Wistar rats weighing 250–500 g are commonly used for the double hemorrhage model.[6,8]

For anesthesia, chloral hydrate (250 mg kg^{-1})[7,10] or ketamine (100 mg kg^{-1}) in combination with xylazine (10 mg kg^{-1})[8] or midazolam (1 mg kg^{-1}) was described.[6] In our experience, the use of intraperitoneal application of midazolam (1 mg kg^{-1} body weight) and ketamine (100 mg kg^{-1} body weight) is safe and good feasible. Animals breathe spontaneously and sufficient, circulation parameters are also relatively stable and a disturbed CBF was not observed by this medication.[11,12] Body temperature of the animals should be maintained at approximately 37°C by a heading pad.

For measurement of blood gas values, for control of the circulation, and for the source of the injected blood, an arterial cannulation is necessary. Additional medication or infusions (crystalloids or hydroxyethyl-starch) to maintain a stable circulation can be administered by a venous approach. Therefore, insertion of a tube (Portex ® polythene tube, luminal diameter: 0.96 mm) in both vessels by one approach is recommended before the procedure. Both tubes can be left inside the vessels until the animals are sacrificed. They can be plugged after irrigation with saline infusion and stored under the skin.

After cannulation of both vessels, the animals are positioned in a stereotactic frame (Fig. 1). The materials necessary for the procedure are demonstrated and described in Fig. 2.

A medial incision should be done after infiltration of the skin and the muscles by local anesthesia (e.g., mepivacaine 1%). The incision should cover the suboccipital region and the arch of C1. Acromiotrapezius muscle has to be cut strictly in the midline to avoid bleeding. Splenius and rhomboid muscle can be laterally displaced with a dissector and kept with a retractor. After that the suboccitital region, the atlanto-occipital membrane, and the C1 arch is exposed. A catheter (Portex ® polythene tube, luminal diameter: 0.28 mm) can be inserted in the cisterna magna after small incision of the atlanto-occipital membrane and the arachnoidea below (Fig. 3). Using this catheter, 0.1 ml of cerebrospinal fluid is withdrawn with a syringe. Thereafter, 0.2 ml of autologous blood are obtained from the femoral artery and slowly injected into the cisterna magna through the catheter to induce the SAH. Leakage of the injected blood can be avoided by fibrillar haemostypticum patches (Tabotamp) and cotton. Animals should be brought in a head down position immediately after the injection and kept in this position for 15 min to ensure a subarachnoid distribution of the blood. After surgical closure of the wounds, application of 5 ml crystalloid solution with 0.0125 mg Fentanyl subcutaneously is recommended.

The identical operative procedure can be repeated 24 h later for induction of the second SAH. Alternatively, the blood injection can be repeated after 48 h, which lowers the mortality rate but results in lower pronounced delayed ischemic deficits.[6,8,9]

During the next days, animals should receive (dependent on their clinical behavior) 5 ml of crystalloid solution with 0.0125 mg Fentanyl subcutaneously twice a day.

3 Evaluation

Delayed consequences of SAH are clinically diagnosed by the so-called delayed ischemic neurological deterioration (DIND). The gold standard in radiological investigations is the digital subtraction angiography (DSA). However, cerebral perfusion measurement by computer tomography (CT) or magnetic resonance (MR) tomography are becoming more important[2] for diagnosis and monitoring of impaired CBF after SAH in the clinical setting. Accordingly, these criteria should also be applied in the animal model.

Fig. 1 Schematic drawing showing a rat positioned in the stereotactic frame

Neurological deterioration of rats after experimental SAH was tried to be objectified by scales and blinded observers. For the rat double hemorrhage model scales by Ryba[13] and Bederson[6,14] were described and used. However, majority of the rats recover only very slowly after the second artificial bleeding,

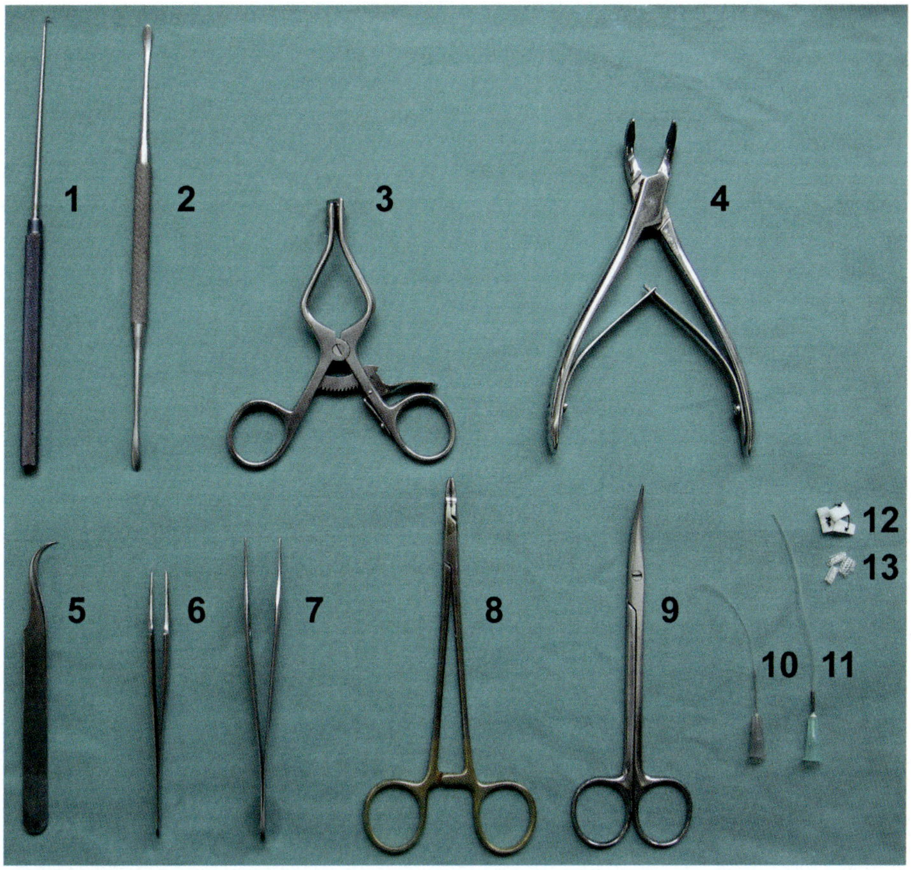

Fig. 2 List of materials necessary for the blood injection SAH model. (1) mikro nerve hook (Codman 80–1532); (2) dissektor (Aesculap OL 165 R); (3) retractor (Fehling ECE-3); (4) rongeur forceps (Aesculap FO410R); (5) forceps, curved (Aesculap BD333R); (6) forceps, fine (Dumont-Met 5); (7) forceps (Aesculap BD513R); (8) needle holder (Fehling KAS-1); (9) scissor (Aesculap BC276R); (10) and (11) Portex ® polythene tube, luminal diameter: 0.28 and 0.96 mm; (12) cotton; (13) fibrillar haemostypticum patches (Tabotamp)

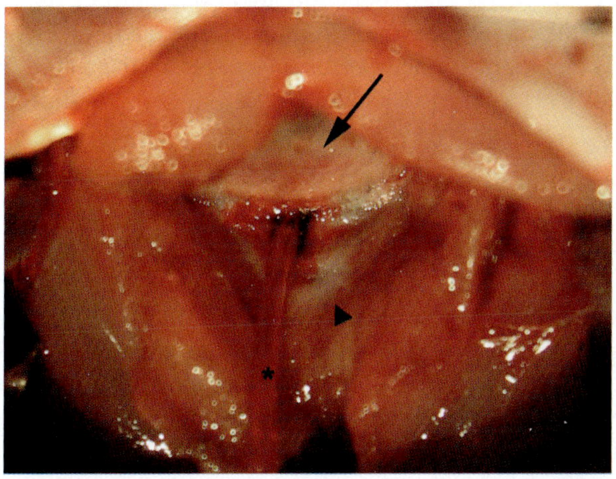

Fig. 3 Intraoperative photography of the suboccipital region showing the suboccipital bone (*arrow*), C1 (*arrowhead*) and the catheter (*asterisk*) inserted in the cistrena magna after penetrating the atlanto-occipital membrane

and a secondary deterioration is difficult to discriminate from the initial effect. Accordingly, neurological signs or the mortality may not be very feasible parameters for the evaluation of the model.

The difficulty in angiographic investigations of rat cerebral arteries because of the small vessel size has been discussed previously, and several techniques have been investigated.[10,15–17] Selective catheterization of the VA shows the highest resolution in imaging the BA but contains a higher risk of cardiac abnormalities caused by the contrast medium usually after multiple angiographies.[16] The average diameter of the rat basilar artery (BA) seems to be a relatively robust value in adult rats. Therefore, the use of one selective DSA at the end of the experiment seems to represent a practical reference parameter for the quantitative determination of delayed CVS in the rat. The time course of the development of angiographic CVS in this model was characterized by our laboratory and we observed a maximum on day 5,[6] which is in line with morphological investigations.[8,9] The maximum reduction of the angiographic BA diameter by the induced CVS amounted to about 50%.[6]

Another method for the evaluation of CVS is the histomorphometrical analysis of cerebral arteries. Commonly used parameters are the inner vessel diameter or the area of the lumen and the thickness of the wall[8,9] in histological sections. Consistently, with angiographic results the reduction of the arterial diameter in histological investigations is about 60–50% of control.[8,9,18] At the same time, the vessel wall thickness increases of about 40–50%.[8,9] Similar quantitative results were observed by analysis of the BA diameter from digital images taken postmortally after excision of the brain,[19] which represents a further feasible evaluation method.

Determination of CBF and/or cerebral blood volume (CBV) is possibly the pathophysiologically most important parameter. In the single hemorrhage rat model, an initial drop of CBF by approximately 40–70% has been shown by laser Doppler flowmetry,[4,20] hydrogen clearance,[21] or microsphere techniques.[7,22,23] These methods are largely not suitable for the detection of delayed effects. Therefore, the application of the noninvasive MR perfusion weighted imaging (PWI) method was the first to characterize the decrease in CBF and CBV in the rat double hemorrhage model in a semiquantitative fashion. Both CBV and CBF are reduced to about one third of the control value on day 5 after the initial bleeding.[6] Similar results may be obtained by more assessable CT angiography and perfusion images in the future but were not established for a rat SAH model so far.

4 Applications and Limitations

Aim of rat subarachnoid blood injection models is mainly to imitate CVS or delayed cerebral perfusion deficits after SAH. Rats are common, relatively robust, and the environment for their keeping and for the present experimental setting is assessable or can be established in the most centers. Furthermore, immunohistochemical or molecular biological equipment like antibodies are more frequently used and available for rats compared with other animals like dogs or monkeys.

The easier single hemorrhage model in the rat has the advantage of a low mortality and complication rate.[4] However, the neurological deterioration and development of CVS seems to be moderate, and the discrimination between acute and delayed effects of the SAH is sometimes difficult.[1,5]

The double hemorrhage model stretches the time course of the pathophysiological sequale. Thus, it comes closer to the clinical setting. Furthermore, the deterioration, the CVS, and the impaired blood flow are more pronounced compared with single models.[6] However, the line between a significant pathological effect on the CBF and the survival of the animals seems to be small in rats. Therefore, it is difficult to induce severe delayed CVS or DIND and associated with a high mortality rate. This mortality even increases by the use of invasive procedures like DSA, which further complicates the monitoring of the pathological effect. The cost effectiveness of the keeping of the rats is in contrast to the extensive and complex evaluation procedures like DSA or MR imaging.

The rat double hemorrhage model represents, therefore, a cost effective small animal model for the development of DIND, which affords considerable stamina of the experimentator. It seems to be predestinated for basic science investigations on the impaired function and morphology of the cerebrovasculature and other tissues during the development of CVS. The extent of this model by in vivo evaluations like MRI or DSA considerably increases the complexity of the model but imitates the clinical course in humans closer than any established rodent model. It may, therefore, additionally be adequate for the use in pharmacological investigations instead of more significant estimated animals like dogs and monkeys. The rat injection models may effect that studies on these animals could be preserved on directly preclinical questions.

References

1. Sehba FA, Bederson JB: Mechanisms of acute brain injury after subarachnoid hemorrhage. Neurol. Res. 28:381–398, 2006
2. Loch MR: Management of cerebral vasospasm. Neurosurg. Rev. 29:179–193, 2006
3. Macdonald RL, Pluta RM, Zhang JH: Cerebral vasospasm after subarachnoid hemorrhage: the emerging revolution. Nat. Clin. Pract. Neurol. 3:256–263, 2007
4. Prunell GF, Mathiesen T, Svendgaard NA: Experimental subarachnoid hemorrhage: cerebral blood flow and brain metabolism during the acute phase in three different models in the rat. Neurosurgery 54:426–436, 2004
5. Megyesi JF, Vollrath B, Cook DA, et al: In vivo animal models of cerebral vasospasm: a review. Neurosurgery 46:448–460, 2000
6. Vatter H, Weidauer S, Konczalla J, et al: Time course in the development of cerebral vasospasm after experimental subarachnoid hemorrhage: clinical and neuroradiological assessment of the rat double hemorrhage model. Neurosurgery 58:1190–1197, 2006
7. Solomon RA, Antunes JL, Chen RY, et al: Decrease in cerebral blood flow in rats after experimental subarachnoid hemorrhage: a new animal model. Stroke 16:58–64, 1985
8. Meguro T, Clower BR, Carpenter R, et al: Improved rat model for cerebral vasospasm studies. Neurol.Res. 23:761–766, 2001
9. Miyagi Y, Carpenter RC, Meguro T, et al: Upregulation of rho A and rho kinase messenger RNAs in the basilar artery of a rat model of subarachnoid hemorrhage. J. Neurosurg. 93:471–476, 2000
10. Delgado TJ, Brismar J, Svendgaard NA: Subarachnoid haemorrhage in the rat: angiography and fluorescence microscopy of the major cerebral arteries. Stroke 16:595–602, 1985

11. Rousselle CH, Lefauconnier JM, Allen DD: Evaluation of anesthetic effects on parameters for the in situ rat brain perfusion technique. Neurosci. Lett. 257:139–142, 1998
12. Lei H, Grinberg O, Nwaigwe CI, et al: The effects of ketamine-xylazine anesthesia on cerebral blood flow and oxygenation observed using nuclear magnetic resonance perfusion imaging and electron paramagnetic resonance oximetry. Brain Res. 913:174–179, 2001
13. Ryba MS, Gordon-Krajcer W, Walski M, et al: Hydroxylamine attenuates the effects of simulated subarachnoid hemorrhage in the rat brain and improves neurological outcome. Brain Res. 850:225–233, 1999
14. Bederson JB, Pitts LH, Tsuji M, et al: Rat middle cerebral artery occlusion: evaluation of the model and development of a neurologic examination. Stroke 17:472–476, 1986
15. Weidauer S, Vatter H, Dettmann E, et al: Assessment of vasospasm in experimental subarachnoid hemorrhage in rats by selective biplane digital subtraction angiography. Neuroradiology 48:176–181, 2006
16. Verlooy J, Van Reempts J, Haseldonckx M, et al: The course of vasospasm following subarachnoid haemorrhage in rats. A vertebrobasilar angiographic study. Acta Neurochir. (Wien.) 117:48–52, 1992
17. Longo M, Blandino A, Ascenti G, et al: Cerebral angiography in the rat with mammographic equipment: a simple, cost-effective method for assessing vasospasm in experimental subarachnoid haemorrhage. Neuroradiology 44:689–694, 2002
18. Vatter H, Konczalla J, Weidauer S, et al: Characterization of the endothelin-B receptor expression and vasomotor function during experimental cerebral vasospasm. Neurosurgery 60:1100–1108, 2007
19. Trandafir CC, Nishihashi T, Wang A, et al: Participation of vasopressin in the development of cerebral vasospasm in a rat model of subarachnoid haemorrhage. Clin. Exp. Pharmacol. Physiol 31:261–266, 2004
20. Naveri L, Stromberg C, Saavedra JM: Angiotensin IV reverses the acute cerebral blood flow reduction after experimental subarachnoid hemorrhage in the rat. J. Cereb. Blood Flow Metab. 14:1096–1099, 1994
21. Jackowski A, Crockard A, Burnstock G, et al: The time course of intracranial pathophysiological changes following experimental subarachnoid haemorrhage in the rat. J. Cereb. Blood Flow Metab. 10:835–849, 1990
22. Saida A, Ito H, Shibuya T, et al: Time-course alterations of monoamine levels and cerebral blood flow in brain regions after subarachnoid hemorrhage in rats. Brain Res. Bull. 43:69–80, 1997
23. Clozel M, Watanabe H: BQ-123, a peptidic endothelin ETA receptor antagonist, prevents the early cerebral vasospasm following subarachnoid hemorrhage after intracisternal but not intravenous injection. Life Sci. 52:825–834, 1993

Chapter 23

Filament Perforation Subarachnoid Hemorrhage: Mouse Model

Hideyuki Kamii* and Teiji Tominaga

Abstract The filament perforation subarachnoid hemorrhage (SAH) mouse model mimics a clinical situation with aneurysmal SAH in humans. In this model, subarachnoid clot induces vasospasm, which has a peak at 3 days after SAH. Moreover, a sudden increase in intracranial pressure immediately after the arterial perforation induces cortical neuronal damage similar to aneurysmal SAH in humans. Therefore, this model is suitable for researches on the various pathophysiologies of SAH. Although uncontrollable volume of SAH basically causes high mortality rate in this model, our techniques perforating the anterior cerebral artery can lead to less mortality and relatively reproducible state after SAH. The complications and limitations in this model are also discussed.

Keywords: Subarachnoid hemorrhage, Mouse, Model, Filament, Perforation

1 Model Selection

The filament perforation subarachnoid hemorrhage (SAH) mouse model has been recently employed to investigate the mechanisms underlying vasospasm and neuronal injury after SAH.[1–5] This model has modified rat SAH model on the basis of endovascular arterial rupture.[6,7] As in the rat SAH model,[8] the mortality rate in the mouse filament model is higher than that in the cisterna magna SAH model, which induces SAH by blood injection into the cirterna magna.[9,10] When only vasospasm after SAH is investigated in the mouse, the cisterna magna SAH model might be suitable for the study because of the low mortality rate. In the filament model, however, SAH causes brain parenchymal damage as well as vasospasm,[2,5] which can mimic the pathological conditions of SAH in humans. With the recent successful development of transgenic and knockout mice, the studies on SAH using a mouse model have become more

*Department of Neuroanesthesia, Kohnan Hospital, Taihaku-ku, Sendai 982-8523, Japan, kamii@kohnan-sendai.or.jp

From: *Animal Models of Acute Neurological Injuries.*
Edited by: J. Chen (eds.) New York, Totowa, NJ

prevalent.[11] The filament perforation SAH mouse model can be used to elucidate molecular mechanisms underlying the pathogenesis of SAH.

2 Material

2.1 Animal

Adult male CD-1 mice, 10–12 weeks of age (35–40 g), are selected in our laboratory. Also C57BL/6J mice are used in several reports.[3,4,9,10] Although the strain of mice is not so an important factor in SAH model as in ischemic model, it seems easier to make SAH model in bigger mice. Since CD-1 mice (35–40 g) are a little bigger than C57BL/6J mice (30–35 g) in 10–12 week of age, we usually use male CD-1 mice for SAH model.

2.2 Anesthesia

The mice are housed under a 12-h light/dark cycle with free access to food and water. Anesthesia is induced in a chamber with 2% halothane in 30% oxygen and 70% nitrous oxide after an intraperitoneal injection of xylazine (10 mg kg^{-1}).[12] The mice are placed in the supine position on an operating table, and the rectal temperature of the animals is maintained at 37°C by a Homeothermic Blanket Control Unit (Harvard Apparatus). Mice are allowed to respire spontaneously, and the anesthesia is maintained with 0.5% halothane in 30% oxygen and 70% nitrous oxide using a face mask during the operation. The left femoral artery is cannulated with a PE-10 catheter for measurement of mean arterial blood pressure, PaO2, PaCO2, and pH.

3 Procedures

3.1 Induction of SAH

1. Make a midline incision of the neck skin, and expose the left common carotid artery (CCA) with careful conservation of the vagus nerve. Then, isolate the external carotid artery (ECA), superior thyroid artery (STA), and occipital artery (OA). The internal carotid artery (ICA) and pterygopalatine artery (PPA) are observed under the OA (Fig. 1a).
2. Electrocoagulate and divide the STA and OA. After placing a microsurgical clip across the ECA origin, ligate the distal end of the ECA with 6-0 silk suture and cut it. Tie another 6-0 suture loosely around the ECA stump. Then, introduce a 2-cm length of 5-0 nylon monofilament, blunted at the tip by heat, into the ECA stump, and tighten the silk suture around it. After removing the microsurgical clip, advance the nylon filament into the ICA through the ECA stump. Take care to avoid introducing the filament into the PPA (Fig. 1b).
3. Advance the filament gently until encountering a resistance. At this point, the tip of the filament is in the left anterior cerebral artery (ACA) near the confluence with the right ACA (Fig. 2a). After pulling back the filament 2 mm (Fig. 2b), advance the filament 5 mm further to perforate the left ACA (Fig. 2c), and withdraw immediately through the ICA into the ECA stump, allowing reperfusion and SAH (Fig. 3a).

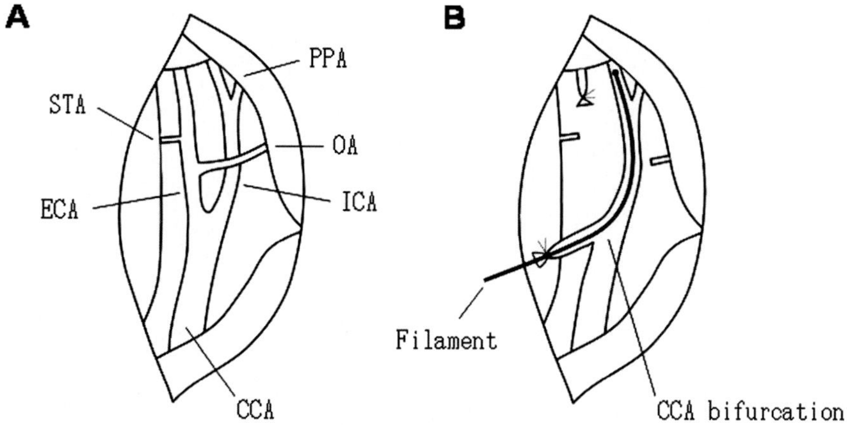

Fig. 1 Schematic representations showing procedures to introduce the filament into the left ICA. (**a**) After a midline incision of the neck skin, the left common carotid artery (CCA) is exposed, and then the external carotid artery (ECA), superior thyroid artery (STA), and occipital artery (OA) are isolated. The internal carotid artery (ICA) and pterygopalatine artery (PPA) are observed under the OA. (**b**) The nylon filament is introduced into the ICA through the ECA stump. Take care to avoid introducing the filament into the PPA

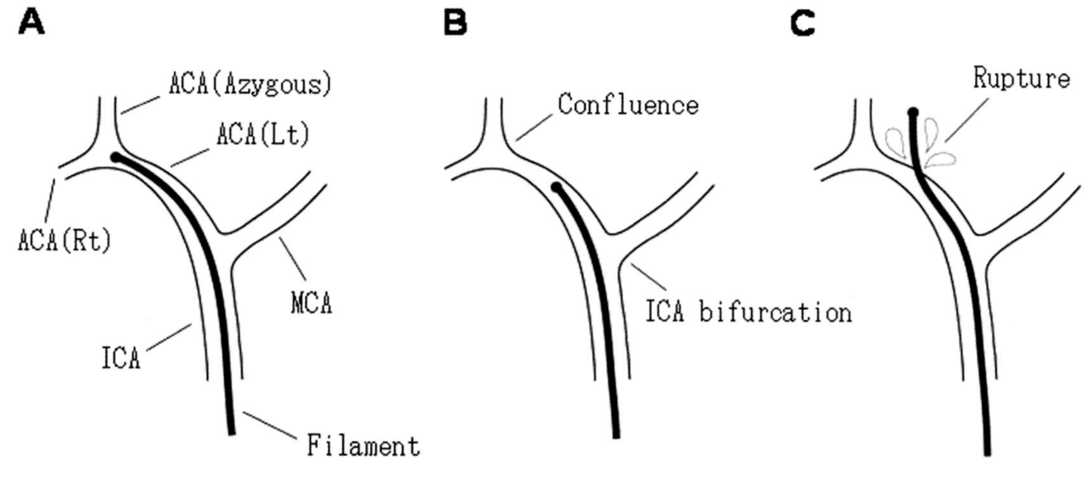

Fig. 2 Schematic representations showing procedures to perforate the left ACA. (**a**) The tip of the filament is in the left anterior cerebral artery (ACA) near the confluence with the right ACA, when a resistance is encountered. (**b**) Pull back the filament 2 mm. (**c**) Then, advance the filament 5 mm further to perforate the left ACA

4. When endovascular SAH occurs, mice show a respiratory failure for several seconds, followed by recovering regular respiration without any treatments. The distance from the CCA bifurcation to the confluence with the right ACA (the origin of the azygous ACA) measures approximately 11 mm. Sham-operated control mice undergo identical procedures except that the filament is withdrawn just after encountering the resistance.
5. After cutting the nylon filament protruding over the ECA stump, close the incision with silk sutures. The entire procedure is completed within approximately 20 min.

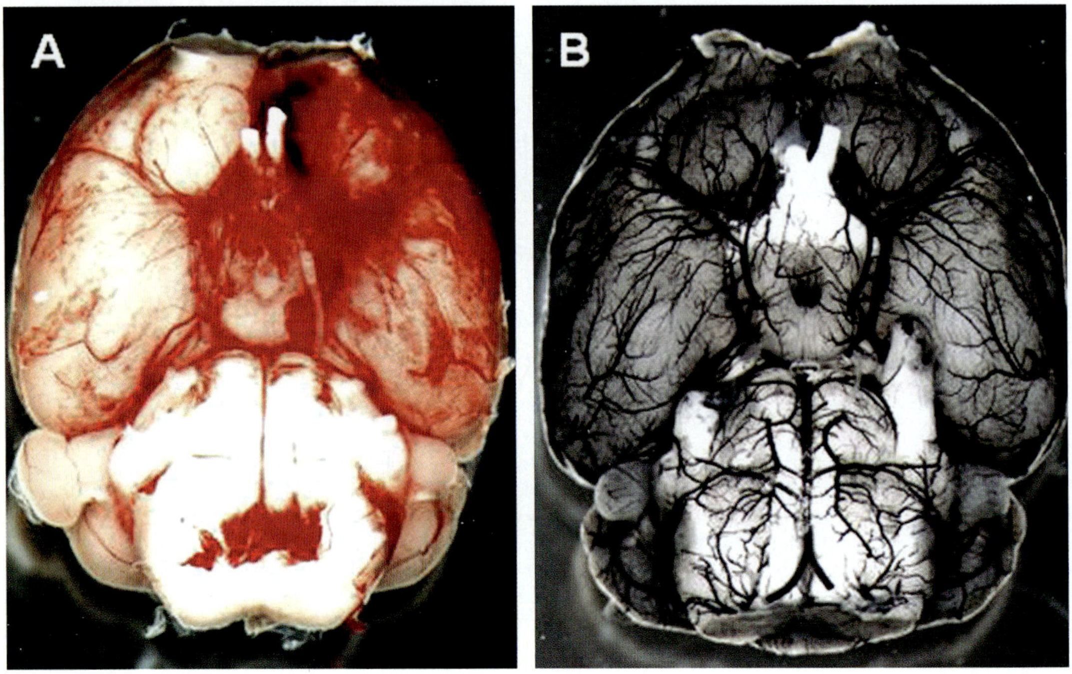

Fig. 3 Representative *photographs* of the brain. (**a**) At 4 h after SAH, blood clots are diffusely observed on the brain surface, especially around the left ACA. (**b**) All vessels of the brain are casted with carbon ink and 10% gelatin

3.2 Perfusion-Fixation and Casting Method

1. Anesthetize the mice with an intraperitoneal injection of 20 mg kg^{-1} pentobarbital. Open the chest, and cannulate the aorta through the left ventricle with a 23-gauge needle. Just after making an incision in the right atrium, begin perfusion with 200 ml of normal saline including heparin, followed by 200 ml of 10% formalin or 4% paraformaldehyde. Perfusion pressure is controlled by setting a container of the solution at 90 cm height from the mice.
2. In case of evaluating the arterial diameter in the mice, casting method is performed following the perfusion-fixation. Gelatin powder (10 g) is dissolved in 100 ml of saline and mixed with 100 ml of carbon ink, and the solution is maintained at 40°C. After the perfusion with 10% formalin, perfuse all vessels with 100 ml of the carbon ink/gelatin solution. Then, refrigerate the mice for 2 h to allow gelatin solidification. Finally, remove the brains, and store them in 10% formalin (Fig. 3b).

4 Outcome Evaluation

4.1 Neurological Examination

It is important to confirm a mouse without hemiparesis after the surgery. When the mice showed right hemiparesis upon the recovery, our preliminary study revealed that the left MCA was often occluded completely, and that a big cerebral infarction in the left MCA area was observed in the brain. Therefore, when a mouse shows right hemiparesis after the surgery, the mouse should be excluded from the study.

4.2 Measurement of the Arterial Diameter

Vasospasm is evaluated by measuring the diameter of proximal portion of the left MCA by a stereomicroscope (Fig. 4). We do not use the ICA and ACA for evaluation of vasospasm, since the filament passes through them, resulting in mechanical injuries to the arteries. In our series, vasospasm peaked at 3 days after SAH (Fig. 5).

4.3 Immunohistochemistry in the Cerebral Artery and Cortex

The mouse brains are embedded in paraffin. Coronal 5-∝m-thick sections near the ICA bifurcation are cut and mounted on slide. The sections can be advanced to the ordinary or immunohistochemical staining for the artery as well as the cortex. In our filament perforation SAH mouse model, iNOS protein was induced in the cerebral arteries as well as in the cerebral cortices (Fig. 6).

5 Advantages, Complications, and Limitations

5.1 Advantages

1. The filament perforation SAH mouse model mimics a clinical patient with aneurysmal SAH in humans, compared with the cisterna magna SAH

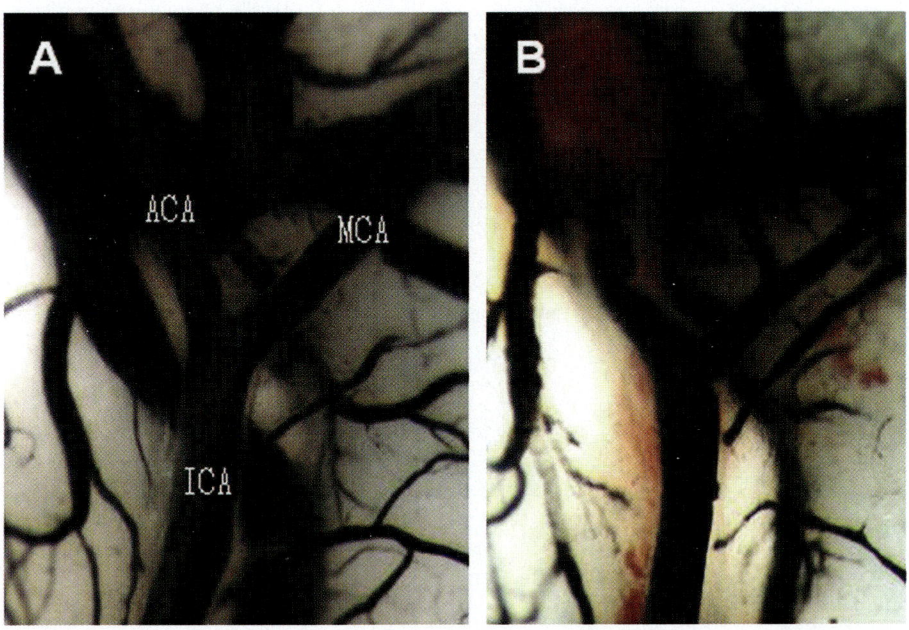

Fig. 4 Representative *photographs* of the cerebral arteries casted with carbon ink and 10% gelatin. (**a**) The left ICA, ACA, and MCA are clearly demonstrated by a stereomicroscope in a sham-operated control mouse. (**b**) At 3 days after SAH, the diameter of the MCA is reduced compared with that in a sham-operated control mouse. Note that the left ACA is perforated between the ICA bifurcation and the confluence with the right ACA

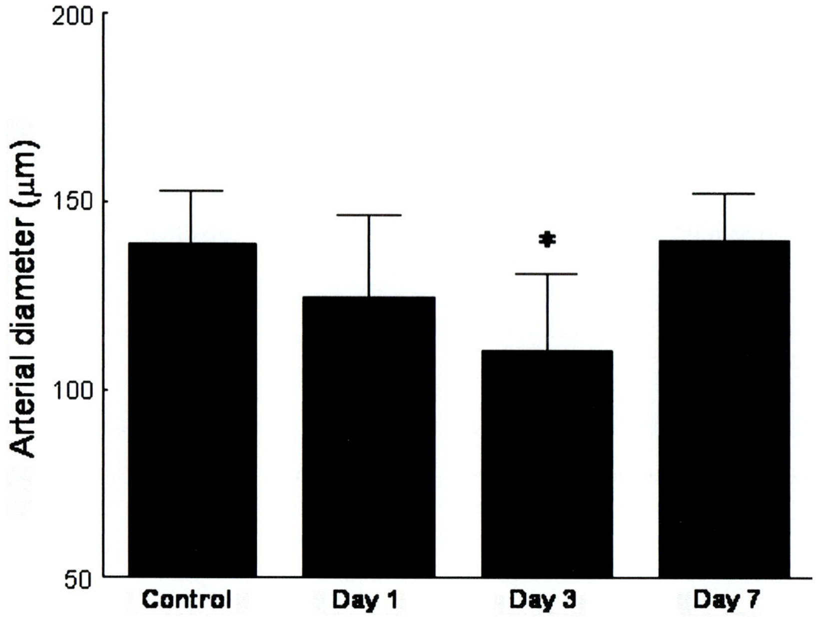

Fig. 5 Sequential changes in the diameter of the MCA after SAH in our series. At 3 days after SAH, the MCA diameter was significantly reduced compared with that in sham-operated control mice ($^{*}p < 0.01$)

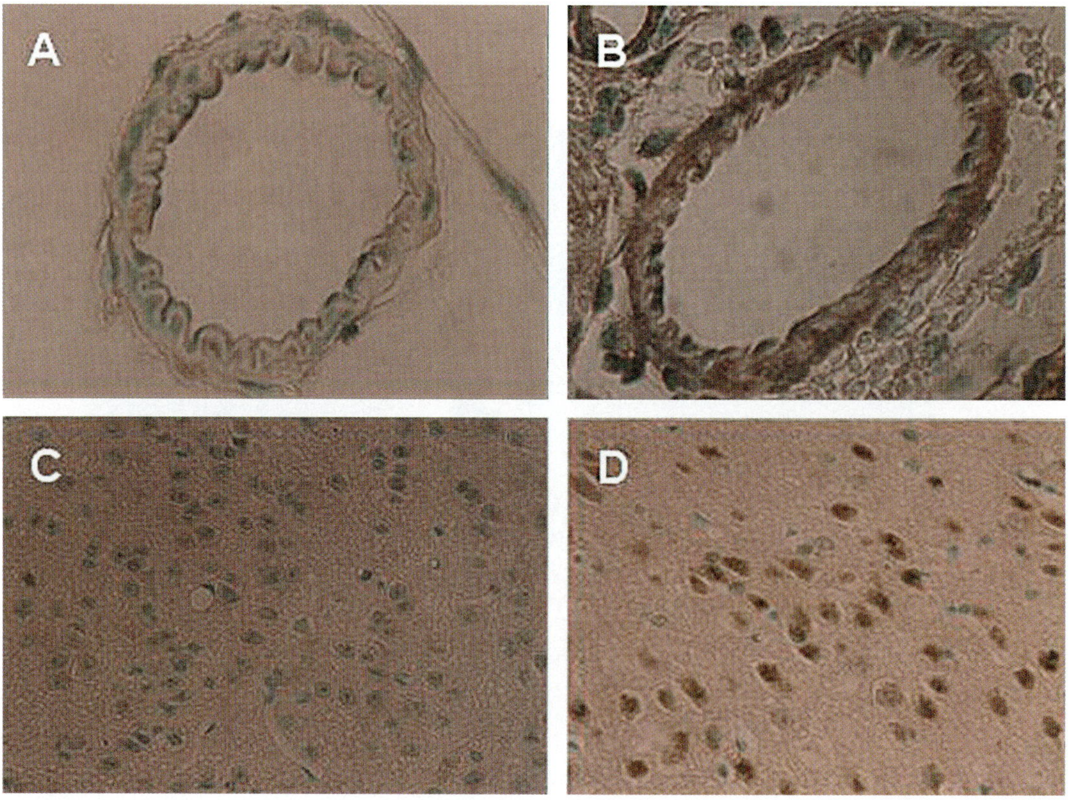

Fig. 6 Immunohistochemical demonstration of iNOS. At 24 h after SAH, iNOS immunoreactivity is observed in the cerebral artery (**b**) and the cortex (**d**). In sham-operated control mice, iNOS immunoreactivity is negative in the cerebral artery (**a**) and the cortex (**c**)

model. A sudden increase and the prolongation of intracranial pressure (ICP) cannot be reproduced in the cisterna magna SAH model.[8] Moreover, the cisterna magna SAH model may not distribute the blood in the subarachnoid space all over the brain. Therefore, the filament perforation SAH model is the most suitable for the researches to investigate vasospasm as well as neuronal injury and neurogenesis following SAH.[2,5]

2. In our filament perforation SAH model, the left ACA is perforated by a nylon filament. That is, since the perforation site is between the ICA bifurcation and the confluence with the right ACA (the origin of the Azygous ACA) (Fig. 2), the filament does not injure the MCA directly. Thus, we can evaluate the vasospasm by SAH itself, not by the mechanical injury, in the proximal portion of the MCA. Although some reports showed that the perforation site was near the ICA bifurcation,[3,4] the arterial rupture near the ICA bifurcation often causes the MCA occlusion resulting in brain infarction of the MCA area in our preliminary studies. Perforating the ACA is a very important technique to avoid ischemic brain damage, and it is a big advantage in our filament perforation SAH mouse model.
3. The filament perforation SAH model is less invasive for mice because of no craniotomy. The techniques are simple, and the entire procedure, skin incision to skin closure, is completed within approximately 20 min in our laboratory.
4. The costs are low to purchase and maintain mice, compared with other species. In addition, the recent successful development of transgenic and knockout mice has made mice more prevalent in numerous researches on stroke including SAH.[11]

5.2 Complications

1. In our SAH model, when typical SAH occurs, mice show respiratory failure for several seconds, while severe SAH often causes long respiratory arrest in the mice. Although they can be resuscitated, they usually die within 24 h due to the high ICP with hypoxic brain injury.
2. When the perforation site is the ICA bifurcation, not the left ACA, the ICA/MCA occlusion often occurs, resulting in brain infarction in the MCA area. Therefore, when the mice show right hemiparesis after the surgery, they should be excluded from the study.
3. At 7 days after SAH, the mice showed hydrocephalus in the brain at approximately 30% in our laboratory.
4. Small intracerebral hematoma was seen in the left frontal cortex at approximately 3% in our series.

5.3 Limitations

1. In the filament perforation SAH model, the blood volume in the subarachnoid space cannot be controlled basically. The blood volume is believed to correlate well with ICP and mortality rate.[8]
2. The mortality rate in our SAH model is approximately 20% within 72 h after SAH. Main cause of the death is respiratory failure because of severe SAH. Although the high mortality rate is inevitable in the filament perforation SAH model even in rats,[6–8] our techniques perforating the left ACA can lead to less mortality rate. In our laboratory, the mortality rate on

perforation of the ICA bifurcation or the left ACA was approximately 45% or 20%, respectively.

3. At 7 days after SAH, approximately 30% of the mice demonstrated hydrocephalus in the brain in our series. The mice with hydrocephalus could not survive for a long time. Thus, the filament perforation SAH mouse model might not be suitable for researches on a chronic phase over a month following SAH.

References

1. Kamii H, Kato I, Kinouchi H, Chan PH, Epstein CJ, Akabane A, Okamoto H, Yoshimoto T. Amelioration of vasospasm after subarachnoid hemorrhage in transgenic mice overexpressing CuZn-superoxide dismutase. Stroke 1999;30:867–872.
2. Saito A, Kamii H, Kato I, Takasawa S, Kondo T, Chan PH, Okamoto H, Yoshimoto T. Transgenic CuZn-superoxide dismutase inhibits NO synthase induction in experimental subarachnoid hemorrhage. Stroke 2001;32:1652–1657.
3. Parra A, McGirt MJ, Sheng H, Laskowitz DT, Pearlstein RD, Warner DS. Mouse model of subarachnoid hemorrhage associated cerebral vasospasm: methodological analysis. Neurol Res 2002;24:510–516.
4. McGirt MJ, Parra A, Sheng H, Higuchi Y, Oury TD, Laskowitz DT, Pearlstein RD, Warner DS. Attenuation of cerebral vasospasm after subarachnoid hemorrhage in mice overexpressing extracellular superoxide dismutase. Stroke 2002;33:2317–2323.
5. Mino M, Kamii H, Fujimura M, Kondo T, Takasawa S, Okamoto H, Yoshimoto T. Temporal changes of neurogenesis in the mouse hippocampus after experimental subarachnoid hemorrhage. Neurol Res 2003;25:839–845.
6. Bederson JB, Germano IM, Guarino L. Cortical blood flow and cerebral perfusion pressure in a new noncraniotomy model of subarachnoid hemorrhage in the rat. Stroke 1995;26:1086–1092.
7. Veelken JA, Laing RJ, Jakubowski J. The Sheffield model of subarachnoid hemorrhage in rats. Stroke 1995;26:1279–1284.
8. Prunell GF, Mathiesen T, Diemer NH, Svendgaard NA. Experimental subarachnoid hemorrhage: subarachnoid blood volume, mortality rate, neuronal death, cerebral blood flow, and perfusion pressure in three different rat models. Neurosurgery 2003;52:165–176.
9. Lin CL, Calisaneller T, Ukita N, Dumont AS, Kassell NF, Lee KS. A murine model of subarachnoid hemorrhage-induced cerebral vasospasm. J Neurosci Methods 2003;123:89–97.
10. Lin CL, Dumont AS, Calisaneller T, Kwan AL, Hwong SL, Lee KS. Monoclonal antibody against E selectin attenuates subarachnoid hemorrhage-induced cerebral vasospasm. Surg Neurol 2005;64:201–206.
11. Cahill J, Calvert JW, Zhang JH. Mechanisms of early brain injury after subarachnoid hemorrhage. J Cereb Blood Flow Metab 2006;26:1341–1353.
12. Kamii H, Kinouchi H, Sharp FR, Chan PH. A model of transient focal cerebral ischemia in the mouse. In: Ohnishi ST, Ohnishi T, eds. Central Nervous System Trauma: Research Techniques. Boca Raton, Fla: CRC Press, 1995:139–146.

Chapter 24

Blood Injection Subarachnoid Hemorrhage Mouse Model

Tarkan Calisaneller, Chih-Lung Lin, Naoya Ukita, Ryo Yoshimura, Yi Wang, Mark Fitzgerald, and Kevin S. Lee

Abstract Cerebral vasospasm is a major complication of subarachnoid hemorrhage (SAH). This chapter describes a reproducible, low-mortality, and low-cost model of SAH-induced vasospasm in the mouse. Autologous blood is injected into the cisterna magna of adult mice, and the diameters of large cerebral arteries are measured postmortem. Acute (6–12 h) and persistent (1–3 days) phases of vascular narrowing are observed in cerebral arteries. Vascular narrowing is associated with thickening of the vascular wall and corrugation of the internal elastic lamina; these histological findings have been described by Lin and associates.[1] A key advantage of this model is the use of a species for which a variety of resources are available, such as a range of genetically-modified animals. This mouse model could therefore assist in identifying and verifying mechanisms contributing to vascular pathology associated with SAH and facilitate the initial screening of candidate therapies for cerebral vasospasm.

Keyswords: Subarachnoid hemorrhage, Cerebral vasospasm, Mouse, Vasoconstriction, Stroke

1 Model Selection

The absence of an effective medical therapy for treating the complications of subarachnoid hemorrhage (SAH) continues to motivate experimental studies to characterize mechanisms of injury and to develop effective therapies. Toward these ends, multiple models of SAH have been developed in a range of species. As described in an excellent review of the available SAH models, none of the models adequately mimics all of the features of SAH and cerebral vasospasm.[2] Nonetheless, most of the models have distinct advantages that can facilitate the elucidation of mechanisms and the development of therapies. Major advantages of the mouse blood injection model[1] include the following:

1. Ability to study SAH-induced vasospasm, albeit over a briefer time course than that observed in larger animals

From: *Animal Models of Acute Neurological Injuries.*
Edited by: J. Chen (eds.) New York, Totowa, NJ

2. Availability of extensive reagents and resources for mouse studies, such as genetically-modified animals
3. High throughput due to low mortality and replicable outcomes
4. Inexpensive because of low costs of animal purchase and housing

2 Materials

Male C57Black/6J mice (Hilltop Labs) weighing 27–32 g are used as experimental animals. In additional studies not described here, we have also used other strains of mice successfully using these techniques.

1. Chloral hydrate (350 mg kg^{-1} i.p) and xylazine (4 mg kg^{-1} i.p.) are used for anesthesia.
2. A heating pad (Harvard Instruments) is used to maintain body temperature 37°C.
3. The right femoral artery is exposed and cannulated with a polyethylene catheter (PE-10).
4. A Zeiss operating microscope (OPMI 1-FC) is used for visualization of all surgical procedures, although any comparable surgical microscope could be used.
5. A 30-gauge syringe needle (Becton Dickenson) is used to puncture the atlanto-occipital membrane.
6. The needle is secured in place using Aron Alpha glue (Ted Pella, Inc).
7. During the surgical procedure, bipolar electrocoagulation (Codman CMC-II-PC) is used for hemostasis.
8. For intracardial perfusion, a 23-gauge butterfly needle is used with 10% formalin perfused at a rate of 5.5 ml min^{-1} for 2 min.
9. For visualizing the vessels, the perfusate is shifted to a mixture of 10% gelatin and India ink at a flow rate of 13.5 ml min^{-1} for 15 s.
10. Vessels are visualized using a steromicroscope (Wild Makroskop 420) with a video camera connected (MTI CCD20).
11. Vascular size is measured using a computer-assisted image analysis system (Metamorph; Universal Imaging). Steps 10 and 11 can use any standard stereo-microscopic and image analysis system.

3 Procedures

All experimental protocols are approved by the University of Virginia Animal Care and Use Committee.

1. Animals are anesthetized with chloral hydrate and xylazine.
2. The right femoral artery is exposed and cannulated.
3. The posterior scalp is incised at the midline and the skull exposed at the junction of the occipital bone and the arch of C1. The occipital muscles are carefully dissected off the occipital bone and retracted infero-laterally by suture. The atlanto-occipital membrane is then exposed.
4. A 30-gauge needle is used to puncture the atlanto-occipital membrane 45° caudally, from the midline toward the left side of the animal, into the cisterna magna (Fig. 1). Aron Alpha glue is used to secure the needle in place.

5. Autologous blood (80–100 μl) is withdrawn from the femoral artery catheter.
6. A volume of 60 μl of freshly-drawn blood is injected through the 30-gauge needle into the cisterna magna. This volume was selected on the basis of pilot studies to optimize the subarachnoid clot, limit leakage from the injection site, and limit mortality.
7. Animals are then immediately placed in a head-down (approximately 75° vertical) position for 10 min to facilitate the diffusion of injected blood in the basal cisterns (Fig. 2).
8. The needle inserted in the cisterna magna and the PE-10 catheter in the femoral artery are then removed.

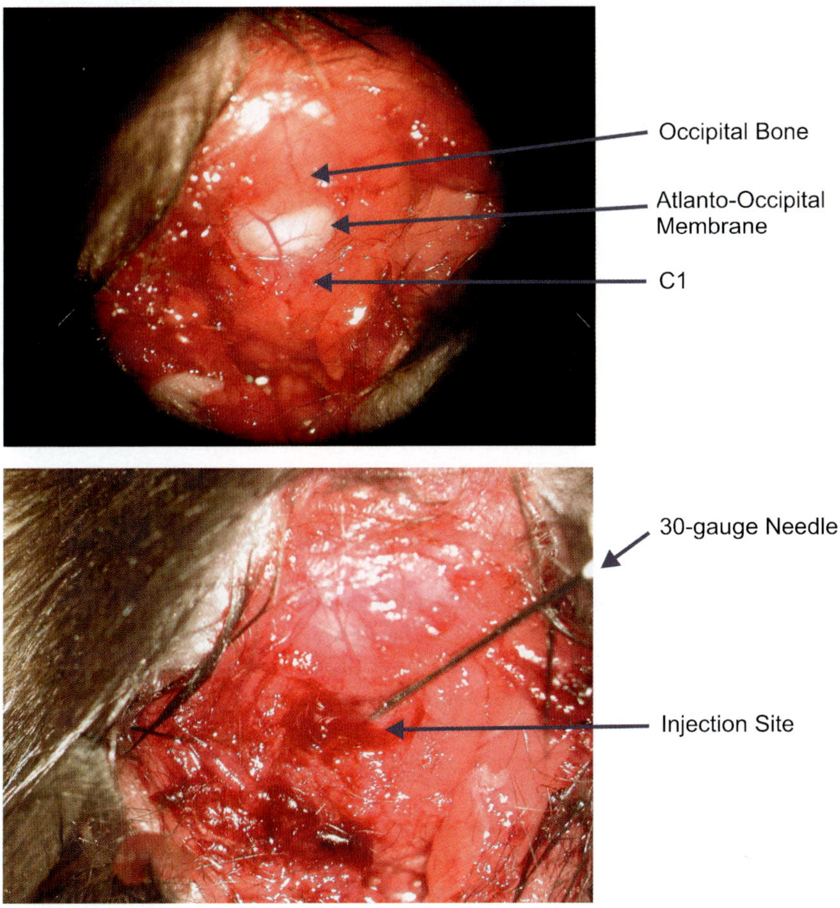

Fig. 1 Anatomical landmarks and injection site for experimental SAH. The upper frame shows the exposed surface of the atlanto-occipital membrane and indicates the occipital bone and C1 as landmarks. The animal's head is toward the top of the frame. The *lower frame* shows the trajectory of the 30-gauge needle as it injects blood through the atlanto-occipital membrane

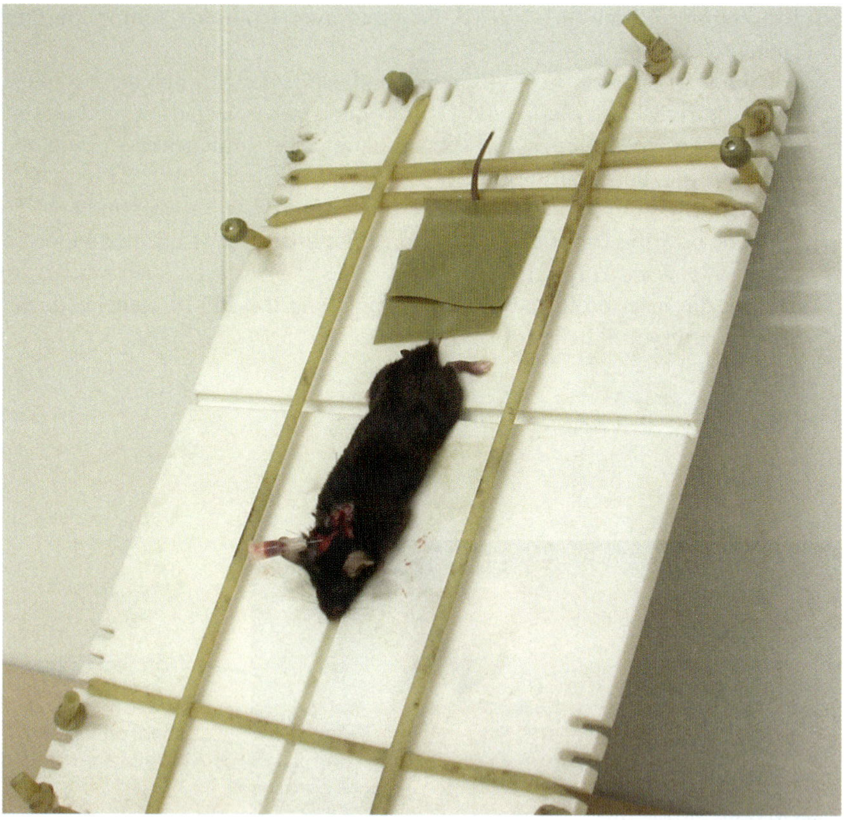

Fig. 2 Positioning of the animal after injection of autologous blood. The animals are placed in a head-down position (approximately 75° vertical) for 10 min to facilitate the diffusion of injected blood in the basal cisterns

9. The neck muscles are approximated, and the incision closed tightly with sutures to prevent cerebrospinal fluid (CSF) or blood leakage.
10. The leg incision is closed with suture.
11. The animal's body temperature is maintained at 37°C until it is fully recovered (about 1 h postoperatively).
12. The postsurgical survival time depends on the purpose of the study. The peak of the persistent phase of vascular narrowing is between 1 and 3 days after SAH; this is an appropriate time frame for measuring the effects of candidate therapies on vascular constriction. The precise time course of vascular narrowing in this model has been described elsewhere.[1]
13. For perfusion-fixation, animals are anesthetized with chloral hydrate and xylazine as described earlier.
14. Animals are perfused by trans-thoracic cannulation of the left ventricle, after clamping the descending thoracic aorta and opening the right atrium with an incision. The first phase of perfusion (2 min) uses 10% formalin (pH 7.4 at 37°C) at a flow rate of 5.5 ml min^{-1}. The second phase of perfusion (15 s) uses a mixture of 10% gelatin and India ink at a flow rate of 13.5 ml min^{-1}. This fills the large cerebral vessels black and allows them to be readily visualized on the surface of the brain, and measured using videomicroscopy (Fig. 3).

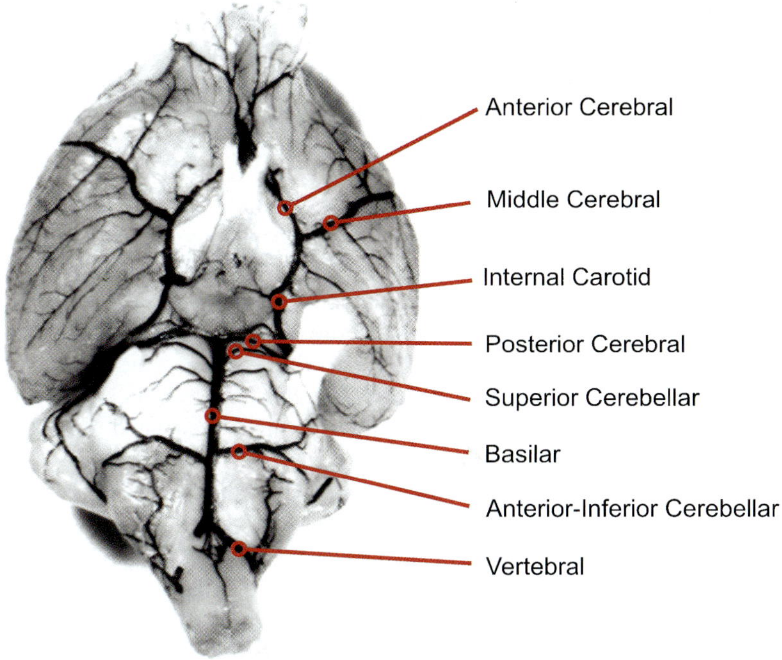

Fig. 3 Vascular labeling for computer-assisted microscopic measurements. The ventral surface of a mouse brain is shown with the large cerebral vessels (*black*) filled with India Ink/gelatin. The major cerebral arteries are labeled, and the vessels that are typically measured (i.e., middle cerebral, anterior cerebral, and basilar arteries) are easily identified for assessment using videomicroscopy

15. Brains are removed from the skulls and stored in the formalin solution until vascular measurement.
16. Vascular measurements are performed by visualizing the large vessels on the surface of the brain with a stereomicroscope and using computer-assisted image analysis.

4 Outcome Evaluation

Vascular Size: Vascular measurements are performed using computer-assisted image analysis by investigators blinded to the group identity of the animals being measured. We typically measure the diameter of three major arteries: the middle cerebral artery at a point 100 μm distal to the bifurcation of the internal carotid artery, the anterior cerebral artery at a point 100 μm distal to the bifurcation of the internal carotid artery, and the basilar artery at three evenly-spaced points along the extent of the artery. Data are expressed as the mean ± the standard deviation of the mean (SD). Group comparisons are performed using an one-way analysis of variance (ANOVA) with the Fisher's LSD posthoc test for a typical study in which a dose-response analysis is performed for a therapeutic candidate. Differences are accepted as significant at the $p < 0.05$ level.

5 Applications, Complications, and Limitations

1. The blood injection model in the mouse is useful for examining mechanisms of, and potential therapies for, vascular pathology after SAH.
2. The animals become alert during the first hour after SAH and no focal neurological deficits are observed.
3. A diffuse blood clot is observed over the major cerebral arteries until approximately Day 2 after SAH, with the clot persisting longest in the anterior circulation.
4. Two phases of reduced arterial diameter are observed. An acute phase peaks around 6–12 h after SAH and second more persistent phase peaks around 1–3 days. The persistence of vascular narrowing is thus considerably briefer than that observed in man or large animal models of SAH. As in other models of SAH, the persistence of vascular narrowing is associated with the rate of clearance of the subarachnoid clot.
5. Our initial series of studies[1] produced a mortality rate of 3.4% (2 of 59) and these deaths were the result of surgical complications.
6. A drawback to this SAH model is that there is no direct rupture of a cerebral artery, as is produced in the mouse endovascular puncture model.[3–5] Consequently, it is not possible to assess the potential complication of rebleeding using the blood injection model. Nonetheless, the low mortality and consistent blood clots produced by direct injection provide a replicable and high throughput approach for examining SAH-induced pathologies.
7. This model is also relatively low in cost. In a typical study, the size of an experimental group is approximately eight animals depending on the outcome measure. Consequently, the blood injection model can be used to rapidly and economically screen candidate therapeutic treatments.[6]

References

1. Lin C-L, Calisaneller T, Ukita N,Dumont AS, Kassell NF, Lee KS. A murine model of subarachnoid hemorrhage-induced cerebral vasospasm. J. Neurosci. Methods 2003;123:89–97.
2. Macdonald RL, Weir B. Ch. 11. Animal models. In: *Cerebral vasospasm* (Macdonald RL, Weir B, eds), San Diego: Academic Press, 2001; pp. 449–475.
3. Kamii H, Kato I, et al. Amelioration of vasospasm after subarachnoid hemorrhage in transgenic mice overexpressing CuZn-superoxide dimutase. Stroke 1999;30: 867–872.
4. Saito A, Kamii H, et al. Transgenic CuZn-superoxide dimutase inhibits NO synthase induction in experimental subarachnoid hemorrhage. Stroke 2001;32:1652–1657.
5. Parra A, McGirt J, Sheng H, Laskowitz DT, Pearlstein RD, Warner DS. Mouse model of subarachnoid hemorrhage associate cerebral vasospasm: Methodological analysis. Neurol. Res. 2002;24:510–516.
6. Lin C-L, Dumont AS, Calisaneller T, Kwan A-L, Hwong S-L, Lee KS. Monoclonal antibody against E-selectin attenuates subarachnoid hemorrhage-induced vasospasm. Surg. Neurol. 2005;64:201–206.

Section II

Cerebral Vasospasm

Chapter 25

Blood Injection Cerebral Vasospasm Rabbit Model

Mario Zuccarello

Abstract Cerebral vasospasm after subarachnoid hemorrhage (SAH) is one of the major causes of cerebral ischemia, and is the leading cause of death and disability following aneurysm rupture. The pathogenesis of cerebral vasospasm is poorly understood, and numerous hypotheses have been proposed to explain its underlying cause. The development of an animal model able to mimic the human condition is extremely useful to try to understand the pathogenesis of post-SAH cerebral vasospasm, ultimately leading to more efficacious treatment. We introduce step-by-step procedures to create a double-hemorrhage model of SAH in rabbits. We also performed angiographic studies of the vertebrobasilar vasculature and measured the diameter of the basilar artery to quantify the post-SAH vasospasm.

Keywords: Rabbit, Subarachnoid hemorrhage, Cerebral vasospasm

1 Model Selection

The rabbit model of subarachnoid hemorrhage (SAH) is useful to study the pathogenesis of post-SAH induced cerebral vasospasm.[1,2] This model was developed to overcome the disadvantage of models using animals that were either too small (rat) or too large (dog). The advantage of the rat model is its economy, thus affording the testing of a wide range of hypotheses. However, its disadvantage is its inconsistency regarding the occurrence of vasospasm. More importantly, in vivo quantification of the vasospasm by angiography is difficult. The advantage of the dog model is the consistency in development of cerebral vasospasm and quantification of vasospasm by angiography. However, its disadvantages include its expense and the relevance of the pathway of canine cerebral circulation to human study; that is the dog receives a large blood supply from the external carotid artery circulation (rete mirabilis).

The rabbit model offers several major advantages. First, rabbits are relatively large animals that are easily handled. Second, the anatomy of the circle of Willis is similar to that in humans. Third, the coagulation

From: *Animal Models of Acute Neurological Injuries.*
Edited by: J. Chen (eds.) New York, Totowa, NJ

cascade is similar to humans; this aspect is particularly valuable because blood and blood components appear to be one of the major causes of vasospasm. Fourth, severe chronic vasospasm is consistently produced. Fifth, this animal model is relatively inexpensive, thus allowing testing of a wide range of hypotheses. Lastly, rabbits are one of the few animals in which in vitro investigations of the basilar artery reactivity following SAH has been extensively performed.[3–8]

2 Materials

Adult immunized and conditioned New Zealand white rabbits weighing 2.5–3.8 Kg are the animals of choice. For the surgical procedures, the following materials are recommended:

1. Rabbit stereotactic frame
2. Small animal respirator (Harvard Apparatus, South Natick, Massachusetts)
3. CO_2 monitor (Ohmeda 5200) and pulse oxymeter (Ohmeda 3740)
4. Spring scissors with especially fine, small blades (F.S.T. 15002-08)
5. Spring scissors with especially fine, small blades (F.S.T. 15005-08)
6. Forceps (Adson, F.S.T. 11027-12)
7. Dumont angled forceps (F.S.T. 11253-20)
8. Dumont forceps (F.S.T. 11251-35)
9. Needle holder (F.S.T. 12500-12)
10. Scalpel handle (F.S.T. 10003-12) and scalpel blades (F.S.T. 10015-00)
11. Spring scissors with especially fine, small blades (F.S.T. 15013-12)
12. Spring scissors with especially fine, small blades (F.S.T. 15370-26)
13. Contrast medium (Angiovist 282)
14. Digital subtraction angiography
15. All purpose sponges (Kendall 9022)
16. Cotton-tipped applicators (Henry Schein 100-6015)
17. Surgical microscope

3 Procedures

3.1 Angiographic Studies

1. Preanesthetize a rabbit with acepromazine (1 mg kg^{-1} IM), then anesthetize with fentanyl-droperidol (0.18–0.23 ml kg^{-1}) and pentobarbital sodium (25 mg kg^{-1} IV).
2. Place the animal on a heating pad to maintain body temperature and insert a rectal thermometer to maintain at 37°C.
3. Lubricate the eyes with Lacrilube ophthalmic ointment (Allergan Pharmaceuticals, Irvine, CA) to prevent drying.
4. Shave one ear, sterilize the skin, and place a 21-gauge IV catheter into the ear vein.
5. Paralyze with Pavulon (0.1 cc kg^{-1} IV), place an endotracheal tube (Portex × 3.5 ETT), and ventilate on room air supplemented with O_2 (0.3 l min^{-1}) with an animal respirator (Harvard Apparatus, South Natick, Massachusetts).
6. Adjust respiratory rate and volume to maintain expiratory pCO_2 (Ohmeda 5200 CO_2 Monitor). Monitor arterial oxygen saturation (SaO_2) and heart rate (Ohmeda 3740 pulse oxymeter).

7. Maintain anesthesia with pentobarbital sodium (25 mg kg^{-1} IV) and Pavulon (0.08 mg/kg/30 min IV).
8. Place a 21-gauge IV catheter into the auricular artery to monitor blood pressure.
9. After shaving the hair and sterilizing the skin with Betadine over the left or right axillary area to expose the subclavian artery, insert a No. 4 French polyethylene catheter directed to the ipsilateral vertebral artery to obtain a selective injection of the vertebrobasilar artery.
10. Obtain arterial blood for blood gas analysis.
11. Inject contrast medium (Angiovist 282) (5 ml s^{-1} for 5 s) and obtain images (4° left anterior oblique projection) of the vertebrobasilar system at two per seconds for 14 s using a rapid sequential angiographic technique. Digital substraction analysis is performed with a small focal spot at 60 kV and 0.8 mA.
12. Remove the subclavian catheter and suture skin in layers.

3.2 Induction of Subarachnoid Hemorrhage

1. Immediately after the Day 0 angiogram, place the rabbit in a stereotactic frame.
2. Shave the hairs and sterilize the neck skin with Betadine.
3. Puncture percutaneously the cisterna magna with a 21-gauge butterfly needle.
4. Maintain blood gas levels by adjusting the respiratory rate and/or tidal volume.
5. Withdraw arterial blood from the central ear artery.
6. Inject arterial blood (0.75 ml kg^{-1}) into the cisterna magna over 3 min.
7. Repeat the blood injection on Day 2 by following the same protocol as Day 0 (Fig. 1)

4 Outcome Evaluations

4.1 Basilar Artery Diameter

Angiograms are placed on a light box above which is fixed a video camera connected to an image analysis computer. The angiographic image is captured by the camera, digitized, and reproduced on the video monitor. An observer blind to the experiment measures basilar artery diameter just below the basilar-posterior cerebral artery junction, just above the basilar vertebral artery junction, and midway between these locations; the means of these three measurements are averaged. Constriction is expressed as a percent of the basilar artery diameter on Day 6 relative to Day 0 (Figs. 2 and 3).[2,5,6]

5 Applications, Complications and Limitations

1. With the relative cost advantage in using rabbits, large numbers of these animals can be more readily used in studies than other more expensive animals. When compared with the rat model, the rabbit model offers a significant advantage because of its larger vessel size, allowing the ability to perform angiography.

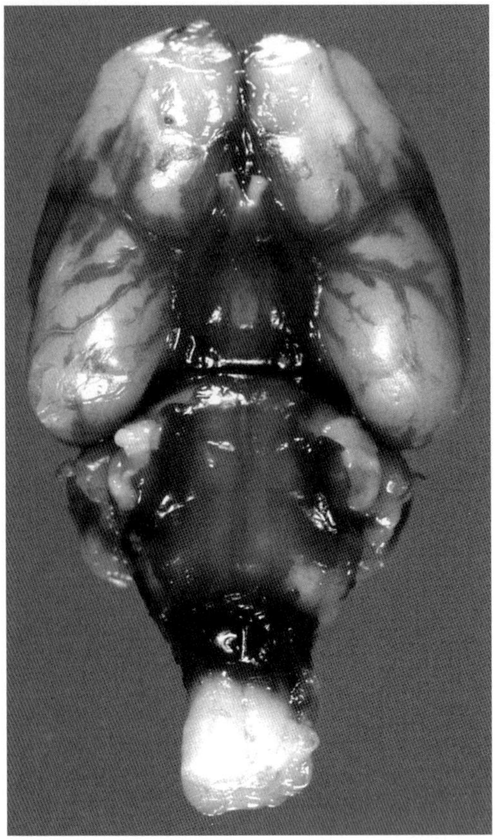

Fig. 1 Rabbit brain showing the presence of a large SAH (with permission from the Mayfield Clinic)

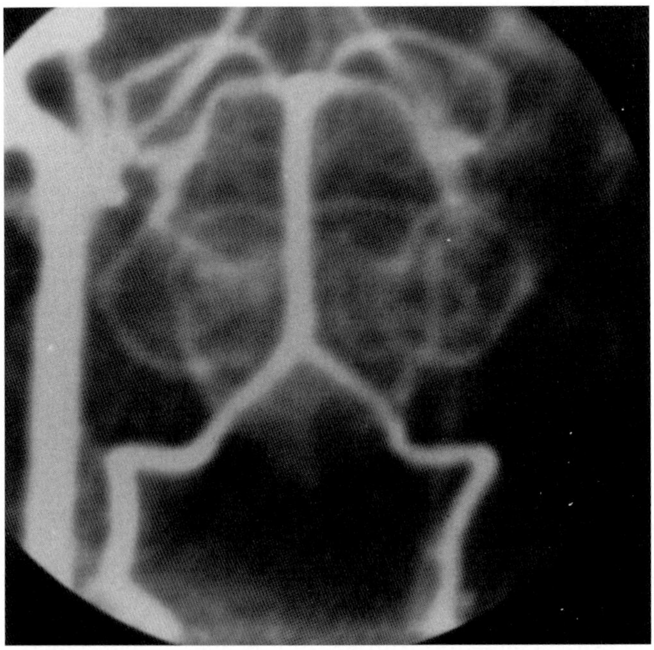

Fig. 2 Digital subtraction angiography of rabbit vertebrobasilar system (Day 0, pre-SAH) (with permission from the Mayfield Clinic)

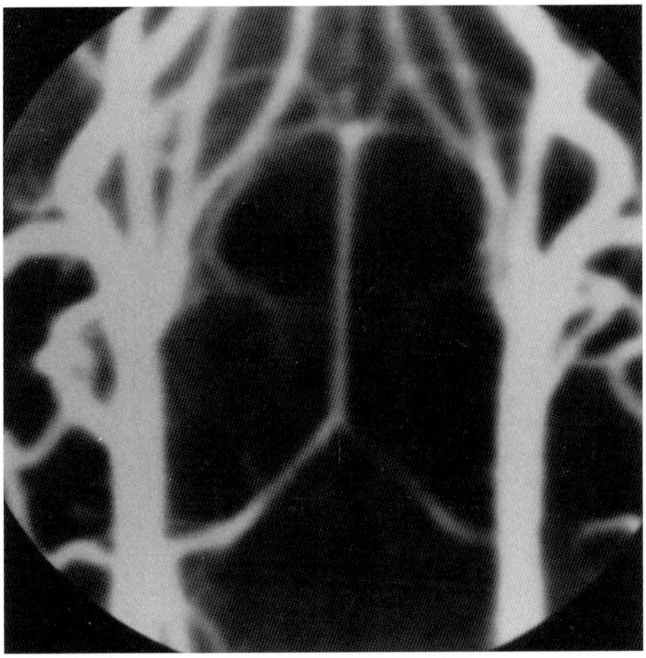

Fig. 3 Digital subtraction angiography of rabbit vertebrobasilar artery after SAH (Day 6). SAH induced approximately 34% constriction of the basilar artery (with permission from the Mayfield Clinic)

2. The rabbit model of SAH is a very versatile animal model. If a single SAH injection is used, cerebral vasospasm develops within 3 days and is not chronic. Therefore, this model may be ideal to provide an initial screening tool for new pharmacological therapies afforded by the large numbers of animals that can be tested before moving to more expensive animal models. If the double-hemorrhage model is used, this model may be best suited to the study of the pathogenesis of SAH-induced cerebral vasospasm because a double injection results in chronic and pharmacologically resistant cerebral vasospasm.[2]
3. The rabbit model is effective to test the acute and delayed effects of SAH on cerebral circulation and metabolism.[2,5,6]
4. This rabbit model has been extensively used in our laboratory to perform pharmacological studies of the basilar artery in situ (Fig. 4). The basilar artery is exposed after creating a cranial window in the rabbit clival area and after removal of the blood clot.[1,3,4,7,8]
5. Mortality rates with the rabbit SAH model are very low in a single-injection model (less than 3%) and acceptably low in the double-injection model (less than 8%).
6. The percutaneous puncture of the cisterna magna to induce SAH requires some practice before it can be performed without complications. The most common complication results from the injection of blood into the brain stem, resulting in the immediate death of the animal. To avoid this complication, the butterfly needle is inserted between the bony ridge of

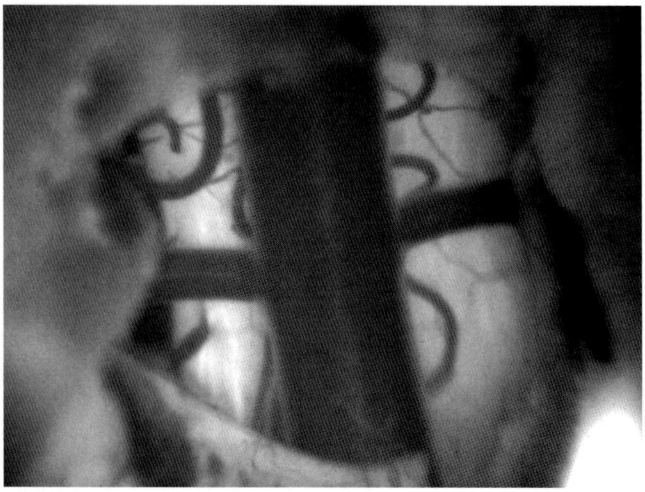

Fig. 4 Exposure of the rabbit basilar artery after transclival cranial window (with permission from the Mayfield Clinic)

the occipital bone and the arch of first cervical vertebra. The needle is then advanced slowly under continuous aspiration and stopped as soon as cerebrospinal fluid is withdrawn.

7. In spite of the significant advantages offered by the rabbit SAH model, the problem still persists as to what constitutes a good model of human vasospasm. One limitation is that the time course of development of cerebral vasospasm in an animal model has an earlier onset than the clinical picture. The pursuit of a better model will undoubtedly continue.

References

1. Zuccarello M, Lewis IA, Rapoport RM. ET_A and ET_B receptors in subarachnoid hemorrhage-induced cerebral vasospasm. *Eur J Pharmacol* 1994;259:R_1–R_2.
2. Zuccarello M, Lewis IA, Upputuri S, Anderson DK. Effect of remacemide on subarachnoid hemorrhage-induced vasospasm in rabbits. *J Neurotrauma* 1994:11:691–698.
3. Zuccarello M, Romano A, Passalacqua M, Rapoport RM. Endothelin-1 induced endothelin-1 release causes cerebral vasospasm. *J Pharm Pharmacol* 1995;47:702.
4. Zuccarello M, Romano A, Passalacqua M, Rapoport RM. Decreased endothelium-dependent relaxation in subarachnoid hemorrhage-induced vasospasm: role of endothelin-1. *Am J Physio*, 1995;269:H1009–H1015.
5. Zuccarello M, Soattin GB, Lewis AI, Breu V, Hallak H, Rapoport RM: Prevention of subarachnoid hemorrhage-induced vasospasm by oral administration of endothelin receptor antagonists. *J Neurosurg*, 1996;84:503–507.
6. Zuccarello M, Boccaletti R, Rapoport RM: Role of extracellular Ca^{2+} in subarachnoid hemorrhage-induced vasospasm of the rabbit basilar artery. *Stroke*, 1996;27:1896–1902.
7. Zuccarello M, Boccaletti R, Rapoport RM: Does blockade of EndothelinB1 receptor activation increase $Endothelin_{B2}$/$Endothelin_A$ receptor -mediated constriction in the rabbit basilar artery. *J Cardiovasc Pharmacol* 1999;3:679–684.
8. Byung Ho Lee, Sherman JD, Zuccarello M, Rapoport RM: Vasospasm following subarachnoid hemorrhage: evidence against functional upregulation of PKC constrictor pathway. *Neurol Res* 2003;25(3):268–270.

Chapter 26

Double Blood-Injection Cerebral Vasospasm Canine Model

Shigeru Nishizawa

Abstract The details of double blood-injection vasospasm canine model are described. In the viewpoints of severity of arterial constriction, time course, and histological changes of cerebral artery exposed subarahcnoid blood clot, it can be a most suitable experimental animal model for the investigation of pathophysilogy of cerebral vasospasm following subarachnoid hemorrhage.

Keywords: Animal model, Basilar artery, Cerebral vasospasm, Dog, Subarachnoid hemorrhage

1 Introduction

The pathophysiological mechanisms of cerebral vasospasm following subarachnoid hemorrhage (SAH) have still been unknown. To investigate the mechanisms of cerebral vasospasm, it is inevitable to use experimental animal models, and various models were developed by many authors. Among them, a canine cerebral vasospasm model has been most widely accepted as a standard model for cerebral vasospasm following SAH.

In the development of the experimental canine model for cerebral vasospasm, a long historical background has to be described. After many trial and error processes, double blood-injection vasospasm canine model was finally established as a most standard experimental animal model.

In this chapter, the historical background of this model is at first introduced. Then, the detail of the present model is described.

2 Historical Background

In 1965, Echlin suggested that cerebral vasospasm was due primarily to some irritation property of blood in the subarachnoid space.[1] He based this theory on his observation that the basilar artery of the dog would constrict following the application of blood to the arterial wall.[1] Other investigators confirmed this

From: *Animal Models of Acute Neurological Injuries.*
Edited by: J. Chen (eds.) New York, Totowa, NJ

evidence in dog, cat, and monkey, and a number of experimental models were developed to permit the study of arterial spasm.[2–10] Using dog, tearing or direct puncture of the artery around circle of Willis was employed to induce vasospasm by many authors,[3,6,7,9] and was reported to be very reliable. However, Echlin experienced high mortality and morbidity rate (uncontrollable bleeding, seizure, respiratory arrest, and death) with this model.[11] He also pointed out that intracranial hypertension and brain swelling, followed by uncontrollable hemorrhage with puncture method, exerted influence to the spasm other than blood itself. In addition, other investigator suggested the release of some vasoactive substances from the damaged brain, which influenced on arterial constriction rather than cerebral vasospasm.[12,13] Finally, those studies were not succeeded.

The basis of the present canine model was primarily introduced by Harvard group. At first, Kuwayama and Zervas et al. developed single injection model using dog, and the study was published in 1972.[14] In that study, only single injection of autologous blood was employed. They injected 2 ml of autologous blood into the cisterna magna of dog by a streotactic method, and carefully analyzed the diameter of the basilar artery following the injection. By the cerebral angiography through the cannulation from the femoral artery to the vertebral artery, they showed early arterial constriction lasting for 30 min, which returned to the subcontrol diameter within 2 h. Since then, arterial constriction restarted 2 days later, and returned to the almost normal size on 14 days. The degree of the arterial constriction was from 25% to 59% with an average of 37% of the control.[14] Although the experimental findings were reproducible and the condition of the dog was acceptable, this method was not used for a standard model. The reason is not clear, but it is probably assumed that the extent of the arterial constriction is not severe. The authors did not mention about the histological changes in this model. Such change might not be compatible enough to that in human cerebral vasospasm.

From the same group, a new canine model of cerebral vasospasm following SAH was published in 1983.[15] In this article, Varsos and Zervas et al. revised the first model (single injection model), and adopted the second injection of autologous blood into the cisterna magna 48 h apart from the first injection.[15] In this new model, "two-hemorrhage" canine model, 4 ml of autologous blood was injected, and chronological angiography was performed. After the first injection, arterial constriction in acute phase lasting for 2 h was detected (Day 1), and the same amount of autologous blood was injected on Day 3. The angiography was performed after the second injection (Day 4), and the following day (Day 5). They investigated the caliber change of the basilar artery, response of three different kinds of drugs (aminophylline, nifedipine, or papaverine) for the treatment of "delayed" cerebral vasospasm, and morphological changes of the artery.[15] Those drugs were standard medications for the treatment of cerebral vasospasm in patients with SAH in those days, and no therapeutic efficacy was observed. The significant caliber change was obtained after the second injection in this model, and such change was not reversible by those drugs. Histological changes of canine basilar artery were also compatible with human cerebral arteries exposed by subarachnoid bleeding.[16–18] On the basis of those evidence, they concluded that any findings of their new experimental model, "two-hemorrhage" canine model resembled the situation found in patients with SAH, and the new model may provide useful data for future evaluation of pharmacotherapy.[15]

3 From Historical Background to the Present Model

It can be emphasized that four important points were described in this paper published by Varsos et al.[15] First, potent and consistent cerebral arterial constriction following experimental SAH was shown using cerebral angiography. Second, the time course of the arterial constriction following SAH was equivalent with that in human cerebral vasospasm after SAH. Third, significant cerebral arterial constriction after SAH was not reversible by conventional drugs inducing vasodilation, such as aminophylline, nifedipine, or papaverine. Fourth, histological changes of long lasting constricting cerebral arteries resembled to those in patients with SAH. As it is well known, cerebral vasospasm in patients with SAH starts 3–4 days after SAH and continues for almost 2 weeks, and significant histological changes occur in vasospastic cerebral arteries.[16,17] To describe that it is a suitable experimental animal model of cerebral vasospasm, these four factors described above are essential to be shown in experimental SAH and cerebral vasospasm animal models.[19] In small animal models such as mouse and rat, histological confirmation can be obtained following SAH. However, visible cerebral vasospasm using cerebral angiography has not been consistently established. To fulfill those factors, canine and monkey models are mostly suitable. An animal model using primates is probably best for investigation of the mechanisms of cerebral vasospasm, because it very resembles to human. However, experimental cost makes unable to use as a most standard model. Alternatively, canine model is reasonable in any aspects.

Varsos et al. mainly emphasized that their model was useful for future evaluation of pharmacotherapy for cerebral vasospasm.[15] Since this paper was published, their model became to be a most standard, widely accepted experimental animal model for various researches of cerebral vasospasm following SAH.

4 Double Blood-Injection Vasospasm Canine Model

4.1 Terminology

Originally, this model is finished 5 days after the first injection.[15] However, various modification of this model have been carried out. Most recently, it becomes as standard to continue the experiment for a week.[20–27] Some literature describes the first day of the experiment as "Day 0", and the final day of the experiment (sacrificing day of the experimental animal) as "Day 6".[26–28] In another literature, the former as "Day 1", and the latter as "Day 7".[20–25] As it makes readers confusing, the exact definition which "Days" are the first day and the final day of the experiment should be clearly described in articles. As already described, double blood-injection vasospasm canine model was developed by Harvard group. Since then, they themselves modified their own model.[20] The author of this chapter (S.N.) also adopts the modified model of this group.[20–25] In this chapter, the details of this model are described according to the Harvard model.

4.2 Animals

The adult dogs of either sex are used. In terms of obtaining consistent data, there were no significant changes between mongrel dogs and one fixed space such as beagle dog. However, to avoid possible variance of the data, it might be better to use single space. The author (S.N.) standardizes the space of dog, and beagle dogs are used for making the model.[24,25]

4.3 Anesthesia

Adult beagle dogs weighing 7–10 kg of either sex are used. Intramuscular ketamine hydrochloride is administered (10 mg kg^{-1}), then dogs are anesthetized by sodium pentobarbital given intravenously (20 mg kg^{-1}). During procedure, a small amount of pentobarbital is additionally injected as necessary. Lactate-Ringer solution is given intravenously during the procedure. Under intravenous anesthesia, a tracheal tube is intubated. Although artificial ventilation is sometimes necessary to maintain the breathing, spontaneous breathing recovers soon most of times. Through intratracheal tube, end-tidal CO_2 is monitored and is kept during 35–40 mmHg during the experiment.

Some papers reported that inhalation of anesthetic gas such as isoflurane, or intravenous injection of propofol is used for the maintenance of anesthesia, and recovery from the anesthesia by these ways are better rather than consecutive injection of pentobarbital.[26,27]

4.4 Cannulation of the Catheter and Aniography

A femoral artery of either side is prepared and a sterile catheter for cerebral angiography is inserted. If repeated angiography is planned, a distal part of the femoral artery should be prepared. When a branch of the femoral artery is available, it is better for use. Then, more proximal part of the artery can be used for following angiography on different days. A protective sheath catheter is useful for the manipulation of the catheter rather than direct insertion of the catheter in the femoral artery for the angiography.

Under fluoroscopic control, a catheter is advanced to a vertebral artery of either side. It is easy to cannulate into the left side of the vertebral artery rather than the right side. In a dog, two trunk arteries are branching from the aortic arch. The first trunk is an innominate artery, and the second is the left subclavian artery. The innominate artery is separating into a common trunk of bilateral carotid arteries and right subclavian artery. A common carotid artery forms right and left common carotid arteries. From right subclavian artery, right vertebral artery is branching. The left vertebral artery directly originates from left subclavian artery with a straight fashion. In these anatomical structures, it is easy to advance a catheter from the descending aorta to the left vertebral artery through the left subclavian artery. A vertebral artery of either side is acutely narrowing at the third or the second vertebral level, then advancement of the catheter above is not accessible and it is wedged. The placement of the tip of the catheter around the fourth vertebral level is optimal.

After advancement of the catheter, a dog is kept in the prone position, and the head of the dog is fixed in the stereotactic frame (Fig. 1). A blood pressure monitoring system is set to the inserted catheter, and the blood pressure is continuously monitored during the next step, angiography. End-tidal CO_2 is

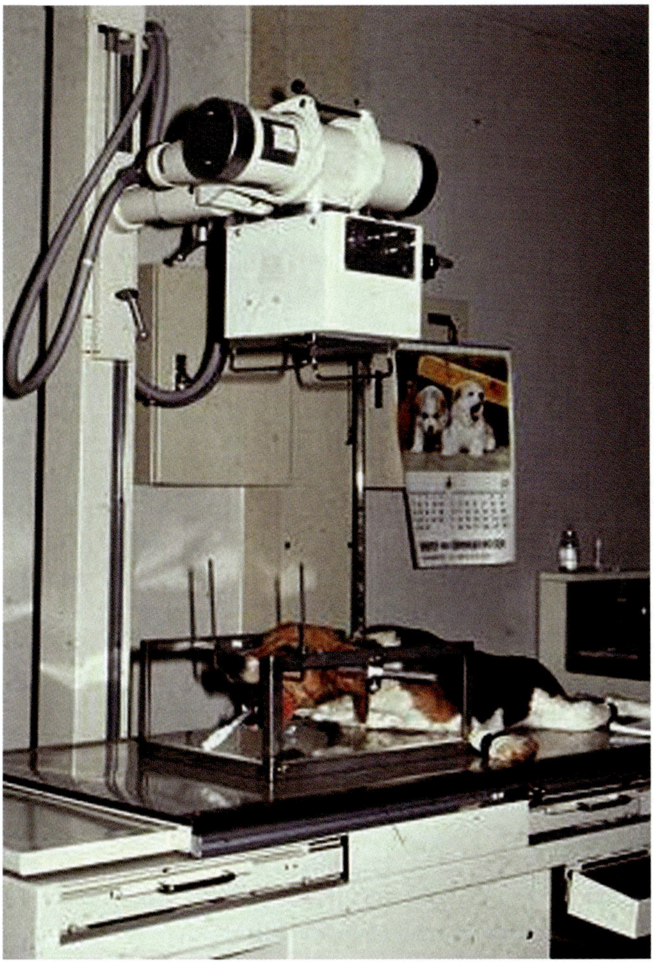

Fig. 1 A *picture* during the angiography. A dog's head is fixed in the streotactic frame. The diameter of the ear-bar is 1.0 cm for the correction of the magnification change among angiographies. End-tidal CO_2 is measured through a tracheal tube, and blood pressure is monitored through a catheter inserted in the vertebral artery

also continuously monitored through an intratracheal tube, and is kept during 35–40 mmHg. After stabilization of the condition of the dog, blood pressure, and end-tidal CO_2 for 10–15 min, angiography is performed. Five to seven milliliters of water-soluble nonionized iodine contrast material is injected manually through a catheter, and an angiogram is taken. It is extremely important to standardize the magnification of angiography for the comparison of the caliber of basilar artery. There are some ways to standardize the magnification. One way is to insert a needle in the subcutaneous tissue of the scalp, and measure the real diameter of the needle by a vernier micrometer and that of an angiogram each time. The difference of the magnification can be known by measurement of both diameters, then such difference can be compensated. Other way is to make ear-bar diameter of the stereotactic frame at 1.0 cm. By measurement of the ear-bar diameter on an angiogram, the difference of the magnification can be detected.

4.5 Cisternal Puncture and Blood Injection

After taking a control angiogram, the next step is injection of autologous arterial blood into the cisterna magna. At first, the neck of the dog is kept flexion. The positioning of the dog's neck can be done either in the stereotactic frame or manually after taking out of the stereotactic frame. In either way, it is important to keep the back-skin of the dog's neck extension by taking a neck-flexion posture. It cannot be overemphasized that successful puncture of the cisterna magna depends on the positioning. Using a 21-gaze needle connecting with a 5 ml-syringe, cisternal puncture is performed. During advancement of the needle, some fine feeling to penetrate the dura can be felt. Thereafter, the syringe is pulled gently and watery-clear cerebrospinal fluid is aspirated. It makes investigators be sure that successful cisternal puncture is carried out. If the needle hit the occipital bone, the needle should be inserted from a different point rather than changing the insertion angle.

Before injection of arterial blood, the same amount of cerebrospinal fluid is gently withdrawn. Three milliliters of nonheparinized arterial blood is collected from the inserted catheter using a new sterilized syringe, and blood is injected into the cisterna magna slowly. A rapid injection may arrest spontaneous breathing, and make a dog very sick. The speed of the injection was reported at 2 ml min^{-1}.[26–28]

The reason why a 21-gaze needle is used for cisternal puncture is to make aspiration of cerebrospinal fluid and injection of blood easy. When a much fine needle is used for the puncture, an investigator feels resistance in the injection of blood and tries to push syringe harder. Such maneuver makes injected blood scattered rather than accumulated around a basilar artery. During the injection of the blood, it is important to make sure that the tip of the needle is exactly in the subarachoid space by aspiration of the cerebrospinal fluid during the injection of the blood, and injection should correctly be done in that space. Otherwise, the blood might be injected into the extradural or intramuscular space, and sufficient blood clot around the basilar artery and significant cerebral vasospasm cannot be obtained.

In terms of amount of arterial blood, some authors reported that 0.5 ml/kg of the blood is injected.[26–28] Although it depends on the investigators, the modification of the total amount of injected blood is acceptable if the condition of the dog after injection is not hazardous.

Following the injection of the blood (the first injection or the first hemorrhage), the dog is tilted with the tail up for 20 min to facilitate setting of the blood around the basilar artery by gravity. After these procedures, the inserted catheter is removed, the femoral artery is sutured, and skin incision is closed. When spontaneous breathing is regular, a tracheal tube is removed, and the dog goes back to the cannel. This is all to be done on "Day 1". In literature, this day is defined as "Day 0".[26–28] In preparation for submission of a paper, clear definition of experimental days is necessary as described before.

4.6 The Second Injection of the Blood

The second injection of the blood was originally performed on "Day 3".[15] The Harvard group modified their original model, and the second injection was performed on "Day 4".[20] It depends on the investigators whether Day 3 or Day 4 is selected for the second injection day. It seems not much difference is seen in the data between the second injection days on "Day 3" and "Day 4".

On the second injection day, the exactly same procedures in anesthesia, cisternal puncture, and blood injection are carried out.

4.7 The Final Day of the Experiment

In the paper published by Varsso et al.,[15] the final day of the experiment is Day 5. However, the experiment was continued until Day 7 in most of recent studies.[20–28] On the final day of the experiment, the dog is anesthetized, and angiography is performed using the same procedure on Day 1 (the first day of the experiment). After taking the final angiography, the dog is sacrificed. The way of sacrifice varies and depends on investigators. Most important point is to sacrifice the dog and take out the basilar artery as quickly as possible. Exanguination from the femoral artery or directly from the heart was reported,[29] but it takes long time to sacrifice. To inject high KCl solution into the heart might make the property of the basilar artery changed. The author adopts the injection of an over-dose of pentobarbital (50 mg kg^{-1}) directly in the heart.[23–25]

After the sacrifice of the dog, a long longitudinal scalp incision is made and bilateral temporal muscles are separated from the cranium. Thereafter, craniotomy is performed and bilateral frontal lobes are incised. The brain is lifted up and the tentorium is bilaterally incised. Then, the whole brain, cerebellum, and the brainstem are exposed. When the brainstem is cut as distally as possible, total specimen can be removed. The brainstem with the basilar artery is separated from other tissues, and it is immersed in the buffer solution. The prepared buffer solutions depend on the investigators, or on the specific purpose of the further experiments. Under a microscope, the basilar artery is dissected, separated from the brainstem, and surrounding blood clots are meticulously removed from the basilar artery. The basilar artery is prepared for use. The surgical specimen obtained from the control dog (no blood injection) and that from the dog in double blood-injection vasospasm canine model are shown in Fig. 2.

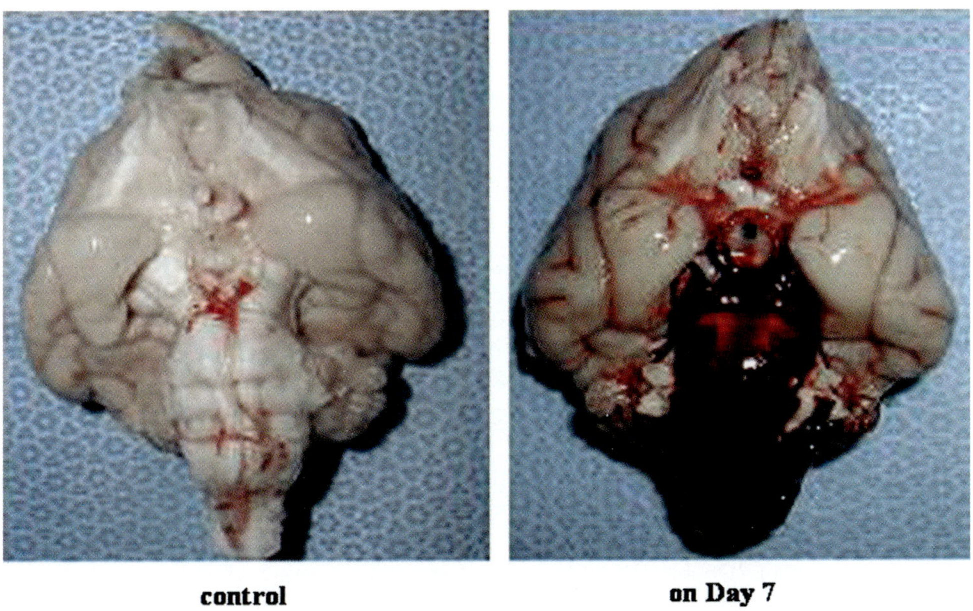

Fig. 2 Surgical specimens of the control (*left*) and double blood-injection canine model (*right*). In the double blood-injection canine model, the ventral surface of the brainstem is densely covered by injected blood, and the basilar artery is completely buried in the blood clot

4.8 Modification of the Model

There are several modifications in the double blood-injection vasospasm canine model. Most of modifications have been described in each step, although such modification in the model induces not much difference in the data.

4.8.1 Angiography

For the special experimental purposes, repeated angiography is carried out. To investigate the efficacy of special drugs for the treatment of cerebral vasospasm, angiography is performed on Day 1 after the first injection of the blood, on Day 4 before and after the second blood injection, on Day 5, on Day 6, and on Day 7 before sacrifice.[24] For the purpose of chronological changes of histological changes,[30] or mechanical contractile property of the artery,[31] the dog is kept alive until Day 28, and angiography is done until Day 28.

These modifications of the double-injection vasospasm canine model are acceptable if they are carefully planned and performed for each specific purpose.

4.9 Measurements of Diameter of Canine Basilar Artery

On an angiogram, the diameter of canine basilar artery is measured. The basilar artery on an angiogram is divided into three segments (upper, middle, and lower), and the midpoint of each segment is measured under a magnified microscope with a scale.[22–25] The diameter of the basilar artery is expressed as the mean value of these three points, and the extent of cerebral vasospasm is expressed as a percentage of the diameter of the control angiogram.[22–25] The margin of the arterial wall is blurred when it is seen under the magnified microscope. The graphic software to measure the diameter of the artery is commercially available. After scanning of the angiogram, the margin of the arterial wall is set first, and then the distance of both margin of the arterial wall is measured.[30]

In any way, it is important that the diameter is measured by at least three different individuals who are not involved in the study to avoid prepossession for the data of the experiment.

5 Results

5.1 Angiographical Cerebral Vasospasm and Its Time Course

In "Double Blood-Injection Vasospasm Canine Model," so called acute cerebral vasospasm occurs, but it resolves within 30 min. The extent of acute vasospasm is 75.2% ± 5.8% of the control diameter (Day 1). The diameters of the basilar artery before and after the second injection are 72.8 ± 2.4% and 49.2 ± 1.9%, respectively. On day 7, the diameter of the artery is 50.1 ± 2.1% (Figs. 3 and 4),[22–25] and the same extent of arterial constriction sustains until day 14.[31] Thereafter, the arterial constriction gradually resolves and returns to the control level on Day 28 (Fig. 5).[30,31] The time course of cerebral vasospasm in "Double Blood-Injection Vasospasm Canine Model" is very similar with that of human cerebral vasospasm following SAH.

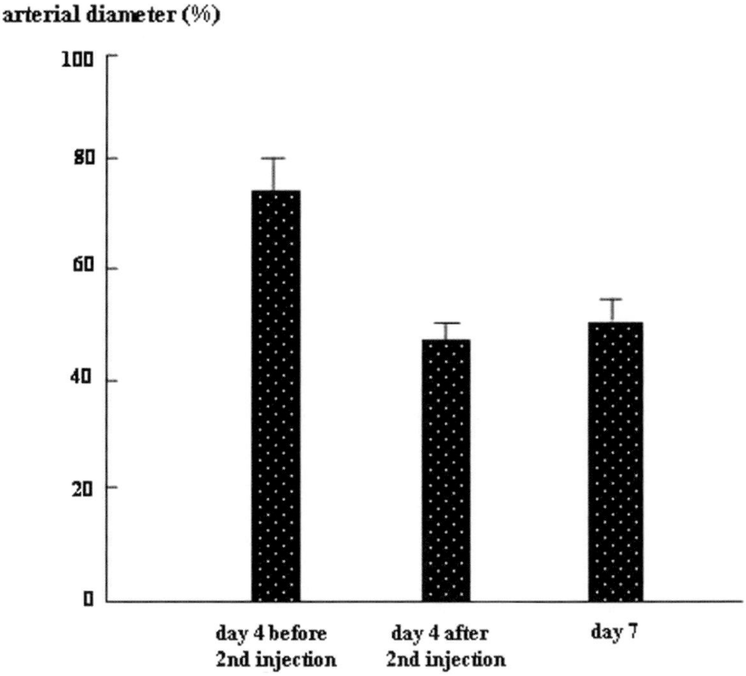

Fig. 3 A graph showing the diameter change of the basilar artery in the double blood-injection canine model. The diameter of the control angiogram on Day 1 is 100%, and the diameter of each experimental day is expressed as percentage of the control

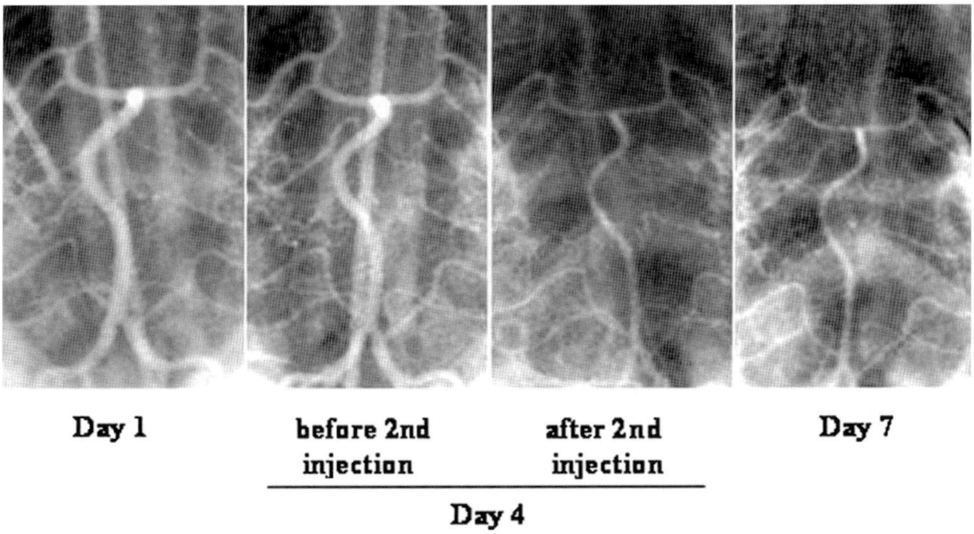

Fig. 4 Angiograms showing the canine basilar artery in the double blood-injection canine model. The angiograms show typical progression of cerebral vasospasm after injection of the blood

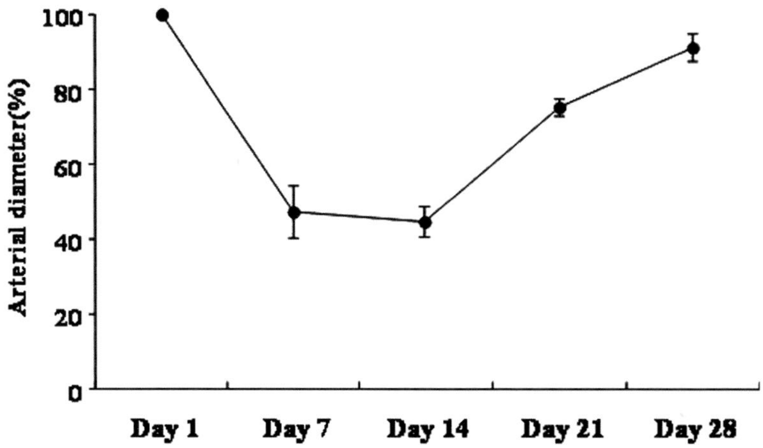

Fig. 5 Time course of the progression and resolution of cerebral vasospasm in the double blood-injection canine model. On Day 28, the diameter of the canine basilar artery returns to the control level

5.2 Histological Changes

In the histological viewpoint, human cerebral arteries exposed subarachnoid hemorrhage show remarkable changes such as endothelial detachment, corrugation of the internal elastic lamina, and thickening of medial smooth muscle cells.[16,17] In "Double Blood-Injection Vasospasm Canine Model," the same changes can chronologically be seen (Figs. 6 and 7),[16–18,30] and such changes are almost compatible with those in human vasospastic cerebral arteries.[16,17]

As the severity, time course of arterial constriction, and histological changes after blood injection in this model are identical with those in human, this model can be considered as a satisfactory model for investigation of the pathophysiological mechanisms of cerebral vasospasm following SAH.

6 Animal Protection

In recent days, animal usage for the experimental purpose is highly emphasized from the viewpoint of animal protection. Especially, it becomes harder and harder to use dogs and cats for any experiment. In near future, it could probably be impossible to use these animals for the experiments. Even it is still possible to use, it cannot be overemphasized about strict following to the guidelines for the care and use of animals, which is approved by each scientific society.

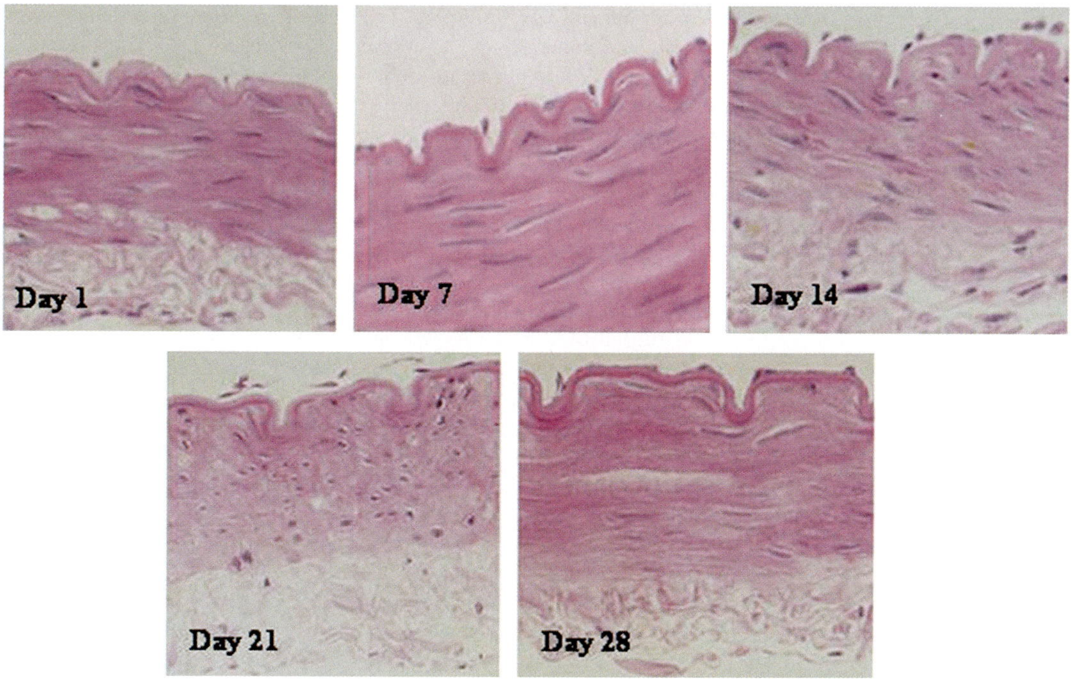

Fig. 6 Histological changes of canine basilar arteries in the double blood-injection canine model (hematoxylin–eosin staining). The medial thickening of the arterial wall is most prominent on Day 7, thereafter, it gradually returns to the control figure on Day 28

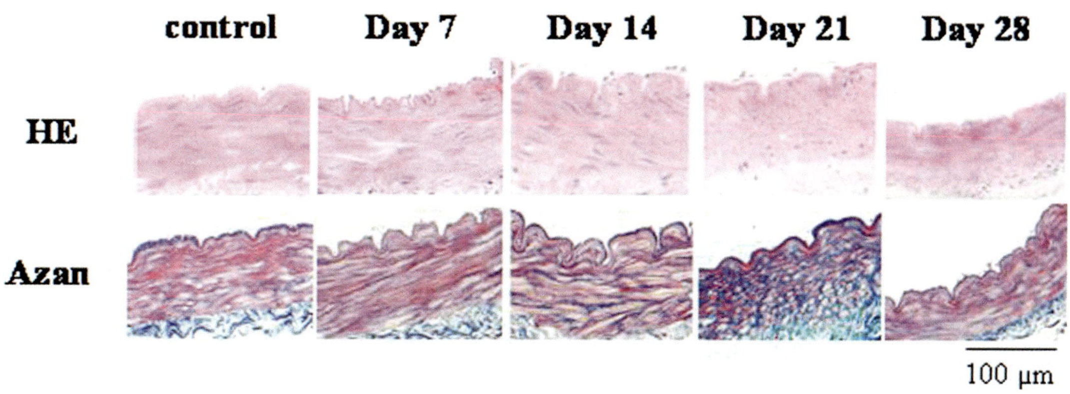

Fig. 7 Histological changes of canine basilar arteries with hematoxylin–eosin and Azan staining in the double blood-injection canine model. The collagen structure of the arterial wall is clearly shown by Azan staining

References

1. Echlin FA. Spasm of basilar and vertebral arteries caused by experimental subarachnoid hemorrhage. J Neurosurg 1965; 23: 1–11.
2. Allcock JM. Arterial spasm in subarachnoid haemorrhage. A clinical and experimental study. Acta Radiol 1966; 5: 73–83.

3. Browley BW, Strandness DE, Kelly WA. The biphasic response of cerebral vasospasm in experimental subarachnoid hemorrhage. J Neurosurg 1968; 28: 1–8.
4. Chow RWB, Newton TH, Smith MC. Cerebral vasospasm induced by subarachnoid blood and serotonin. An angiographic study. Invest Radiol 1968; 3: 402–407.
5. Kapp J, Mahaley MS Jr, Odom GL. Cerebral arterial spasm. Part I: Evaluation of experimental variables affecting the diameter of the exposed basilar artery. J Neurosurg 1968; 29: 331–338.
6. Landau B, Ransohoff J. Prolonged cerebral vasospasm in experimental subarachnoid hemorrhage. Neurology (Minneap) 1968; 18: 1056–1065.
7. Simeone FA, Ryan KG, Cotter JR. Prolonged experimental cerebral vasospasm. J Neurosurg 1968; 29: 357–366.
8. Arutiunov AI, Baron MA, Majorova NA. Experimentla and clinical study of the development of spasm of the cerebral arteries related to subarachnoid hemorrhage. J Neurosurg 1970; 32: 617–625.
9. Weir B, Erasmo R, Miller J. Vasospasm in response to repeated subarachnoid hemorrhages in the monkey. J Neurosurg 1970; 23: 395–406.
10. Wilkins RH, Levitt P. Intracranial arterial spasm in the dog. A chronic experimental model. J Neurosurg 1970; 33: 260–269.
11. Echlin FA: Current concepts in the etiology and treatment of vasospasm. Clin Neurousurg 1968; 15: 133–160.
12. Osterholm JL, Meyer R. Experimental effects of free serotonin on the brain and its relation to brain injury. Part 2: Trauma-induced alterations in spinal fluid and brain. J Neurosurg 1969; 31: 413–416.
13. Wilkins RH, Odom GL. Intracranial arterial spasm associated with craniocerebral trauma. J Neurosurg 1970; 32: 626–633.
14. Kuwayama A, Zervas NT. Belson R, Shintani A, Pickren K. A model for experimental cerebral arterial spasm. Stroke 1972; 3: 49–56.
15. Varsos VG, Liszczak TM, Han DH, Kistler JP, Vielma J, Black PM, Heros RC, Zervas NT. Delayed cerebral vasospasm is not reversible by aminophylline, nifedipine, or papaverine in a "two-hemorrhage" canine model. J Neurosurg 1983; 58: 11–17.
16. Eldevik OP, Krisiansen K, Torvik A. Subarachnoid hemorrhage and cerebrovascular spasm. Morphological study of intracranial arteries based on animal experiments and human autopsies. J Neurosurg 1981; 55: 869–876.
17. Findlay JM, Weir BKA, Kanamaru K, Espinosa F. Arterial wall changes in cerebral vasospasm. Neurosurgery 1989; 25: 736–746.
18. Seifert V, Stolke D, Reale E. Ultrastructural changes of the basilar artery following experimental subarachnoid haemorrhage. Acta Neurochir (Wien) 1989; 100: 164–171.
19. Nishziawa S. Controversial issues regarding pathophysiology of vasospasm: A review. In Macdonald RL, ed. Cerebral Vasospasm. Advances in Research and Treatment. Thieme, 2004: 103–105.
20. Peterson JW, Nishizawa S, Hackett JD, Bun T, Teramura A, Zervas NT: Cyclosporine A reduces cerebral vasospasm after subarachnoid hemorrhage in dogs. 1990; 21: 133–137.
21. Nishizawa S, Nezu N, Uemura K. Direct evidence for a key role of protein kinase C in the development of vasospasm after subarachnoid hemorrhage. J Neurosurg 1992; 76: 635–639. 103–105.
22. Nishizawa S, Peterson JW, Shimoyama I, Iwasaki K, Uemura K. Therapeutic effect of a new immunosuppressant, FK-506, on vasospasm after subarachnoid hemorrhage. Neurosurgery 1993; 32: 986–992.
23. Nishizawa S, Yamamoto S, Yokoyama T, Ryu H, Uemura K. Chronological changes of arterial diameter, cGMP, and protein kinase C in the development of vasospasm. Stroke 1995; 26: 1916–1921.

24. Nishizawa S, Obara K, Nakayama K, Koide M, Yokoyama T, Yokota N, Ohta S. Protein kinase C δ and α are involved in the development of vasospasm after subarachnoid hemorrhage. Eur J Pharmacol 2000; 398: 113–119.
25. Nishizawa S, Obara K, Koide M, Nakayama K, Ohta S, Yokoyama T. Attenuation of canine cerebral vasospasm after subarachnoid hemorrhage by protein kinase C inhibitors despite augmented phosphorylation of myosin light chain. J Vasc Res 2003; 40: 168–179.
26. Yamaguchi M, Zhou C, Heistad DD, Watanabe Y, Zhang JH. Gene transfer of extracellular superoxide dismutase failed to prevent cerebral vasospasm after experimental subarachnoid hemorrhage. Stroke 2004; 35: 2512–2517.
27. Yatsushige H, Yamaguchi M, zhou C, Calvert JW, Zhang JH. Role of c-Jun N-terminal kinase in cerebral vasospasm after experimental subarachnoid hemorrhage. Stroke 2005; 36: 1538–1543.
28. Zhou C, Yamaguchi M, Coloban AR, Zhang JH. Role of p53 and apoptosis in cerebral vasospasm after subarachnoid hemorrhage. J Cereb Blood Flow Metab 2005; 25: 572–582.
29. Nishizawa S, Peterson JW, Shimoyama I, Uemura K. Relation between protein kinase C and calmodulin systems in cerebrovascular contraction: nvestigation of the pathogenesis of vasospasm after subarachnoid hemorrhage. Neurosurgery 1992; 4: 711–716.
30. Yamaguchi-Okada M, Nishizawa S, Koide M, Nonaka Y. Biomechanical and phenotypic changes in the vasospastic canine basilar artery after subarachnoid hemorrhage. J Appl Physiol 2005; 99: 2045–2052.
31. Koide M, Nishizawa S, Ohta S, Yokoyama T, Namba H. Chronological changes of the contractile mechanism in prolonged vasospasm after subarachnoid hemorrhage: from protein kinase C to protein tyrosine kinase. Neurosurgery 2002; 51: 1468–1476.

Chapter 27

Nonhuman Primate Blood-Clot Placement Cerebral Vasospasm Model

Ayako Kawashima and R. Loch Macdonald

Abstract Cerebral vasospasm is delayed narrowing of the large cerebral arteries in response to blood clots left in the subarachnoid space after subarachnoid hemorrhage (SAH) from any cause, although it is most common after aneurysmal SAH because this process deposits the most dense and voluminous blood in the subarachnoid space, and it is this blood clot that causes vasospasm. Vasospasm is one of the main causes of morbidity and mortality in patients with aneurysmal SAH. Animal models of vasospasm have been developed because of the difficulties modeling delayed vasospasm in vitro. These models use one of three techniques to simulate SAH: (1) an intracranial artery is punctured allowing blood to surround the artery; (2) an artery is surgically exposed and clotted autologous blood obtained from another site is placed around the artery; or (3) autologous blood from another site is injected into the subarachnoid space and allowed to surround the intracranial arteries. Each technique has advantages and disadvantages. In this chapter, we discuss the advantages and disadvantages of and methods for creation of SAH and vasospasm in nonhuman primates. Arterial diameters are easily quantified by angiography, and SAH is created by frontal craniectomy, exposure of the intracranial arteries of the anterior circle of Willis, and placement of autologous clotted blood around these arteries.

Keywords: Cerebral angiography, Cerebral vasospasm, Nonhuman primate, Subarachnoid hemorrhage

1 Model Selection

Vasospasm has been modeled in vitro and in vivo. It may be difficult to generalize results of acute studies of arterial segments studied in vitro[1] with vasospasm. Arteries in vitro are denervated, not subjected to pulsative pressure and flow, and may not have normal endothelial function. Changes in gene expression can occur within hours and may influence the responses.[2] In most but not all monitoring systems in vitro the intra and extraluminal surfaces are exposed to vasoactive substances, which probably do not occur in vivo. Pathological changes occur in vasospastic

From: *Animal Models of Acute Neurological Injuries.*
Edited by: J. Chen (eds.) New York, Totowa, NJ

arteries but are not known to occur in experiments in vitro, which may be important assuming that the pathological changes are important and are not simply an epiphenomenon.

Mice, rats, rabbits, cats, dogs, and monkeys have been subjected to SAH to create vasospasm.[3] In general, injection of blood into the cisterna magna is used in rodents and results in mild, short-lived vasospasm. Vasospasm can be prolonged by repeating the injection a day or two later. An endovascular perforation model also is used in rats. The only models that produce prolonged, severe vasospasm are in dogs and monkeys.[3,4] Cisternal injections are typically done in dogs. The monkey model described here was invented by Bryce Weir in Edmonton, Alberta, in the 1980s. His experiments began with transcranial blood injections into the Rhesus monkey subarachnoid space but these produced only short-lived and mild vasospasm.[5] Injections of large volumes of blood did worsen the vasospasm but he conceived the intracranial clot placement model, which was adopted, because it was more reliable. The use of an operating microscope to perform a frontal craniectomy in the cynomolgus monkey, to open the arachnoid and expose the internal carotid, posterior communicating, middle and anterior cerebral arteries, and pack clotted autologous arterial blood around them was first reported in 1984.[6]

1.1 Advantages of This Model[4]

1.1.1 The model reliably creates severe, delayed vasospasm most like those occurring in humans. No other model produces vasospasm of the same severity and time course as this model. The dog double-hemorrhage model involves two injections of autologous blood into the cisterna magna several days apart, and this does cause severe vasospasm. However, the time course is somewhat longer than that occurring in man and injecting blood twice complicates the interpretation of time course studies. Rodents do not develop vasospasm that is severe or long-lasting, and this prolonged, severe course may be associated with processes that are pathophysiologically important and that are not even detected in these other models.

1.1.2 The knowledge that treatments that work in this model work in humans and those that do not work in this model are very unlikely to affect vasospasm in humans. This cannot be said of rodent models. The nonhuman primate model demonstrated that sodium nitroprusside,[7] nimodipine,[8] and tirilazad[9] had no or very modest effects on vasospasm whereas clot lysis with tissue plasminogen activator[10] and endothelin antagonists[11] were effective, corresponding to results in humans.[12,13]

1.1.3 There are clot-side arteries exposed to SAH that can be compared within animals to the contralateral, non clot-side arteries.

1.1.4 Some animals develop cerebral ischemia and infarction, which may be important in the vasospasm syndrome.

1.1.5 Intracranial arteries and a subarachnoid space similar to that in humans are present so experiments with clot placement, replacement, and removal as well as detailed time-course studies and systemic and intracranial drug delivery can be done easily.

1.2 Disadvantages of the Nonhuman Primate Model

1.2.1 There is a relative dearth of information on gene sequences of cynomolgus monkeys and paucity of specific molecular biological supplies such as antibodies for monkeys.

1.2.2 Variability is greater. The animals are not an inbred strain like rodents used in biological research. Therefore, interanimal variations occur, which affect the sizes of intracranial arteries, anatomy, composition of the blood, and of the arteries that may affect the degree of vasospasm produced.

1.2.3 The cost, time, and equipment required and complexity are substantial compared with rodents.

1.2.4 There are no genetically-modified monkeys. Manipulation of proteins has to be done pharmacologically at present. Some work has been done altering protein expression in cerebral arteries but this is mainly in the endothelial and adventitial cells, and it is very difficult to accomplish in the smooth muscle cells.[14]

1.2.5 The early effects of SAH are not modeled. The blood clot is placed surgically without increasing the intracranial pressure, so the aneurysm rupture and intracranial pressure changes that occur during this, if any, do not occur. The extent to which these processes contribute to vasospasm is unclear; however, obviously this model cannot determine this. Unfortunately, no model can. The closed-skull endovascular perforation model of SAH in rats mimics the acute effects of SAH[15,16] but has important disadvantages in that there is great variability in the volume of SAH, mortality is high, severe, delayed vasospasm is not produced, and there is frequently a permanent middle cerebral artery occlusion associated with the initial SAH, making this a model of SAH plus acute ischemia and not SAH alone.

1.2.6 Young female animals without other diseases are used. Few investigators have used old animals for SAH research or even stroke studies, so this is a problem common to the entire field.[17]

1.2.7 Many old and new-world monkeys, including the cynomolgus monkey, carry simian B virus (Cercopithecine herpesvirus 1).[18] They also can be infected with tuberculosis, numerous other tropical diseases, and even filoviruses such as Ebola and Marburg viruses.[19] Infection of humans with simian B virus is fatal in about 70% of cases, and the hemorrhagic fever associaed with filoviruses is worse. Most animals used now in medical research are repeatedly tested and used only if B and filovirus-negative.

1.2.8 Radical animal rights activists such as People for the Ethical Treatment of Animals (PETA) have killed humans in support of their cause and threatened researchers and their families. The PETA Web site lists how to support stopping research on animals, although members avoiding use of medical treatments developed through animal research does not seem to be necessary.

2 Materials

Juvenile female cynomolgus monkeys (*Macaca fasicularis*) weighing 2.5–4.0 kg are used. Regulatory requirements vary but generally over 1 month quarantine is required. Substantial preplanning is required because animals also usually cannot be shipped when the ambient temperature is too low or high. The animals are housed in crush cages and rooms complying with all federal requirements. The procedures are carried out in a veterinary operating room equipped in much the same fashion as a human operating room. The following general equipment is needed:

1. Small animal ventilator to deliver a tidal volume of approximately 8–10 ml kg^{-1} at a rate of 10–20 breaths per minute
2. Anesthetic machine equipped to deliver isoflurane and oxygen
3. Thermostatically-controlled heating blanket or warmed-air blanket to maintain body temperature
4. Angiography machine such as a standard, high-resolution fluoroscope machine capable of performing digital subtraction angiography
5. Radiolucent animal operating table
6. Physiological monitoring machines for percutaneous oxygen saturation, end-tidal PCO_2, heart rate, noninvasive blood pressure
7. An operating microscope, more than a simple dissecting microscope, such as at least the older Zeiss models. Extra light bulbs must be available.
8. Monopolar surgical coagulator, grounding pad and bipolar coagulation machines
9. Suction
10. Tape measure and magnification standard for angiography
11. Electronic scale to weigh the blood clot

For conduct of the angiography and creation of SAH, the following sterile instruments are recommended (Fig. 1):

1. Four small hemostats
2. One regular size needle driver
3. One curved, 4 or 5″ micro needle driver
4. One Mayo scissors
5. One short Metzenbaum scissors
6. Two fine vascular forceps
7. Two Adson toothed forceps
8. Scalpel handle
9. Self-retaining brain retractor and retractor blade
10. Small rongeur for making craniectomy
11. Trephine and pointed tip to make initial burr hole. An air-powered surgical drill also would work
12. #5 Frazier suction tube
13. Piece of angiography guide wire to fit an 18 guage angiography catheter
14. Two small metal beakers
15. Penfield microdissector
16. Pin head holder
17. Arachnoid knife
18. Bipolar forceps with cord

The disposable sterile supplies that are used include (Fig. 2):

1. Sterile weighing boat to weigh blood clot
2. Red-top tubes to put arterial blood from the angiography catheter into to clot
3. Sterile gauze (4 × 4)

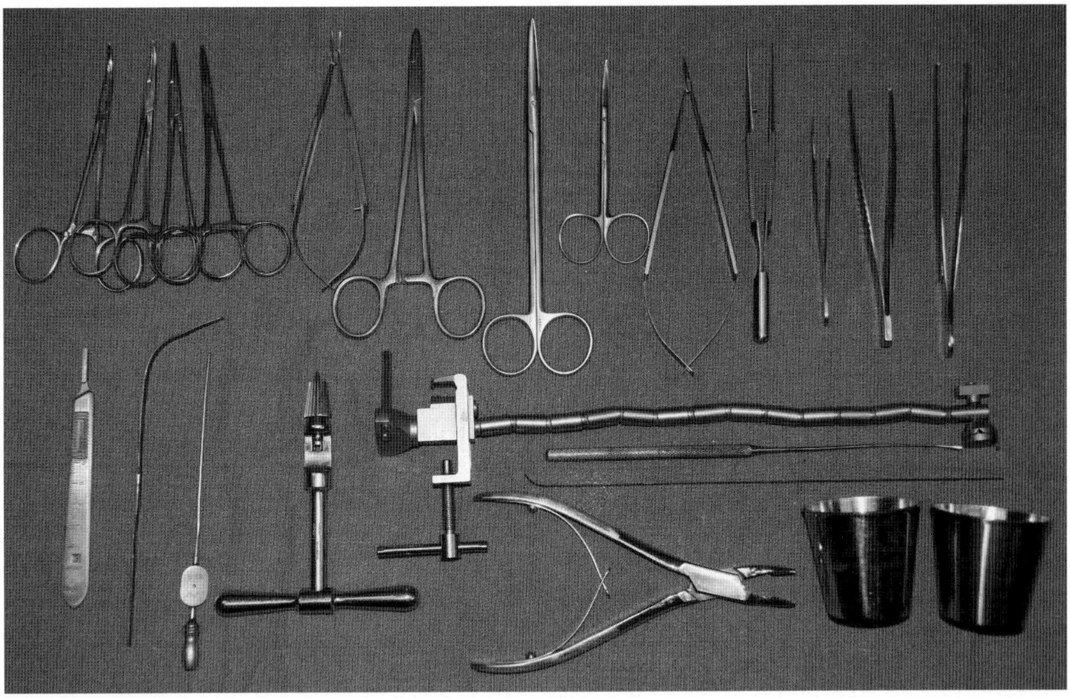

Fig. 1 *Photograph* of surgical instruments used for this model. From left to right on the *top row* are 4 hemostats, micro needle driver, standard needle driver, three types of scissors and four types of forceps. The *bottom row* from left to right is a scalpel handle, brain retractor, suction, trephine, self-retaining retractor arm, microdissector, angiography guide wire, ronguer and two metal cups. Bipolar and monopolar cautery and a sterile suction tubing are packaged separately

4. Cotton patties
5. Sterile drapes
6. 18 gauge angiography catheter
7. Three-way stopcock
8. Syringe (1 ml) with 25 gauge needle for blood gases, needle is bent and used as a dural hook
9. #10 scalpel blade
10. Syringes (2–10 ml) for catheter work and blood for SAH
11. Nonabsorbable monofilament 3-0 suture on a cutting needle for skin. Perfect scalp closure is required. We have not found that stapling devices can provide this.
12. 6-0 monofilament nylon or polypropylene suture on a noncutting needle for suturing dura
13. 4-0 silk suture for ligating the brachial artery
14. Absorbable 3-0 braided absorbable suture (such as polyglactin or an equivalent) on a noncutting needle for suturing the muscle and fascia
15. Bone wax
16. Surgical gloves
17. Heparin (1:1,000 solution for blood gas syringe only), mannitol (20% for intravenous injection), physiological saline and contrast agent such as 60% Hypaque (diatrizoate meglumine USP, 60%, Amersham Health, Princeton, NJ)

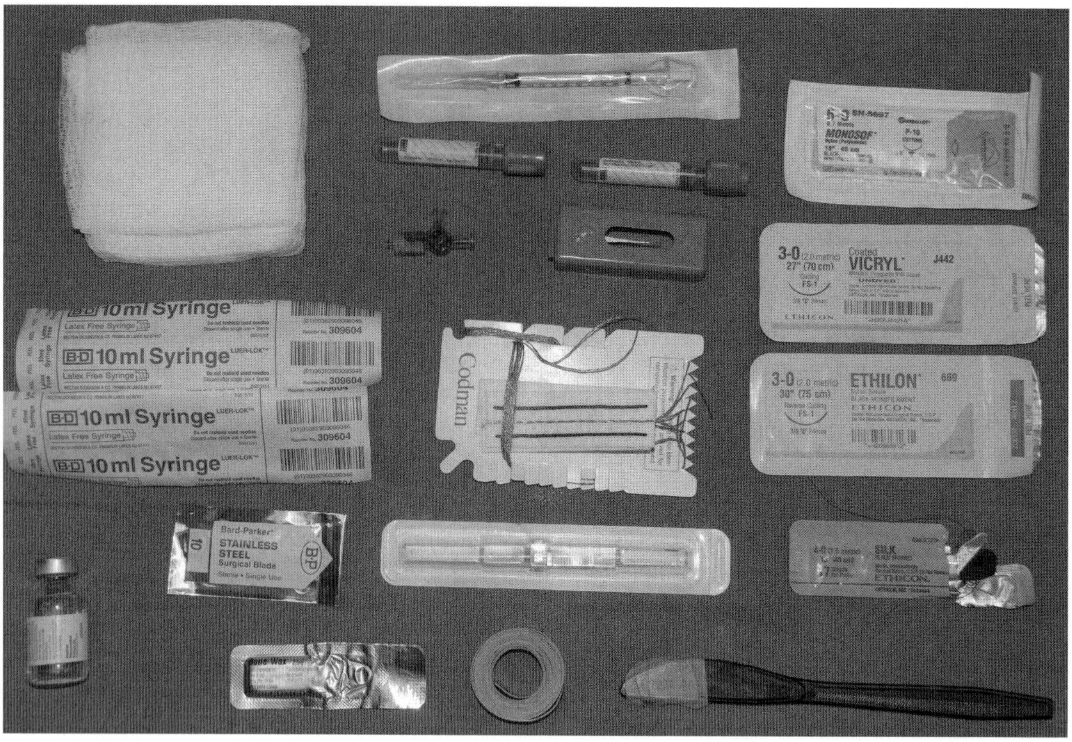

Fig. 2 Disposable supplies for angiography and creation of SAH by clot placement. Starting from the *top left*, there is guaze, 1 ml syringe with 25 guage needle, red-top tubes for blood, 3-way stopcock, arachnoid knife, 6-0 monofilament nonabsorbable suture, 3-0 absorbable suture, 2–10 ml syringes, patties, 3-0 monofilament nonabsorbable suture, heparin solution, scalpel blade, bone wax, 18 gauge 1.8 in. angiocatheter, tape measure, silk ligatures, and a coin taped to a stick to be placed behind the monkey's head during angiography as a magnification standard. Not shown are mannitol, contrast, weighing boat, surgical gloves and drapes and sterile surgeon's gown

3 Procedures

3.1 Anesthesia

The monkey is sedated in the crush cage with ketamine hydrochloride, 10 mg kg^{-1}, intramuscularly. It is weighed and an intravenous catheter is placed in a calf vein or other accessible superficial vein. The animal is intubated transorally with an unarmoured number 4 pediatric endotracheal tube. An antibiotic such as enrofloxacin, 2.5 mg kg^{-1}, is administered intramuscularly. The head, both groins, and the right axilla are shaved with a clipper, and the animal is then transported to the operating room. Ventilation is with oxygen and 1–3% isoflurane, using a small animal respirator. End-tidal PCO_2, body temperature, transcutaneous oxygen saturation, blood pressure, and heart rate are monitored and maintained at physiological levels. For the first stages when angiography is performed, it is important to maintain normal arterial PCO_2. Body temperature will drop precipitously if the animal is not kept covered. It is best maintained with a forced-air warming blanket.

3.2 Angiography

The animal is placed supine and the head is taped straight up with the neck flexed to permit an anteroposterior angiogram projection. A magnification standard is placed under the neck and the right hand is taped in an abducted and externally rotated posture to the operating table. A rolled gauze under the triceps will rotate the arm externally and expose the grove between the biceps anteriorly and triceps posteriorly, where the neurovascular bundle resides. The right axilla and upper arm are prepared with povidone iodine and a disposable, sterile drape with a 2 cm hole cut in the middle of it placed over the proximal neurovascular bundle. The surgeon puts on a lead apron and sterile gloves. Under the operating microscope, a 1 cm incision is made over the artery, which is exposed using microscissors and fine vascular forceps. With experience, this tedious dissection can be accomplished in minutes. The venous plexus around the artery is fragile but easily freed and does not need to be cauterized. Excess manipulation or cauterization will make the brachial artery spastic so it is best not to start any bleeding to begin with. The artery should be exposed proximal to the profunda brachii artery, since distal to this it is quite small and difficult to catheterize. The artery is tied distally with a 4-0 silk suture. A 4-0 silk suture is placed around the artery proximally, and it is catheterized with a sterile 18 gauge polypropylene catheter connected to a stop cock through which a guidewire is preplaced.

The guidewire is removed and a 10 ml syringe is connected to the stopcock. Blood (8 ml) is withdrawn from the catheter, placed in red-top tubes, and allowed to clot for unilateral SAH. Do not put heparin in or even near the catheter or stopcock as the minutest wiff of it will prevent the blood from clotting. At this point, a 1 ml syringe with a small amount of heparin in it is used to withdraw 300 μl blood for analysis of blood gases and hematological values.

The operating microscope is removed, and the fluoroscope is brought in and centered over the head. The isoflurane concentration may be below 1% but should be increased because we have routinely observed that injection of contrast causes the animal to move or vomit, and if the animal vomits the procedure has to be aborted otherwise it will die. The distance from the radiographic source to the operating table is measured and kept constant for every angiogram. The closer the head is to the radiographic detector, the clearer the image will be. Maximum magnification is used and exposure factors for the catheter runs should always be the same. We obtain a single, anteroposterior midarterial phase angiogram by manual injection of 5–8 ml 60% diatrizoate meglumine USP under digital subtraction video fluoroscopy.

The catheter is withdrawn from the brachial artery and the artery ligated proximally with 4-0 silk. We have not found it necessary to reconstruct the artery and have cannulated it more proximally for subsequent angiograms that have been done up to three more times as survival procedures. The incision is closed with interrupted, monofilament 3-0 nylon sutures. The catheter can be left in place, heparinized, and used to continuously monitor blood pressure for nonsurvival procedures if desired.

3.3 Craniectomy and Clot Placement

Mannitol (0.5 g kg^{-1}) is administered intravenously and anesthetic depth monitored to make sure pin placement and craniectomy can be safely performed.

The head is fixed in a 3-point pin fixation device and the right scalp prepared with povidone-iodine solution and then draped in sterile fashion (Fig. 3). The head position is the key to being able to see down the fissure as well as along the anterior cerebral artery. Ventilation is adjusted to lower the endtidal PCO_2 to 25–35 mmHg. Mono and bipolar cautery and suction are readied as well as all of the equipment including patties cut into small strips 5 mm wide and 2 cm long and 2–3 mm balls of bone wax.

Different incisions are possible but the point to keep in mind is that the temporalis muscle is very thick and the purpose of this operation is to get exposure to retract the brain as much as possible and put as much clot next to the arteries as can be done. Instead of a standard pterional, frontotemporal incision, we tend to use a reverse C incision based posteriorly (Fig. 3). The skin is incised and then monopolar cautery is used to incise the temporalis muscle along the same incision. The muscle is reflected posteriorly. The total

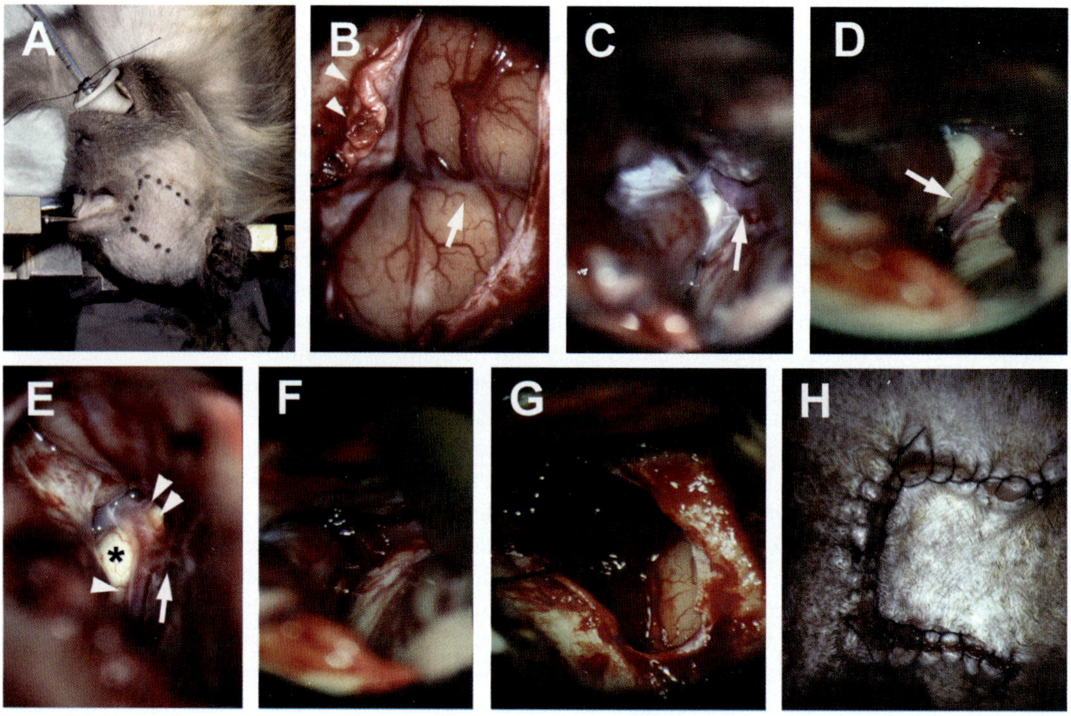

Fig. 3 Step-by-step procedures for creation of SAH. (**A**) The head position in a 3-point pin fixation device viewed from directly above the monkey. The incision is marked. (**B**) The craniectomy has been performed and the dura opened (*arrow heads* show free edge of dura next to the sphenoid ridge. The *arrow* shows the Sylvian fissure with the temporal lobe above and the frontal lobe inferior. (**C**) A retractor has been placed to retract the temporal lobe posteriorly. The *arrow* shows the middle cerebral artery exposed by opening of the arachnoid down to the carotid bifurcation. (**D**) After complete opening of the arachnoid, the anterior cerebral artery is well seen (*arrow*) as the frontal lobe is elevated with a microdissector. (**E**) The intradural segment of the internal carotid (*double arrow heads*, the *upper arrow head* is over the posterior communicating artery), the precommunicating segment of the anterior cerebral (*single arrow head*) and the sphenoidal segment of the middle cerebral artery (*arrow*) are exposed. The optic nerve (*asterisk*) is relatively larger than the arteries in this species compared with those in humans. (**F**) The clotted autologous blood has begun to be placed against the arteries of the right side of the circle of Willis. (**G**) Clot placement is complete. (**H**) The wound was closed in layers and in this case with a running monofilament 3-0 nylon suture

blood volume in these animals is about 75 ml kg^{-1} so excess blood loss has to be avoided. Pay attention to the deep temporal arteries that traverse up under the zygomatic arch deep to the deep temporal fascia. They need to be coagulated. The incision courses along the top of the zygomatic arch and then superiorly along the orbital rim for about 2 cm before turning posteriorly. A cuff of fascia is left attached to the bone for resuturing but all the muscle has to be removed from behind the orbit. The muscle is covered with moist gauze and held posteriorly. We place it under the self-retaining retractor blade that is mounted on the head holder at this stage. The muscle under the deep temporal fascia usually is excised with the fascia, and the fascia can be used as a dural patch.

The operating microscope is brought in. A craniotomy is made in the shape of a kidney, about 2 cm long, centered over the pterion and extending to the floor of the anterior and close to the floor of the middle cranial fossa. Take care not to push the dura away too much as extradural bleeding is difficult to control and perfect hemostasis is best. Also the brain underneath is like the most fragile, mushy neonatal brain ever and will bruise easily by compression through the dura. The sphenoid ridge is relatively flat compared with the human, but should be ronguered or drilled away for several millimeters. Bleeding from the diploic space can be marked from the inferior part of the craniectomy where the bone thickens, and is controlled with bone wax.

The dura mater is opened with a bent, 25 gauge needle used a hook and a scalpel blade (Fig. 3). The opening is completed in a C-shape based on the sphenoid ridge, and the dura is held anteriorly with a 6-0 suture. In some cases, a large venous sinus will be seen in the temporal dura. This heralds the presence of a draining vein from the Sylvian fissure to the dura of the sphenoid ridge. The temporal lobe should not be retracted or the vein will be avulsed and one will have to struggle to control the bleeding. Bridging veins should be looked for, coagulated, and cut.

Next, we place strips of patty 5 mm wide by about 2 cm long on the frontal and temporal lobes. Early deep retraction contuses the brain and causes intracerebral hematomas so we start the dissection laterally in the Sylvian fissure. The arachnoid is first incised at the lateral end of the fissure and craniectomy. We use an arachnoid knife and Jeweler's forceps. The brain is softer than toothpaste and overly aggressive suction will disrupt the pia and brain instantly. Under high magnification, the frontal and temporal operculae are separated and the distal middle cerebral artery exposed. The initial exposure will usually be beyond the bifurcation. Cerebrospinal fluid (CSF) has to be suctioned and small veins bridging across the fissure are coagulated. As the dissection progresses deeper, two fine vascular forceps can be used to pull apart the arachnoid and tease out the middle cerebral artery. At this point, enough CSF will have been suctioned to relax the brain. A $5 \times 5\,mm^2$ patty segment is put distally between the frontal and temporal operculae, and the frontal lobe is elevated with the retractor. The arachnoid is opened widely down over the intradural segment of the internal carotid, the precommunicating segment of anterior cerebral and the sphenoidal segment of the middle cerebral artery. It usually is possible to follow the posterior communicating artery back to the membrane of Lilliequist. The retractor is repositioned to retract the temporal lobe posteriorly.

The clotted autologous blood is poured out of the red-top tubes onto a sterile gauze and cut into small pieces. It is weighed and then placed piece-meal along the exposed cerebral arteries of the right side of the circle of Willis. Pieces can be put under the middle cerebral artery where it is free of perforators and can be lifted off the brain. The retractor and patties are removed as more clot is packed in.

Once as much clot as possible is packed in, the dura is closed with interrupted 6-0 sutures. The temporalis fascia is closed with interrupted absorbable 3-0 sutures and the skin with interrupted monofilament sutures (Fig. 3).

4 Postoperative Treatment and Euthanasia

The isoflurane is turned off and the animal ventilated and then extubated when awake. Buprenorphine, 10–30 μg kg^{-1}, is given intramuscularly at this time and then repeated every 12 h for 48 h and then after that as needed. Enrofloxacin, 2.5 mg kg^{-1}, intramuscularly is continued twice daily for 24 h unless there are signs of infection in which case it is continued. The animals are returned to their cages where they are fed their usual diet of biscuits, fruit, and water. They are observed every 12 h and all monitoring and interventions recorded in the animal's records. If they are ill, monitoring may be more frequent as determined by severity of illness.

After 7 days or whatever time is scheduled for the experiment, the monkey is placed under general anesthesia again and cerebral angiography is repeated. Euthanasia is then done according to what endpoints are needed. Perfusion through the left ventricle with phosphate-buffered saline (PBS) followed by paraformaldehyde is used for histologic endpoints, and with ice-cold PBS for molecular endpoints or isometric tension studies.

The brain is removed by reflecting the scalp and removing the cranium with an autopsy saw. The brain is immersed in ice-cold PBS and the cerebral arteries and clot removed. The remaining clot can be weighed. Dissect the cerebral arteries (internal carotid, precommunicating segments of the anterior cerebral, sphenoidal segment of middle cerebral, basilar artery) free under the operating microscope and place in ice-cold Krebs-Henseleit buffer. The brain can be placed in a custom-built brain mold (Fig. 4) and sectioned into 12–13 coronal sections of equal thickness.

5 Data Analysis

5.1 Angiography

To assess arterial diameters from cerebral angiograms (Fig. 5), an optical micrometer can be used to measure diameters at predetermined points along the extradural and intradural internal carotid, anterior and middle cerebral and basilar arteries.[11,20,21] When we use digital subtraction angiography, the images are saved electronically and can be measured using edge-detection programs written in Igor and NiDAQ Tools 1.5 (Wavemetrics, Lake Oswego, OR), although the key aspect is determining where in the penumbral area of the dye column on the angiogram that the edge of the artery will be defined, whether it is done electronically or visually. Thus, the reader should familiarize themselves with the literature on this subject.[22] Measurements should be conducted by an

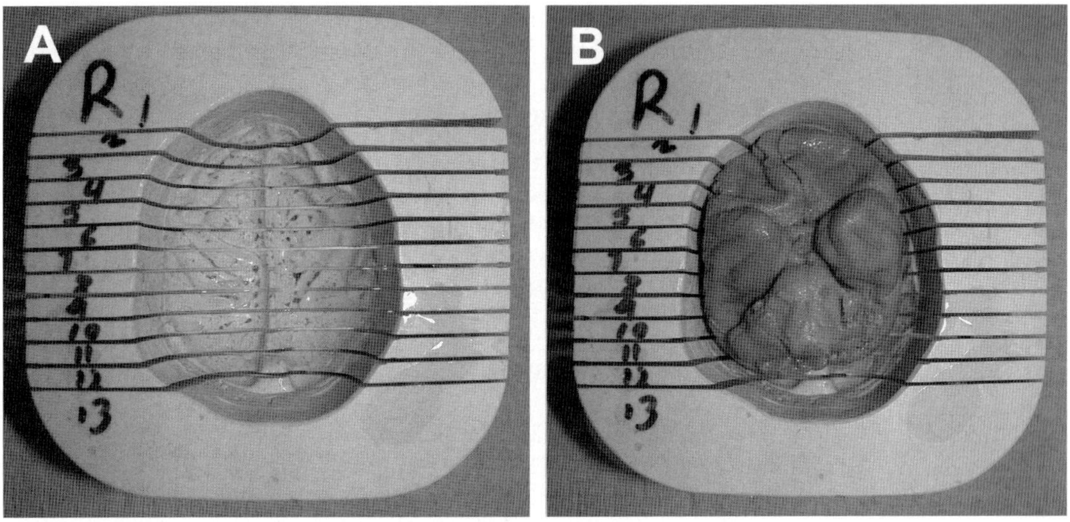

Fig. 4 (**A**) A custom-built brain mold. (**B**) Fixed monkey brain in the mold. The brain is smaller than the mold due to fixation; the fresh brain fills the mold

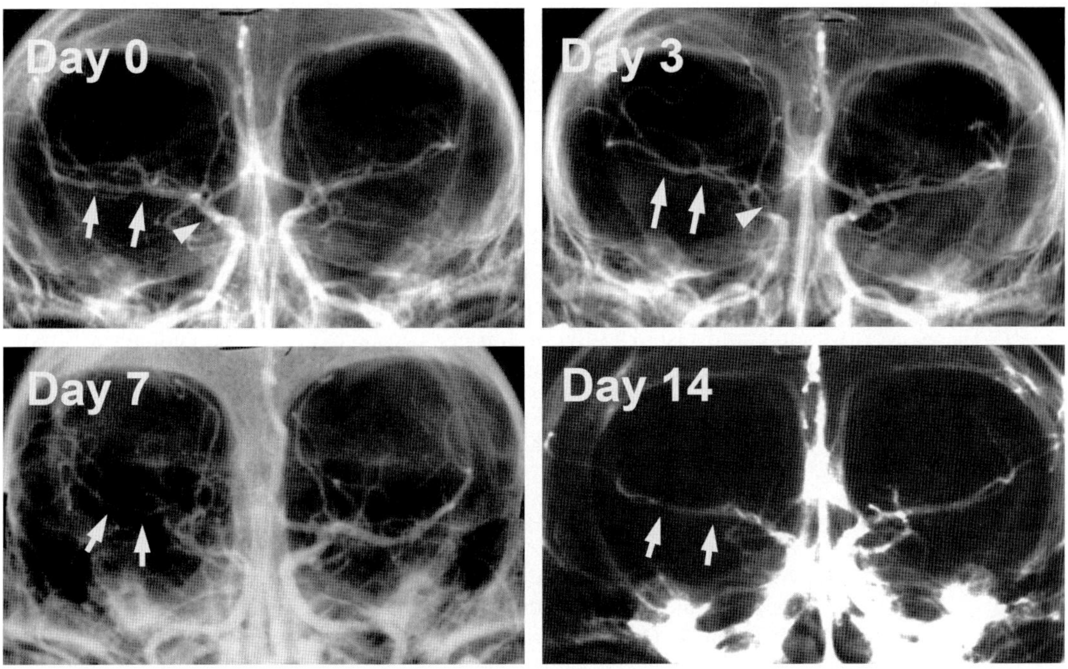

Fig. 5 Representative anteroposterior angiograms from baseline before SAH (**A**) and then 3 (**B**), 7 (**C**), and 14 (**D**) days after SAH showing development and resolution of right-sided vasospasm (*arrows* point to right middle cerebral artery and *arrow head* to right internal carotid artery)

investigator without knowledge of the monkey's group and in random order. Vasospasm can be assessed by comparing angiograms taken before and at some time after induction of SAH and treated as a continuous variable or categorized as absent (0–10% reduction in diameter), mild (10–25% reduction), moderate (25–50% reduction), or severe (>50% reduction). One also can compare within animals at each time the SAH-side and control-side arteries.

5.2 Additional Endpoints

These are limited only by the imagination and needs of the investigator. The authors have used isometric tension studies,[23] arterial protein and messenger ribonucleic acid levels,[24] second messenger concentrations,[25] immunohistochemistry,[26] morphometric measurements,[27] and studied the brain and blood clot.[20] We have not used behavior assessment yet.

6 Applications, Complications, and Limitations

The senior author is perhaps the person with the largest experience in the world with this model. This is a technically-demanding procedure where errors are punished by surgical fatality. For craniotomy and SAH, we recently have conducted 129 unilateral SAH, 58 bilateral SAH, and 29 clot removals on 187 monkeys. There was unexpected early death in 19 (10%), early illness necessitating planned euthanasia in 8 (4%) and morbidity in 13 (7%). Overall total morbidity and mortality was 41/187 = 22%. Neurological deficits and seizures develop in less than 6% of animals and infection develops in less than 5% of animals.

The model is suitable for a vast array of studies such as intracranial drug delivery by Ommaya reservoir or continuous infusion,[28] continuous intravenous drug delivery,[21] and measurement of intracranial pressure and cerebral blood flow.

Normally no elaborate catheterization techniques are needed but occasionally the anatomy of the aortic arch makes visualization of both carotids impossible by injection of the brachial artery so a transfemoral approach is necessary. This is always an option but we have found the brachial approach simple and fast. The femoral artery can be exposed and catheterized with a 4F angiography catheter that is advanced into the aortic arch or selectively into each common carotid artery. The risks of this procedure are that the iliac artery can become spastic and make removal of the catheter difficult or impossible.

References

1. Wilkins RH: Attempts at prevention or treatment of intracranial arterial spasm: an update. Neurosurgery 18:808–825, 1986.
2. Motterlini R, Gonzales A, Foresti R, Clark JE, Green CJ, Winslow RM: Heme oxygenase-1-derived carbon monoxide contributes to the suppression of acute hypertensive responses in vivo. Circ Res 83:568–577, 1998.
3. Macdonald RL, Weir B: *Cerebral Vasospasm.* San Diego, Academic, 2001.
4. Megyesi JF, Vollrath B, Cook DA, Findlay JM: In vivo animal models of cerebral vasospasm: a review. **Neurosurgery** 46:448–460, 2000.
5. Weir B, Erasmo R, Miller J, McIntyre J, Secord D, Mielke B: Vasospasm in response to repeated subarachnoid hemorrhages in the monkey. J Neurosurg 33:395–406, 1970.
6. Espinosa F, Weir B, Shnitka T, Overton T, Boisvert D: A randomized placebo-controlled double-blind trial of nimodipine after SAH in monkeys. Part 2: Pathological findings. J Neurosurg 60:1176–1185, 1984.

7. Rothberg CS, Weir B, Overton TR: Treatment of subarachnoid hemorrhage with sodium nitroprusside and phenylephrine: an experimental study. Neurosurgery 5:588–595, 1979.
8. Espinosa F, Weir B, Overton T, Castor W, Grace M, Boisvert D: A randomized placebo-controlled double-blind trial of nimodipine after SAH in monkeys. Part 1: Clinical and radiological findings. J Neurosurg 60:1167–1175, 1984.
9. Steinke DE, Weir BK, Findlay JM, Tanabe T, Grace M, Krushelnycky BW: A trial of the 21-aminosteroid U74006F in a primate model of chronic cerebral vasospasm. Neurosurgery 24:179–186, 1989.
10. Findlay JM, Weir BK, Gordon P, Grace M, Baughman R: Safety and efficacy of intrathecal thrombolytic therapy in a primate model of cerebral vasospasm. Neurosurgery 24:491–498, 1989.
11. Hino A, Weir BK, Macdonald RL, Thisted RA, Kim CJ, Johns LM: Prospective, randomized, double-blind trial of BQ-123 and bosentan for prevention of vasospasm following subarachnoid hemorrhage in monkeys. J Neurosurg 83:503–509, 1995.
12. Findlay JM, Kassell NF, Weir BK, Haley ECJ, Kongable G, Germanson, Truskowski L, Alves WM, Holness RO, Knuckey NW: A randomized trial of intraoperative, intracisternal tissue plasminogen activator for the prevention of vasospasm. Neurosurgery 37:168–176, 1995.
13. Macdonald RL, Kakarieka A, Mayer SA, Pasqualin A, Rufenacht DA, Schmiedek P, Kassell NF: Prevention of cerebral vasospasm after aneurysmal subarachnoid hemorrhage with clazosentan, an endothelin receptor antagonist. Neurosurgery 59:453 (Abstract), 2006.
14. Saitoh Y, Kato A, Hagihara Y, Kaneda Y, Yoshimine T: Gene therapy for ischemic brain diseases. Curr Gene Ther 3:49–58, 2003.
15. Veelken JA, Laing RJ, Jakubowski J: The Sheffield model of subarachnoid hemorrhage in rats. Stroke 26:1279–1283, 1995.
16. Bederson JB, Germano IM, Guarino L: Cortical blood flow and cerebral perfusion pressure in a new noncraniotomy model of subarachnoid hemorrhage in the rat. Stroke 26:1086–1091, 1995.
17. The Stroke Therapy Academic Industry Round Table (STAIR): Recommendations for standards regarding preclinical neuroprotective and restorative drug development. Stroke 30:2752–2758, 1999.
18. Eberle R, Hilliard J: The simian herpesviruses. Infect Agents Dis 4:55–70, 1995.
19. DeMarcus TA, Tipple MA, Ostrowski SR: US policy for disease control among imported nonhuman primates. J Infect Dis 179 Suppl 1:S281–S282, 1999.
20. Zhang ZD, Yamini B, Komuro T, Ono S, Johns L, Marton LS, Weir B, Macdonald RL: Vasospasm in monkeys resolves because of loss of and encasement of subarachnoid blood clot. Stroke 32:1868–1874, 2001.
21. Macdonald RL, Curry DJ, Aihara Y, Zhang ZD, Jahromi BS, Yassari R: Magnesium and experimental vasospasm. J Neurosurg 100:106–110, 2004.
22. Elisevich K, Cunningham IA, Assis L: Size estimation and magnification error in radiographic imaging: implications for classification of arteriovenous malformations. Am J Neuroradiol 16:531–538, 1995.
23. Macdonald RL, Zhang J, Marton LS, Weir B: Effects of cell-permeant calcium chelators on contractility in monkey basilar artery. J Neurotrauma 16:37–47, 1999.
24. Hino A, Tokuyama Y, Kobayashi M, Yano M, Weir B, Takeda J, Wang X, Bell GI, Macdonald RL: Increased expression of endothelin B receptor mRNA following subarachnoid hemorrhage in monkeys. J Cereb Blood Flow Metab 16:688–697, 1996.
25. Aihara Y, Jahromi BS, Yassari R, Sayama T, Macdonald RL: Effects of a nitric oxide donor on and correlation of changes in cyclic nucleotide levels with experimental vasospasm. Neurosurgery 52:661–667, 2003.
26. Ono S, Zhang ZD, Marton LS, Yamini B, Windmeyer E, Johns L, Kowalczuk A, Lin G, Macdonald RL: Heme oxygenase-1 and ferritin are increased in cerebral

arteries after subarachnoid hemorrhage in monkeys. J Cereb Blood Flow Metab 20:1066–1076, 2000.

27. Macdonald RL, Weir BK, Grace MG, Martin TP, Doi M, Cook DA: Morphometric analysis of monkey cerebral arteries exposed in vivo to whole blood, oxyhemoglobin, methemoglobin, and bilirubin. Blood Vessels 28:498–510, 1991.
28. Macdonald RL, Zhang ZD, Curry D, Elas M, Aihara Y, Halpern H, Jahromi BS, Johns L: Intracisternal sodium nitroprusside fails to prevent vasospasm in nonhuman primates. Neurosurgery 51:761–768, 2002.

Section III

Intracerebral Hemorrhage

Chapter 28

Blood Injection Intracerebral Hemorrhage Rat Model

Thomas O'Lynnger, Ya Hua, and Guohua Xi

Abstract The autologous blood injection intracerebral hemorrhage (ICH) model has been used successfully in rats because they are large enough to accommodate all of the procedures and protocols, and that measurement of pathophysiological parameters in the rat brain can be extrapolated to humans. In this chapter, we present a step-by-step procedure to model ICH by injecting autologous whole blood. This model can be readily modified to investigate the effects of infusing different compounds other than autologous whole blood.

Keywords: Intracerebral hemorrhage, Brain edema, Blood–brain barrier permeability, Brain atrophy

1 Model Selection

Blood injection is a suitable model for studying intracerebral hemorrhage (ICH). Autologous blood obtained from the femoral artery is infused into the right caudate nucleus to simulate hemorrhage (Fig. 1). Advantages of this model include the following:

1. In vivo observation of the pathological changes of ICH and the results of therapeutic manipulations
2. A relative homogeneity within strains owing to inbreeding
3. A small brain size well-suited to immunohistochemical and biochemical studies
4. The ability to study a variety of physiologic parameters that are altered by the hematoma, including intracranial pressure, cerebral blood flow, cerebral perfusion pressure, and neurological function
5. Ease of reproducibility and rarity of complications
6. The ability to measure blood–brain permeability, brain edema, and brain atrophy and extrapolate those measurements to the human situation
7. Modification of the procedure to test the neurotoxic and neuoprotective effects of a virtually unlimited number of infused compounds

From: *Animal Models of Acute Neurological Injuries.*
Edited by: J. Chen (eds.) New York, Totowa, NJ

Fig. 1 Rat positioned in a stereotactic frame prepared for injection in the right caudate. The scalp is incised to expose the area to be injected, and a 1 mm cranial burr hole is drilled to create an entry point for the needle

2 Materials

Sprague-Dawley (SD) rats weighing 250–300 g are used in this model of ICH. The materials recommended for the surgical procedure include the following:

1. Kopf stereotactic frame
2. Infusion pump (Harvard)
3. Dremel drill with 1 mm drill bit
4. Moria spring scissors with very sharp points
5. Scalpel handle with scalpel blades
6. Fine iris scissors
7. Curved Mayo scissors
8. Dumont forceps
9. Dumont angled forceps
10. Adson forceps with teeth
11. Heiss retractor
12. Hemostats
13. Needle holder
14. Cotton-tipped applicators
15. Syringes (1 mL)
16. Polyethylene tubing (PE-50)
17. 23-gauge and 26-gauge needles
18. 5–0 black braided silk (Deknatel)
19. 3–0 Dermalon suture

3 Procedure

1. Inject sodium pentobarbital (50 mg kg^{-1}, i.p.) to anesthetize the rat.
2. Apply ophthalmic ointment to both eyes.

3. Use a heating pad with feedback regulator to maintain body temperature at 37°C.
4. Use a hair clipper to remove hair from the scalp and right groin regions and apply Betadine to the exposed skin.
5. Make a 2 cm incision in the groin at the region of the right femoral artery.
6. Use the curved Mayo scissors to cut away fat to expose the artery, being careful to avoid cutting any major blood vessels.
7. Expose the femoral artery and vein using cotton-tipped applicators and fine iris scissors to cut away fascia.
8. Separate 1.5 cm of the femoral artery from the femoral vein using angled forceps and a cotton-tipped applicator, being careful not to tear the artery.
9. Ligate the distal end of the artery using 5–0 black braided silk and clip one end of the silk with a hemostat and place to the side to help expose the artery.
10. Place a thread of silk under the proximal end of the artery and clip both ends with a hemostat to temporarily occlude blood flow.
11. Using spring scissors, prick the artery to create an opening for the catheter without severing the blood vessel.
12. Make sure the PE-50 catheter is trimmed to a point and attached to a 23-gauge needle and 1 mL syringe filled with saline.
13. Insert the tubing into the opening with the pointed side down using two pairs of forceps to hold the tubing and the artery in place.
14. Use a loop of silk to cinch the tubing in place.
15. Inject a small amount of saline to ensure that blood is being pumped into the tubing.
16. Exchange the saline-filled syringe for an empty syringe and release the tension on the proximal occluding thread to fill the syringe with 0.2 mL of blood.
17. Using another syringe that has been filled with heparin and then emptied completely, withdraw a 0.2 mL blood sample to analyze glucose, hematocrit, and blood gases.
18. After sampling the blood and measuring blood pressure if appropriate, put the tension back on the proximal end of the artery, then gently slide the PE-50 out enough so that the loop originally cinching the tubing to the artery slides off, then tighten it to ligate the artery.
19. Remove the silk on the proximal end and cut all the loose ends of silk, then suture the skin using 2–3 3–0 Dermalon sutures.
20. Place the rat into the Kopf stereotactic frame and make a 2 cm sagittal incision over the midline of the scalp.
21. Using cotton-tipped applicators, clear away the fascia to expose the bregma, then position the Heiss retractor under the skin to keep the area exposed.
22. Measure 4.0 mm laterally to the right and 0.1 mm anterior to the bregma and mark the area for the cranial burr hole, making sure to stay posterior to the coronal suture.
23. Holding the Dremel drill with 1 mm drill bit vertically over the mark, slowly drill through the cranium, stopping when the pressure is no longer felt in order avoid tearing the dura.

24. Place the 0.2 mL sample of nonheparinized blood in the infusion pump with a 26-gauge needle with the beveled edge pointing away from the midline.
25. Position the needle over the burr hole and insert the tip of the needle into the brain just until the beveled edge is no longer visible, then insert the needle 5.5 mm ventrally into the right caudate.
26. Use the infusion pump to infuse 100 μL of autologous whole blood into the right basal ganglia at a rate of 10 μL min^{-1}, leaving the needle in place for 2 min once the infusion is complete.
27. Retract the needle and fill the burr hole with bone wax.
28. Suture the skin using two to three 3–0 Dermalon sutures.

4 Outcome Evaluations

1. Brain water content: To examine brain edema formation, which is a major determinant of survival in patients with ICH. It is also a simple, reproducible, and inexpensive measure of brain injury.[1–3]
2. Blood–brain barrier (BBB) permeability: The commonly used assessors of BBB disruption are radioisotopes and Evans blue.[4,5]
3. Brain genomics: To determine changes of gene expression.[6]
4. Histology: To evaluate the size, location, extent of the hemorrhage,[7] neuronal death,[8,9] and brain atrophy,[9,10] coronal sections should be prepared (Figs. 2 and 3).
5. Behavioral testing: Forelimb placing test, the forelimb use asymmetry test, and the corner turn test are sensitive measures of neurological deficits induced by ICH.[7]

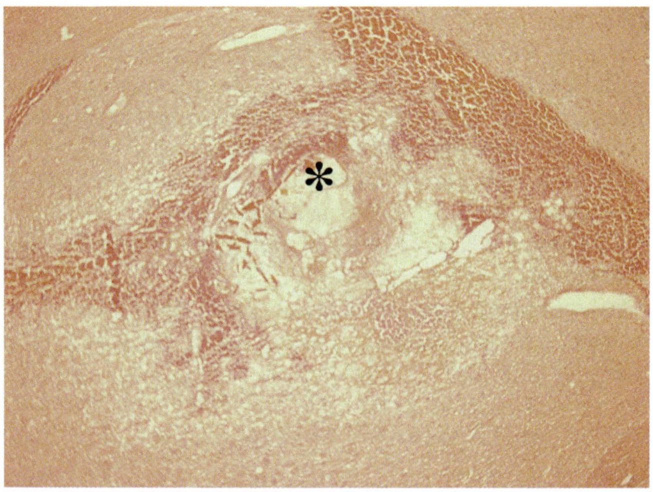

Fig. 2 H&E staining of a coronal slice of the right caudate 3 days after injection of whole blood. The a*sterisk* denotes hematoma

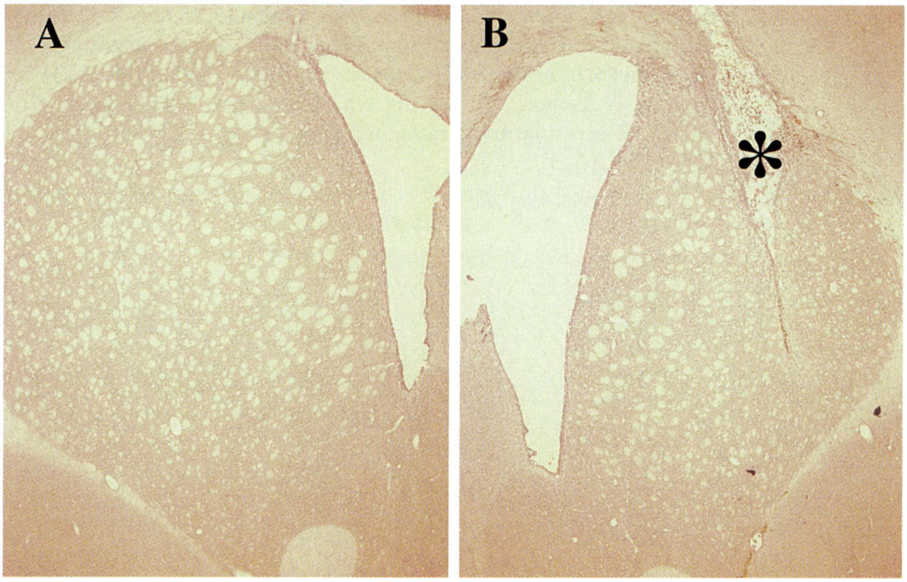

Fig. 3 Brain atrophy after ICH in rats. Coronal brain sections (**A**: the contralateral caudate; **B**: the ipsilateral caudate) two month after ICH. Note enlargement of ventricle and atrophy of striatum in the ipsilateral side. The a*sterisk* denotes hematoma cavity

5 Limitations

1. Lack of a ruptured blood vessel as the source of the ICH
2. Lacks the hematoma enlargement that occurs in ICH patients
3. Very low mortality rate

References

1. Yang GY, Betz AL, Chenevert TL, Brunberg JA, Hoff JT. Experimental intracerebral hemorrhage: Relationship between brain edema, blood flow, and blood-brain barrier permeability in rats. *J Neurosurg*. 1994;81:93–102
2. Xi G, Wanger KR, Hua Y, deCourten-Myers GM, Myers RE. Plasma but not packed red blood cells infused in pig cerebral white matter induces rapidly developing, marked edema. *Soc Neurosci Abs*. 1996;22:1424
3. Xi G, Keep RF, Hoff JT. Erythrocytes and delayed brain edema formation following intracerebral hemorrhage in rats. *J Neurosurg*. 1998;89:991–996
4. Xi G, Hua Y, Bhasin RR, Ennis SR, Keep RF, Hoff JT. Mechanisms of edema formation after intracerebral hemorrhage: Effects of extravasated red blood cells on blood flow and blood-brain barrier integrity. *Stroke* 2001;32:2932–2938
5. Belayev L, Saul I, Busto R, Danielyan K, Vigdorchik A, Khoutorova L, Ginsberg MD. Albumin treatment reduces neurological deficit and protects blood-brain barrier integrity after acute intracortical hematoma in the rat. *Stroke* 2005;36:326–331
6. Lu A, Tang Y, Ran R, Ardizzone TD, Wagner KR, Sharp FR. Brain genomics of intracerebral hemorrhage. *J Cereb Blood Flow Metab*. 2006;26:230–252
7. Hua Y, Schallert T, Keep RF, Wu J, Hoff JT, Xi G. Behavioral tests after intracerebral hemorrhage in the rat. *Stroke* 2002;33:2478–2484
8. Wu J, Hua Y, Keep RF, Schallert T, Hoff JT, Xi G. Oxidative brain injury from extravasated erythrocytes after intracerebral hemorrhage. *Brain Res*. 2002;953:45–52

9. Felberg RA, Grotta JC, Shirzadi AL, Strong R, Narayana P, Hill-Felberg SJ, Aronowski J. Cell death in experimental intracerebral hemorrhage: The "black hole" model of hemorrhagic damage. *Ann Neurol*. 2002;51:517–524
10. Hua Y, Nakamura T, Keep R, Wu J, Schallert T, Hoff J, Xi G. Long-term effects of experimental intracerebral hemorrhage: The role of iron. *J Neurosurg*. 2006;104:305–312

Chapter 29

Bacterial Collagenase Injection Intracerebral Hemorrhage Rat Model

Mark Grossetete and Gary A. Rosenberg

Abstract Intracerebral hemorrhage (ICH) has been modeled in the laboratory with the intracerebral injection of autologous blood, injection of emboli, or proteolysis of the basal lamina surrounding cerebral capillaries by bacterial collagenase (BC). Selection of the model depends on the concept of the pathophysiology behind the bleeding. Modeling a rapidly progressive bleed without excessive tissue destruction can be done by the injection of autologous blood. Injection of small blood clots into the carotid artery of the rabbit mimics an embolic stroke with hemorrhagic transformation. BC produces a highly reproducible model of intracerebral bleeding that evolves over several hours, produces inflammation around the lesion, and slowly resolves, simulating the process that has been observed to occur in human studies of serially imaged patients with ICH. BC initiates an injury that is propagated by the production of endogenous proteolytic enzymes of the matrix metalloproteinase (MMP) gene family. The inflammatory response, which is seen in both the BC model and the autologous blood model, is an important part of the pathological changes produced in the brain by the disruption of the blood vessels and brain edema. MMP inhibitors reduce the damage to the brain by the hemorrhagic mass lesion. A number of other therapeutic agents reduce injury in BC-induced ICH.

Keywords: Apoptosis, Brain edema, BB-94, Matrix metalloproteinases, Matrix metalloproteinase inhibitors

1 Introduction

Intracerebral hemorrhage (ICH) is a common neurological disease for which an effective treatment is lacking.[1] Studies of the pathophysiology have been done in several experimental models. The selection of the model depends on the concepts of the origin and propagation of the intracranial bleeding. Two major theories have been debated since the middle of the nineteenth century: in one, a damaged blood vessel ruptures and releases a large quantity of blood into the tissue, while in the other, ischemia initiates a molecular cascade with proteases causing hemorrhagic transformation (HT).[2] Because of the extensive

From: *Animal Models of Acute Neurological Injuries.*
Edited by: J. Chen (eds.) New York, Totowa, NJ

destruction of tissues as seen at autopsy, prior to the modern era of imaging, the origin and timing of the bleed was obscured. However, computed tomography (CT) scanning has helped resolve this controversy. Through the use of serial imaging, it has been observed that in a significant number of patients, the hemorrhagic mass grew over several hours, suggesting release of a tissue disrupting substance.[3] This growth occurred in about 30% of patients with the rest having a static mass lesion. This heterogeneity in the onset and course of ICH implies that a single model will be inadequate to determine the usefulness of a therapy and that several models will be needed.

Two models are currently being used most extensively to study ICH. The first to be developed was the direct injection of autologous blood into the caudate, and the second involves the intraparenchymal injection of bacterial collagenase (BC). The blood infusion model rapidly creates a mass lesion with blood extravasation into the tissue. The collagenase-induced hemorrhage model simulates the formation of a mass lesion over several hours because of damage to the blood vessels as would be seen with secondary bleeding from HT. This review will focus on the pros and cons of the BC-ICH model.

2 BC-Induced ICH

BC is produced by certain bacteria, providing them with a mechanism for pathological damage. The BC used in most studies comes from *Clostridium histolyticum*, and is encoded in two distinct chromosomal genes, *ColH* and *ColG*. The collagenases produced by the bacteria degrade both native and denatured collagen. Bacterial collagenases are in the zinc metalloproteinase family, and are referred to as zincins, because they share a zinc-containing HEXXH motif in the active site, whereas the metzincins contain a conserved methionine residue situated under the active site; the gluzincins contain a conserved glutamate residue toward the C-terminal region that probably acts as the third zinc ligand (Fig. 1).

Cleavage of C-terminal peptides of ColH results in the formation of a 98 kDa gelatinase, with decreased substrate capacity; the 98 kDa form can only degrade denatured but not native collagen. In an early study, several proteases considered destructive to the blood-brain barrier (BBB) were injected into the cerebral ventricles of rats to study their effect on capillary permeability; of the proteases tested BC proved to be the most disruptive to the BBB.[4] During a study of the effect of brain edema on the movement of radio-labeled sucrose through the extracellular matrix, we infused BC into the caudate nucleus along with the radio-labeled tracer.[5] The BC produced a hemorrhage at the site of the injection, which proved to be highly reproducible. This unexpected finding formed the basis for the current model of BC-induced ICH.[6]

3 Rodent Selection in BC-Induced ICH

The BC-induced ICH model was developed in rat, and this species remains the most widely used for induction of the hemorrhage. More recently, mouse models have been utilized for studies in genetically manipulated mice. A systematic

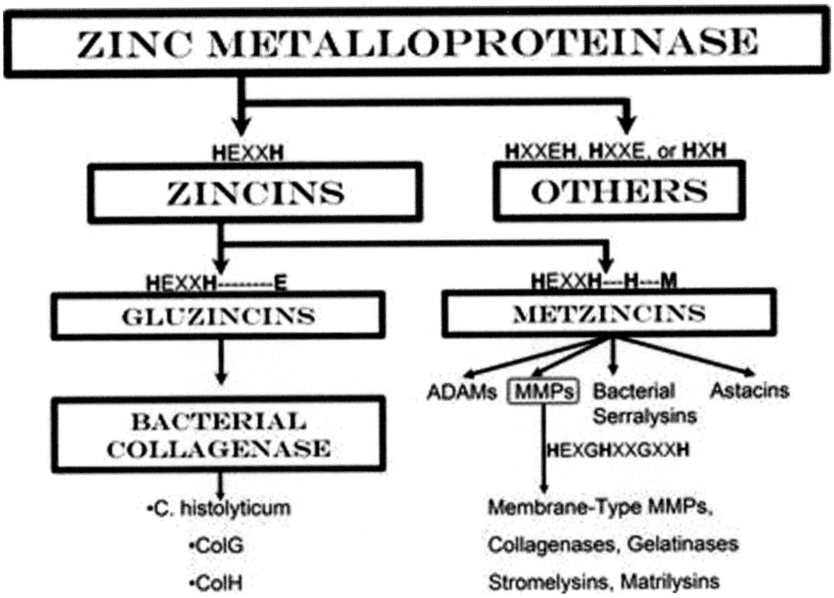

Fig. 1 Classification of bacterial collagenases and the matrix metalloproteinases (MMPs)

study has not been done on the differences between hemorrhages in the two species. In the rat, the larger brain size allows for the study of the flow of edema fluid through the brain. Edema forms around the mass lesion and in the white matter fiber tracts around the hippocampus; MRI shows the excess fluid in both locations.[7,8] The distribution of the edema in the posterior regions provides a convenient location to monitor the effects of drugs on the edema.

The injection of BC results in the endogenous production of MMP-9, which is induced in inflammation. A dose study was performed to determine an appropriate concentration and injectate volume of BC that provided a consistent hemorrhage size and a less than 10% death rate. Initially, two types of BC were evaluated for use in the study. Type XI (Sigma-Aldrich) was initially used at a concentration of 0.075 U in a 0.5 μL injectate volume; however, Type XI also contains other contaminating proteases in addition to the BC activity. Utilization of this collagenase type resulted in a large hemorrhage and a mortality rate greater than 10%. A switch was made to cell culture tested, Type VII-S BC (Sigma-Aldrich), from *Clostridium histolyticum*, which is sterile and highly purified. The same dose still resulted in a mortality rate that was higher than 10%; a dose of 0.0375 U in 0.5 μL was found suitable and produced approximately a 10 mm^3 hemorrhage volume.

The analysis of the endogenous production of the gelatinases, especially of MMP-9, required the accurate determination of the appropriate molecular weights of the exogenously administered BC vs. the endogenous production of MMPs. Naïve BC that is diluted and run on a zymogram shows two predominant bands at the apparent molecular weights of 110 and 103 kDa. Lower molecular weight species are also present in bacterial collagenase;

however, when a 1-h-post CIH brain sample is analyzed by zymography, the only BC bands detectable by zymography are the 110, 103, and 84 kDa bands; murine pro-MMP-9 (98 kDa) and pro-MMP-2 (72 kDa) are present as well (Fig. 2).

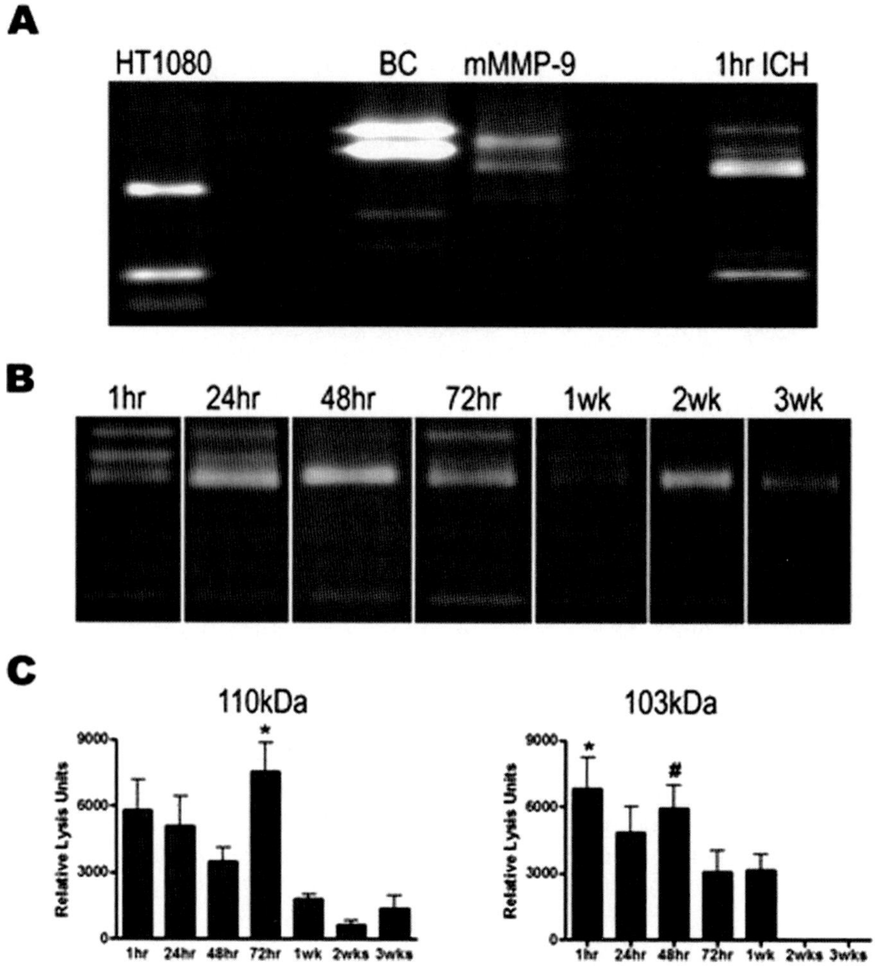

Fig. 2 Molecular weights of MMP-2, MMP-9, and Type VII bacterial collagenase (BC). (**A**) Gelatin-substrate zymography showing human fibrosarcoma cell conditioned media (HT1080, proMMP-9, 92 kDa; MMP-2, 72 and 62 kDa), the determined molecular weights of the injected BC components (110, 103, 93, 84, and 76 kDa), and 50 pg of 4-aminophenylmercuric acetate (APMA)-activated murine MMP-9 (Chemicon; 105, 98, and 88 kDa). One-hour post-ICH of the ipsilateral striatum is shown for comparison. Note that the 110 and 103 kDa bands of BC are the predominant forms visible. (**B**) Composite time course zymography showing endogenous MMP production and the reduction in the levels of the exogenously administered BC. Only the 98 kDa intermediate pro-form of murine MMP-9 was detected following ICH. The 105 kDa band was not detected. (**c**) Time course measurement of BC in injected striata. Exogenously administered BC is greatly decreased by 48 h (110 kDa) and 72 h (103 kDa). The 72 h time point is significantly elevated over 1, 2, and 3 weeks for the 110 kDa band, whereas the 1 h time point is statistically different from 2 and 3 weeks, and 48 h is statistically elevated over the 3-week time point. The elevation by 72 h of the apparent 110 kDa band is likely an endogenous source that may be the result of a compound conjugating with MMP-9. ($^{*}P < 0.01$, $\#P < 0.05$)

4 Materials and Methods in the Rodent

This description is based on the mouse model, which is used for testing of genetic models. The same method with minor modifications can be used in the rat.

1. Mice are initially anesthetized with 2.5% halothane delivered in 70% nitrous oxide/30% oxygen and maintained with 1.75% halothane in the above gas mixture.
2. Mice are then injected subcutaneously with 0.1 mL of 0.027 mg mL^{-1} of the analgesic, buprenorphine, and placed in a Kopf (Kopf Instruments, Tujunga, CA) stereotaxic head-holder. Utilizing sterile technique, a midline incision is made, and a burr hole is drilled through the skull 2.5 mm lateral from bregma over the left hemisphere.
3. A 5 µL syringe (Hamilton Company, Reno, NV) with a 32-gauge noncoring needle is inserted 3.0 mm from the ventral surface of the dura into the region of the caudoputamen.
4. Bacterial collagenase (0.5 µL of 0.0375 U Type VII-S; Sigma-Aldrich Corp., St. Louis, MO) or vehicle is injected over a 2-min period, and the needle is left in place another 2 min to help prevent sample reflux prior to needle withdrawal.
5. The burr hole is sealed with bone wax, and the skin incision is closed with 6–0 silk.
6. The mice are warmed with a heat lamp and provided with water and food ad libitum.
7. At the selected time point, euthanasia is performed following administration of 2.5% halothane delivered in 70% nitrous oxide/30% oxygen by cervical dislocation and the brain is harvested.

5 Tissue Collection and Processing for Zymography

1. The brains are frozen in −85°C cooled 2-methylbutane (Fisher Scientific Company, Pittsburgh, PA). Tissue collection of the ipsilateral and contralateral striata immediately follows.
2. The brain is sectioned so as to collect a 4 mm coronal section of tissue encompassing the hemorrhage; two millimeters rostral and 2 mm caudal of the injection site at bregma are collected.
3. From this coronal section, the respective striata, of approximately 10 mg, are collected into 200 µL ice-cold 100 mM phosphate buffer (Sigma), pH 7.2 with 0.2% Triton X-100 (Bio-Rad Laboratories, Hercules, CA, USA) and homogenized.
4. The supernatants are collected and a separate aliquot is saved at −85°C for later bicinchoninic (BCA) protein content analysis performed in triplicate (Pierce Co., Rockford, IL, USA).
5. The supernatants are incubated for 1 h at 4°C with 50 µL of prepared gelatin-coated Sepharose 4B beads (GE Healthcare Bio-Sciences, Piscataway, NJ, USA) to concentrate the gelatinases, after which, the supernatant is discarded, and the beads washed and incubated with eluting buffer containing phosphate buffer plus 10% dimethylsulphoxide (DMSO; Sigma).
6. The elution supernatant is collected and aliquoted for later analysis.

7. To measure relative gelatinase levels, the gelatin-bead extracted samples, diluted 1:5 in Laemmli sample buffer (Bio-Rad), are separated on 15-well 10% SDS-PAGE gels embedded with 1% gelatin (Sigma).
8. To control for between gel differences, the same sample is run across three different gels and the average of the three is utilized. A protein standard (Bio-Rad), 187.5 μU BC (Sigma), conditioned media from HT1080 human fibrosarcoma cells, which constitutively express MMP-9 (92 kDa) and MMP-2 (72 kDa), and activated murine MMP-9 (Chemicon, Temecula, CA, USA) are all included on the gels to verify the molecular weights of the bands.
9. The gels are run for 72 min at 150 V, after which, the gels are washed for 20 min in 2.5% Triton X-100 (Sigma) at 80 rpm to remove the sodium dodecylsulfate (SDS; Fisher). Removal of the SDS allows for reconformation of the enzymes, and their proteolytic activity is then allowed to progress for 70 h at 37°C in incubation buffer (500 mM Tris, pH 7.6, 200 mM $NaCl_2$, 5 mM $CaCl_2$, 0.02% Brij-35, and 0.02% NaN_3; Sigma).
10. Proteolytic bands are visualized by staining with Coomassie R-250 dye (Sigma) for 1 h with gentle shaking followed by 3 days of destaining in 10% acetic acid. Gels are dried between two sheets of cellophane (Bio-Rad) and scanned with an AGFA Duoscan (AGFA Corporation, Ridgefield Park, NJ, USA), followed by densitometric analysis using AlphaEase (Alpha Innotech Corporation, San Leandro, CA, USA).

6 Edema Measurement

1. For the determination of brain water content following BC injection, experimental and control mice, weighing 25–30 g, are analyzed by calculating the percent water present from dry weight vs. wet weight of treated animal's brains.
2. Twenty-four hours after treatment, the mice are sacrificed and brains are sectioned coronally into 2 quadrants each of the left and right hemispheres. Coronal section 1 contains regions from bregma +1.0 to −1.0 for samples L-1 and R-1, while coronal section 2 contains regions from approximately bregma −1.0 to −4.5.
3. Each section is weighed before and then again after drying for 24 h at 100°C. Water content is expressed as the percentage change between wet weight (WW) and dry weight (DW): $[(WW - DW)/WW] \times 100$.

7 Pathological Sequelae

Following the intracerebral injection of BC, blood vessel damage and tissue disruption begins. Multifocal damage occurs at the site of injection and 4 h after the injection a clot begins to form. Histopathological studies show that the initial damage leads to extravasation of intact red blood cells. By approximately 24 h, lysis of the red blood cells begins, and there is further damage to the BBB. By 48 h, edema is present in the region of the injury and in the posterior white matter tracts next to the hippocampus (Fig. 3).[6] Brain injury occurs not only as a result of the mass effect of the expanding hemorrhage,

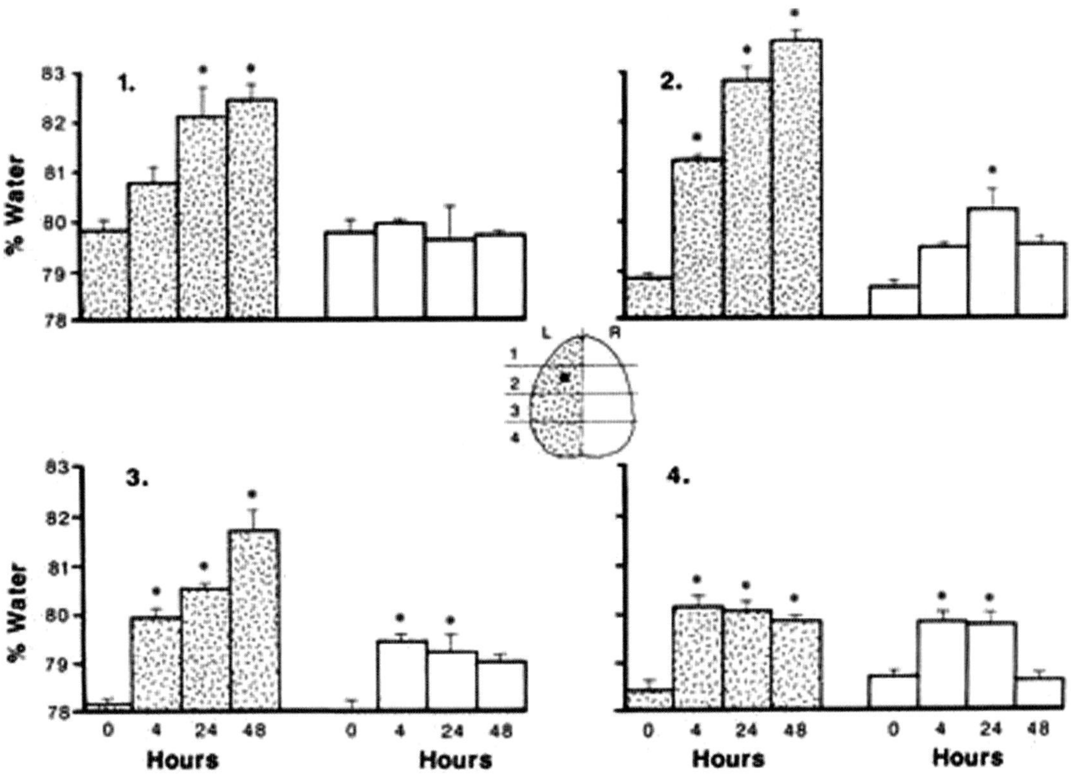

Fig. 3 Water content in four brain sections of 13 rats 4, 24, and 48 h after infusion of 0.5 U of collagenase into the caudoputamen. The small drawing in the center shows the regions sampled and the needle puncture site. ($^*P < 0.05$ difference from control by analysis of variance). Permission pending

but also occurs following an inflammatory reaction due to the infiltration of such blood.

BC bands are reduced within several hours of injection, and the endogenous production of MMP-9 is observed in the gelatin zymograms. Because of the differences in molecular weights of the BC and the mouse or rat MMP-9, it is possible to separate the exogenous from the endogenous. The MMP-9 causes disruption of the basal lamina around the blood vessel. As MMP-9 is disruptive to the BBB, it would be assumed that the MMP-9 knockout would have less damage. Paradoxically, the injection of BC into the MMP-9 knockout mouse caused greater damage than in the wild-type.[9] Recently, we found that a broad-spectrum MMP inhibitor, BB-94, increased the damage in the mouse model of BC-induced bleeding, which is consistent with the MMP-9 knockout studies.[10] However, those data are inconsistent with the beneficial effects seen with other broad-spectrum MMP inhibitors, including BB-1101 and GM6001.[11,12] These seemingly contradictory findings will require further studies to be resolved. The BC model has been used extensively to test drugs. Agents tested include matrix metalloproteinase inhibitors, free radical scavengers, cyclooxygenase inhibitors, as well as other agents (Table 1).

Table 1 Agents shown to improve various outcome measures in rodent collagenase-induced ICH.

Treatments	Outcome	References
Atrial natriuretic peptide or arginine vasopressin V_1-receptor antagonist	Decreased posterior edema	13
Ca^{2+} channel blocker and serotonin S_2 antagonist [(S)-emopamil]	Decreased posterior edema	15
MMP inhibitors (BB-1101 and GM6001)	Decreased posterior edema	11, 12
GABA agonist (Muscimol)	Neurobehavioral improvement Tissue sparing	16
Citicoline	Neurobehavioral improvement Decreased infarct size	17
TNF-α antisense oligodeoxy-nucleotide (ORF4-PE)	Neurobehavioral improvement Reduced cell death	18
Free radical scavenger (NXY-059[a])	Neurobehavioral improvement	19
FK-506 (Tacrolimus)	Neurobehavioral improvement	20
Minocycline	Neurobehavioral improvement Reduced cell death, microglia/ macrophage activation	(21)
Macrophage/microglial inhibitory factor (MIF; tuftsin fragment 1–3)	Neurobehavioral improvement Reduced injury volume, edema, and degenerating Neurons	22
COX-2 inhibitor (Celecoxib)	Reduced inflammation, brain edema, and perihematomal cell death	23
Statins – HMG-CoA[b] reductase inhibitors (Atorvastatin)	Neurobehavioral improvement, reduced perihematomal cell death, inflammation, hemispheric atrophy, and edema	24

[a]NXY-059: disodium 4-[(*tert*-butylimino) methyl] benzene-1,3-disulfonate-N-oxide
[b]HMG-CoA: 3-hydroxy-3-methyl-glutaryl-coenyme A

8 Progressive Enlargement of the Hemorrhage and Recovery of Function

Primary brain injury results from the mass effect of the space occupying lesion; continued expansion of the hematoma occurs in approximately one-third of the patients with ICH due to cell death, edema, and rebleeding.[3] Mechanical disruption of the surrounding tissue as well as physiological parameters such as hypertension or a coagulation deficit may all lead to the enlargement of the hematoma following the initial ictus. Early work suggested ischemia in the penumbral zone of the ICH; however, human diffusion-weighted magnetic resonance imaging (MRI) and positron emission tomography (PET) failed to show an ischemic zone.[13] Cerebral blood flow (CBF) studies in rats with BC-induced ICH showed reduced CBF in the site of injection, but in no other locations (Fig. 4).[13]

An advantage of the BC-ICH model is the ability to follow the animals over an extended period with behavioral measures. In the antologous blood method, there is more rapid absorption of the blood after injection, and the rate of recovery is faster than seen in the clinical situation. Many agents have long-term benefits, while others are only useful for the early stages of

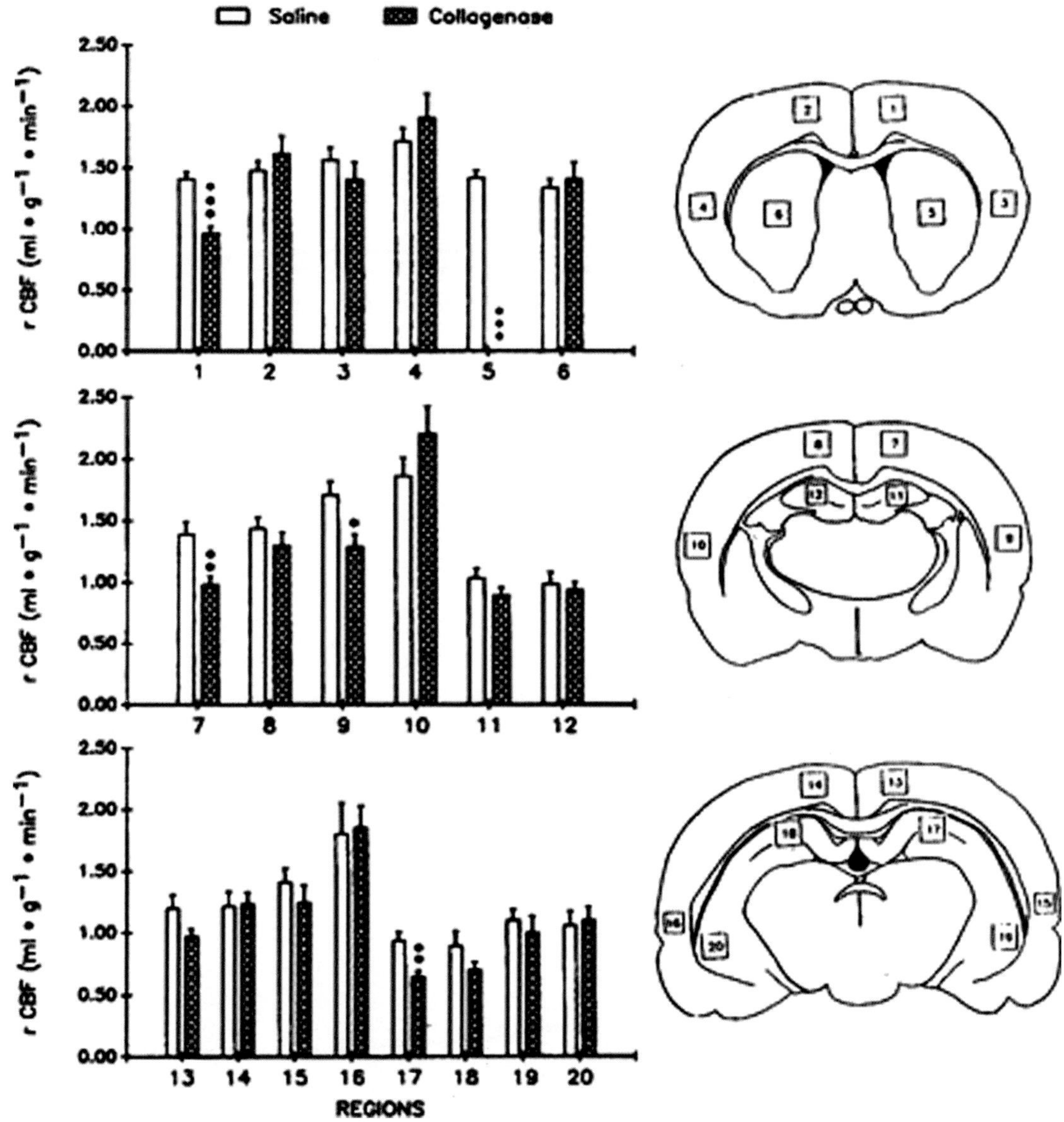

Fig. 4 Bar graphs showing regional cerebral blood flow (rCBF) measured by [^{14}C]-iodoantipyrine autoradiography in rats injected with 2 μL saline or 0.5 IU bacterial collagenase in left caudoputamen (region 5). Regions are identified in brain slice illustrations beside each graph and are different from sites of water and electrolyte measurement. ($^{*}P < 0.05$, $^{**}P < 0.01$, $^{***}P < 0.001$ difference for saline vs. collagenase). Permission pending

the injury, making it important to have the ability to observe both the short and long-term effects. A grading scale was developed in the initial description of the BC method. The components that went into the final composite score are shown in Fig. 5.[6]

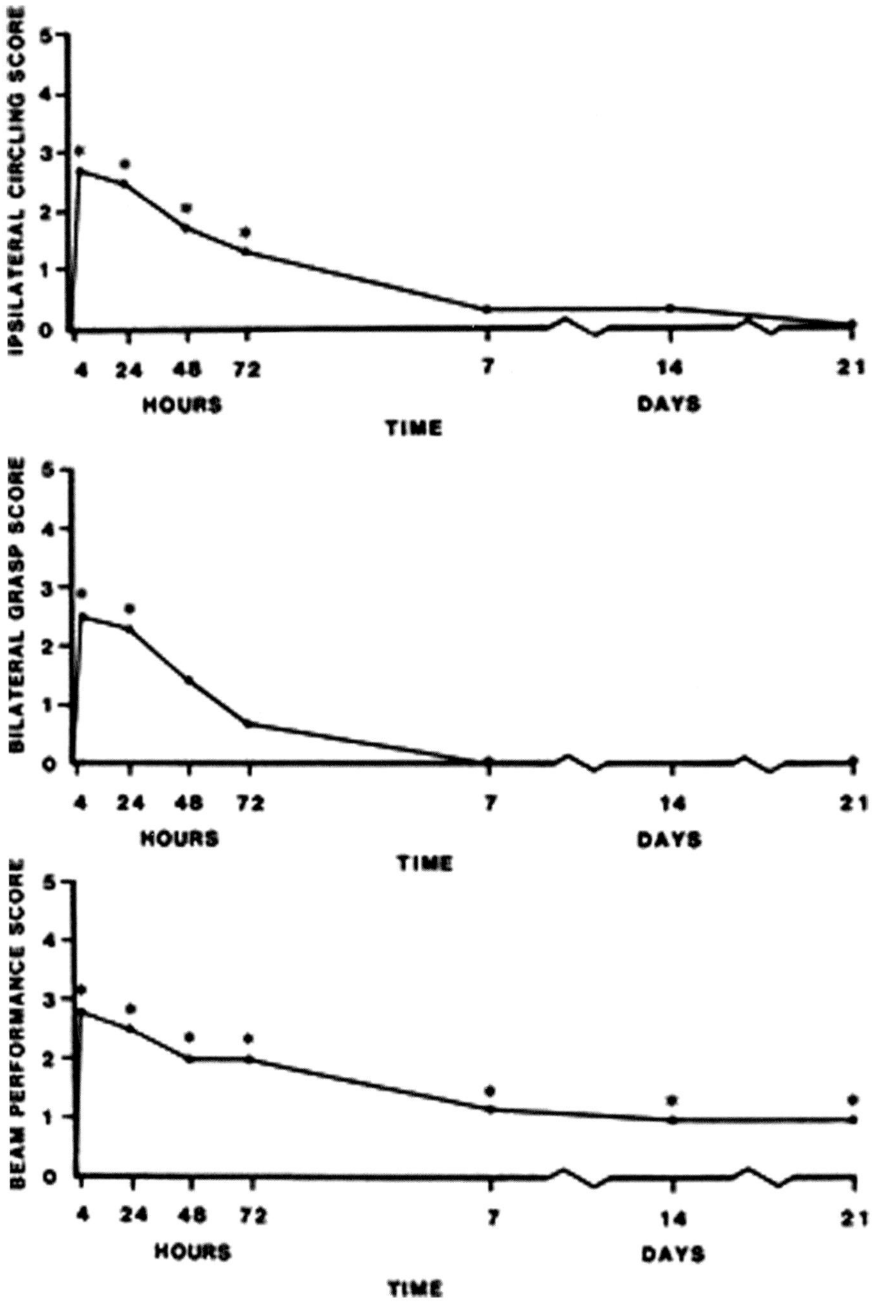

Fig. 5 Injury scores over three weeks are shown for three different neurobehavioral tests for rats infused with 0.5 U of collagenase into the caudoputamen. Prior to collagenase infusion, the rats all performed without deficit. Ipsilateral circling score was significantly impaired for 72 h. Bilateral grasp score recovered most rapidly. The beam walking task was the slowest to recover. These three tests can be scored to form a composite score to grade recovery. (*$P < 0.05$ level)

9 Conclusions

The selection of the appropriate model for the study of the pathophysiology of ICH depends to a large extent on the concept of the origin of the bleed. For a rapid release of blood into the parenchyma with dissection of the blood products into the brain, the use of the autologous blood infusion method may be the most relevant model. In the case of a slower release of blood, as is now known to occur in a significant number of patients with ICH, the BC model more closely simulates the growth of the lesion over time. HT cannot be modeled with the injection of blood because of the multifocal nature of the injury within the ischemic regions. To this end, the use of BC is more appropriate. As the lesion matures, the major contributing factors to the pathology are the result of the blood and debris in the brain. At later stages, the BC model is more reliable in that it mimics the effects of the naturally occurring process, whereas injected blood is more rapidly resolved producing a more artificial reaction. Ultimately, the selection of the model will be dependent upon the type of injury, but optimally both models can be used to increase reliability of the results.

References

1. Caplan LR. Intracerebral haemorrhage. Lancet 1992;339:656–658.
2. Zulch KJ, Minckler J. Hemorrhage, thrombosis, embolism. In: Pathology of the Nervous System. New York: McGraw-Hill Book Corp., 1971:1499–1535.
3. Brott T, Broderick J, Kothari R, et al. Early hemorrhage growth in patients with intracerebral hemorrhage. Stroke 1997;28:1–5.
4. Godeau G, Robert AM. Mechanism of action of collagenase on the blood-brain barrier permeability. Increase of endothelial cell pinocytotic activity as shown with horse-radish peroxidase as a tracer. Cell Biol Int Rep 1979;3:747–751.
5. Rosenberg GA, Estrada E, Wesley M, Kyner WT. Autoradiographic patterns of brain interstitial fluid flow after collagenase-induced hemorrhage in rat. Acta Neurochirurgica 1990;51:280–282.
6. Rosenberg GA, Mun-Bryce S, Wesley M, Kornfeld M. Collagenase-induced intracerebral hemorrhage in rats. Stroke 1990;21:801–807.
7. Brown MS, Kornfeld M, Mun-Bryce S, Sibbitt RR, Rosenberg GA. Comparison of magnetic resonance imaging and histology in collagenase-induced hemorrhage in the rat. J Neuroimaging 1995;5:23–33.
8. Del Bigio MR, Yan HJ, Buist R, Peeling J. Experimental intracerebral hemorrhage in rats. Magnetic resonance imaging and histopathological correlates. Stroke 1996;27:2312–2319; discussion 2319–2312.
9. Tang J, Liu J, Zhou C, et al. Mmp-9 deficiency enhances collagenase-induced intracerebral hemorrhage and brain injury in mutant mice. J Cereb Blood Flow Metab 2004;24:1133–1145.
10. Grossetete M, Rosenberg GA. Matrix metalloproteinase inhibition facilitates cell death in intracerebral hemorrhage in mouse. J Cereb Blood Flow Metab 2008;28:752–763.
11. Rosenberg GA, Navratil M. Metalloproteinase inhibition blocks edema in intracerebral hemorrhage in the rat. Neurology 1997;48:921–926.
12. Wang J, Tsirka SE. Neuroprotection by inhibition of matrix metalloproteinases in a mouse model of intracerebral haemorrhage. Brain 2005;128:1622–1633.
13. Rosenberg GA, Scremin O, Estrada E, Kyner WT. Arginine vasopressin V1-antagonist and atrial natriuretic peptide reduce hemorrhagic brain edema in rats. Stroke 1992;23:1767–1773.

14. Zazulia AR, Diringer MN, Videen TO, et al. Hypoperfusion without ischemia surrounding acute intracerebral hemorrhage. J Cereb Blood Flow Metab 2001;21:804–810.
15. Rosenberg GA, Navratil MJ. (S)-emopamil reduces brain edema from collagenase-induced hemorrhage in rats. Stroke 1994;25:2067–2071.
16. Lyden PD, Jackson-Friedman C, Lonzo-Doktor L. Medical therapy for intracerebral hematoma with the gamma-aminobutyric acid-A agonist muscimol. Stroke 1997;28:387–391.
17. Clark W, Gunion-Rinker L, Lessov N, Hazel K. Citicoline treatment for experimental intracerebral hemorrhage in mice. Stroke 1998;29:2136–2140.
18. Mayne M, Ni W, Yan HJ, et al. Antisense oligodeoxynucleotide inhibition of tumor necrosis factor-alpha expression is neuroprotective after intracerebral hemorrhage. Stroke 2001;32:240–248.
19. Peeling J, Del Bigio MR, Corbett D, Green AR, Jackson DM. Efficacy of disodium 4-[(tert-butylimino)methyl]benzene-1,3-disulfonate *N*-oxide (NXY-059), a free radical trapping agent, in a rat model of hemorrhagic stroke. Neuropharmacology 2001;40:433–439.
20. Peeling J, Yan HJ, Corbett D, Xue M, Del Bigio MR. Effect of FK-506 on inflammation and behavioral outcome following intracerebral hemorrhage in rat. Exp Neurol 2001;167:341–347.
21. Power C, Henry S, Del Bigio MR, et al. Intracerebral hemorrhage induces macrophage activation and matrix metalloproteinases. Ann Neurol 2003;53:731–742.
22. Wang J, Tsirka SE. Tuftsin fragment 1–3 is beneficial when delivered after the induction of intracerebral hemorrhage. Stroke 2005;36:613–618.
23. Chu K, Jeong SW, Jung KH, et al. Celecoxib induces functional recovery after intracerebral hemorrhage with reduction of brain edema and perihematomal cell death. J Cereb Blood Flow Metab 2004;24:926–933.
24. Jung KH, Chu K, Jeong SW, et al. HMG-CoA reductase inhibitor, atorvastatin, promotes sensorimotor recovery, suppressing acute inflammatory reaction after experimental intracerebral hemorrhage. Stroke 2004;35:1744–1749.

Chapter 30

Blood Injection Intracerebral Hemorrhage Mouse Model

Ludmila Belayev

Abstract A major limitation to intracerebral hemorrhage (ICH) research is the lack of a reproducible animal model. Several animal models of ICH have been developed in mice. In this chapter, we introduce a step-by-step procedure to perform a double-injection method of ICH in the mouse. The present ICH model produces a consistent neurological deficit, hematoma volume, and brain swelling. This model closely mimics human hypertensive basal ganglionic ICH and can be useful to evaluate the pathophysiological, biochemical, and molecular processes after ICH as well as potential new treatments.

Keywords: Animal models, Behavior, Blood, Brain edema, Mice, Stroke

1 Experimental Model Selection

Intracerebral hemorrhage (ICH) accounts for 10–15% of all strokes, and is associated with high morbidity and mortality. The most frequent sites of primary spontaneously ICH in humans are the putamen (50%), thalamus (15%), pons (10–15%), and cerebellum (10%).[1] Several animal models of ICH have been developed in mice, rats, rabbits, cats, and primates.[2] The study of intracerebral hemorrhage in the mouse has attracted increasing attention, particularly because the availability of genetically modified (transgenic and knockout) mouse strains provides an unique opportunity to study specific pathways or injury processes. A widely-used ICH model, developed by Rosenberg and colleagues[3] in the rat, uses a local injection of bacterial collagenase into the basal ganglia and was subsequently studied in the mouse.[4] Collagenase dissolves the extracellular matrix and leads to blood vessel rupture and an intracerebral bleed. Although the collagenase method is a simple means of producing hemorrhaging and is reproducible, several reports demonstrated that bacterial collagenase causes a significant inflammatory reaction and likely differs from the mechanism that produces ICH in humans.[5]

Another commonly used ICH model employs the injection (or infusion) of autologous blood into the brain parenchyma of mice.[6,7] This model was designed to mimic the natural events that occur with spontaneous ICH in humans. Intraparenchimal injection of blood is an effective technique for

From: *Animal Models of Acute Neurological Injuries.*
Edited by: J. Chen (eds.) New York, Totowa, NJ

producing an intracerebral hematoma. A disadvantage of the blood infusion model is the potential for ventricular rupture as well as backflow of infused blood along the needle track.[8] A double-injection ICH model in rat and mouse has been developed, in which a small amount of blood is infused into the striatum at a slow rate to allow blood clotting along the needle track; the remaining blood is then infused to generate the hematoma.[9–11] Major advantages of this model include reproducible hematoma volume, consistent neurological deficit, hypoperfusion, and brain swelling.

2 Materials

Male C57BL/6 mice weighing 25–30 g are the animals of choice.

The following instruments and materials are recommended for the ICH experiment:

1. GAS anesthesia instruments (Stoelting): Open isoflurane system (50251), isoflurane/halothane scavenger (50206), and induction chamber (50216)
2. Cranial temperature controller (Omega Engineering): Monogram benchtop controller CSC32T (unit set for type T only) and temperature probe: HYP-30–1/2 in. (30-gauge)
3. Rectal temperature controller (CMA/microdialysis AB): CMA/150 temperature controller (8315000), heating pad (8315001), temperature probe (8315002), and insulation pad (8315003)
4. Digi-Med blood pressure analyzer, model 400, Micro-Med Inc
5. Blood gas system (model ABL 5, Radiometer America Inc)
6. Infusion pump (KDS-100, kd Scientific)
7. Stereotactic instrument (model 900), mouse adaptor (model 926), mouse anesthesia mask (907) from David Kopf Instruments
8. Micromotor drill (51453, Stoelting)
9. Stainless steel cannula (30-gauge, 3/4″, W/30° PT, 7460, Popper & Sons Inc) and Luer stub adaptor, 23-gauge
10. Intramedic polyethylene tubing (Becton Dickinson): PE 10 (427401) and PE 20 (427406)
11. Glass syringe (250 μL, Hamilton company)
12. Gelfoam sponges (9083300, Henry Schein Medical)

3 Procedures

1. Anesthetize a mouse with isoflurane (3% for induction and 1% for maintenance of anesthesia during surgical procedure) in a mixture of 70% nitrous oxide and 30% oxygen delivered by face-mask. Shave the hairs over the head and the right groin area.
2. Insert temperature probes into the rectum and the left temporalis muscle and use separate warming lamps to maintain rectal and temporalis muscle temperature at 36.0–37.5°C.
3. Insert PE 10 catheter into the right femoral artery for continuous blood pressure monitoring and periodic blood sampling for arterial gases and pH.
4. Place mouse on the stereotactic frame (Fig. 1).
5. Make a midline scalp incision.

6. Make a burr hole in the skull (2 mm lateral to midline, 1 mm anterior to bregma, depth 4 mm below the surface of the skull) using a drill.
7. Take blood from the heart of the donor mouse with a 1 mL syringe, which was flushed with heparin before blood withdrawal.
8. Introduce a 30-gauge stainless steel cannula through a burr hole into the left striatum.
9. Inject 5 µL of whole blood from the donor mouse over 3 min, followed 7 min later by 10 µL injected over 5 min, using a microinfusion pump.
10. Slowly withdraw the injection cannula 10 min after the second injection.
11. Seal the bone hole with bone wax.
12. Suture the wounds and place the animal in a temperature controlled incubator with free access to food and water.

4 Outcome evaluations

4.1 Neurological Deficit

Animals began to exhibit neurological signs of an ICH within 60 min. A standardized battery of behavioral tests is recommended to quantify sensorimotor neurological function after ICH. The battery consists of two tests that have been used previously to evaluate various aspects of neurologic function: (1) the postural reflex test, to examine upper body posture while the animal is suspended by the tail[12]; and (2) the forelimb placing test, to examine sensorimotor integration in forelimb placing responses to visual, tactile, and proprioceptive stimuli.[13] Behavioral tests should be conducted by an observer blinded to the identity of the individual animals.

4.1.1 Postural Reflex Test

Mouse is suspended by the tail 1 m above the floor. Intact mouse extends both forelimbs toward the floor; animal displaying this behavior is assigned a score of 0. A score of 1 is given if the mouse flexes one or both forelimbs. The animal is then given the lateral push test, which involves placing the mouse on a sheet of plastic-coated paper and applying lateral pressure behind the shoulders in the left and right directions. If the mouse is unable to resist the force equally in both directions, it receives a score of 2.

4.1.2 Elicited Forelimb Placing

For visual placing, the animal is cupped in the experimenter's hands with the forelimb hanging free and is slowly tilted and lowered from each side toward a tabletop. Intact mouse reaches for the table with both forelimbs extended. By moving the animal laterally toward the table edge, sideways visual placing could be assessed as well. Tactile placing is judged by lightly contacting the dorsal and then the lateral surface of the mouse's forepaw to the table edge while slightly elevating its head to obscure its view of the table. Intact mouse immediately places the paw on the tabletop, while impaired mice either were slow to place or did not place at all. Proprioceptive placing is also assessed by pressing the mouse's paw against the table edge to stimulate limb muscles. For each of these tests, a score of 0 is given for normal, immediate placing. A score of 1 is given if the placing was delayed or incomplete. A score of 2 indicated placing was absent.

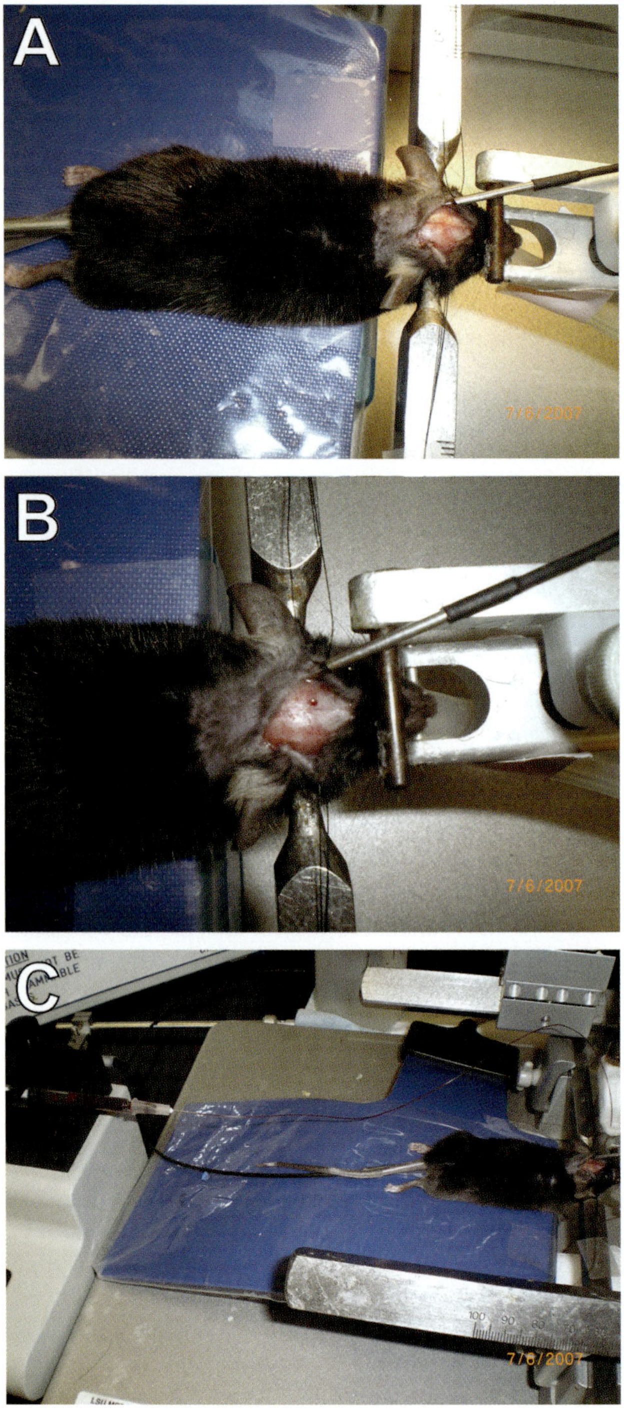

Fig. 1 Step-by-step procedures. (**A**) Place mouse on the stereotactic frame, insert temperature probes into the rectum and the left temporalis muscle, and make a midline scalp incision. (**B**) Make a burr hole in the skull using a drill. (**C**) Introduce a stainless steel cannula through a burr hole into the left striatum and inject blood using a micro-infusion pump

4.1.3 Total Neurological Score

Total neurological score is calculated as the sum of scores on forward visual placing (range: 0–2), lateral visual placing (0–2), dorsal (0–2) and lateral (0–2) tactile placing, proprioceptive placing (0–2), and postural reflex (0–2). Thus, the maximal possible score is 12, as previously described by Belayev et al.[10]

4.2 Histopathological Examination

After completion of the experiment, animal is perfused with a mixture of 40% formaldehyde, glacial acetic acid, and methanol (FAM, 1:1:8 by volume), as previously described.[10] Then brain blocks are embedded in paraffin, and 10-μm-thick sections are cut in the coronal plane and stained with hematoxylin and eosin. Histological sections are digitized at 8 standardized coronal levels (bregma levels: +1.7, +1.1, +0.62, +0.26, +0.46, −1.22, −1.94, and −2.92 mm).[14] The following areas are measured on each section: ipsilateral and contralateral hemisphere areas, area of hematoma, and total affected area (Fig. 2). Hematoma size and brain swelling is determined using image analysis system.

5 Applications, Complications, and Limitations

5.1 Applications for Animal Model of ICH

The blood-injection method replicates the localization and size of ICH observed clinically but differs in the respect that, in patients with ICH, abnormal microvascular fragility probably plays an important predisposing role. The brain tissue responses to ICH include blood-brain barrier (BBB) disruption, edema development, inflammation, tissue necrosis, clot absorption, and scar or cavity formation. Animal models provide a new understanding of the pathophysiology underlying ICH-induced injury, including information on the roles of mass effect, elevated intracranial pressure, alterations in cerebral blood flow and metabolism, and impact of specific blood components on brain edema formation.[15]

We present a double-injection ICH model that satisfies the need of highly reproducible hematomas in the mouse. Major advantages of this model include reproducible hematoma volume, consistent neurological deficit together with consistent BBB dysfunction and brain swelling. This model closely mimics human hypertensive basal ganglionic ICH and should be useful for the evaluation of pharmaceutical therapies.

5.2 Rapid Injection vs. Slow Infusion

As opposed to the rapid injection of blood used in previous models, this model uses a slow infusion of fresh blood into the brain parenchyma. The presence of a 15-μL blood clot produces a localized striatal hematoma in mice. We think that the slow infusion of blood limits its extravasation into the subarachnoid and ventricular spaces and more closely mimics the natural process. Furthermore, it also avoids undesired nonphysiological pressure injury to adjacent tissue.

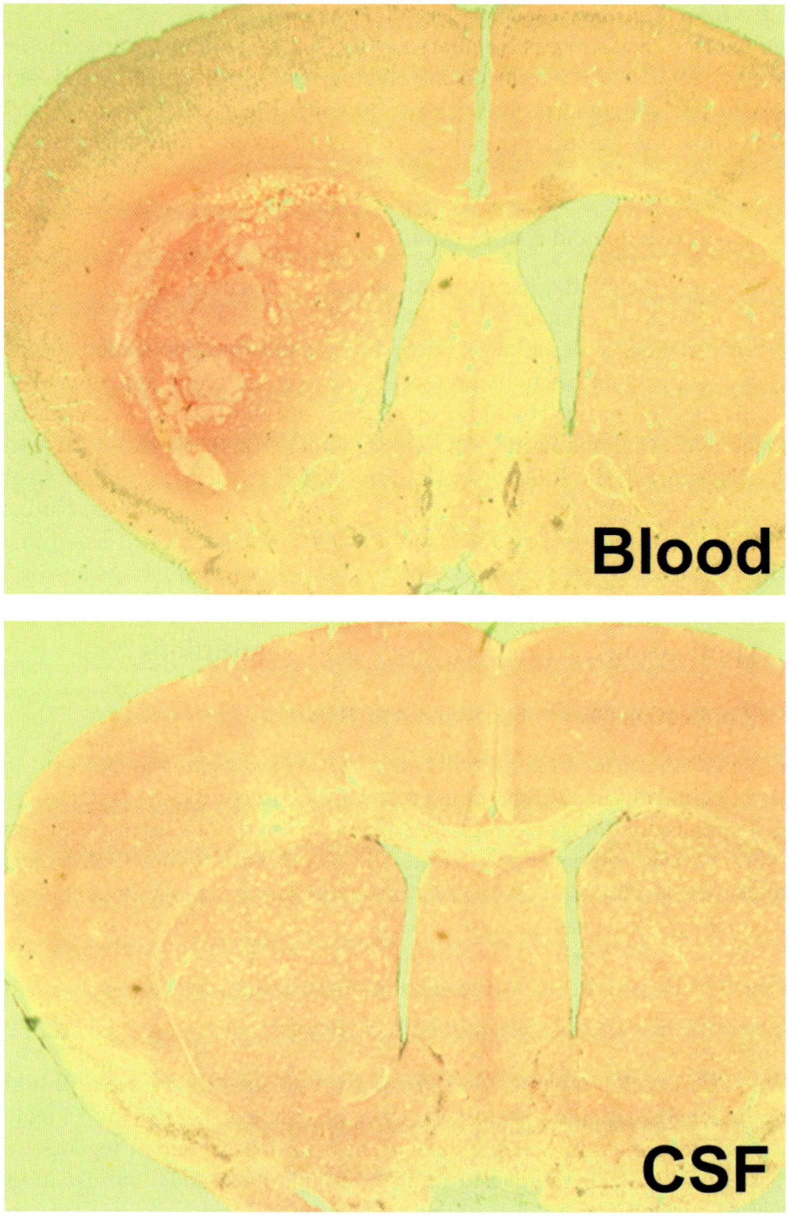

Fig. 2 Paraffin-embedded coronal brain sections, stained with hematoxylin and eosin 2 days after infusion of blood (*top*) or CSF (*bottom*) into the mouse brain. A moderate-sized striatal hematoma is evident after blood injection. In contrast, the CSF injected mouse showed only a small, nonhemorrhagic lesion

5.3 Autologous vs. Donor Blood

Brain edema is an important clinical endpoint for ICH studies. Hematoma within brain parenchyma triggers a series of events leading to brain-barrier disruption and edema. Brain water content is significantly increased in the ipsilateral basal ganglia and cortex in C57BL/6 mice at 1 and 3 days after injection of autologous blood.[6] Interestingly, donor blood induces significantly more brain edema than autologous blood on day 1 after ICH.[6] This model can be recommended for the further investigations of brain edema and to test agents that reduce edema formation in the ICH model.

5.4 Application for Genetically Modified Mice

The development of the mouse ICH model has attracted increasing attention because the availability of gene knockout or transgenic animals may assist in elucidating the mechanisms of brain injury after ICH. Considerable evidence demonstrates that ICH induces alterations in gene expression, enzyme, and protein degradation, which contribute to cellular injury after ICH. Studies using knockout and transgenic mice in ICH models include the heme oxygenase-1 (HO-1-/-)[16] and HO-2-/-[17] mice, complement C3-deficient mice,[18] matrix metalloproteinase-9 deficient mice,[19] tPA-deficient mice,[20] and many others.

5.5 Influence of Gender

Male mice are the animals of choice. In ischemic stroke models, behavioral score and brain infarcts are significantly reduced in females as opposed to males.[21] In the ICH mouse model, female mice showed a faster recovery for behavioral tests and reduced brain edema compared with male animals.[6] The greater neuroprotection afforded to females in ischemia, and ICH is likely because of the effects of circulating estrogens and progestins.[6]

5.6 Importance of Temperature Regulation

It is now well-recognized that even small decreases in brain temperature protect the brain from ischemic injury.[22] Mice after ICH fail to regulate their temperatures, becoming severely hypothermic during the postoperative period, and that these temperature changes may have a profound effect on the outcome of ICH. We recommend close monitoring of cranial and body temperature during surgical procedure and also allowing mice to recover from surgery in a temperature-controlled incubator for 48 h.

5.7 Importance of Neurological Evaluation

Observations of neurological deficits are important, not only in clinical stroke patients, but also in animal models of ICH. Two sensorimotor tests (the postural reflex and the forelimb placing tests) applied here appear to be particularly sensitive to this model. They can help to detect a neurological deficit and also to identify animals in which ICH model was not successfully reproduced.

5.8 Bleeding Problems

Bleeding may occur during the surgical procedure. To help with this problem, keep the incision of the skin over the skull to the midline and use a haemostatic sponge or wax to stop bleeding.

References

1. Broderick JP, Adams HP, Jr., Barsan W, Feinberg W, Feldmann E, Grotta J, Kase C, Krieger D, Mayberg M, Tilley B, Zabramski JM, Zuccarello M. Guidelines for the management of spontaneous intracerebral hemorrhage: A statement for healthcare

professionals from a special writing group of the Stroke Council, American Heart Association. Stroke 1999;30:905–915.

2. Wagner KR, Broderick JP. Hemorrhagic stroke: pathophysiological mechanisms and neuroprotective treatments. In: Lo, EH, Marwah J, eds. Neuroprotection. Scottsdale: Permanent Press, 2002:471–508.
3. Rosenberg GA, Mun-Bryce S, Wesley M, Kornfeld M. Collagenase-induced intracerebral hemorrhage in rats. Stroke 1990;21:801–807.
4. Clark W, Gunion-Rinker L, Lessov N, Hazel K. Citicoline treatment for experimental intracerebral hemorrhage in mice. Stroke 1998;29:2136–2140.
5. Del Bigio MR, Yan HJ, Buist R, Peeling J. Experimental intracerebral hemorrhage in rats. Magnetic resonance imaging and histopathological correlates. Stroke 1996;27:2312–2319.
6. Nakamura T, Xi G, Hua Y, Schallert T, Hoff JT, Keep RF. Intracerebral hemorrhage in mice: model characterization and application for genetically modified mice. J Cereb Blood Flow Metab 2004;24:487–494.
7. Zhao X, Sun G, Zhang J, Strong R, Song W, Gonzales N, Grotta JC, Aronowski J. Hematoma resolution as a target for intracerebral hemorrhage treatment: role for peroxisome proliferator-activated receptor gamma in microglia/macrophages. Ann Neurol 2007;61:352–362.
8. Andaluz N, Zuccarello M, Wagner KR. Experimental animal models of intracerebral hemorrhage. Neurosurg Clin N Am 2002;13:385–339.
9. Deinsberger W, Vogel J, Kuschinsky W, Auer LM, Boker DK. Experimental intracerebral hemorrhage: description of a double injection model in rats. Neurol Res 1996;18:475–477.
10. Belayev L, Saul I, Curbelo K, Busto R, Belayev A, Zhang Y, Riyamongkol P, Zhao W, Ginsberg MD. Experimental intracerebral hemorrhage in the mouse: histological, behavioral, and hemodynamic characterization of a double-injection model. Stroke 2003;34:2221–2227.
11. Ma B, Zhang J. Nimodipine treatment to assess a modified mouse model of intracerebral hemorrhage. Brain Res 2006;1078:182–188.
12. Bederson JB, Pitts LH, Tsuji M, Nishimura MC, Davis RL, Bartkowski H. Rat middle cerebral artery occlusion: evaluation of the model and development of a neurologic examination. Stroke 1986;17:472–476.
13. De Ryck M, Van Reempts J, Borgers M, Wauquier A, Janssen PJ. Photochemical stroke model: flunarizine prevents sensorimotor deficits after neocortical infarcts in rats. Stroke 1989;20:1383–1390.
14. Franklin KBJ, Paxinos G. The Mouse Brain in Stereotaxic Coordinates. San Diego: Academic Press, 1997.
15. Wagner KR. Modeling intracerebral hemorrhage: glutamate, nuclear factor-kappa B signaling and cytokines. Stroke 2007;38:753–758.
16. Wang J, Dore S. Heme oxygenase-1 exacerbates early brain injury after intracerebral haemorrhage. Brain 2007;130:1643–1652.
17. Qu Y, Chen-Roetling J, Benvenisti-Zarom L, Regan RF. Attenuation of oxidative injury after induction of experimental intracerebral hemorrhage in heme oxygenase-2 knockout mice. J Neurosurg 2007;106:428–435.
18. Yang S, Nakamura T, Hua Y, Keep RF, Younger JG, Hoff JT, Xi G. Intracerebral hemorrhage in complement C3-deficient mice. Acta Neurochir Suppl 2006;96:227–231.
19. Tang J, Liu J, Zhou C, Alexander JS, Nanda A, Granger DN, Zhang JH. Mmp-9 deficiency enhances collagenase-induced intracerebral hemorrhage and brain injury in mutant mice. J Cereb Blood Flow Metab 2004;24:1133–1145.
20. Thiex R, Mayfrank L, Rohde V, Gilsbach JM, Tsirka SA. The role of endogenous versus exogenous tPA on edema formation in murine ICH. Exp Neurol 2004;189:25–32.

21. Alkayed NJ, Harukuni I, Kimes AS, London ED, Traystman RJ, Hum PD. Gender-linked brain injury in experimental stroke. Stroke 1998;29:159–166.
22. Busto R, Ginsberg MD. The influence of altered brain temperature in cerebral ischemia. In: Ginsberg MD, Bogousslavsky J, eds. Cerebrovascular Disease: Pathophysiology, Diagnosis and Management. Malden: Blackwell Science, 1998.

Chapter 31

Blood Injection Intracerebral Hemorrhage Pig Model

Ya Hua, Yuxiang Gu, and Guohua Xi

Abstract The autologous blood injection ICH model has been used successfully in pigs. The well-developed white matter in the pig brain (in contrast to the rat and the mouse) is good for studies of white matter injury after ICH. Swine provides an excellent model for the study of edema formation and white matter brain damage. Of the nonprimate mammals, pigs have a cardiovascular system, which is most similar to humans. In this chapter, we present a step-by-step procedure to produce ICH by injecting autologous whole blood.

Keywords: Intracerebral hemorrhage, Brain edema, White matter injury, Swine

1 Model Selection

Experimental models of intracerebral hemorrhage (ICH) have been available since the 1960s and commonly involve the injection of autologous blood into the frontal lobe of dogs, cats, pigs, or monkeys.[1–5] Intracerebral blood injection model is a useful model for studying ICH-induced brain injury in pigs (Fig. 1).[6,7] Autologous blood obtained from the femoral artery is infused into the white matter in the frontal lobe to simulate hemorrhage (Figs. 2 and 3). Mentioned below are the advantages of this model:

1. Well-developed white matter to determine white matter brain damage induced by ICH[4]
2. A nice ICH model well suited to clot aspiration studies[8–10]
3. Large brain size suitable to magnetic resonance imaging studies[11]
4. The ability to study a variety of physiologic parameters that are altered by the hematoma, including intracranial pressure, cerebral blood flow, and cerebral perfusion pressure
5. Ease of reproducibility and rarity of complications
6. Modification of the procedure to test the neurotoxic effects of thrombin and red blood cells[12]

From: *Animal Models of Acute Neurological Injuries.*
Edited by: J. Chen (eds.) New York, Totowa, NJ

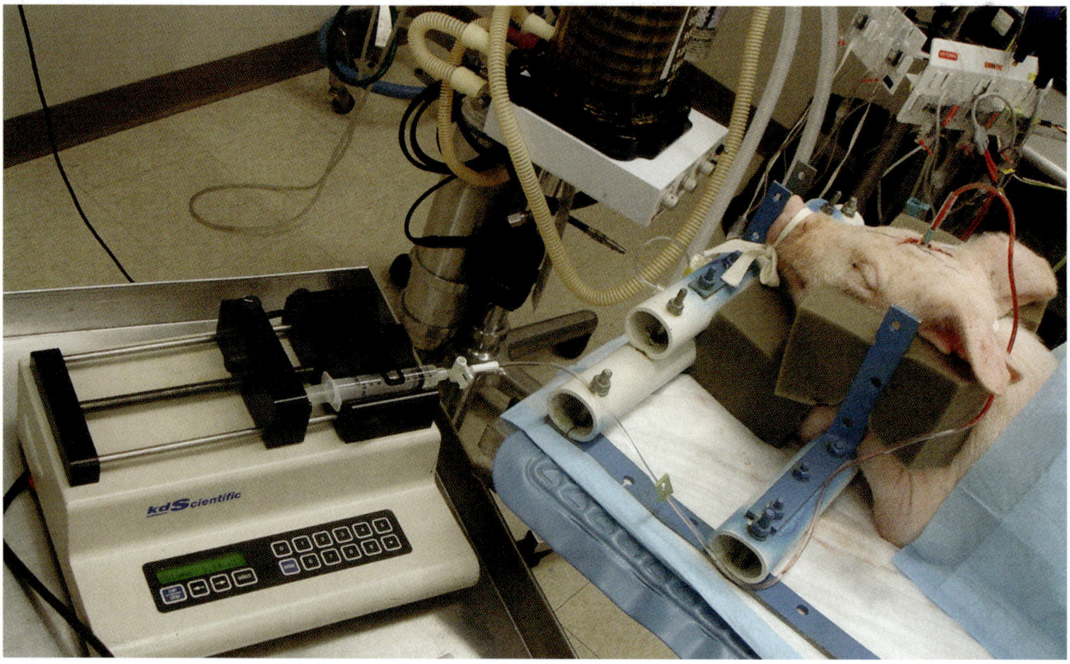

Fig. 1 Autologous whole blood is injected into the white matter of the right frontal lobe in the pig

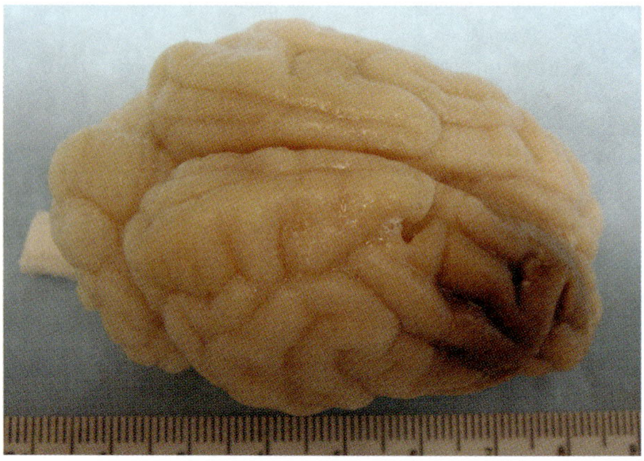

Fig. 2 A pig brain 1 week after an intracerebral injection of 2.5 ml autologous whole blood

2 Materials

Immature pigs weighing ~10 kg are used in this model of intracerebral hemorrhage. The materials recommended for the surgical procedure include the following:

1. Anesthesia machine
2. Respirator
3. Biophysical recorders with pressure transducers
4. Blood gas analyzer

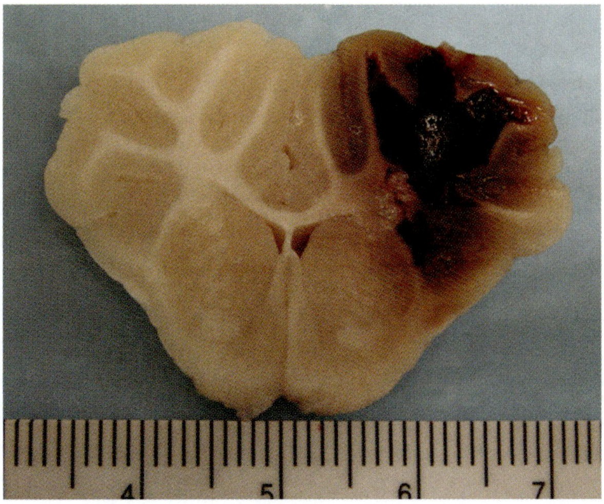

Fig. 3 A coronal section of pig brain 7 days after an intracerebral injection of 2.5 ml autologous whole blood. Note red blood cell lysis and reddish brain tissue around the clot

5. Warm water blanket
6. Glucose analyzer
7. Cuffed endotracheal tube (6.0 mm I.D.)
8. Silicone elastomer tubing (Sil-med)
9. Infusion pump
10. Dremel drill with 1.5 mm drill bit
11. Moria spring scissors with sharp points
12. Scalpel handle with scalpel blades
13. Fine iris scissors
14. Curved Mayo scissors
15. Dumont forceps
16. Dumont angled forceps
17. Adson forceps with teeth
18. Heiss retractor
19. Three-way stopcocks
20. Hemostats
21. Needle holder
22. Cotton-tipped applicators
23. Syringes (1 and 5 ml)
24. Polyethylene tubing (PE-160)
25. Needles (20G)
26. Sterile plastic catheters (20G)
27. 4–0 black braided silk (Deknatel)
28. 3–0 Dermalon suture

3 Procedure

1. Immature swine (~10 kg) will be fasted overnight with water ad libitum.
2. Inject Telazol (6 mg kg^{-1}, IM) and xylazine (2.2 mg kg^{-1}, IM) for induction of anesthesia and endotracheal intubation.

3. Intubate the animal orotracheally when a surgical plane of anesthesia is reached.
4. Use isoflurane in oxygen as the gas anesthetic for swine.
5. Use a warm water blanket to maintain body temperature at 37°C.
6. Use a hair clipper to remove hair from the scalp and right groin regions and apply Betadine to the exposed skin.
7. Make a 4 cm incision in the groin at the region of the right femoral artery.
8. Expose the femoral artery and vein using hemostats.
9. Separate 2 cm of the femoral artery from the femoral vein using angled forceps and a cotton tipped applicator, being careful not to tear the artery.
10. Ligate the distal end of the artery using 4–0 black braided silk and clip one end of the silk with a hemostat and place to the side to help expose the artery.
11. Place a thread of silk under the proximal end of the artery and clip both ends with a hemostat to temporarily occlude blood flow.
12. Using spring scissors prick the artery to create an opening for the catheter without severing the blood vessel.
13. Make sure the PE-160 catheter is trimmed to a point and attached to a three-way stopcock.
14. Insert the tubing into the opening with the pointed side down using two pairs of forceps to hold the tubing and the artery in place.
15. Use a loop of silk to cinch the tubing in place.
16. Inject a small amount of saline to ensure that blood is being pumped into the tubing.
17. Exchange the saline-filled syringe for an empty syringe and release the tension on the proximal occluding thread to fill the syringe with 5 ml of blood.
18. Using another syringe that has been filled with heparin and then emptied completely, withdraw a 0.2 ml blood sample to analyze glucose, hematocrit, and blood gases.
19. After sampling the blood and measuring whether the blood pressure is appropriate, put the tension back on the proximal end of the artery, then gently slide the PE-160 out enough so that the loop originally cinching the tubing to the artery slides off, then tighten it to ligate the artery.
20. Remove the silk on the proximal end and cut all the loose ends of silk, then suture the skin using 3–0 Dermalon sutures.
21. Make a 4 cm sagittal incision over the midline of the scalp.
22. Using cotton-tipped applicators, clear away the fascia to expose the coronal suture, then position the Heiss retractor under the skin to keep the area exposed.
23. Measure 10 mm laterally to the right and 10 mm anterior to the coronal suture and mark the area for the cranial burr hole.
24. Holding the Dremel drill with 1.5 mm drill bit vertically over the mark, slowly drill through the cranium, stopping when the pressure is no longer felt in order avoid tearing the dura.
25. A 20-gauge sterile plastic catheter cut to a length of 14 mm will be placed stereotaxically (guided by the pig brain atlas) into the center of the left frontal cerebral white matter (centrum semiovale) at the level of the caudate nucleus and cemented in place.

26. A 30-cm-long silicone elastomer tubing (Sil-med) connected to the arterial catheter will be filled with 10 ml of arterial blood by opening a three-way stopcock.
27. Use the infusion pump to infuse 2.5 ml of autologous whole blood into the brain tissue over 20 min.
28. Retract the plastic catheter and fill the burr hole with bone wax.
29. Suture the skin using 3–0 Dermalon sutures.

4 Outcome Evaluations

1. Brain edema: This can be determined by brain water content[4,10,13,14] and magnetic resonance imaging measurements.[11,15,16]
2. Blood-brain barrier (BBB) permeability: Evans blue is used to determine BBB disruption after ICH in pigs.[4,10]
3. Histology: Brain sections are stained with hematoxylin and eosine (H&E) and Luxol fast blue (Fig. 4). Using a computerized image analysis system (NIH image), the areas of white matter, ventricle, and hemisphere on each section are measured. The total volumes of white matter, hemisphere, and ventricle are calculated by multiplying area in each section by the distance between sections.
4. Brain metabolic changes: Adenosine triphosphate, phosphocreatine, glycogen, glucose, and lactate levels in the perihematomal zone can be determined.[17]

5 Limitations

1. There is a lack of a ruptured blood vessel as the source of the ICH.
2. This model lacks the hematoma enlargement, which occurs in ICH patients.
3. Functional outcome is very important in ICH studies. However, minimal neurological deficits are present in this pig ICH model.
4. The mortality rate is very low.

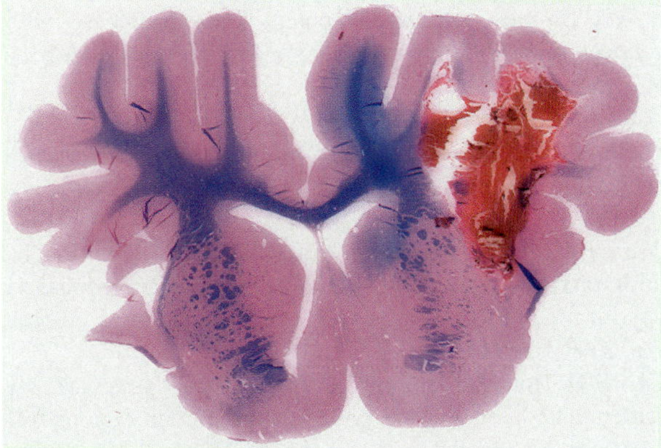

Fig. 4 Luxol fast blue staining showing ICH-induced white matter injury in pigs

References

1. Whisnant JP, Sayer GP, Millikan CH. Experimental intracerebral hematoma. *Arch Neurol.* 1963;9:586–592
2. Sussman BJ, Barber JB, Goald H. Experimental intracerebral hematoma. Reduction of oxygen tension in brain and cerebrospinal fluid. *J Neurosurg.* 1974;41:177–186
3. Takasugi S, Ueda S, Matsumoto K. Chronological changes in spontaneous intracerebral hematoma–An experimental and clinical study. *Stroke.* 1985;16:651–658
4. Wagner KR, Xi G, Hua Y, Kleinholz M, de Courten-Myers GM, Myers RE, Broderick JP, Brott TG. Lobar intracerebral hemorrhage model in pigs: Rapid edema development in perihematomal white matter. *Stroke.* 1996;27:490–497
5. Xi G, Keep R, Hoff J. Mechanisms of brain injury after intracerebral hemorrhage. *Lancet Neurol.* 2006;5:53–63
6. Andaluz N, Zuccarello M, Wagner KR. Experimental animal models of intracerebral hemorrhage. *Neurosurg Clin N Am.* 2002;13:385–393
7. Xi G, Wagner KR, Keep RF, Hua Y, de Courten-Myers GM, Broderick JP, Brott TG, Hoff JT. The role of blood clot formation on early edema development following experimental intracerebral hemorrhage. *Stroke.* 1998;29:2580–2586
8. Wagner KR, Dwyer BE. Hematoma removal, heme, and heme oxygenase following hemorrhagic stroke. *Ann New York Acad Sci.* 2004;1012:237–251
9. Wagner KR. Modeling intracerebral hemorrhage: Glutamate, neclear factor-kappa b signaling and cytokines. *Stroke.* 2007;38:753–758
10. Wagner KR, Xi G, Hua Y, Zuccarello M, de Courten-Myers GM, Broderick JP, Brott TG. Ultra-early clot aspiration after lysis with tissue plasminogen activator in a porcine model of intracerebral hemorrhage: Edema reduction and blood-brain barrier protection. *J Neurosurg.* 1999;90:491–498
11. Rohde V, Rohde I, Thiex R, Ince A, Jung A, Duckers G, Groschel K, Rottger C, Kuker W, Muller HD, Gilsbach JM. Fibrinolysis therapy achieved with tissue plasminogen activator and aspiration of the liquefied clot after experimental intracerebral hemorrhage: Rapid reduction in hematoma volume but intensification of delayed edema formation. *J Neurosurg.* 2002;97:954–962
12. Wagner KR, Packard BA, Hall CL, Smulian AG, Linke MJ, De Courten-Myers GM, Packard LM, Hall NC. Protein oxidation and heme oxygenase-1 induction in porcine white matter following intracerebral infusions of whole blood or plasma. *Dev Neurosci.* 2002;24:154–160
13. Wagner KR, Dwyer B, Hua Y, Cody M, Broderick JP, Brott TG. Tin-mesoporphyrin reduces intracerebral"mass"in a pig intracerebral hemorrhage. *Stroke.* 1998;29:332
14. Wagner KR, Hua Y, de Courten-Myers GM, Broderick JP, Nishimura RN, Lu SY, Dwyer BE. Tin-mesoporphyrin, a potent heme oxygenase inhibitor, for treatment of intracerebral hemorrhage: In vivo and in vitro studies. *Cell Mol Biol.* 2000;46:597–608
15. Thiex R, Kuker W, Muller HD, Rohde I, Schroder JM, Gilsbach JM, Rohde V. The long-term effect of recombinant tissue-plasminogen-activator (rt-pa) on edema formation in a large-animal model of intracerebral hemorrhage. *Neurol Res.* 2003;25:254–262
16. Thiex R, Weis J, Krings T, Barreiro S, Yakisikli-Alemi F, Gilsbach JM, Rohde V. Addition of intravenous *n*-methyl-d-aspartate receptor antagonists to local fibrinolytic therapy for the optimal treatment of experimental intracerebral hemorrhages. *J Neurosurg.* 2007;106:314–320
17. Wagner KR, Xi G, Hua Y, Kleinholz M, de Courten-Myers GM, Myers RE. Early metabolic alterations in edematous perihematomal brain regions following experimental intracerebral hemorrhage. *J Neurosurg.* 1998;88:1058–1065

Part IV

Traumatic CNS Injury Animal Models

Section I

Experimental Brain Injury Models

Chapter 32

Fluid Percussion Injury Model

Jonathan Lifshitz

Abstract Hallmarks of traumatic brain injury in man can be faithfully reproduced in the laboratory using fluid percussion injury. Variations in the surgical procedure provide the ability to induce focal diffuse or mixed focal and diffuse brain injury in various laboratory species. Being fully scalable, fluid percussion can induce mild, moderate, or severe brain injury in subjects of any age. This chapter outlines the procedures for midline (diffuse) fluid percussion brain injury in adult male Sprague-Dawley rats and lateral (mixed) fluid percussion brain injury in adult male mice. With these procedures, it becomes possible to generate brain-injured laboratory animals for studies of injury-induced pathophysiology and behavioral deficits, for which rational therapeutic interventions can be implemented.

Keywords: Fluid percussion, Rat, Mouse, Brain injury, Trauma

1 Model Selection

Fluid percussion permits the study of experimental traumatic brain injury (TBI) in a model that is reproducible, clinically relevant, and scalable between species and injury severities. Brain injury is induced by a rapid (~20 ms) fluid pulse through a craniotomy onto the intact dura that follows the inner curvature of the skull and creates an elastic decompression of the brain.[1,2] The mechanical forces disrupt cell membranes, blood vessels, and neuronal processes. By increasing the angle from which the pendulum hammer falls, greater pressures can be generated to travel through the fluid-filled cylinder and impact the brain. At a moderate level of injury, 20–25% of animals die as a result of the injury within the acute posttraumatic period (15 min), generally from respiratory failure and pulmonary edema. This is a normal and desired feature of TBI models, as it reflects human TBI.

In laboratories worldwide, subtle variations in surgical and injury procedures reproduce the spectrum of brain injuries found in the human population. Primarily, the location of the injury site determines the major features of the injury, where a <u>midline</u> location induces a *diffuse injury*

From: *Animal Models of Acute Neurological Injuries.*
Edited by: J. Chen (eds.) New York, Totowa, NJ

and a lateral location induces a *focal injury with a diffuse component*.[3–5] Fluid percussion injury reproduces the acute reflex suppression, functional deficits, and histopathology evident after TBI in man.[6] The model continues to be implemented to evaluate pathophysiological mechanisms underlying posttraumatic degeneration, behavioral deficits, and therapeutic interventions to mitigate the degeneration and promote recovery of function.[7]

This chapter focuses on the streamlined procedures employed in the author's laboratory, which are a result of years training in several different laboratories in these techniques. In either rats or mice, the cranial surgery is performed under inhaled isoflurane anesthesia, allowing a rapid recovery from the surgery. Injury is induced later the same day under light isoflurane anesthesia to mimic, as close as possible, the human condition at the time of injury.

2 Materials

2.1 Animals

Fluid percussion brain injury has been successfully performed on various species, including cats, rabbits, pigs, rats, and mice. The adaptation of fluid percussion to rats[8–10] was followed by its implementation in mice.[11] The procedures outlined in this chapter focus on adult male Sprague-Dawley rats (~400 g) and 8-week-old adult male C57Bl/6 mice (~25 g). To maximize the success of the brain injury, examine all animals for any signs of ill health (e.g., rough coat, bleeding or dirty eyes, runny or bleeding nose, and scratches around eyes or nose area). Weigh all animals prior in order to surgery to follow injury-induced weight loss.

2.2 Equipment

2.2.1 Anesthesia

Rodent nose cone for inhaled anesthesia

2.2.2 Surgery

Microwave heating pads (BrainTree Scientific, #39DP)
Stereotactic head holder
Scalpel
Delicate bone scraper (Fine Science Tools, #10075–16)
Bulldogs (Fine Science Tools, #18050–28, #18051–28)
Chisels Wedelstaedt 3/4 DE (Henry Schein, #600–4972)

2.2.2.1 Rat Surgery
Dremel tool with engraving cutter #106
Trephine (4.7 mm) (Miltex, #26–140)
Fingernail drill with 5/64″ drill bit (Miltex, #33–232)
Stainless steel skull screws (2–56 × 3/16″) (Small Parts Inc., #MX-0256-03B-25)

2.2.2.2 Mouse Surgery
Custom trephine (3 mm)

Research Instrumentation Shop
University of Pennsylvania, Philadelphia, PA
http://www.uphs.upenn.edu/biocbiop/ris/ris.html
Weed whacker line for cranial discs (1.7 mm diameter)
Side-grasping forceps (7 × 7) (Henry Schein, #6–124XL)
3M Vetbond tissue adhesive (Henry Schein, #700–3449)

2.2.3 Injury Hub

1 1/2″ needle (20 g) (Becton Dickinson, #305176)
Syringe (1 cc)
Razor blades
Cyanoacrylate superglue gel (Loctite, #01–30379)
Tissue forceps (Henry Schein, #6–114)
Methyl methacrylate (Permanent Repair Acrylic Powder, Henry Schein, #5470526; Repair Acrylic Liquid, Henry Schein, #1255142)
Saline-filled syringe, blunted needle bent 90°

2.2.3.1 Rat Injury Hub
Cosmetic pencil sharpener

2.2.3.2 Mouse Injury Hub
Baxter extension tubing (Baxter, #2C5643)

2.2.4 Fluid Percussion Injury

Fluid percussion injury device
Custom Design and Fabrication
Virginia Commonwealth University, Richmond, VA
http://www.radiology.vcu.edu/newradsite/customdesign/FPI.html
The Website contains an FPI assembly and operation manual.
Recording oscilloscope (Tektronics, Model 1001B)
1 1/2″ Heavy duty Velcro to secure injury device to the bench
High vacuum grease (Fisher Scientific, #14–635–5D)
Dish washing solution
Jet dry spotless water softener to minimize air bubbles in cylinder
Gelfoam (Henry Schein, #908–3300)

3 Procedure

3.1 Administer Anesthesia and Secure in Head Holder

- Follow a standard surgery sheet, recording the progress of the surgery (see Appendix)
- Anesthetize animals with 5% isoflurane for 5 min in an induction chamber
- Shave or remove hair from the scalp, as appropriate
- Secure animal in a stereotaxic frame equipped with a specialized nose cone for continuous inhalation of isoflurane (2.0–2.5%). The subsequent surgical procedure is performed within a hooded area that prevents any fumes from leaking out into the laboratory.
- Apply ophthalmic ointment to the eyes to keep them moist during the surgery
- Wash the scalp with 70% alcohol and betadine solution (antiseptic)

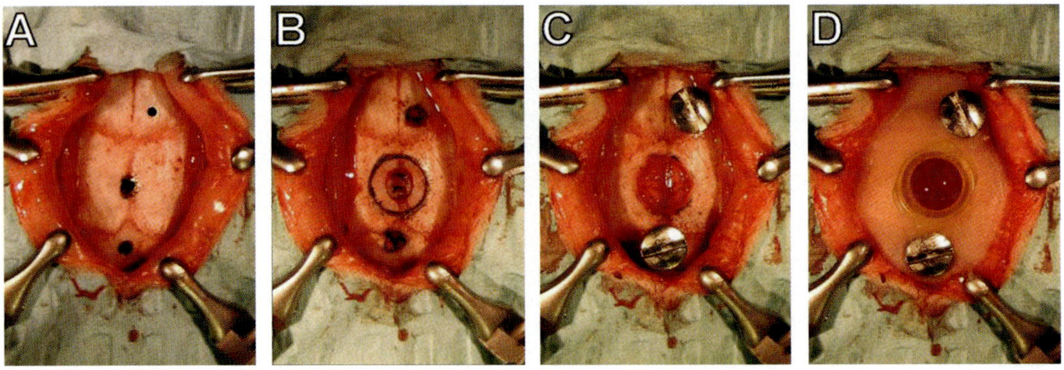

Fig. 1 Photographs of the midline craniotomy procedure in the rat. (**A**) After a midline incision and cleaning the skull, three marks are placed for the anterior skull screw, craniotomy center, and posterior skull screw, from top to bottom. (**B**) The burr holes for the skull screws have been expanded with a fingernail drill, and the trephination has been completed to the point of removing the craniotomy. (**C**) The craniotomy has been removed exposing the intact dura and two skull screws have been inserted. (**D**) The injury hub has been secured with cyanoacrylate to the craniotomy and encased in methyl methacrylate

- Continuously monitor anesthesia by observing muscle relaxation, in addition to assessing the toe pinch reflex every 10 min. Animals under appropriate anesthesia have a steady respiration rate.

3.2 Cranial Surgery for Hub Placement

- Make a midline incision in the scalp, from between the eyes past the ears
- Expose the skull and scrape the fascia from the skull using the delicate bone scraper, cotton swabs, and gauze. Clear away the temporal muscle as necessary
- Attach the Bull Dog clips to the edges of the incision (two anterior, two posterior) to expose the surgical field. Allow the Bull Dogs to fall down and hold the incision open (Fig. 1A, 2A).

3.2.1 Cranial Surgery for Hub Placement: Rat

- Mark the locations on the skull of the two skull screws (far rostral and caudal) and craniotomy center (midway between bregma and lambda, over the sagittal suture) using a felt marker (Fig. 1A).
- Drill pilot holes at all three markings using a Dremel tool.
- Expand the two skull screw holes with a finger nail drill and 5/64″ drill bit (Fig. 1B).
- Place the centering pin inside the 4.7 mm diameter trephine. Anchor the centering pin in the pilot hole at the craniotomy center.
- Perform the craniotomy by repeatedly turning and spinning the trephine (Fig. 2C). Be diligent to clean the bone debris from the trephine with a brush. Apply saline to moisten the bone and aid in trephination. As needed, angle the trephine to work on regions that remain thicker.
- Repeatedly test the progress of the craniotomy. Apply mild pressure (tap) around the circumference of the craniotomy with forceps. As the skull thins, the craniotomy will begin to move independently of the skull (observed by a change in reflection of fluid in the groove). The craniotomy is complete when the bone moves independently in all directions.

- Pry the bone loose with the Wedelstaedt chisels, scalpel or forceps (Fig. 2E), working around the circumference, without disturbing the underlying dura.
- Secure stainless steel screws in the skull screw holes. Hold the screws with tissue forceps, while advancing the screw with a screwdriver (Fig. 1C).

3.2.2 Cranial Surgery for Hub Placement: Mouse

- Trim (shave) weed whacker line with a razor blade as thin as possible to make circular cranial discs (Fig. 2B, inset).
- Pick up the cranial disc with the side-grasping forceps. Dip the cranial disc into a drop of Vetbond tissue adhesive placed on a nonabsorbent surface.

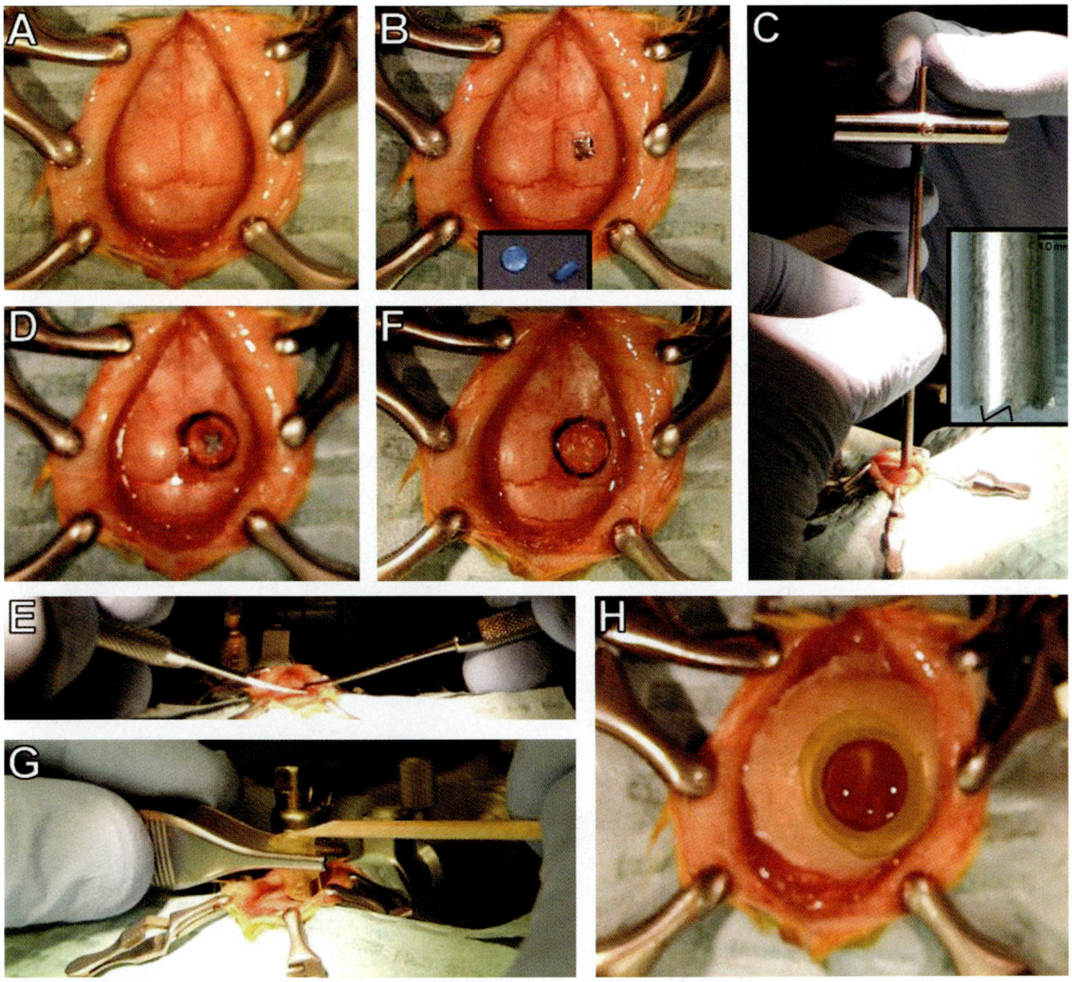

Fig. 2 Photographs of the lateral craniotomy procedure in the mouse. (**A**) A midline incision exposes the skull to be cleaned of overlying fascia. (**B**) The cranial disc (inset) is attached to the skull at the craniotomy center using Vetbond® and colored for clarity. (**C**) The craniotomy is performed with a custom-made mouse trephine (inset, 3.0 mm outer diameter), supporting the shaft while spinning the top. (**D**) As the trephination proceeds, a groove forms outside the cranial disc. (**E**) Using two Wedelstaedt chisels, the craniotomy is lifted and pried from the skull. (**F**) The craniotomy has been removed exposing the intact dura. (**G**) The injury hub is positioned with tissue forceps and an applicator stick, making sure the marked long side is facing lateral. (**H**) The injury hub has been secured with cyanoacrylate to the craniotomy and encased in methyl methacrylate

- Place the cranial disc at the location of the craniotomy, release the forceps and position disc with a wooden applicator stick (Fig. 2B). Wick away excess Vetbond using twisted Kimwipe tissues.
- Place the 3.0 mm diameter trephine over the cranial disc and perform the craniotomy by repeatedly turning and spinning the trephine (Fig. 2C). Be diligent to clean the bone debris from the trephine with a brush. Apply saline to moisten the bone and aid in trephination. As needed, angle the trephine to work on regions that are thicker.
- Repeatedly test the progress of the craniotomy under magnification (Fig. 2D). Apply mild pressure (tap) around the circumference of the craniotomy with forceps. As the skull thins, the craniotomy will begin to move independently of the skull (observed by a change in reflection of fluid in the groove). The craniotomy is complete when the bone moves independently in all directions.
- Under magnification, pry the bone loose with the Wedelstaedt chisels, scalpel, or forceps (Fig. 2E, F), working around the circumference, without disturbing the underlying dura.

3.3 Injury Hub

3.3.1 Injury Hub Construction

- Attach a 22 g, 1 1/2″ needle to a 1 cc syringe. Place the needle into a laboratory bench pad (Fig. 3A).
- Using a razor blade, cut the female Luer-Loc hub from the needle (Fig. 3B). For the rat, the cut is parallel to the Luer-loc and the outer diameter will

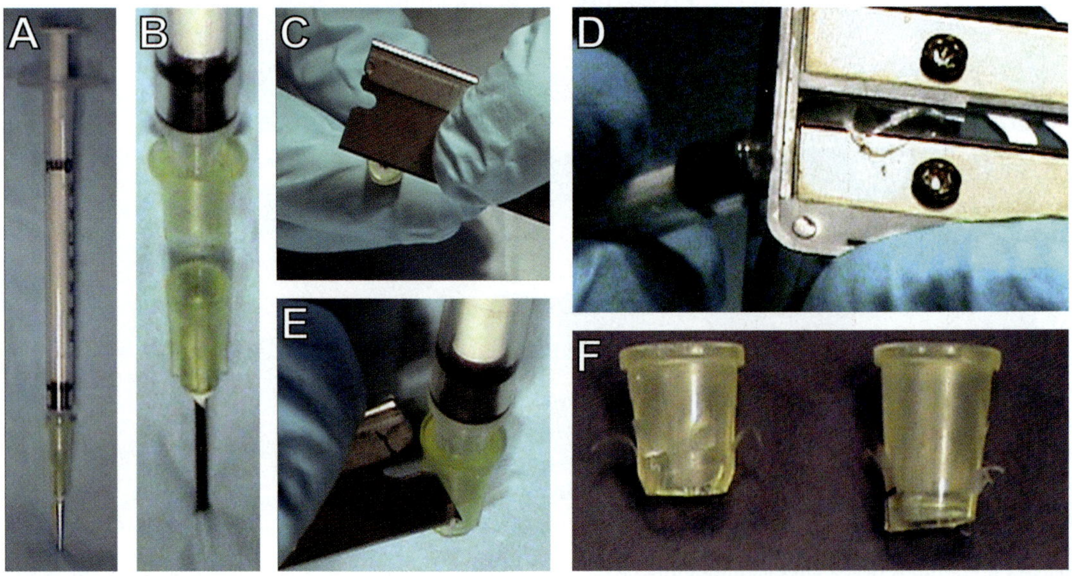

Fig. 3 Photographs of the injury hub construction. (**A**) A 22 g needle is attached to a 1 cc syringe and inserted into laboratory bench paper. (**B**) Using a razor blade, the needle is cut between the Luer-loc and the needle. (**C**) The injury hub is trimmed flat with a razor blade. (**D**) For the rat, the injury hub is beveled using a cosmetic sharpener, holding the syringe at a steep angle. (**E**) Burrs are shaved into the injury hub with a razor blade, bracing the hub on a table. (**F**) Examples of a rat injury hub (*left*, note the beveled edge) and a mouse injury hub (*right*, note the angled tip for a lateral placement)

be ~4.7 mm. For the mouse, the cut is angled slightly from the Luer-loc. Confirm that the inside diameter is 3.0 mm by trephining through the injury hub. Mark the long side with a felt marker.
- Inspect the cut edge of the injury hub. Trim as necessary (Fig. 3C).
- For the rat, bevel the cut edge of the injury hub with the cosmetic pencil sharpener (Fig. 3D).
- Shave thin burrs around the injury hub from the Luer-Loc edge to the cut edge using a razor (Fig. 3E).

3.3.2 Injury Hub Placement

- For a lateral craniotomy, elevate the stereotaxic frame on the side of the craniotomy to minimize the likelihood of the injury hub falling off the skull.
- Hold the injury hub in tissue forceps (not with the teeth). Apply three drops of cyanoacrylate gel on the outside of the hub, above the cut end.
- Position the hub over the craniotomy (with magnification if necessary), using a beveled wooden applicator stick to hold it in place, while the forceps are released (Fig. 2G). For the rat, the injury hub is normal to the dural surface and fits *inside* the craniotomy. For the mouse, the injury hub is parallel to the ground and fits *outside* the craniotomy.
- Gently scrape the cyanoacrylate gel down the injury hub and onto the skull.
- Apply more cyanoacrylate gel to the junction between the injury hub and the skull, spreading it with the beveled wooden applicator stick. The cyanoacrylate gel adheres the injury hub to the skull, in addition to sealing the skull.
- After the cyanoacrylate gel dries, encase the injury hub (and screws) in methyl methacrylate cement. Apply the methyl methacrylate from a 1 cc syringe, when it begins to hold shape in the mixing container (Fig. 1D, 2H).
- Fill the injury hub with saline.
- Suture the anterior and posterior edges of the incision to prevent tearing of the wound.
- Remove the animal from the stereotaxic frame and anesthesia. Place the animal in a normothermic warming box until awake and alert.
- Monitor animals for any untoward signs of pain or distress.

3.4 Injury

- Follow a standard injury sheet, recording the parameters and outcome (see Appendix).
- Assess animals for regained activity, inquisitiveness, grooming, defecation, urination, appearance of eyes, breathing rate and gait.
- Reanesthetize animal after a 30–60 min period of recovery.
- Clean out the injury hub using a cotton tip applicator (thinned with a razor and twisted). Irrigate with saline if necessary. Fill injury hub with saline.
- Drop the pendulum hammer several times to prime the device (Fig. 4).
- Connect the female Luer-loc injury hub on the animal to the male Luer-loc fitting on the fluid percussion injury device. To avoid air, tap the plunger on the device to expel fluid and create continuity between the fluid of the cylinder and the injury hub. For rats, hold them in your left hand, with their head between your extended thumb and index finger, lying on their right

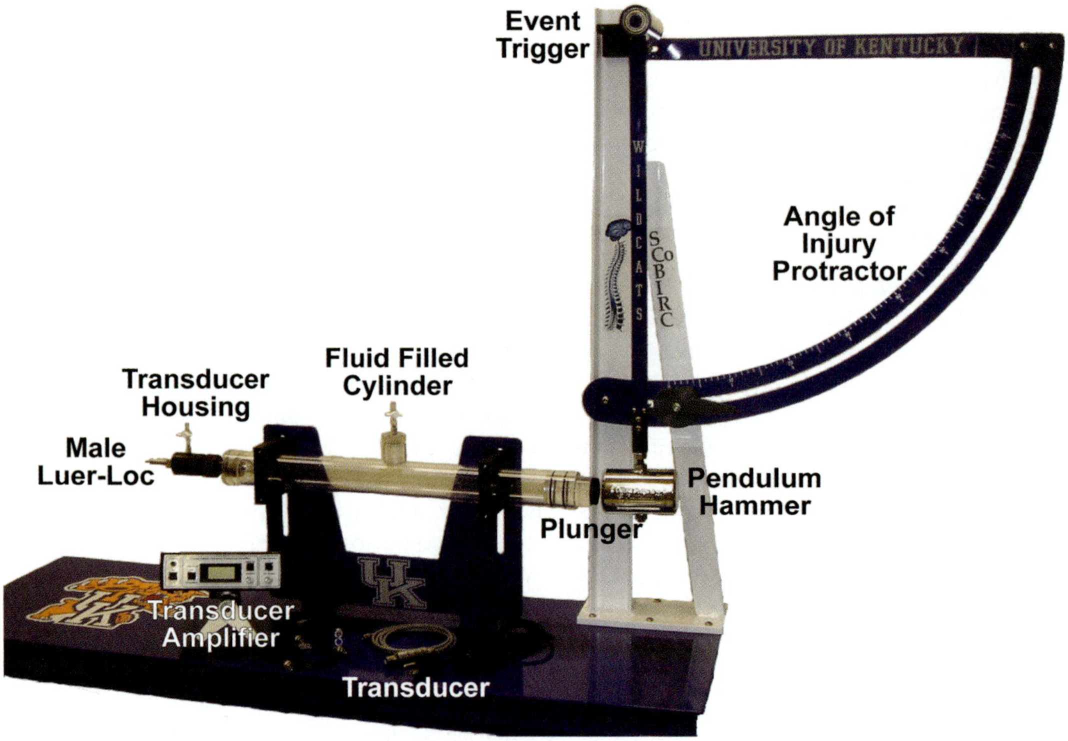

Fig. 4 Photograph of the University of Kentucky Spinal Cord & Brain Injury Research Center fluid percussion device. The pendulum can be raised to fall from angles ranging from 0–75°. The pendulum strikes the plunger, generating a fluid pulse through the fluid filled cylinder and past the transducer housing. The transducer measures the pressure pulse when an event is triggered, and is amplified and displayed on an oscilloscope (not shown). The pulse exits through the male Luer-loc at the left-hand side, where the fluid impacts the intact dural surface

side. Attach the rat directly to the device. For mice, attach them to the device using a Luer-loc extension tube. Mice should lie on their right side on a heating pad.

- Once a normal breathing pattern resumes (1–2 breaths per second), before sensitivity to stimulation, induce the injury by releasing the pendulum hammer. Catch the hammer after it strikes the plunger and return it to the catch.
- Start the timer to record the righting reflex time.
- Remove the injury hub and methyl methacrylate *en bloc*, by applying pressure to the bridge of the nose, while the hub is still attached to the device. Visually inspect the injury hub for obstructions.
- Observe and record the extent of apnea, seizure, brain herniation and hemorrhage. If the dura is compromised, the animal should be euthanized.
- Suture the scalp incision and apply topical lidocaine. Control bleeding with Gelfoam if necessary. Leave the craniotomy open.
- Place the animal in a supine position on a warming pad. The time elapsed until the animal spontaneously rights is recorded as the righting reflex time.
- Upon return of righting reflex, return animals to an incubator equipped with a heating pad.

3.4.1 Oscilloscope Trace

The transducer amplifier currently available is calibrated for 10 mV per PSI (pound per square inch). Figure 5 shows a representative injury pressure trace, with a peak magnitude of 180 mV, which is equivalent to 18 PSI or 1.22 atmospheres. Injury traces should have a smooth rise, plateau, and fall components. Sharp changes in the trace indicate air bubbles in the system.

3.5 Postoperative Care

- Monitor animals visually to ensure a safe recovery every 10 min postinjury. Within 15–20 min after injury, surviving animals are alert. Within 1 h after injury, animals are ambulating. Brain-injured and uninjured control animals typically show no untoward effects once they have recovered from anesthesia, and resume normal eating, drinking, and grooming patterns.
- Follow the Postoperative Evaluation Sheet (see Appendix) to record the external examination, physical examination, suture site, and a pain evaluation. Typically, animals require no special supportive care after surgery. These injuries do not produce overt signs of post-operative pain, and do not call for pain monitoring or drugs to manage pain.
- Weigh animals daily. Animals can lose up to 20% of their body weight after surgery and injury. Weight loss exceeding 20% indicates significant injuries that require intensive post-operative care or euthanasia.

4 Outcome Evaluation

4.1 Mortality and Acute Reflex Suppression

A moderate injury severity results in 20% mortality within the acute posttraumatic period (15 min), resulting from respiratory failure and pulmonary

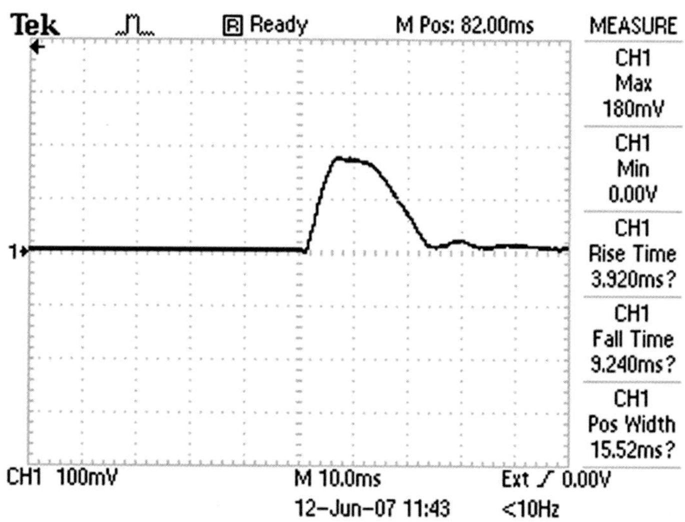

Fig. 5 Sample injury pressure trace recorded at 100 mV and 10 ms per division. The trace shows a linear increase in pressure, a declining plateau and a return to baseline. Rapid and sharp deviations in this shape indicate air bubbles within the transducer housing. The oscilloscope can measure the maximum voltage (180 mV), minimum (0 mV), rise time (3.92 ms), fall time (9.24 ms), and positive width of the trace (15.52 ms)

edema. This is a normal and desired feature of brain injury models, because it reflects the mortality in human brain injury. In our experience, midline fluid percussion can be induced at 1.9 and 1.7 atm in the rat and mouse, respectively. Lateral fluid percussion can be induced at 2.4 and 1.4 atm in the rat and mouse, respectively. The righting reflex (see Sect. 3.4) is an acute neurological evaluation of injury severity. For a moderate injury, the righting reflex time for brain injured rodents is approximately 6 min.

4.2 Behavioral Evaluation

Extensive reviews cover many aspects of post-traumatic behavioral deficits.[12,13] Almost all behavioral tests developed for rodents can be applied to brain injury studies, including a neurological score, Morris water maze and beam balance.

4.3 Histology

At acute time points, petechial hemorrhages can be found in multifocal sites throughout the brain. Depending on the craniotomy location and injury severity, the cortical lesion continues to develop into a cavity over days to weeks post-injury. In this period, the blood brain barrier is compromised, with significant glial responses (e.g., infiltrating macrophages and reactive astrocytes). Numerous forms of neuronal damage and death are observed rapidly and progress over time.

5 Advantages, Complications, Limitations, Alternatives

5.1 Advantages

The clinical relevance of the model allows the study of brain injury phenomena, from behavior to molecular biology. The model achieves face validity by the nature of the mechanical forces applied to the brain, and the ensuing pathophysiological responses. The recovery period after surgery before injury enhances the face validity of the brain injury model by returning the animals to a condition that more closely resembles the human condition (in which brain injury occurs in the absence of anesthesia). Also, the open craniotomy reduces the effects of injury-induced rises in intracranial pressure, like a decompressive craniotomy.

The midline injury results in bilateral diffuse pathology that globally suppresses function, such that therapeutic interventions can be tested against an injured control. The lateral injury results in ipsilateral focal injury and contralateral diffuse injury, such that the injury is distinct for every brain region. Only specific lateralized functions may be suppressed after a lateral brain injury. However, the ensuing cavity permits targeted therapeutic intervention, especially for cell transplantation and bioengineered matrices.

5.2 Complications

During the cranial surgery, the dura can be compromised by the trephination or craniotomy removal. With the pressure generated by the injury, the dura will tear and the brain will herniate through the craniotomy. The compromised dura changes the injury by extending the opening of the blood brain barrier and displacing neural tissue. Therefore, the animals with dural tears following injury must be identified and excluded from the study.

When placing the injury hub over the craniotomy, cyanoacrylate can spread on to the dura. Glue on the dura will change its mechanical properties and alter the injury. After removing the head assembly, visually inspect the injury hub for glue or other obstructions.

The plunger on the fluid percussion injury device is secured with high pressure vacuum grease, which maintains a water-tight cylinder. Overnight, the vacuum grease and plunger stick together. Therefore, fluid must be repeatedly removed and injected into the cylinder to loosen the plunger (without introducing air). A reproducible injury pressure trace indicates that the device is working according to specification. With few moving parts and no electronics, no additional calibration is needed.

5.3 Limitations

To date, most studies using fluid percussion injury have used a moderate level of injury. Concerns about injury variability are based on the extent of damaged tissue and range of physiological responses among animals. In contrast to the tissue *destruction* caused by other CNS injury models, the tissue *disruption* following fluid percussion is not constrained by a ceiling effect. The injury to the brain occurs within a broad range, necessitating larger group sizes to detect significant effects.

5.4 Alternatives and Variations

As a fully scalable injury model, fluid percussion injury can be adapted to a variety of species over a wide range of injury severities, depending on the angle from which the pendulum is dropped. For each species-severity combination, the parameters of the injury must be determined empirically to produce a desired level of mortality and acute neurological reflex suppression. Additional variations include age, gender, and craniotomy location. Please note that variations in species will change scalp, skull, and dura thickness, necessitating slight adjustments in surgical technique to perform a successful craniotomy.

Other laboratories employ alternate anesthesia paradigms. Every anesthetic has side effects that influence the progression of the injury. Inhaled anesthetics, such as isoflurane, are cleared rapidly, thereby minimizing side effects. Some laboratories induce the injury immediately after surgery, whereas others wait until the next day. The 60–90 min period described in this chapter allows the anesthetic to be cleared and the animals to rest before injury.

References

1. Thibault LE, Meaney DF, Anderson BJ, Marmarou A. Biomechanical aspects of a fluid percussion model of brain injury. J Neurotrauma 1992; 9(4):311–322.
2. Dixon CE, Lighthall JW, Anderson TE. Physiologic, histopathologic, and cineradiographic characterization of a new fluid-percussion model of experimental brain injury in the rat. J Neurotrauma 1988; 5(2):91–104.
3. Floyd CL, Golden KM, Black RT, Hamm RJ, Lyeth BG. Craniectomy position affects morris water maze performance and hippocampal cell loss after parasagittal fluid percussion. J Neurotrauma 2002; 19(3):303–316.

4. Vink R, Mullins PG, Temple MD, Bao W, Faden AI. Small shifts in craniotomy position in the lateral fluid percussion injury model are associated with differential lesion development. J Neurotrauma 2001; 18(8):839–847.
5. Iwamoto Y, Yamaki T, Murakami N et al. Investigation of morphological change of lateral and midline fluid percussion injury in rats, using magnetic resonance imaging. Neurosurgery 1997; 40(1):163–167.
6. Graham DI, Gennarelli TA, McIntosh TK. Trauma. In: Graham DI, Lantos PL, editors. Greenfield's Neuropathology. 7th ed. London: Arnold Publishers; 2002 pp. 823–898.
7. Thompson HJ, Lifshitz J, Marklund N et al. Lateral fluid percussion brain injury: a 15-year review and evaluation. J Neurotrauma 2005; 22(1):42–75.
8. Dixon CE, Lyeth BG, Povlishock JT et al. A fluid percussion model of experimental brain injury in the rat. J Neurosurg 1987; 67(1):110–119.
9. McIntosh TK, Noble L, Andrews B, Faden AI. Traumatic brain injury in the rat: characterization of a midline fluid-percussion model. Cent Nerv Syst Trauma 1987; 4(2):119–134.
10. McIntosh TK, Vink R, Noble L et al. Traumatic brain injury in the rat: characterization of a lateral fluid-percussion model. Neurosci 1989; 28(1):233–244.
11. Carbonell WS, Maris DO, McCall T, Grady MS. Adaptation of the fluid percussion injury model to the mouse. J Neurotrauma 1998; 15(3):217–229.
12. Fujimoto ST, Longhi L, Saatman KE, McIntosh TK. Motor and cognitive function evaluation following experimental traumatic brain injury. Neurosci Biobehav Rev 2004; 28(4):365–378.
13. Hamm RJ. Neurobehavioral assessment of outcome following traumatic brain injury in rats: an evaluation of selected measures. J Neurotrauma 2001; 18(11):1207–1216.

Appendix

FLUID PERCUSSION BRAIN INJURY

Spinal Cord & Brain Injury Research Center

Animal ID#:____________ Species: rat mouse Animal Marking:___________

Protocol #____________________________

Investigator: ______________ Strain:___________________

Barcode #: _______________ Emergency Ph#: ____________

Contact Person:_______________ Lab Ph#: ________________

Surgery Date:______________ Injury Date:_______________

Experiment:____

Date	Clinical Notes

Euthanized: Date: ____________

Substance Used: ___________ Amount: _______ Route: ____________

Perfusion

Perfusion Date: ______________ Surgeon: ______________

Sodium Pentobarbital: Route: _________ Amount: ______________

Perfusion Technique: ______________

Saline: ______________ __________ml

Fixative: ______________ __________ml

Perfusion Quality: Good Moderate Marginal Poor

Post-Fix: Brain / SkullDays:________ Fixative:______________

Brain Sectioning

Sectioning Date: ___________

Cutting Scheme: ____

____

1

SURGERY

Animal ID#: ______________

Pre-Surgical Observation*

Good Poor

**Animals assessed for normal activity, inquisitiveness, condition of hair coat, eating, drinking, defecation, urination, appearance of eyes, breathing rate, gait, and bodyweight.*

Surgical Preparation

Weight: ________

Anesthesia Type: ____Isoflurane____ ______________

Percentage:______ Duration:__________ Time:______

- Shave ___

Surgical Procedure

Anesthesia Type:______________

Percentage:______ Duration:__________ Time:______

- Secured in stereotaxic frame ___
  Set stereotaxic frame: Earbar: ___ Nose cone unit: ___
- Lacri-Lube ointment ___
- Betadine & Alcohol Scrub ___
- Sterile Gloves ___
- Midline incision and skull exposure ___
- Holes drilled in frontal and occipital bones and expanded with pin-set ___
- Midline craniotomy between bregma and lambda ___
- Screws placed within frontal / occipital bone holes ___
- Cut needle hub at lower white line & fashion injury hub___
- Needle hub placed over craniotomy with cyanoacrylate ___
- Dental cement poured around hub and over screws ___
- Wound closure with : suture staples
- Lidocaine ___
- Bacitracin ___
- Animal Returned to Holding Cage___ Time:______

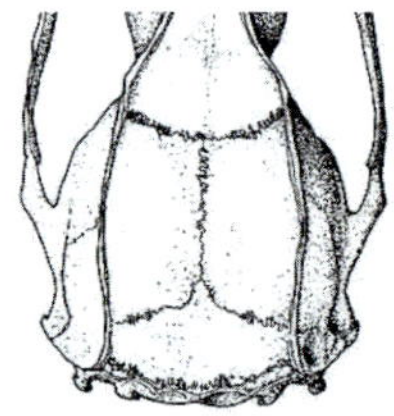

Surgical notes: ____

____

____

____

____

Procedure was completed: ❑ As expected ❑ Better than expected ❑ Worse than expected

Surgeon Signature: ______________________________ Date: ______________

FLUID PERCUSSION INJURY

Animal ID#: ______________

INJURY **SHAM**

Pre-Injury Observation*

Good Poor

**Animals assessed for normal activity, inquisitiveness, condition of hair coat, eating, drinking, defecation, urination, appearance of eyes, breathing rate, gait, and bodyweight.*

Anesthesia Type:______Isoflurane______ ______________

Percentage: _______ Duration: _____________ Time: _______

Pre-Injury Inspection

- Visual inspection of injury hub ______________________________
- Visual inspection of the dura ____ ______________________________

Hammer Angle: __________ Injury Time: _______

Magnitude of injury: __________ ATM __________mV (147.28 mV = 1 ATM)

MAX: __________

MIN: __________

Rise Time: __________

Fall Time: __________

Positive Width: __________

Oscilloscope Filename:______________________________

Animal Status Post-Injury

Apnea: __________ secs	❑ massage of thorax		❑ no intervention		
Seizure: __________ secs					
Herniation:	❑ none	❑ small	❑ moderate	❑ large	
Hematoma:	❑ none	❑ small	❑ moderate	❑ large	❑ bloody
Dura Breach:	❑ none	❑ small	❑ extensive		
Breathing:	❑ normal	❑ deep	❑ regular	❑ rapid	
	❑ labored	❑ shallow	❑ irregular	❑ slow	

Righting Reflex Time: _______________

Notes following injury / sham: ____

____

____

- Wound closure with: suture staples
- Lidocaine ___
- Bacitracin ___

Time returned to incubator:________

Time returned to colony:_________

Date / Time animal is to be perfused: __________________ Survival Time __________

Procedure was completed: ❑ As expected ❑ Better than expected ❑ Worse than expected

Surgeon Signature: ____ Date: ____________________

Post-Operative Evaluation

Animal ID#: ______________

Date					
Days post-procedure					
Time					
Observer's Initials					

EXTERNAL EXAMINATION

Active?	Y N	Y N	Y N	Y N	Y N
Inquisitive?	Y N	Y N	Y N	Y N	Y N
Crusty eyes?	Y N	Y N	Y N	Y N	Y N
Eating?	Y N	Y N	Y N	Y N	Y N
Drinking?	Y N	Y N	Y N	Y N	Y N
Feces?	Y N	Y N	Y N	Y N	Y N
Urine?	Y N	Y N	Y N	Y N	Y N

PHYSICAL EXAMINATION

Type of Breathing?					
Gait?					
Coat Appearance?					
Dehydration?*	Y N	Y N	Y N	Y N	Y N
Body Appearance?					
Body Weight?					

SUTURE SITE

Wound edges red?	Y N	Y N	Y N	Y N	Y N
Swelling around incision?	Y N	Y N	Y N	Y N	Y N
Swelling under incision?	Y N	Y N	Y N	Y N	Y N
Sutures missing?	Y N	Y N	Y N	Y N	Y N
Exudate from incision?	Y N	Y N	Y N	Y N	Y N

PAIN EVALUATION**

Behavior					
Locomotion					
Appearance of incision					
Pain on palpation of surgery site					
Total					

N = normal L = labored R = rapid S = shallow Y = yes N = no
G = good F = fair P = poor
*Gently pinch up a fold of skin. Skin of dehydrated animals will stay pinched up.
**Scale 0-3

4

Chapter 33

Controlled Cortical Impact Injury Model

C. Edward Dixon and Anthony E. Kline

Abstract Controlled focal mechanical cortical deformations in rodents have been widely used to model histological, physiological, neurochemical, and functional aspects of human traumatic brain injury. The most common technique termed controlled cortical impact (CCI) utilizes a mechanically actuated piston to control the velocity and depth of cortical impact. In this chapter, we introduce step-by-step procedures to perform a typical CCI in rodents. Details regarding anesthesia, surgery and CCI device operation are included.

Keywords: Rat, Traumatic brain injury, Contusion, Craniotomy

1 Model Selection

Experimental traumatic brain injury (TBI) employing a pneumatic impactor was first introduced for use in laboratory ferrets by Lighthall and colleagues[1,2] and subsequently adapted for rat by Dixon and colleagues in an attempt to better control the biomechanical parameters of brain injury.[3] Unlike the fluid-percussion injury device that disperses a stream of solution intracranially that cannot be readily quantified, the controlled cortical impact (CCI) model of experimental brain injury takes advantage of biomechanical events contributing to injury. These biomechanical events can be analyzed by establishing a quantifiable relationship between measurable engineering parameters, such as force, velocity, and tissue deformation, and the magnitude of tissue damage and/or functional impairment. These controlled mechanical variables enable accurate, reliable, and independent control of the deformation parameters over a wide range of contact velocities.

The CCI injury device (Fig. 1, Pittsburgh Precision Instruments, Inc.) consists of a small (1.975 cm) bore, double-acting stroke-constrained pneumatic cylinder with a 5.0 cm stroke. The cylinder is rigidly mounted on a crossbar in either an angled (perpendicular to the dura surface) or vertical position. The lower rod end has an impact tip attached, which varies in geometry (rounded or flat edge) and diameter (5–6 mm), and the upper rod end is attached to a sensor system, which measures velocity. The impactor tip is pneumatically driven at a predetermined velocity, depth, and dura-

From: *Animal Models of Acute Neurological Injuries.*
Edited by: J. Chen (eds.) New York, Totowa, NJ

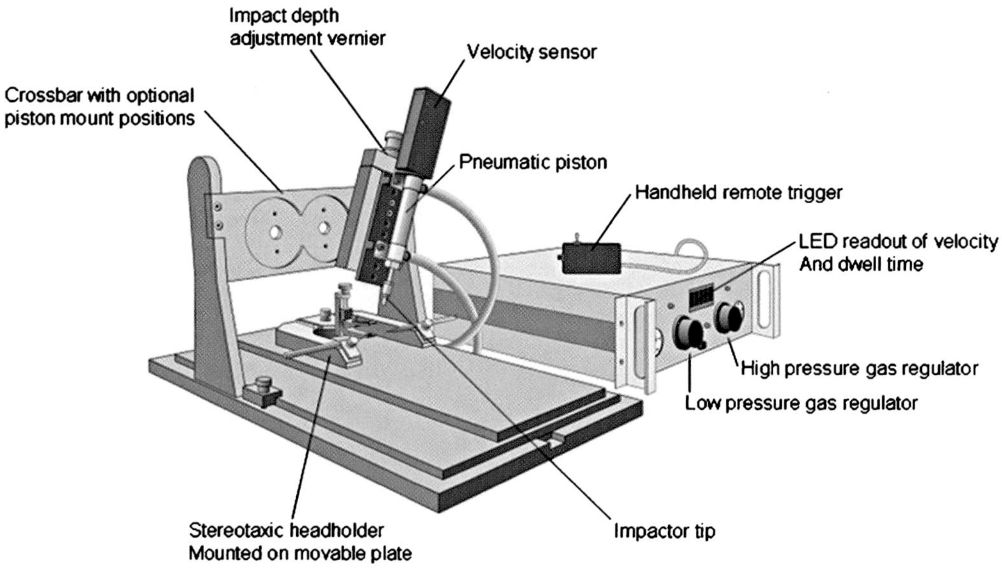

Fig. 1 Diagram of a rat controlled cortical impact injury device with the major components labeled

tion of tissue deformation. A typical penetration depth for this device of 2.6–2.8 mm with a velocity of 4.0 m s^{-1} and a dwell time of 50–150 ms consistently produces an injury of moderate severity. However, exact injury parameters are laboratory-dependent. The velocity of the impacting shaft is controlled by varying the gas pressure. Other CCI devices have been recently developed, which utilize electromagnetic actuators to drive the impactor tip in an effort to provide greater control of the impact parameters.[4–6]

The CCI model also produces morphological and cerebrovascular injury responses that resemble certain aspects of human TBI. Commonly observed are graded histologic and axonal derangements,[1–3,7,8] as well as disruption of the blood brain barrier,[9,10] subdural and intraparenchymal hematoma, edema, inflammation, and alterations in CBF.[10] Similarly, the CCI model also produces neurobehavioral and cognitive impairments like those observed in human patients. In contrast to other TBI models, the CCI device produces a pronounced cortical contusion.

2 Materials

Adult male Sprague-Dawley rats (Harlan, Indianapolis, IN) weighing 300–325 g on the day of surgery are typically used in our experiments. The following surgical tools (listed below and shown in Fig. 2) are utilized when producing a CCI injury.

1. Laryngoscope handle and Miller stainless steel blade (Harvard Apparatus, BS4 59–6582 and BS4 59–6573)
2. Heart holding forceps, 6″ (Roboz, RS-5255)

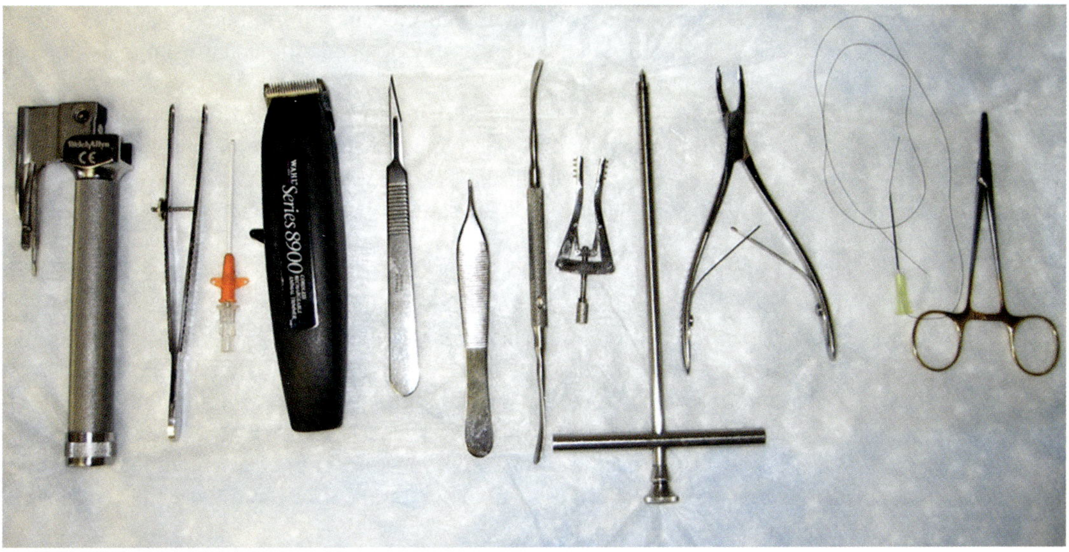

Fig. 2 Tools used during the production of a CCI injury in rats are listed from *left* to *right*: Laryngoscope handle and Miller stainless steel blade, 6 in. heart holding forceps, 14 G catheter for intubation, cordless small animal hair clipper, 4 in. scalpel handle and sterile surgical blade #11, 4 $^{3}/_{4}$ inch micro dissecting forceps, 7 in. periosteal elevator, microdissecting retractor, Michele trephine, 5 $^{1}/_{2}$ inch curved bone rongeurs, 20 G suture needle, silk black braided suture, and 4 $^{1}/_{2}$ inch Olsen-Hagar needle holder

3. Intubation tube, 14 G catheter (Baxter, 2N1110)
4. Cordless small animal hair clipper (Harvard Apparatus, NP 52–9719)
5. Scalpel handle, 4″ and sterile surgical blade 11 (Roboz, RS-9800 and RS-9843)
6. Microdissecting forceps, 1 × 2 teeth, 4 3/4″ (Roboz, RS-5238)
7. Periosteal elevator, 7″ (Roboz, RS-8820)
8. Microdissecting retractor, 4 × 4 prongs (Roboz, RS-6510)
9. Michele trephine (Miltex, 26–140)
10. Bone rongeurs, 5 $^{1}/_{2}$″ curved (Roboz, RS-8303)
11. Suture needle, B-D 20 G, 1 1/2″ (Becton Dickinson, 305176) and 4.0 silk black braided suture (Harvard Apparatus, BS4 51–7698)
12. Olsen-Hegar needle holder, 4 $^{1}/_{2}$″ (Roboz, RS-7880)

3 Procedures

1. Prior to surgery, test the CCI device to confirm that it is set to produce the desired impact velocity and dwell time.
2. Induce anesthesia with an inspired concentration of 4% isoflurane (IsoFlo®, Abbott Laboratories, North Chicago, IL) in 2:1 N_2O/O_2 in a vented anesthesia chamber for 60 s.
3. Using the heart holding forceps, grasp the tongue and insert the laryngoscope until the trachea is visible. Once visible, intubate with a 14 G catheter.
4. After endotracheal intubation, secure the rat in a stereotaxic frame and ventilate mechanically (Harvard Rodent Ventilator, Model 683, Harvard Apparatus, Inc., Holliston, MA) during surgery.

5. Maintain anesthesia with an inspired concentration of 2% isoflurane in 2:1 N_2O/O_2 mixture. Adjustments may be needed to ensure suppression of the pedal response to toe pinch.
6. Apply sterile ophthalmic ointment to each eye to compensate for the decrease in lacrimation during anesthesia.
7. Shave the scalp and scrub the skin with betadine.
8. Apply a sterile drape over the rat with an opening to expose the scalp.
9. Utilizing aseptic technique, make a 1 in. midline scalp incision.
10. Separate the muscle from skull with the microdissecting forceps and periosteal elevator.
11. Reflect the skin and fascia to expose the skull.
12. Make a craniectomy (6-mm in diameter) encompassing bregma and lambda and between the sagittal suture and the coronal ridge (Fig. 3) with the hand-held Michele trephine.
13. Remove the resulting bone flap and further enlarge the craniectomy with rongeurs until there is sufficient clearance for the impact tip, making sure not to disrupt the dura mater.
14. Extend the impacting shaft and center and lower the impact tip over the craniectomy until it just contacts the dura mater.
15. Zero the impact tip to the cortical surface.

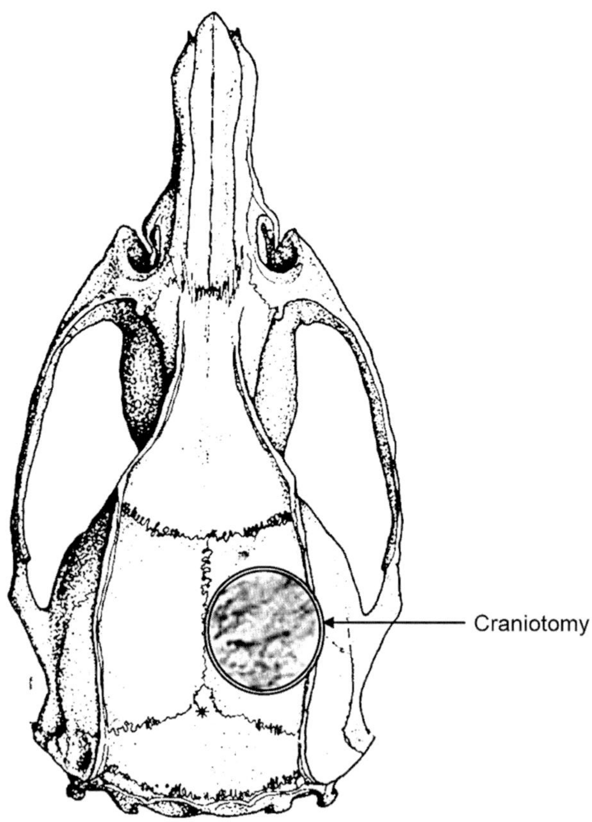

Fig. 3 Illustration of a rat skull showing the location of the craniotomy for a parasagittal impact

16. Retract the impacting shaft and advance the impact tip 2.8 mm to produce a brain injury of moderate severity (2.8 mm tissue deformation at 4 m s^{-1}).
17. Produce the cortical impact and discontinue the anesthesia. Note that the bone flap is not replaced after the CCI injury because it may produce a secondary injury by exerting pressure on the traumatized brain.
18. Suture the wound closed and apply lidocaine jelly to the incision site to minimize acute discomfort.
19. Remove the rat from the stereotaxic holder and extubate.
20. Perform acute neurological assessments (limb withdrawal reflex after a brief pinch to the footpad and righting reflex latency).
21. Place rat in a holding cage until the effects of anesthesia have abated (as evidenced by spontaneous movement) and then return to the colony.

4 Outcome Evaluations

CCI produces graded morphological and functional responses that can be used to monitor injury severity and evaluate therapies. Lesion volumes produced by CCI are calculated by measuring tissue loss in serial coronal sections stained by cresyl violet or hematoxylin and eosin. Hippocampal neuronal loss/survival is measured semiquantitatively by counting healthy-appearing neurons or by employing unbiased stereological methods.[11] A variety of tests are used to evaluate neurological and cognitive function after CCI. Gross vestibulomotor function is assessed in the rat using a beam-balance task.[12–14] Finer components of vestibulomotor function and coordination are assessed using a beam-walking task.[12–15] Neurological status can also be measured in mice using a "grip test", adapted from Hall.[16] Cognitive function in rats and mice after CCI has been primarily assessed using the Morris water maze test.[17–22] Conditioned fear response can also be used to assess cognitive function after TBI.[23]

5 Advantages, Complications, and Limitations

1. The CCI model can be scaled to other species by varying the impactor tip size. Mice,[24] rats,[3] ferrets,[1] pigs,[25] and primates[26] have employed 3, 10, 6, 15, and 15 mm diameter impact tips, respectively. The volume of tissue deformation is also related to the geometry (spherical vs. flat) of the impactor tip.
2. Accurate tissue deformations rely upon precise zeroing of the impact tip to the cortical surface prior to setting the impact depth. The zero point can be determined by visual inspection, force sensors, or using a continuity circuit. Variability between reported depths of deformation and injury severities between various labs may be due to differences in the zeroing method.
3. Although pathophysiology is not limited to the site of injury,[27] CCI is not a model of pure diffuse brain injury.
4. Studies suggest that the early post-CCI use of isoflurane, despite practical logistical issues, could potentially provide clinical benefits in TBI vs. other commonly used sedatives or analgesics.[28] Furthermore, the choice of postinjury sedation and analgesia could have important implications on attempts to translate novel therapies from bench to field or bedside.

5. The site of injury may influence the experimental response to CCI. CCI has been investigated using midline,[3] parasagittal,[29] and frontal sites of impact.[30]
6. In our opinion, the size of the craniotomy can influence the biomechanics of injury and should be kept as consistent as possible.

References

1. Lighthall JW. Controlled cortical impact: A new experimental brain injury model. J Neurotrauma 1988;5(1):1–15.
2. Lighthall JW, Goshgarian HG, Pinderski CR. Characterization of axonal injury produced by controlled cortical impact. J Neurotrauma 1990;7(2):65–76.
3. Dixon CE, Clifton GL, Lighthall LW, Yaghmai AA, Hayes RL. A controlled cortical impact model of traumatic brain injury in the rat. J Neurosci Meth 1991;39:253–262.
4. Bilgen M. A new device for experimental modeling of central nervous system injuries. Neurorehabil Neural Repair 2005;19(3):219–226.
5. Bayly PV, Dikranian KT, Black EE, Young C, Qin Y-Q, Labruyere J, Olney JW. Spatiotemporal evolution of apoptotic neurodegeneration following traumatic injury to the developing rat brain. Brain Res 2006;1107(1): 70–81.
6. Brody DL, Mac Donald C, Kessens CC, Yuede C, Parsadanian M, Spinner M, Kim E, Schwetye KE, Holtzman DM, Bayly PV. Electromagnetic controlled cortical impact device for precise, graded experimental traumatic brain injury. J Neurotrauma 2007;24(4):657–673.
7. Goodman JC, Cherian L, Bryan RM, Robertson CS. Lateral cortical impact injury in rats: Pathologic effects of varying cortical compression and impact velocity. J Neurotrauma 1994;11(5):587–597.
8. Meaney DF, Ross DT, Winkelstein BA, Brasko J, Goldstein D, Bilston LB, Thibault LE, Gennarelli TA. Modification of the cortical impact model to produce axonal injury in the rat cerebral cortex. J Neurotrauma 1994;11(5):599–612.
9. Dhillon HS, Donaldson D, Dempsey RJ, Prasad MR. Regional levels of free fatty acids and Evans blue extravasation after experimental brain injury. J Neurotrauma 1994;11(4):405–415.
10. Kochanek P, Marion DW, Zhang W, Schiding JK, White M, Palmer AM, Clark RS, O'Malley ME, Styren SD, Ho C, et al. Severe controlled cortical impact in rats: Assessment of cererbral edema, blood flow, and contusion volume. J Neurotrauma 1995;12(6):1015–1025.
11. Baldwin SA, Gibson T, Callihan CT, Sullivan PG, Palmer E, Scheff SW. Neuronal cell loss in the CA3 subfield of the hippocampus following cortical contusion utilizing the optical disector method for cell counting. J Neurotrauma 1997;14(6):385–398.
12. Kline AE, Massucci JL, Dixon CE, Zafonte RD, Bolinger BD. The therapeutic efficacy conferred by the 5-HT(1A) receptor agonist 8-Hydroxy-2-(di-*n*-propylamino) tetralin (8-OH-DPAT) after experimental traumatic brain injury is not mediated by concomitant hypothermia. J Neurotrauma 2004;21:175–185.
13. Kline AE, Wagner AK, Westergom BP, Malena RR, Zafonte RD, Olsen AS, Sozda CN, Luthra P, Panda M, Cheng JP, Aslam HA. Acute treatment with the 5-HT_{1A} receptor agonist 8-OH-DPAT and chronic environmental enrichment confer neurobehavioral benefit after experimental brain trauma. Behav Brain Res 2007;177:186–194.
14. Dixon CE, Ma X, Kline AE, Yan HQ, Ferimer H, Kochanek PM, Wisniewski SR, Jenkins LW, Marion DW. Acute etomidate treatment reduces cognitive deficits and histopathology in traumatic brain injured rats. Crit Care Med 2003;31(8):2222–2227.
15. Feeney DM, Gonzalez A, Law WA. Amphetamine, haloperidol, and experience interact to affect rate of recovery after motor cortex injury. Science 1982;217: 855–857.

16. Hall ED. High-dose glucocorticoid treatment improves neurological recovery in head-injured mice. J Neurosurgery 1985;62: 882–887.
17. Morris RG, Garrud P, Rawlins JN, O'Keefe J. Place navigation impaired in rats with hippocampal lesions. Nature 1982;297:681–683.
18. Smith DH, Okiyama K, Thomas MJ, Claussen B, McIntosh TK. Evaluation of memory dysfunction following experimental brain injury using the Morris water maze. J Neurotrauma 1991;8(4):259–269.
19. Hamm RJ, Dixon CE, Gbadebo DM, Singha AK, Lyeth BG, Jenkins LW, Hayes RL. Cognitive deficits following traumatic brain injury produced by controlled cortical impact. J Neurotrauma 1992;9(1):11–20.
20. Kline AE, Bolinger B, Kochanek PM, Carlos TM, Yan HQ, Jenkins LW, Marion DW, Dixon CE. Systemic administration of interleukin-10 suppresses the beneficial effects of moderate hypothermia following traumatic brain injury in rats. Brain Res 2002;937:22–31.
21. Kline AE, Massucci JL, Ma X, Zafonte RD, Dixon CE. Bromocriptine reduces lipid peroxidation and enhances spatial learning and hippocampal neuron survival in a rodent model of focal brain trauma. J Neurotrauma 2004;21:1712–1722.
22. Kline AE, Massucci JL, Zafonte RD, Dixon CE, DeFeo JR, Rogers EH. Differential effects of single versus multiple administrations of haloperidol and risperidone on functional outcome after experimental brain trauma. Crit Care Med 2007;35:919–924.
23. Dash PK, Mach SA, Blum S, Moore AN. Intrahippocampal wortmannin infusion enhances long-term spatial and contextual memories. Learn Mem 2002;9(4):167–177.
24. Smith DH, Soares HD, Pierce JS, Perlman KG, Saatman KE, Meaney DF, Dixon CE, McIntosh TK. A model of parasagittal controlled cortical impact in the mouse: cognitive and histopathologic effects. J Neurotrauma 1995;12(2):169–178.
25. Manley GT, Rosenthal G, Lam M, Morabito D, Yan D, Derugin N, Bollen A, Knudson MM, Panter SS. Controlled cortical impact in swine: Pathophysiology and biomechanics. J Neurotrauma 2006;23(2):128–139.
26. Nemoto EM, Rao G, Robinson T, Saunders T, Kirkman J, Davis D, Kuwabara H, Dixon CE. Effect of local cooling (15 degrees C for 24 hours) with the Chillerpad after traumatic brain injury in the nonhuman primate. Adv Exp Med Biol 2006;578: 311–315.
27. Thompson SN, Gibson TR, Thompson BM, Deng Y, Hall ED. Relationship of calpain-mediated proteolysis to the expression of axonal and synaptic plasticity markers following traumatic brain injury in mice. Exp Neurol 2006;201(1):253–265.
28. Statler KD, Alexander HL, Vagni V, Dixon CE, Clark RSB, Jenkins L, Kochanek PM. Comparison of seven anesthetics on outcome after traumatic brain injury by controlled cortical impact in adult, male rats. J Neurotrauma 2006;23(1):97–108.
29. Dixon CE. The application of rodent models to the study of brain injury biomechanics. In: Hoerner EF, ed. Head and Neck Injuries in Sports, ASTM STP 1229. Philadelphia: American Society for Testing and Materials, 1994:154–167.
30. Hoffman SW, Fülöp Z, Stein DG. Bilateral frontal cortical contusion in rats: Behavioral and anatomic consequences. J Neurotrauma 1994;11(4):417–431.

Chapter 34

Marmarou Weight Drop Injury Model

Christina R. Marmarou, Ruth Prieto, Keisuke Taya, Harold F. Young, and Anthony Marmarou

Abstract The impact acceleration model of diffuse traumatic brain injury (DTBI), commonly referred to as the "Marmarou" weight drop model is widely utilized to replicate DTBI, without focal lesion to characterize changes that closely parallel abnormalities characteristic of human DTBI cause by motor vehicle accidents or falls. In this chapter, we describe in detail procedures used to replicate DTBI of varying degrees of severity; the clinically relevant condition of posttraumatic hypoxia and hypotension; and experimental paradigms utilized for intraoperative multiparameter, physiological monitoring. We provide brief summaries of reproducible outcome evaluations characteristic of this model inclusive of (1) intraoperative physiological responses including mean arterial blood pressure (MABP), pH, pO_2, and pCO_2, and intra-cranial pressure (ICP); (2) gross and neurohistopathological outcomes; (3) functional studies that highlight somatomotor and cognitive behavioral changes; (4) in vitro outcome evaluations employing magnetic resonance imaging (MRI), ^{1}H Magnetic resonance spectroscopy (^{1}H MRS); and (5) high-performance liquid chromatography (HPLC) for the analysis of changes in cellular function, brain edema and cerebral metabolites including *N*-acetyl-aspartate (NAA) and adenosine tri-phosphate. Finally, we describe the complications associated with the use of this model, mainly technical in nature, easily overcome with repetitious practice and state limitations of this model, notably the use of multiparameter intraoperative monitoring techniques that can be costprohibitive due to the expense of the required equipment. We conclude by stating the major advantage of this model; it has been used extensively in the laboratory to evaluate the effect of pharmacological intervention to treat DTBI. Importantly, these treatments have recently reached Phase II & III clinical trial status.

Keywords: Diffuse traumatic brain injury (DTBI), Neuronal damage, Traumatic axonal damage, Rat, Secondary injury, Neurobehavioral deficits, Cellular edema, Intracranial pressure (ICP)

From: *Animal Models of Acute Neurological Injuries.*
Edited by: J. Chen (eds.) New York, Totowa, NJ

1 Model Selection

The impact-acceleration model of diffuse traumatic brain injury (DTBI)[1,2] has been widely used to investigate the pathobiological mechanisms that are similar to features of the human condition following closed head "impact acceleration" type of DTBI. In humans, these injuries occur most frequently during motor vehicle accidents (MVA) and falls[3,4] and are problematic as the pathology of injury is diffusely scattered throughout the cerebrum and brain stem,[3] and patients often present with normal computer tomography (CT) scans prohibiting diagnosis and acute, preventative intervention.[4] This injury model generates reproducible neuronal and axonal pathology[2,5–7] concomitant with blood–brain barrier (BBB) disruption[2,6,8,9] within discrete loci throughout the cerebrum and brain stem. Similar to that described in man,[3,4] this model can be utilized to reproduce graded DTBI injury; mild, moderate, or severe DTBI, by altering the height at which the weights are allowed to accelerate (1–2.1 m) prior to closed skull impact, or implementing secondary insults of hypoxia and hypotension (10 or 30 m) immediately after TBI.[2,5,6,8–11] Economical and technically simple the Marmarou weight drop model, developed in 1991, has been instrumental in characterizing the neurophysiological, histopathological, functional, imaging and biochemical changes that closely parallel human DTBI.[1–12] Taken in concert, studies replicating these experimental procedures are critically important when implementing drug delivery to measure the efficacy of pharmaceutical compounds important for translational research targeting pharmacological intervention for the treatment of DTBI in humans.[12,13]

2 Materials

2.1 Animals

Adult male Sprague Dawley (SD) rats weighing 350–400 g are the preferred species, and weight range utilized for replication of consistent outcome measures reported previously using this injury model.

2.2 Anesthesia

Isoflurane (Henry Schein, Melville, NY) is the anesthetic of choice used during surgical procedures for replication of consistent outcome measures reported previously using this injury model.

2.3 Equipment

2.3.1 Anesthesia Equipment

Narkomed, North American Drager ventilator (Louisville, KY); Drager 19.1 isoflurane vaporizer (Louisville, KY); Induction chamber – VAC-RITE 1,500 cc, (Becton Dickinson, Franklin Lakes, NY), rodent intubation stand (Harvard Apparatus, Holliston, MA), rodent laryngoscope with fiber optic bulb (Henry Schein, Melville, NY), intubation tube – 9 cm precut length of PE-205 tubing (VWR, Bridgeport, NJ) intratracheal end cut at a 45° angle, 20 cm precut length of 3–0 black monofilament nylon spool suture (Henry Schein, Mellville, NY) used to secure intratracheal intubation tube, cone mask – Envelope the distal end of 6 cm in diameter ventilator hub (Narkomed,

North American Drager, Louisville, KY) using 4 cm^2 piece of latex (cut from a latex surgical glove), secure with surgical tape and make a 1 cm slit in the center of latex.

2.3.2 Surgery Equipment

2.3.2.1 Surgical Instruments: Forceps (1), spring scissors with fine, small blades (1), hemostats (3), tissue forceps (1) fine tipped forceps (2), fine-tipped scissors (1), needle holders (1), hemostats, curved end (1), dissector (1), scalpel handle (1) and scalpel blades (2), micro rongeur (3) hand-held pen vice and preset 1.5 mm drill bit.

2.3.2.2 General Surgical Supplies: Surgical absorbent under-pads, latex surgical gloves, 20 gauge (G) needles (3), 1 cc syringes (10), 5 cc syringes (5), 3–0 black monofilament nylon spool suture (Ethilon), 3–0 black monofilament sterile sutures with needle (2), all-purpose sponges (Kendall), cotton-tipped applicators (VWR, Bridgeport, NJ).

2.3.2.3 Intraoperative Surgical Supplies: PE-50 tubing (VWR, Bridgeport, NJ), ophthalmic ointment, antiseptic (Betadine), antibacterial (Bactitracin), topical anesthetic (lidocaine), intradermal injectable anesthetic (Marcaine), (Henry Schein, Melville, NY), 0.2% heparin in 0.9% sterile normal saline solution, 0.9% sterile normal saline.

2.3.2.4 Intraoperative Surgical Equipment: Surgical dissecting microscope, rodent temperature control system (Harvard Apparatus, Holliston, MA), blood gas machine (ABL700 series, Radiometer, Copenhagen, Denmark), rat stereotactic frame with stereotactic arm (Knopf, Instruments, Tajunga, CA) Codman ICP probe and monitor box (Codman, Inc, New Bedford, MA.), Power Lab on-line physiological monitoring system and pressure transducers (AD Instruments, Inc. Mid Atlantic, US), Dell computer, carpenters level and cyano-acrylic (crazy glue) (Ace Hardware, Richmond, VA).

2.3.2.5 Trauma Equipment for Induction of DTBI: The Marmarou weight drop model (Fig. 1) apparatus was developed by Dr. Anthony Marmarou and colleagues in 1991 at the Medical College of Virginia, Richmond, VA.[1,2] The devices can be ordered by contacting the corresponding author. The models are priced modestly, based only on the cost of manufacturing, labor and shipping and include the metal stand (1 metal base with wheels, 1 metal rod, and 3 clamps); plexi-glass tube (1 per unit); sectioned 500 g brass weight set; rectangular, plexi-glass bed,[1] 4–0 spring constant foam (Foam to Size, Ashland, VA) inserts (10 per unit); stainless steel discs (helmets), 10 mm in diameter and 2 mm in depth (10 per unit). Several laboratories have constructed their own version of this simple weight drop model, and the results are relatively consistent with the manufactured model.

3 Procedures

3.1 Anesthesia

Preferentially, anesthesia is delivered using a Narkomed, North American Drager Ventilator, initially set to deliver 200–250 ml gas volume, at 15 cycles per minute, using a gas/air mixture of N_2O (70%) and O_2 (30%) (Fig. 2A). The

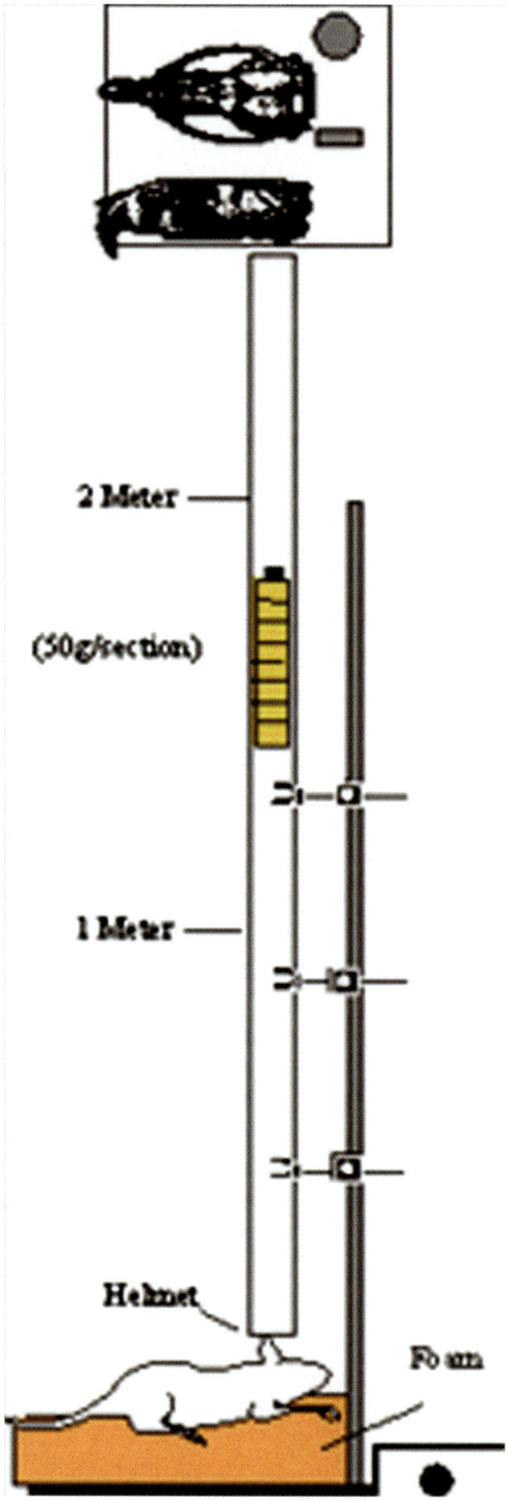

Fig. 1 The Marmarou weight drop model. The anesthetized rat is disconnected from the respirator and placed prone on the foam bed under the hollow Plexiglass tube. Trauma is induced by dropping the cylindrical column (50 g per section) of segmented brass weighing (450 g) through the Plexiglas tube from a distance of 1–2 m onto the metallic disc (helmet) fixed to the skull vault of the animal (*top figure*)

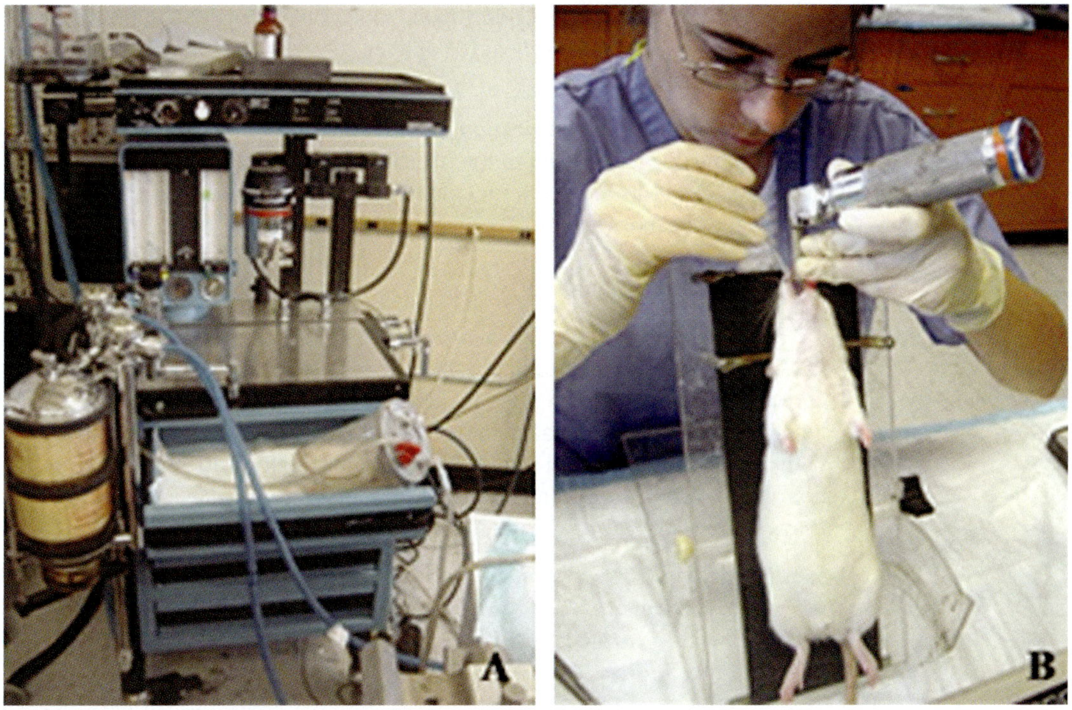

Fig. 2 Step-by-step procedures utilized for induction of anesthesia and endo-tracheal intubation. Rats are initially induced with isoflurane (4.0%) and a mixture of N_2O (70%) and O_2 (30%) for a period of 4 min in a ventilated anesthesia induction chamber (**A**). The rat is then positioned on the rodent intubation stand by hanging in a supine position via the front incisor teeth, inserting the laryngoscope into the oral cavity and inserting the intubation tube into the trachea (**B**)

percent of isoflurane delivered is adjusted using the "dial" on the Drager 19.1 Isoflurane Vaporizer from 0.5% to 4.0%, described in detail, below (Fig. 2a).

3.1.1 Induction of Anesthesia

Rats are initially induced with isoflurane (4.0%) and a mixture of N_2O (70%) and O_2 (30%) for a period of 4 min in a ventilated anesthesia induction chamber (Fig. 2a, rat shown in induction chamber).

3.1.2 Maintenance of Anesthesia

Anesthesia is maintained with isoflurane, 1.5–2.0% during the preinjury surgical procedures and then at 0.5–1.5% during the postinjury surgical procedure in a mixture of N_2O (70%) and O_2 (30%).

3.1.3 Delivery of Anesthesia

Depending on experimental design, anesthesia is delivered throughout the surgical period via an endo-tracheal intubation tube where animals are mechanically ventilated (highly recommended, Fig. 2b) or a cone-mask delivery system where animals are allowed to breathe spontaneously (procedures described in detail in the next two sections).

3.1.3.1 Mechanical Ventilation via Intratracheal Intubation: (a) Position the rat on the rodent intubation stand (Harvard Apparatus, Holliston, MA) (Fig. 2b) by hanging in a supine position via the front incisor teeth; (b) insert the laryngoscope into the oral cavity, and visualize the periodicity of the opening and closing of the vocal cords (c) during an "open phase" insert 6 mm of the 9 mm intubation tube into the trachea; (d) deliver anesthesia as previously described in Sect. 3.1.2.(Fig. 2b)

3.1.3.2 Spontaneous Respiration, via Cone-Mask Delivery of Anesthesia: (a) Insert rat muzzle to the level of the fibrisia into the precut slit in the cone-mask; (b) allow the rat to breathe spontaneously throughout the surgical period; (c) deliver anesthesia as previously described in Sect. 3.1.2.

3.2 Surgical Preparation

3.2.1 General Surgical Preparation

(a) Apply ophthalmic ointment to the eyes to prevent eyes desiccation; (b) prepare two surgical incision sites for cranial impact and cannulation of the left femoral artery and vein; shave the skin over the cranial vault and swab the area with an antiseptic (bentadine) and topical anesthetic (lidocaine), then reposition the rat from the supine to the prone position and shave the left femoral triangle areas and swab with bentadine and lidocaine.

3.2.2 Surgical Preparation for Intraoperative, Multiparameter Physiological Monitoring: Temperature, Mean Arterial Blood Pressure (MABP), Blood Gas Analysis (pH, pCO_2, pO_2) & ICP (Figs. 3 and 4).

3.2.2.1 Temperature: To maintain normal core body temperature at 37°C throughout the experimental procedures, (a) place the anesthetized rat in the supine position, (b) insert the rectal probe, (c) connect to thermostat that automatically controls the heating pad (Harvard Apparatus, Holliston, MA) (Figs. 3a and 4a).

3.2.2.2 MABP and Blood Gas Analysis: To monitor and maintain MABP, pH, pCO_2, pO_2 it is necessary to cannulate the femoral vessels (Fig. 3a, b).

3.2.2.2.1 Prepare Two Cannulae Lines for Insertion into Femoral Artery and Vein: (a) Firmly grasp with needle forceps the distal end of a 20 G needle and remove 1/3 of the distal pointed tip; (b) insert the blunted needle into a 25 cm length of PE50 tubing (VWR, Bridgeport, NJ); (c) attach the distal, hub of the 20 G needle to a 3 way stop-cock attached to a preloaded 5 cc syringe containing 0.2% heparinized; (d) to prepare for easy insertion into the vessel cut the opposing free end of the cannulae line at a 45° angle.

3.2.2.2.2 Isolate the Femoral Artery and Vein (Fig. 3a, b): (a) Pre-cut six 10 cm lengths of 3–0 surgical spool thread, (b) using tissue forceps and fine-tipped surgical scissors make a 4 cm incision, from medial to lateral (45° angle), roughly tracing the course of the femoral artery and vein. (c) With the aid of a dissection microscope, breach the fascia, separate the femoral muscles, and expose the femoral sheath that encapsulates the femoral vein, artery, and nerve (VAN) anatomically positioned from medial to lateral, (d) using two fine-tipped dissection forceps; firmly grasp the thin, transparent, femoral

Fig. 3 (continued) Note in the enlargement of the boxed area in (**A**), the anatomical positioning of the larger femoral artery and the reestablishment of blood flow in relation to the cannulated femoral vein, located medial (**B**)

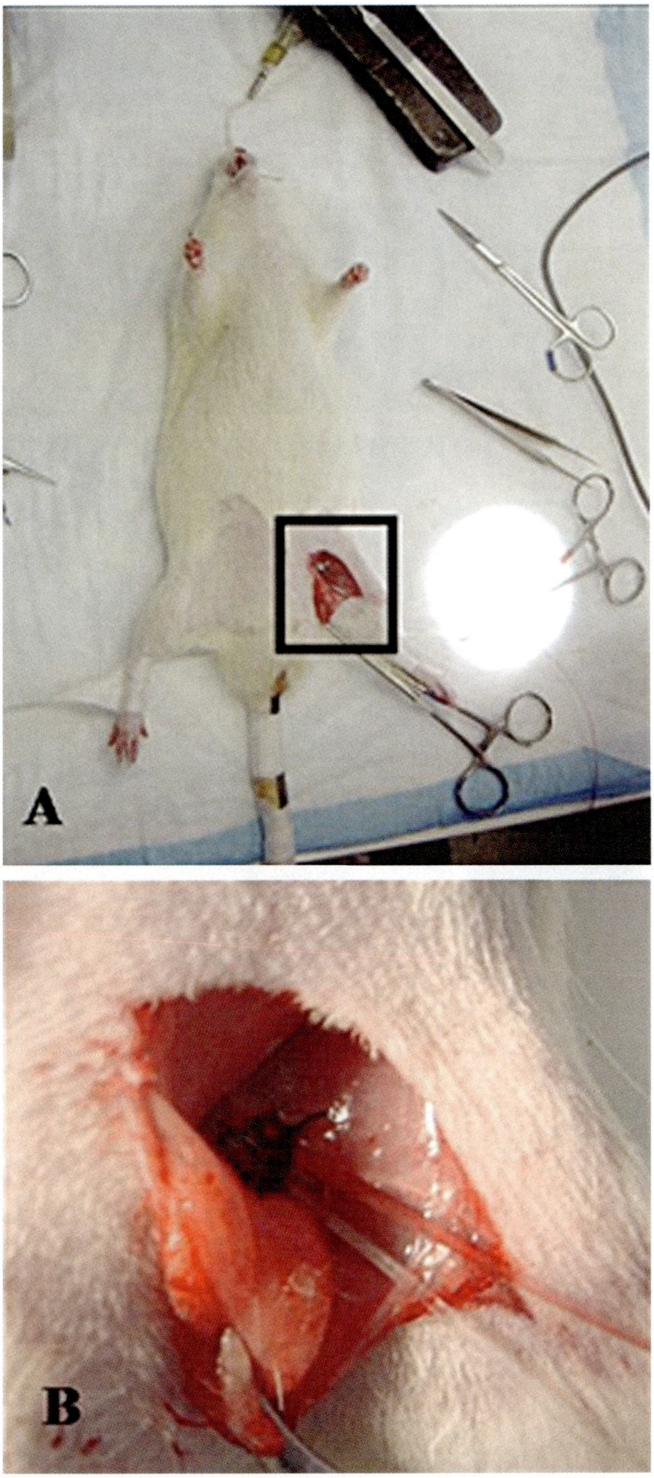

Fig. 3 Step-by-step procedures utilized for surgical preparation for intraoperative physiological monitoring of temperature and MABP. The anesthetized rat is placed in the supine position and normothermia is maintained via a rectal probe. MABP and venous fluids are monitored and delivered by cannulation of the artery and vein (**A**).

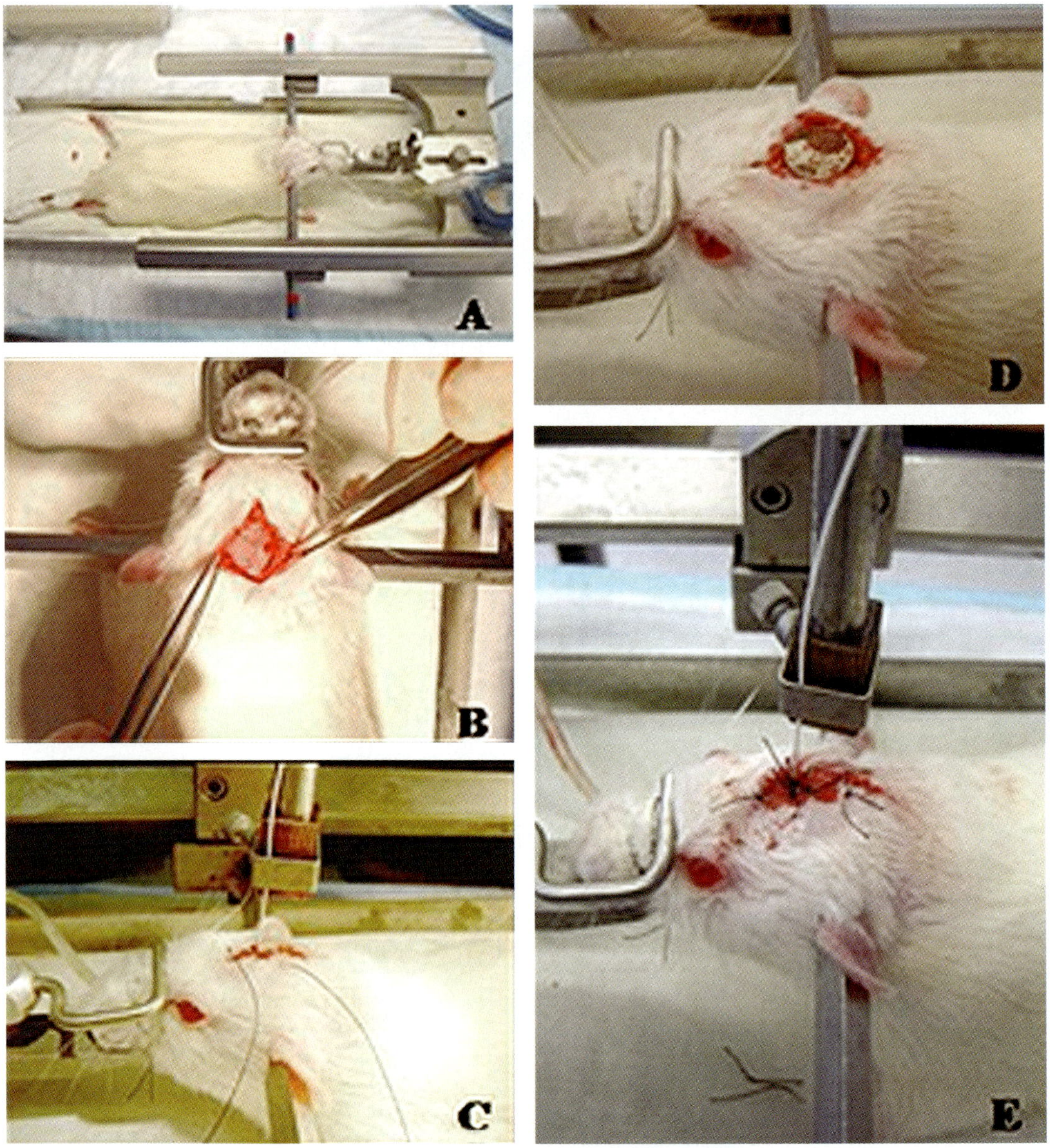

Fig. 4 Step-by-step procedures utilized for intraoperative physiological monitoring of intracranial pressure before and following DTBI. The anesthetized rat is positioned on stereotactic frame and secured with ear-bars inserted bilaterally into right and left ear canals (**A**). A 20 mm midline scalp incision is used to expose the skull, remove the periosteom and a burr-hole is drilled 3 mm lateral and 2 mm caudal to bregma (**B**). The ICP fiber-optic probe is inserted using stereo-tactic placement through the burr-hole (**C**. To prepare for DTBI, the metallic disc is mounted on the parietal bone midline between bregma and lambda, straddling the sagittal suture (**D**). Following DTBI, the ICP fiber-optic probe is repositioned using the same sterotactic co-ordinates (**E**)

sheath at two adjacent points at midline and in a continuous motion pull in medial and lateral directions, stripping the femoral sheath and exposing the femoral vessels; (e) isolate the larger femoral artery (located lateral and pink in color) from the smaller femoral vein (located medial and blue in color); (f) remove blood with surgical Q-tips avoiding direct contact with the vessels to avoid potential mechanically induced vasospasm, (g) using a pair of adson forceps, loop 3 of the pre-cut sutures, around the isolated artery at medial and lateral positions, approximately 10 mm apart, then place 1 suture at the midline position; (h) repeat step (g) using the isolated vein; (i) loosely ties all sutures, maintaining ligatures 2 mm from the surface of the vessels.

3.2.2.2 Cannulate the Femoral Vessels: Cannulate the artery first (a) clamp hemostats to the two free ends of the medial placed ligatures (the pre-cut sutures, now "looped" around the vessel and loosely tied); (b) clamp hemostats to the two free ends of the laterally placed ligatures; (c) position the hemostats so that the ligatures looped around the vessels are pulled, gently, just until the point that they occlude blood flow to the vessel; (d) gently remove all blood from the surgical field using cotton-tipped swabs (e) at the midline position, 2 mm distal to the centrally located ligature use fine-tipped forceps to firmly grasp the superior wall of the vessel (f) using spring scissors with especially fine blades make a "V" shaped incision in the anterior wall of the vessel taking care to not bisect the artery; (g) insert 5 mm of the angled end of the pre-prepared cannulae into the intra-luminal area of the vessel; (h) tightly secure the centrally placed median ligature by pulling into a knot, encircling the catheter and extra-luminal surface of the vessel; (i) release the tension on the most medially placed ligature by releasing the free ends of the suture from the hemostat and note the reestablishment of blood flow; (j) secure the ligature by pulling into a knot, encircling the catheter and extra-luminal surface of the vessel (k) "tie off" the most laterally placed suture; (l) flush the line with 0.2 cc 2% heparanized saline periodically to prevent coagulation or 2 cc per hour to prevent dehydration. Repeat steps (a) through (l) to cannulate the femoral vein (Fig. 3a, c).

3.2.2.3 Intracranial Pressure (ICP): For real-time assessment of ICP, (a) position rat on stereotactic frame and secure with ear-bars inserted bilaterally into right and left ear canals; (b) make a 20 mm midline scalp incision using a scalpel; (c) expose the skull by using two cotton-tipped applicators to gently push the periosteom of the skull to the most lateral edges of the incision; (d) apply gentle pressure to absorb bleeding and dry exposed skull; (e) make one small 1.5 mm burr hole using a hand-held pen vice and pre-set 1.5 mm drill bit through the skull bone over the right frontal lobe, 3 mm lateral and 2 mm caudal to bregma (f) using stereo-tactic placement insert the 1.2 mm diameter Codman fiber-optic probe, 3 mm deep into brain tissue; (g) attach distal end of probe to Codman ICP monitor attached to a data acquisition system.

3.3 Intraoperative, Multiparameter Monitoring

MABP and ICP are monitored and recorded continuously using a data acquisition system (ADInstruments, Colorado Springs, CO). For blood gas assessment withdraw a total of four to six, 0.2 cc blood gas samples are utilized to monitor pH, pO_2 and pCO_2 (15 min intervals) and deliver replacement fluids (0.9% sterile NaCl, 5 cc per hour).

3.4 Surgical Preparation and Induction of Moderate DTBI

After stabilization of blood gases, MABP and ICP, (a) reopen the 20 mm midline scalp incision using a scalpel; (b) expose the skull by using two cotton-tipped applicators to gently push the periosteom of the skull to the most lateral edges of the incision; (c) apply gentle pressure to absorb bleeding and dry exposed skull; (c) coat the scored side of the stainless steel disk with cyanoacrylic glue (crazy glue) and mount on the parietal bone midline between bregma and lambda, straddling the sagittal suture; (d) allow glue to dry for 3 min. (e) disconnect rat from the respirator and place prone on the foam bed under the hollow Plexiglass tube; (f) secure rodent by strapping surgical tape over the dorsal surface attached to either side of the Plexiglas bed (g) induced by trauma by dropping the cylindrical column of segmented brass weighing (450 g) through the Plexiglas tube from a distance of 2 m onto the disc fixed to the skull vault of the animal (h) rapidly reconnect to anesthesia with mechanical ventilation; (i) remove the metallic helmet; (j) check for skull fracture; (k) reinsert ICP catheter as described earlier and suture incision using sterile 3–0 suture with needle.

3.4.1 Induction of Mild, Moderate, and Severe DTBI and/or Induction of Secondary Insults of Hypoxia and Hypotension

The biomechanics of this model allow for the reproduction of graded severities of DTBI, reproduced by simply altering the height at which the weights are dropped (1–2.1 m), and changing the amount of weight that is dropped (500 or 450 g). To produce mild injuries, allow 450 g of weight to fall from 1 m; for moderate injury, allow 450 g of weight to fall 2 m; for severe injury, allow 500 g of weight to fall 2.1 m.[10] Secondary insults of hypoxia and hypotension are induced by manipulation of anesthesia levels and anesthesia delivery, immediately following DTBI, and maintained depending on experimental design for a period of 10 or 30 min post-injury. (a) To induce hypoxia, reduce inspired oxygen to 12% of pretrauma values; (b) to induce hypotension, increase isoflurane level to 2% above baseline while increasing the inspired nitrous oxide to 88%; (c) observe the gradual reduction in MABP to 30–40 mmHg; (d) maintain for the duration of the secondary insult (10–30 m).

4 Outcome Evaluations

4.1 General Physiological Observations

We recommended that for each experimental series, two sham-operated rats be used to calibrate anesthesia equipment and therefore establish anesthesia delivery settings effective at maintaining the rat within normal physiological range (MABP, pH, paO_2, and $PaCO_2$) throughout the surgical period. Normothermia should be maintained at 37°C. MABP should be maintained at 100 ± 10 mmHg. Blood gas analysis should be applied during the surgical period at 15 min intervals for the first 1 h and then 1 time for each additional 30-min period. At these time points, rats sustaining DTBI will show no alteration of changes in MABP, pH, pO_2, pCO_2, or body temperature either at 30 and 15 min preinjury or 15 and 30 min post-TBI when compared with sham-operated controls.[1,2,5–13] Following DTBI, a classic hypertensive surge in MABP (150 ± 10 mmHg) is followed by a brief, transitory period of hypotension (60± mmHg). MABP returns to baseline, pre-DTBI within 5–10 min post insult.[1,2,6] Secondary insults of hypoxia and hypotension sustained for either 10 or 30 min employed following DTBI will result in hypotension measured by a drop in MABP to 40 ± 5 mmHg and a reduction in pO_2 (determined by blood gas analysis) to 47 ± 5 mmHg.[5,6] Normal preinjury ICP valued range between 5 and 10 mmHg. DTBI will result in a rise in ICP to 28 ± 3 mmHg at 30 min postinjury followed by a gradual decline, but elevated, above baseline, measured over 4 h.[6] DTBI with secondary insults of hypoxia and hypotension result in an increased and enduring rise in ICP (68 ± 27 mmHg) at 4 h postinjuries.[6] Seizure is often positively correlated to the severity of insult. Following a 1 m drop, no seizures are observed. However, following a 2 m drop, seizure occurs in roughly 50% of animals.[1,2]

4.2 Gross Pathological Changes

To evaluate gross pathological changes, perfusion fixed brain should be examined for subarachnoid, intraventricular, and brain stem petechial hemorrhage, focal lesion, ventricular dilation and subcistern congestion. Typically in this diffuse model, there is no focal lesion. Further, only in the severe injury paradigm does subarachnoid, intraventricular, and brain stem petechial hemorrhage occur. Ventricular dilation is not significantly different than sham until 24 h post-TBI.[2]

4.3 Neurohistopathological Changes

The response of neurons, axons, astrocytes, and small blood vessel to this injury model is remarkably rich. Cortical and hippocampal necrosis and apoptosis; diffuse traumatic axonal injury particularly in the corpus callosum and white matter tracts of the brain stem; and BBB dysfunction accompanied by edema are hallmark features of the resulting histopathological change commonly used for endpoints to validate the variations of this model, either mild, moderate, severe, or with secondary insults of hypoxia and hypotension.[2,5,7,12] Following transcardial fixation, we recommend coronal sectioning of the cerebral cortex or sagittal sectioning of the brain stem using a vibratome to produce 40 μm sections or paraffin embedding and a microtome to produce 10 μm sections. Successful outcome evaluations using routine H & E staining, and/or multiple-label immuohistochemical (IHC) protocols, followed by analysis using light, electron or confocal microscopy have demonstrated a significant and enduring number of damaged neurons in supraventricular cortex and hippocampus[2,5,7,12]; hypertrophic and hyperplasic response of astrocytes; microvascular changes, consistent with BBB dysfunction and edema and diffuse axonal injury including the sequential stages and types of reactive axonal change consistent with an aggressive development of white matter degeneration.[2,7,12]

4.4 Behavior Changes

This model has been utilized to successfully demonstrate that much like the human condition of DTBI; injury severity predicts motor and cognitive performance.[10] Using a battery of well-validated tests including acute neurological testing, somatomotor evaluations including the beam walk and beam balance test and cognitive assessment via the Morris Water Maze, this model is useful to delineate the effects of injury severity on motor and cognitive deficits, and can be used to determine the effect of treatment measured by these functional outcome measures.[10,11] To our knowledge, the Barnes Maze test has not yet been reported as an outcome evaluation, not the paw withdraw to pain paradigm. Both of these tests may also prove to be sensitive outcome measures.

4.5 Magnetic Resonance Imaging (MRI), Spectroscopy (^{1}H MRS) – MRI and ^{1}H MRS

The use of MRI, ^{1}H MRI – MRI, and ^{1}H MRS have opened a new era for in vivo evaluation of the cellular injury, brain edema, and metabolic derangement associated with DTBI. These techniques and endpoints have been utilized following experimental DTBI in rat resulting in translational studies currently underway in traumatically injured patients.[6,8,9,12]

4.5.1 T1-Weighted MRI

It is utilized to study BBB breakdown via the administration of Gadolinium-DTPA. Following DTBI, an increment in the signal intensity demonstrates that BBB opening occurs immediately after TBI, but is transitory and by 30 m postinjury is completely closed.[8] Signal intensity is increased in TBI with hypoxia and hypotension supporting an enduring BBB opening.[8]

4.5.2 T2-Weighted MRI

It is utilized to demonstrate the presence of brain edema (hyper-intense areas). Rats subjected to DTBI show symmetric hyper-intense areas confined in both hippocampal areas at 1 h post-injury and fade in the following 72 h.[4]

4.5.3 Diffusion Weighted Imaging (DWI)

Out come evaluations demonstrate that DTBI results in a significant but transient increase in ADC indicative of vasogenic edema (30–60 min postinjury), followed by a continuing decrease in ADC suggesting a predominately cellular edema, maintained over 7 days postinjury.[6,8,9]

4.5.4 ^{1}H Magnetic Resonance Spectroscopy (^{1}H MRS)

It is a powerful tool for measuring changes in cerebral metabolites. Novel data recently reported that following DTBI, the principal metabolites associated with the progression of brain edema are the organic osmolytes, taurine and myo-inositol, which were elevated between 24–48 h.[14]

5 Biochemical Analysis

5.1 High-Performance Liquid Chromatography (HPLC)

It is a method of in vitro analysis of metabolites. DTBI results in a reduction in *N*-acetyl-aspartate (NAA) and adenosine tri-phosphate (ATP). NAA and ATP levels recover following moderate DTBI, but are enduring following severe DTBI. These techniques are useful as they demonstrate that metabolic derangement including mitochondrial dysfunction and energetic impairment are predictors of the severity of DTBI.[13]

6 Complications, Advantages and Limitations

6.1 Complications

6.1.2 Skull Fracture

Skull fracture may present as a common complication and occur on the average of 1 in 20 rats. Because of the potential correlation with focal injury models, these animals are always excluded from the study. The incidence of skull fracture is reduced by using rats weighing between 350 and 400 g. As this is a "closed head" injury model of DTBI, the impact is delivered to the intact skull; therefore, the thickness of the skull is a major factor in replicating the biomechanics capable of producing graded DTBI. Rats weighing less than 350 g have a thinner skull and result in a higher incidence of skull fracture, death, and aberrant outcome measures (Sect. 4), uncharacteristic of previously reported, historical results. Alternatively, the incidence of skull fracture may be related to poor maintenance of the DTBI model, i.e., faulty equipment. To obviate these problems, prior to each injury, a carpenter's level should be used to insure the vertical angle of the plexiglass tube. The user should adjust the clamps and ensure an unrestricted drop of the weights. In addition, the foam should be replaced after the 20th traumatized rat. This will prevent the occurrence of degenerating foam that can alter the replication of the DTBI model.

6.2 Advantages

This model of DTBI replicates a clinical condition that there is presently no cure. The DTBI model in rat is well established. The species is large enough to allow multiparameter intraoperative, physiological monitoring that allows for the modeling of graded severities of human DTBI. We are currently in the process of developing a mouse model of DTBI with the promise of the potential scientific application of knock-out strains, the rat, however, has historically been used for this model, so novel users can compare historic results from previous studies. Finally, low mortality rates reduce the number of rats utilized and minimize cost.

6.3 Limitations

6.3.1 Technical Difficulties

For some laboratories, intratracheal intubation and cannulation of the femoral vessels represent difficult tasks to master. Notably, we emphasize that it is essential, to visualize the "open" phase of the vocal cords prior to attempting to insert the intubation tube. Also avoid insertion of the intubation tube "to deep" where the air flow and anesthesia delivery is impeded and rests against the carina of the trachea. Pulling back the intubation tube 2 mm will obviate this technical difficulty. The most commonly made error is to insert the intubation tube within the posterior positioned esophagus. A correctly positioned intubation tube will inflate the lungs, thus expanding the thoracic cavity. An ill placed intubation tube within the esophagus will result in an obvious expansion of the abdominal cavity. In this scenario, simply withdraw the intubation tube and reposition correctly. Also, as the intubation tube often becomes filled with secretions generated by the lungs, we recommend suction and removal of these fluids applied by inserting an aspiration device into the intubation tube, easily made by placing a 20 G blunt end needle on the end of an empty 5 cc syringe and aspirating fluid by dry withdrawal. Further, as the lung tissue becomes more compliant with delivery of anesthesia and artificial ventilation, a reduction in air delivery volume by 20–50 ml may reduce the amount of edema in the lungs. Finally, although cone-mask anesthesia may present as a simple solution to intubation, the mortality rate is near 50%,[1,2] and the majority of historical studies have utilized intratracheal intubation and mechanical ventilation.

The more difficult task, to some new investigators is the successful cannulation of the femoral artery and vein for the intraoperative assessment of physiological parameters (MABP and blood gas analysis, i.e., pH, pCO_2, and pO_2). In our laboratory, this is an essential step to insure the reliability and reproducibility of the model. Cannulation also allows for physiological assessment of the effect of various pharmacological treatment compounds, the essence of translation research using TBI animal models. Cannulation, therefore, represents a routine part of the surgical preparation. To master this surgical skill, we suggest utilizing both the right and left femoral artery and vein to practice and therefore perfect this vitally important surgical technique.

6.3.2 Mortality

Using adult male rats, weight is inversely correlated with mortality. Using rats weighing 350–400 g, the mortality rate is approximately 10% in the artificially ventilated, anesthetized rats and 50% in the spontaneously breathing, anesthetized rats. This percentage of rats will die within 10 min following severe (2 m, 450 g) TBI. The mortality rate for less severe injuries (1 m) is negligible. Brain stem injury due to traumatic axonal injury is purportedly the major complication that contributes to animal death following TBI.[2,7,12] Death after resuscitation may occur in few instances if the animal is not hydrated by daily intraperitoneal injections of saline or provided with softened rat chow. Perforia and weight loss commonly occur days to weeks following injury, and in severe cases humane sacrifice is recommended.

6.3.3 Variation

The reproducible outcomes of the experimental paradigms of this model are optimal if anesthesia, ventilator support, fluid replacement, and postoperative care maintain the animal within normal physiological homeostasis throughout the surgical period and are maintained to the time of sacrifice. Our laboratory has provided the opportunity for new investigators to purchase the impact acceleration weight drop device, at cost value. In this case, reproducibility is assured. In some cases, others have replicated the model with good results. In few instances, however, the biomechanics have not been assured, leading to slight variability but comparable data sets. To avoid these confounds, we have emphasized the used of intraoperative, multiparameter monitoring to ensure the reproducibility of consistent results and obviate experimental variables that may contribute to inconsistency in the various injury paradigms characteristic of severe, moderate, or mild TBI or TBI coupled with hypoxia and hypotension.

Acknowledgments The authors recognize Ms. Lori Davis and Ms. Joy Galloway for help with the preparation of this document along with Mr. Charlie Kim and Mr. Travis Faber for technical assistance with photography. Special thanks to Dr. Andrew Beaumont, M.D, PhD for editorial input and scientific evaluation.

References

1. Marmarou A, Foda MA, Van Den Brink W, Campbell J, Kita H, Demetriadou K. A new model of diffuse brain injury in rats Par I: Pathophysiology and biomechanics. J Neurosurg 1994; 80: 291–300
2. Foda MA, Marmarou A. A new model of diffuse brain injury in rats Part II: Morphological characterization. J Neurosurg 1994; 80: 301–13.

3. Adams JH, Doyley D, Ford I, Gennarelli TA, Graham DI and McLellan DR: Diffuse axonal injury in head injury: definition, diagnosis and grading. Histopathology 1989; 15: 49–59
4. Marshall LF, Gautille T, Klauber MR, Eisenberg HM, Jane J, Luessen TG, Marmarou A, Foulkes MA. The outcome of sever closed head injury. J Neurosurg 1991; 71: S28–S26.
5. Yamamoto M, Marmarou CR, Stiefel MF, Beaumont A, Marmarou A. Neuroprotective effect of hypothermia on neuronal injury in diffuse traumatic brain injury coupled with hypoxia and hypotension. J Neurotrauma 1999;16: 487–500.
6. Ito J, Marmarou A, Barzo P, Fatouros P, Corwin F. Characterization of edema by diffusion-weighted imaging in experimental traumatic brain injury. J Neurosurgery 1999; 84: 97–103
7. Marmarou CR, Walker SA, Davis CL, Povlishock JT. Quantitative analysis of the relationship between intra-axonal neurofilament compaction and impaired axonal transport following diffuse traumatic brain injury. J Neurotrauma 2005; 22 : 1066–80.
8. Barzo P, Marmarou A, Fatouros P, Hayasaki K and Corwin F. Contribution of vasogenic and cellular edema to traumatic brain swelling measured by diffusion-weighted imaging. J Neurosurg 1997; 87: 900–907
9. Beaumont A, Marmarou A, Hayasaki K, Barzo P, Fatouros P, Corwin F, Marmarou C, Dunbar J. The permissive nature of blood brain barrier (BBB) opening in edema formation following traumatic brain injury. Acta Neurochir Suppl 2000;76: 125–9.
10. Beaumont A, Marmarou A, Czigner A, Yamamoto M, Demetriadou K, Shirotani T, Marmarou C, Dunbar J. The impact-acceleration model of head injury: Injury severity predicts motor and cognitive performance after trauma. Neurol Res 1999; 21: 742–54
11. Beaumont A, Marmarou C, Marmarou A. The effects of human corticotrophin releasing factor on motor and cognitive deficits after impact acceleration injury. Neurol Res 2000; 22: 665–73.
12. Marmarou CR, Povlishock JT. Administration of the immunophilin ligand FK506 differentially attenuates neurofilament compaction and impaired axonal transport in injured axons following diffuse traumatic brain injury. Exp Neurol 2006; 197: 353–62.
13. Signoretti S, Marmarou A, Tavazzi B, Lazzarino G, Beaumont A, Vagnozzi R. *N*-Acetylaspartate reduction as a measure of injury severity and mitochondrial dysfunction following diffuse traumatic brain injury. J Neurotrauma 2001; 18: 977–991
14. Pascual JM, Solivera J, Prieto R, Barrios L, Lopez-Larrubia P, Cerdan S and Roda JM. Time Course of early metabolic changes following diffuse traumatic brain injury in rats as detected by ^{1}H NMR Spectroscopy. J Neurotrauma 2007; 24: 944–959

Section II

Experimental Spinal Cord Injury Models

Chapter 35

MASCIS Spinal Cord Contusion Model

Wise Young

Abstract The impactor is the most widely used rodent spinal cord injury model. The Multicenter Animal Spinal Cord Injury Study (MASCIS) adopted the model in 1993 and standardized protocols for anesthesia, surgery, spinal cord contusion, and post-injury care of rats, as well as specific outcome measures. The MASCIS model stipulates Long-Evan's hooded rats that are 77 ± 1-day old and anesthetized with pentobarbital (intraperitoneal 45 mg kg^{-1} female, 6 mg kg^{-1} male). After removing the T10 and half of T9 lamina, the rat is suspended by clamping the T8 and T11 dorsal vertebral processes. The Impactor is used to drop a 10-gram rod 12.5, 25.0, or 5.0 mm onto T11 spinal cord. The Impactor tracks the trajectory of the falling rod (±10 μm and ±10 μs) and measures impact velocity (Vi), cord compression distance (Cd), time (Ct), and rate (Cr = Cd/Ct). The rats get tracheal suction for respiratory depression and subcutaneous saline (5–10 ml) for dehydration. The rats are treated with cephalosporin (25 mg kg^{-1} day^{-1}) for 7 days and daily bladder expression. Recurrent urinary tract infections are treated with a 10-day course of enrofloxacin (2.5 mg kg^{-1} day^{-1}). If the rats develop autophagia, they receive daily oral acetaminophen (65 mg kg^{-1} Children's Tylenol). Spinal cord lesion volume is estimated from Na and K concentrations of the spinal cord. Walking recovery is assessed with the BBB locomotor scale. Spared white matter is used to assess spinal cord damage at 6–12 weeks.

Keywords: Spinal cord injury, Contusion, Surgery, Care, Walking, Lesion volume

1 Introduction

The Multicenter Animal Spinal Injury Study (MASCIS) Impactor was first developed in the early 1990s, in response to the demand for a standardized rat spinal cord injury model. In 1993, the National Institutes of Health funded the first multicenter animal spinal cord injury study (MASCIS) to validate the model for multicenter studies of therapies. Eight leading spinal cord injury centers worked together to test approximately 800 rats per year,

From: *Animal Models of Acute Neurological Injuries.*
Edited by: J. Chen (eds.) New York, Totowa, NJ

testing and validating all aspects of the model. The model has been described by several papers.[1–4]

The MASCIS model is more than a device for contusing the spinal cord. The model included detailed criteria for the animals, including strain, body weight, and preoperative care. It defines the anesthetic procedure, the surgery, the timing of the injury after anesthesia induction, antibiotics, resuscitation methods, fluid replacement, and postoperative care. In addition, the model defines the care and methods of assessing tissue damage and locomotion of rats after the injury. The model differs from all others in its definition of not only the mechanical but biological parameters of the injury.

Briefly, the model is carried out in 77 ± 1-day-old rats (Long-Evan's hooded) anesthetized with pentobarbital (45 mg kg^{-1} female, 65 mg kg^{-1} male). The injury site is exposed with a T9-10 laminectomy, removing half of T9 and all of T10. The injury is carried out by dropping a 10 g rod 6.25, 12.5, 25.0, or 50.0 mm onto the spinal cord between T9 and T10. The impactor head is 2.1 mm. Animals are monitored for respiratory depression by observing the color of their feet and revived by tracheal suction if cyanosis is present.

Postoperatively, the rats all receive a prophylactic antibiotic, i.e., cephalozolin (daily 25 mg kg^{-1} injections subcutaneous) for 7 days. If recurrent urinary tract infections occur, the rats get a 10-day course of Baytril (2.5 mg kg^{-1} day^{-1} subcutaneous). The rats are inspected twice daily and their bladders are expressed once a day. If the rats show evidence of autophagia (biting of dermatomes below the injury site), they are treated with 67.5 mg kg^{-1} of acetaminophen (oral). The animals are weighed and scored weekly for BBB locomotor scores. Depending on the experimental plan, the rats are euthanized either by decapitation or intracardiac perfusion of formaldehyde.

The following sections will summarize the preoperative preparation, the surgery, the injury, postinjury care, and various outcome measures that have been developed for the model, including the BBB locomotor score and spared white matter measurements. Some of the information has been published in various papers and chapters but this is the first article that summarizes the procedures in detail.

2 Animals

The MASCIS model stipulates Long-Evans hooded rats that are 77 ± 1-day old at the time of injury. The animals should be obtained at least two weeks before surgery and accomodated to the animal holding facility. The rats are not fasted before the injury. The animals are anesthetized with intraperitoneal pentobarbital. Isoflurane is used for other procedures such as transplantation.

Rat strain: Long-Evan's hooded rats are robust, outbred, and widely available from many commercial sources. Although other rat strains can be used, investigators should recognize that inflammation, immune response, and responses to injury and treatment differ amongst rat strains. Appropriate controls should be used.

Age: The size of the spinal cord of rats correlate linearly with age of the rat and not to body weight. Male and female rats differ substantially in their body weight as they mature. Male Long-Evans hooded rats reach 320 g by 77-days old while female rats tend to be smaller at 240 g. However, their spinal cords are very similar in size and weight.

Accomodation: Shipping is quite stressful to rats. Most rats do not eat during the shipping period and they tend to respond to injuries differently. Therefore, rats should be ordered and delivered at least 2 weeks before surgery. They must be accomodated to animal facility.
No fasting: The rats should not be fasted before surgery. They should be allowed access to food and water ad libitum. The reason is because rats become hypoglycemic (blood glucose $<50\,\text{mg dl}^{-1}$) if they do not feed for 12 h. Hypoglycemia increases severity of injuries. Rats do not vomit or aspirate. There is no rationale of preoperative fasting.
Pentobarbital: The rats are anesthetized with intraperitoneal pentobarbital ($45\,\text{mg kg}^{-1}$ for female and $65\,\text{mg kg}^{-1}$ for males). To inject the rats, it is useful to cover the rat's head with the end of a sock and gently immobilized until the rat is relaxed. The injection is then made on the right, two fingers above the pelvis.
Isoflurane: For procedures to inject cells or other manipulations of the spinal cord besides spinal cord injury, isoflurane is recommended. Anesthesia is induced by exposing the animals to 5% isoflurane. When the animal is deeply anesthetized, the anesthesia is maintained with 2% isoflurane by spontaneous breathing.
Testing anesthesia: The rat should become deeply anesthetized within 5 min. The anesthetic depth is tested by manual pinching of the middle of the tail. A well-anesthetized rat shows a transient response of chest muscles and breathing response but no limb movement to a firm mid-tail pinch.

3 Surgery

The surgery should be carried out under sterile conditions. The anesthetized animals are first shaved and prepared for surgery. A standardized sterile instrument pack should contain everything necessary for the surgery. A laminetomy is carried out to expose the spinal cord. After the spinal cord has been hit with impactor, the wound is closed.
Preparation: The rats are shaved in a hood with an electrical hair cutter, exposing the back for the laminectomy and, if necessary, the neck for placement of intravenous lines. Care should be taken to avoid exposure to rat hair to prevent development of allergies. The skin is prepared with a betadine solution. The surgeon is masked, gowned, and gloved.
Surgical Instruments: The following surgical instruments are part of the essential instrument pack, as shown in Fig. 1. These include scalpel, large and small forceps, mini-rongeur for removing bone, mosquito clamps, a dental elevator for detaching muscle from the vertebral processes, microscissors and scissors, and stainless steel clips and applicator. In addition, sterile gauze pads, cotton pads, and drapes should be available, as well as squirt bottles of sterile saline solution and 95% alcohol solutions. All instruments that touch the inside of the wound must be sterile. The surgeon should wear sterile gloves. To use the instruments in sequential surgeries, they are resterilized by inserting the tips into a hot beads sterilizer.
Laminectomy: The skin is incised with the scalpel at the midline between T5 and L1 (identified by counting the ribs). Muscles are detached sharply from the dorsal processes while the gauze pads are used to apply pressure to bleeding points. The bone elevator is used to clear the muscle from the lateral processes

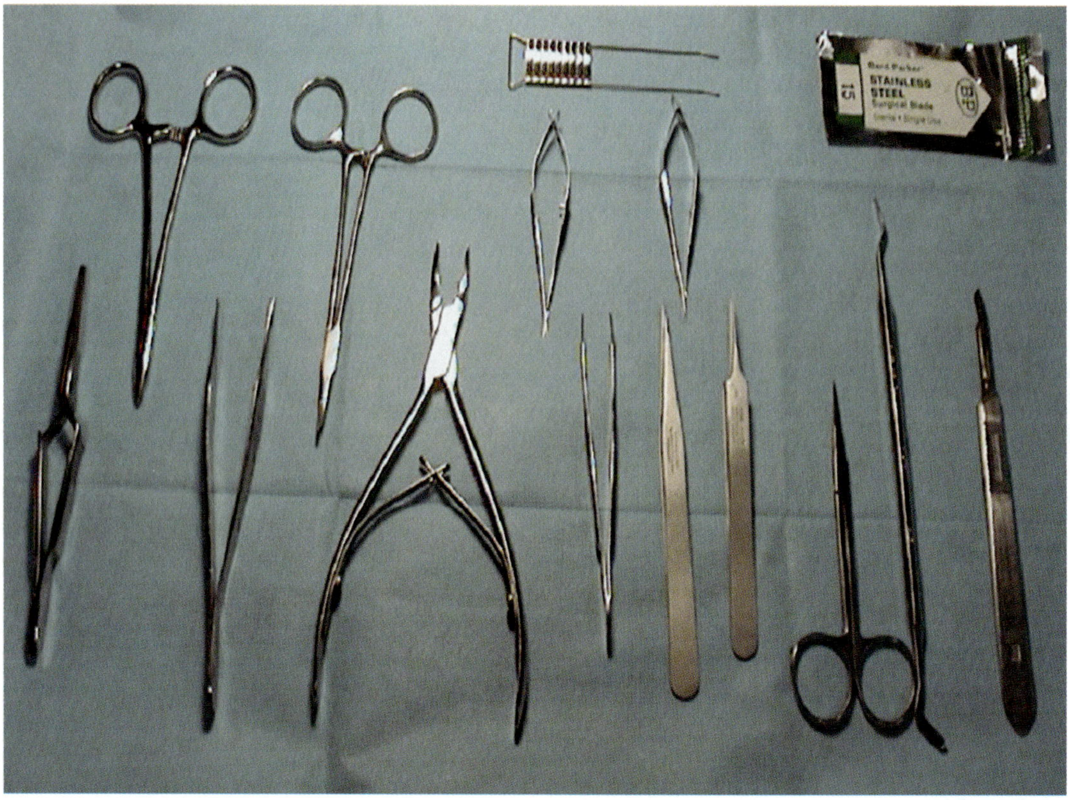

Fig. 1 Surgical instruments used in the operation of the rat. From left to right, they are stainless steel clip applier, mosquito clamp, large toothed tissue forcep, mosquito clamp, mini-rongeur, stainless steel clips (top), microscissors, fine-toothed forceps, fine forceps, microscissor, fine forceps, small scissor, scalpel blade, bone elevator, scalpel handle

of T9-T10. Using the forcep with teeth to grasp, the T9 dorsal process, the microrongeur is used to nibble at bone between T9 and 10, to remove the half of T9 and all of T10. The laminectomy opening should be rectangular and 2.5 mm in width. Care is taken to avoid damaging the spinal cord and the T8 and T11 dorsal vertebral processes during the laminectomy. The dura is left intact. The T12 cord is exposed (Fig. 2).

Wound Closure: After hitting the spinal cord with the Impactor (see next section), the rat is unclamped. Using cotton balls moistened with sterile saline, the dural surface of the spinal cord and the laminectomy site should be cleaned so that the cord can be clearly seen through the dura. Muscle is sutured with silk, and the skin is closed with stainless steel clips. The rat is then placed in a small transparent placed on a heating pad and monitored until it wakes up. If the rats show any signs of cyanosis, tracheal suction will be carried out to remove secretions and to stimulate respiration.

4 Impactor

The rat is suspended with clamps holding T9 and T11. Two types of clamps are available. The first type of clamp is held by right angle screw blocks, and

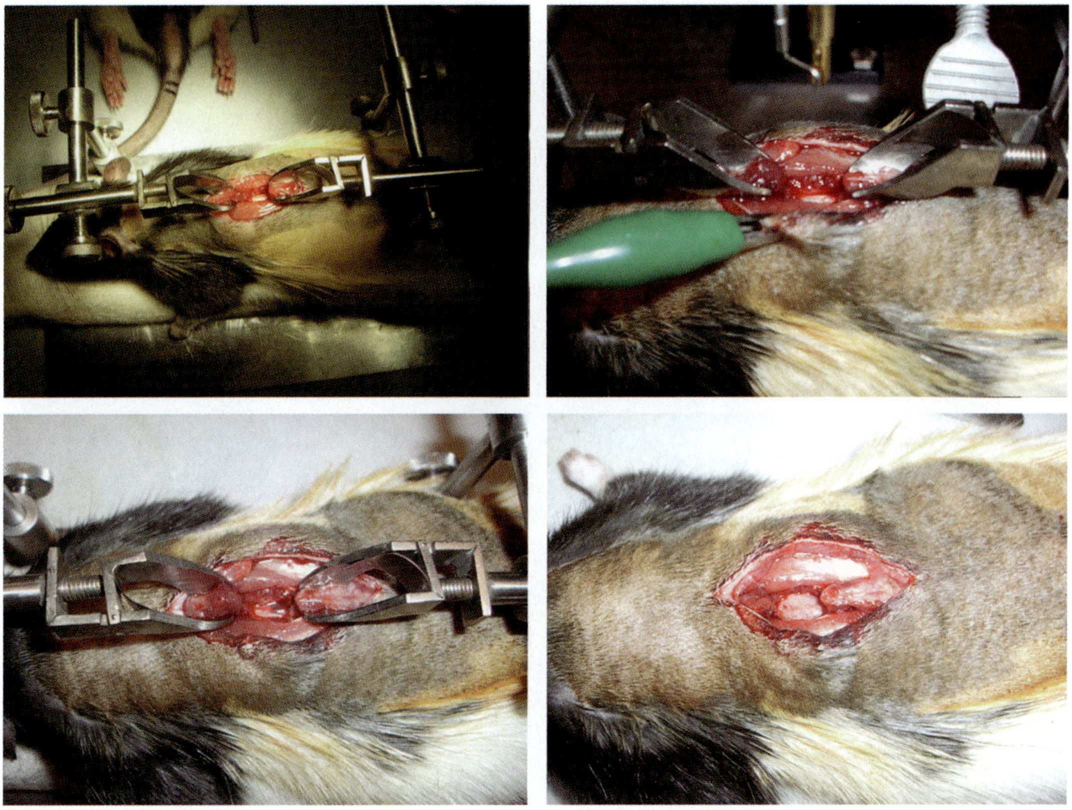

Fig. 2 Laminectomy and impact site: *Top left*: the T11 spinal cord has been exposed by a T9–10 laminectomy. *Top right*: closeup picture of the surgery site before placement and centering of the impactor rod. *Bottom left*: the spinal cord shortly after impact. *Bottom right*: the spinal cord covered with a piece of fat, just before closure

is designed to hold the T9 and T11 vertebral processes. If the laminectomy is done at the incorrect levels, these clamps may not fit. The second type is an angled clamp that allows room for instruments on top and can be used to clamp other vertebral processes. A foam block (covered with a clean material) is used to support the chest of the rat. Figure 3 shows pictures of the Impactor.

Zero the Impactor: The Impactor is positioned with the Impactor rod placed at zero mm and the Impactor head centered on the spinal cord between T9 and T10. The ground alligator is placed on the clamps (in continuity with the animal) to detecting contact between the impator rod and spinal cord. Lower the impactor till the head touches the spinal cord, producing a buzzing sound and a light. The impactor rod is raised slightly until the sound and light stops. The impactor is then lowered slowly with micromanipulator until the sound just occurs.

Set the Impactor: The computer program for the Impactor should have been started. Note the computer runs on MS-DOS and should be a 486 computer. The program name is MAS. All commands are single key click commands. Set the height of the weight drop, the name of the file for recording the impact, and set the computer to record the impact. The impactor rod is raised to the

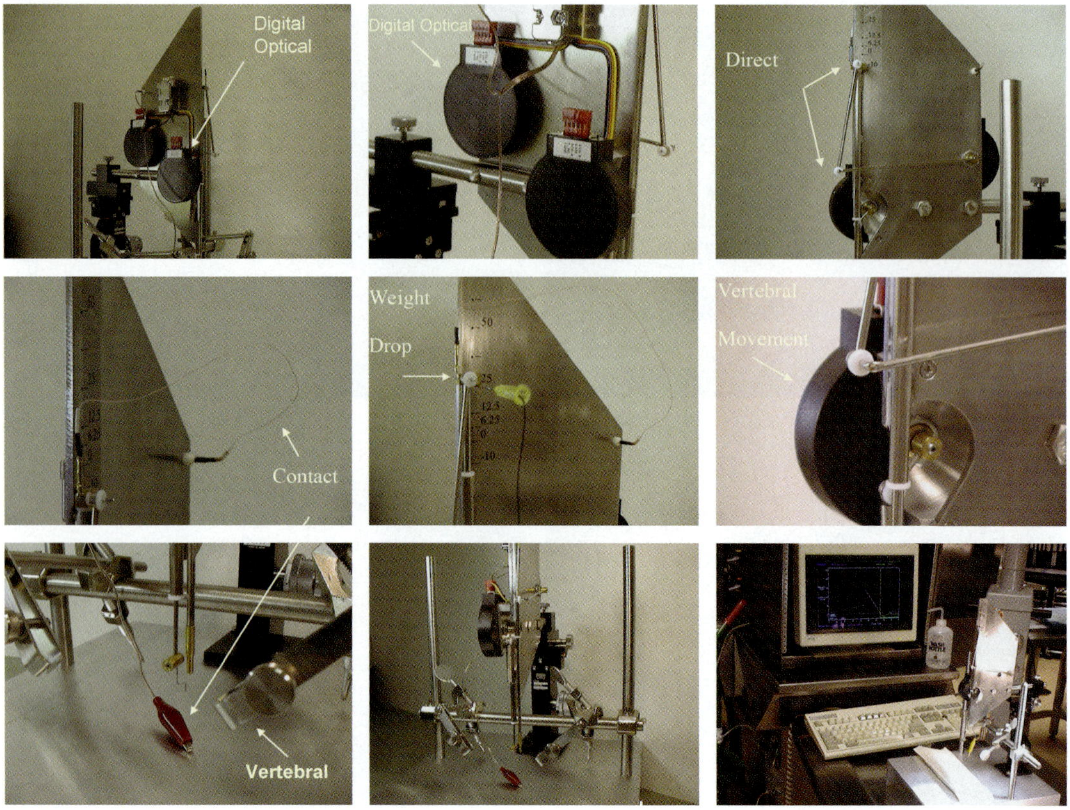

Fig. 3 Impactor: *Top*: digital optical potentiometers and linkage to the impactor rod. *Middle*: contact detector wire, weight drop release, vertebral movement detector. *Bottom*: vertebral clamps, frontal view of the impactor, and the setup in the laboratory

intended level for weight drop, i.e., 6.25, 12.5, 25.0, or 50.0 mm weight drop. The pin is inserted into the hole to hold the rod up, and the tip is pushed so that it is flush to the Impactor steel plate.

Contusion: The injury should take place at 60 min after induction of anesthesia. An impact done 15 min early can result in 20% greater lesion volume. Ensure that the impactor is centered on the spinal cord. Pull the pin and let the rod drop. The rat should jump slightly, the legs should go into extension, and the tail should whip back and forth several times. Remove the impactor rod from the cord surface. The computer should display the impactor parameters. Inspect the impact site. There may be subdural hemorrhage evident. Release the clamps. Bring the rat back to the operating table.

Skin Closure: Using saline-soaked cotton balls, wipe the dural surface clean. Pinch a small piece of subcutaneous fat and place it on the dural surface to inhibit adhesions and formation of a collagenous scar over the laminectomy site. If long term inhibition of adhesions is desired, stretch place a piece of Parafilm® over the fat. Using silk and a cutting needle, suture the muscles at the midline at two points above and below the impact site while avoiding the injury site. Use three stainless steel clamps to close the skin, one at the middle and the other two between the middle and the ends of the wound.

5 Impact Parameters

The Impactor provides a detailed record of the falling rod and the displacement of the spinal cord. The measurements are accurate to ±10 µm and ±10 µs. In addition, the impact records the time and position of contact, the deepest displacement of spinal cord. It does linear regression to find the slope of the line during the 2 ms just before contact of the spinal cord, giving the impact velocity (Vi). The Impactor Program provides three other parameters: the compression depth (Cd), compression time (Ct), and compression rate (Cr). Figure 4 shows a graph of the Impactor trajectory printed by the program. Figure 5 shows the impact parameters.

Impact Velocity (Vi): Vi is the what the impactor delivers. Under ideal circumstances, the impact velocity can be predicted by Galileo's formula where $H = (g{\bullet}T^2)/2$, where H is the height of the weight drop and T is the time required for the rod to fall that height. If so, T = !($2H/g$), $g = 9{,}800\,\mathrm{m\ ms}^{-2}$, and $V = gT$. From these equations, we can readily calculate that the idealized impact velocities should be 0.3571, 0.4945, 0.6993, and 0.9889 for the rod

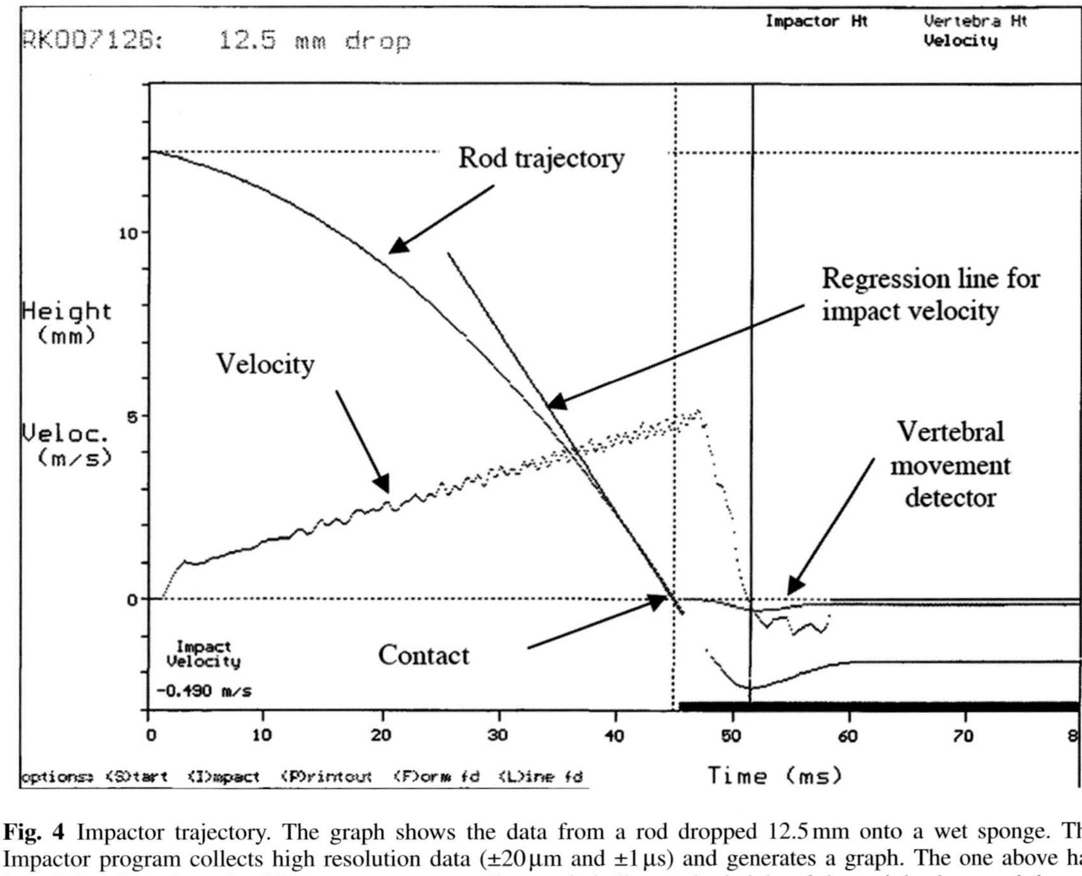

Fig. 4 Impactor trajectory. The graph shows the data from a rod dropped 12.5 mm onto a wet sponge. The Impactor program collects high resolution data (±20 µm and ±1 µs) and generates a graph. The one above has been labeled to show the different components. The *y*-axis indicates the height of the weight drop, and the rod trajectory shows the rod falling until it contacts the spinal cord. The program estimates impact velocity from a linear regression of the rod position during the 2 ms before contact

RK00725H PARAMETER CALCULATIONS Impactor v7.0

Basic Impact Parameters (relative to baseline / t=0 ms)

Variable	Sample	MM	Time (ms)
Baseline	7457	-0.002	71.979
Drop start	1	24.823	0.010
Impact	7451	0.038	71.921
Cord_Rev	7926	-2.565	76.506
Vert_Rev	8058	-0.468	77.780
Crd-Surf_Rev	7919	-2.270	76.438
Compression		2.308	

Impact Height and Velocity Calculations

Variable	Height (mm)	Velocity (m/sec)	Time (msec)
Expected :	24.903	-0.700	71.392
Actual :	24.785	-0.682	71.911
Error :	0.118	0.018	0.519
Error (%):	0.5	2.6	0.727

Fig. 5 Impact parameters: The *picture* shows the graphic display of the impact parameters. The top set of numbers have three columns named Sample, MM, and Time (ms). Sample is the number of pulses received by the computer. MM is the conversion of those pulses to millimeter. Time is in milliseconds. Cd is estimated from MM (Impact-Crd-Surf Rev). Ct is estimated from Time (Impact-Crd-Surf). The lower table lists the expected and actual height, velocity and time of the impacts. The absolute and percentage errors are listed

dropped from 6.25, 12.5, 25.0, 50.0 mm heights. If the impact velocity is close to the ideal velocities, this means that the impactor is working.

Compression Depth (Cd): Cd is how much the dropping rod compressed the spinal cord. It is typically in the range of 2–3 mm and increases with greater height of the weight drop. If Cd is much greater than 3 mm, this usually means that the spinal cord was not well clamped. If it is much less than 2 mm, it usually means that the impactor rod has hit bone along the way.

Compression Time (Ct). Ct is the time required for the dropping rod to compress the spinal cord to the deepest point. Usually expressed in milliseconds (ms), Ct is typically on the order of several milliseconds. The time of compression tends to increase with greater weight drop heights. The greatest utility of Ct, however, is its use for calculating the cord compression rate.

Compression Rate (Cr). Cr is the average rate at which the spinal cord is being compressed by the falling weight, calculated from Cd/Ct. Cr correlates linearly with Vi, as well as lesion volume, spared white matter, and BBB locomotor scores. In general, it is 70–80% of the value of Vi. The difference in Vi and Cr is proportional to the energy absorbed by the spinal cord.

The Impactor program is designed to give a warning the impact parameters deviate from expected amounts by more than 5%. This is an arbitrary criteria. The best measure of impactor consistency is the relationship between Vi and Cr. During their first 15–30 impacts, most investigators have correlation coefficients on the order of 0.60 but they often reach 0.80 after 30 impacts. Very experienced surgeons have correlation coefficients of 0.95 or better.

6 Postoperative Care

The rats should be kept on a warm pad and placed in a prominent place in the operating room where it can be seen by many people. During the immediate postinjury, the rat is most susceptible to respiratory depression and therefore must be frequently inspected.
Respiratory Depression: While we use to use arterial blood gases (from a tail artery) to determine respiratory status of the animal, we discovered that the color of the feet is just as good or better determinant of the oxygenation status of the rats. If the rat is well oxygenated, the feet are bright pink. If they are dusky or even blue, this means the rats must get respiratory stimulation.
Tracheal Suction: We use a thin soft catheter on a 5 ml plastic syringe. After inserting the catheter into the trachea, we pull back on the 5 ml syringe, producing a vacuum. As the catheter is slowly withdrawn, it sucks up mucus and also stimulates the rat to breathe. This method is by far the most effective of any that we have tried. If the rat is not responsive to the tracheal suction, it may be necessary to use the fingers to provide manual ventilation and then do the tracheal suction again.
Fluid Replacement: The need for fluid replacement depends on the rat. To test the hydration status of the rats, pinch the skin on the back and let go. Normally, the skin goes flat within a second. If the skin remains elevated and takes several minutes to fall, this means that the rat is dehydrated. We usually give 5 ml of sterile saline subcutaneously for females and 10 ml for males.
Prophylactic Antibiotic: To prevent wound and urinary tract infections, all rats receive a daily dose of cephalosporins (25 mg kg^{-1} per day) for 7 days. This is a broad spectrum antibiotic that eliminates most common bacteria. When recurrent infections occur, they are usually of the gram-negative variety, particularly hemophilus strains that are sensitive to fluroquinolones.
Daily Bladder Expression: The rats should receive daily bladder expression. This is done by holding the rats around the thorax with one hand, and the thumb and first two fingers of the other hand are used to gently express urine from the bladder. The urine is collected in a test-tube to measure the volume of the urine expressed every morning.
Hemorrhagic Cystitis: This is a major cause of death in the postoperative period. It manifests with bloody urine during the first day after injury. This condition results from stress and may be associated with excessive hemorrhage during surgery and bladder infection. The animal is already on antibiotics at the time. When there is blood, some urine should be left in the bladder to avoid clot formation.

7 Chronic Animal Care

The most common complications that cause morbidity and mortality in rats with chronic spinal cord injury are recurrent urinary tract infections and autophagia.
Recurrent Urinary Tract Infections: Recurrent urinary tract infections occur in about 25% of the animals and usually at 1–2 weeks after injury. We treat the animals with a 10-day course of Baytril (enrofloxacin 2.5 mg kg^{-1} day^{-1}).

About 10% of the animals get recurrent infections after the Baytril treatments. These animals usually have a bladder or renal stone. This drug is also effective against mycoplasma and is used primarily for veterinary purposes.
Flaccid bladders: A third or more rats may develop a flaccid bladder after a 25 mm weight drop, perhaps due to spread of the damage from the T9–10 impact site to the sacral sympatheic and parasympathetic nuclei. In any case, the first sign of a flaccid bladder is difficulty expressing them. The bladders often become quite big and thin-walled and must very carefully expessed or else the bladder will burst.
Autophagia: One third of the rats will develop a behavior called autophagia, usually starting 2–3 weeks after injury. They bite themselves on dermatomes below the injury site. The first sign is reddening of the skin and hair loss (Stage 1). This is soon followed by biting that penetrates skin exposing subcutaneous layers (Stage 2), exposing muscle (Stage 3), and sometimes viscera (Stage 4).
Autotomy: About 5% of the rats will bite themselves in foot and the toes. The first manifestation of this is often bloody toes and may rapidly progress to the point that the rat will eat its foot. Acetaminophen may stop this behavior but we usually euthanize the rat if there is significant damage to the foot because this would interfere with evaluation of its locomotor function.
Autophagia treatment: An effective treatment of autophagia is 65 mg kg^{-1} of acetaminophen (oral, children's Tylenol) per day until the lesion is completely healed. The rats like the cherry-flavored fluid and will readily eat it from a dropper. Daily treatment, if started as soon as level 1 autophagia is seen and continued until the lesion is completely healed, is effective in stopping autophagic behavior. If the treatment is stopped prematurely, the rats may engage in a rebound autophagia that rapidly progressed to stage 4.
Failure to Thrive: Animals may fail to thrive for a variety of reasons. The condition can be documented by weekly weighing. From ages 77- to 100-days old, rats are gaining weight from 220 to 350 g for females and 320–600 g for males. However, despite the huge difference body weight between males and females, spinal cord weights remain the same.

8 Outcome Measures

We developed several outcome measures to assess spinal cord injuries in rats.
Lesion volume: To measure lesion volume, we cut the rat spinal cord into five 5-mm long pieces. The first piece is centered on the impact site, called IMP. Adjacent proximal and distal 5-mm pieces are P1 and D1. The pieces proximal and distal to P1 and D1 are P2 and D2. We homogenize the samples with ultrasound in 10 ml of deionized water and analyze 10 µl aliquots of the tissue solution with atomic absorption spectroscopy for Na and K.[1,5,6] Differences of tissue sodium ($[Na]_t$) and potassium ($[K]_t$) concentration linearly predict the ratio of intracellular and total tissue volume (Vi/Vt) or cell volume fraction. If intracellular and extracellular fluids are isotonic and Na and K account for >90% of tissue fluid osmolarity, the sums of tissue, extracellular, and intracellular Na and K concentrations should be similar, i.e. [Na]t + [K]t = [Na]e + [K]e = [Na]i + [K]i and [Na]e-[Na]i = [K]i-[K]e = G. If so, [Na]t-[K]t is linearly related to Vi/Vt with a slope of 2G and a *y*-intercept of [Na]e-[K]e, i.e. [Na]

t-[K]t = [Na]e-[K]e + 2G(Vi/Vt). We estimate change in cell volume by subtracting normal Vi/Vt from Vi/Vt of the injured cord, i.e. ΔVi/Vt. Since 1 mg of wet tissue is about 1 μl, multiplying ΔVi/Vt by tissue wet weight (*W* in mg) gives loss of cell volume in microliters. The lesion volume can also be estimated from $[K]_t$, i.e. $V_{ii}/V_t = ([K]_t\text{-}4)/120$ and is expressed in microliters of cells lost from the spinal cord within 1.25 cm from the contusion center.

BBB Locomotor Score: The BBB score was developed to evaluate locomotor recovery in rats after spinal cord contusion.[7] On the basis of the observations of 10 behaviors and the rate at which they recovered after a contusion injury, the score ranks behaviors by how late they occurred, generating a progressive recovery curve. The score is reliable[3] and correlates with spared white matter.[8] Team scores were significantly more reliable and closer to those given experienced scorers.[3] Thus, the method recommends a pair of observers situated opposite of each other, observing the rat for 4 min, and coming up with a consensus score. In case of disagreement, the lower score is used. The components of the BBB scores are recorded as well as the final consensus score. Although some of the behaviors are based on one side, the final BBB score is a single score that reflects both left and right side.

Spared White Matter: In general, gray matter is lost at the contusion site, and a thin rim of spared white matter typically remains. We measure spared white matter by staining coronal sections of the injury site with Luxol Fast Blue. We usually make the measurements from the coronal section representing the contusion epicenter with the greatest white matter loss. Spinal cord stained with luxol fast blue reflects the presence of myelinated axons. The spared white matter is expressed in square millimeter.

References

1. Constantini S, Young W. The effects of methylprednisolone and the ganglioside GM1 on acute spinal cord injury in rats. Journal of Neurosurgery 1994;80(1):97–111.
2. Beattie MS, Bresnahan JC, Komon J, et al. Endogenous repair after spinal cord contusion injuries in the rat. Exp Neurol 1997;148(2):453–63.
3. Basso DM, Beattie MS, Bresnahan JC, et al. MASCIS evaluation of open field locomotor scores: effects of experience and teamwork on reliability. Multicenter Animal Spinal Cord Injury Study. J Neurotrauma 1996;13(7):343–59.
4. Young W. Spinal cord contusion models. Prog Brain Res 2002;137:231–55.
5. Young W, Kume-Kick J, Constantini S. Glucocorticoid therapy of spinal cord injury. [Review] [243 refs]. Annals of the New York Academy of Sciences 1994;743:241–63; discussion 63–5.
6. Young W, Constantini S. Ionic and water shifts in injured central nervous tissues. In: Salzman S, Faden AI, eds. Neurobiology of Central Nervous System Trauma. Oxford: Oxford University Press; 1993:Chapter 11, pp. 221–33.
7. Basso DM, Beattie MS, Bresnahan JC. A sensitive and reliable locomotor rating scale for open field testing in rats. J Neurotrauma 1995;12(1):1–21.
8. Basso DM, Beattie MS, Bresnahan JC. Graded histological and locomotor outcomes after spinal cord contusion using the NYU weight-drop device versus transection. Exp Neurol 1996;139(2):244–56.

Chapter 36

Infinite Horizon Spinal Cord Contusion Model

Stephen Scheff and Kelly N. Roberts

Abstract The IH impactor is capable of inflicting a consistent contusion injury at various spinal cord levels in the rat. The device is capable of producing graded magnitudes of injury as a result of user-defined changes in the amount of force applied to the rigidly stabilized spinal column and exposed spinal cord. Following a laminectomy, a small diameter impounding tip displaces the exposed spinal cord at a preset amount of force. This contusion results in an injury to both the gray matter and the white matter of the spinal cord at T10, the extent to which depends on the applied force. The methods for performing this injury with the IH impactor are described in detail.

Keywords: Spinal cord injury, Contusion, Laminectomy, Spinal vertebral column clamping

1 Model Selection

There are numerous devices available for modeling the human contusion spinal cord injury. The Infinite Horizon spinal cord impactor (IH 0400) (Fig. 1) is a force-based contusion device that is commercially available from Precision Systems and Instrumentation, LLC (PSI, Fairfax Station, VA, http://www.presysin.com). The device was designed to produce a reliable contusion injury to the exposed rodent spinal cord.[1] The basic design of the device uses a stepping motor to apply a controlled impact defined in terms of force (kilo dynes). A rigid housing contains the stepping motor that drives a mechanical impactor rack with an inline force sensor and impactor tip attached. The force sensor uses a calibrated load cell to directly measure the amount of force applied by the stepping motor and rack to the spinal cord. In this way, movement artifacts, such as those associated with respiration of the animal, are eliminated. Another important feature of the IH device is that it is not necessary to touch the exposed spinal cord with the impactor tip prior to the injury sequence to provide a displacement reference. The rack is connected to the stepping motor at all times, which means that the motor torque defines the maximum force level. The rack displacement is determined by a very sensitive linear encoder that uses a 3 μm step resolution. The operating principal is simple. The user defines the force to be applied to the exposed spinal cord

From: *Animal Models of Acute Neurological Injuries.*
Edited by: J. Chen (eds.) New York, Totowa, NJ

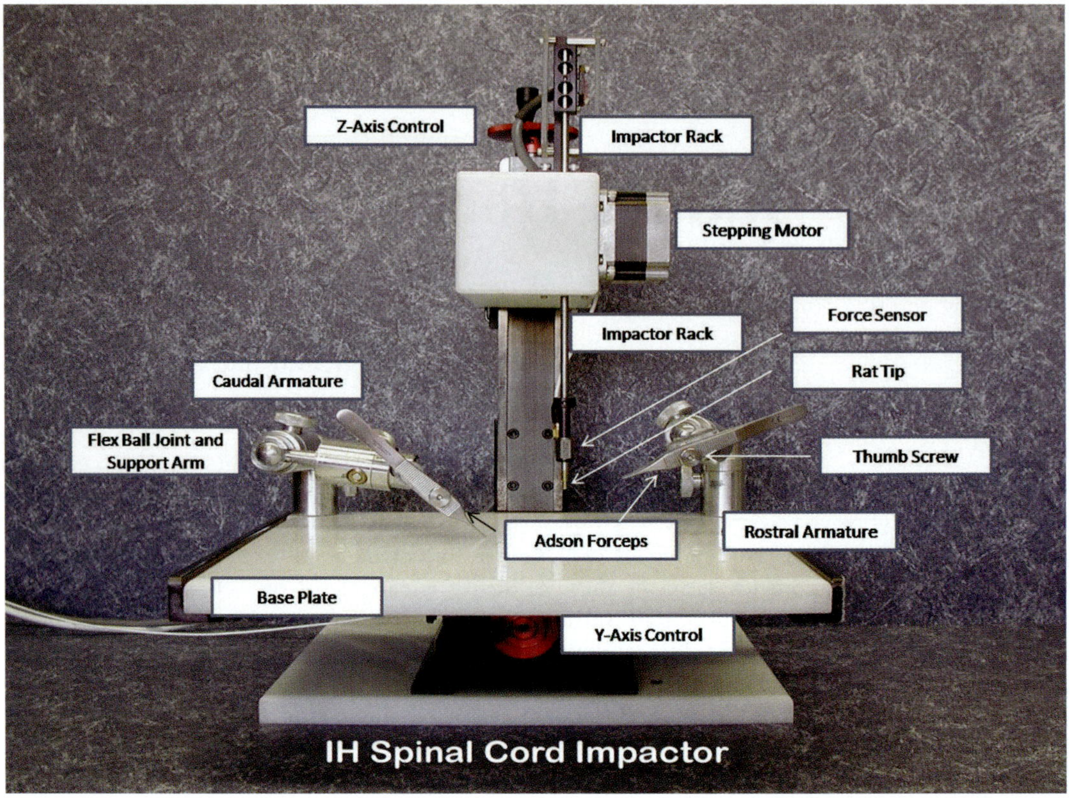

Fig. 1 The IH impactor as viewed from the front. A stepping motor drives the impactor rack toward the exposed spinal cord. The force sensor on the impactor determines when a user-defined force has been applied by the impactor tip to the spinal cord. The spinal column is rigidly held in place with the stabilizing Adson forceps that are attached to the rostral and caudal armatures. Flexible ball joints on the armature allow the forceps to be adjusted during the clamping process. The impactor rack, force sensor, and impactor tip can be raised or lowered with the Z-axis control. The base plate is removed from the IH impactor for the clamping procedure

using the system's software. When initiated, the motor takes a step by applying torque to the rack and moves the rack with attached force sensory and impactor tip in a vertical downward direction. A microcontroller continuously samples the force between impactor tip and exposed spinal cord. If the detected force does not exceed the user-defined force, another step is required from the stepping motor. This process continues until the user-defined force is achieved. The steps are very small and very fast such that the rack appears to move in one continuous motion. When the sampled force equals the user-defined force, the stepping motor stops and the opposite torque is applied resulting in the rack with impactor tip moving away from the exposed spinal cord. In this way, the impactor tip is immediately and quickly withdrawn from the injury site. The software immediately displays the actual force applied to the spinal cord, the amount of tissue displacement and the velocity of the impactor tip at peak force measurement. The displacement is taken from the point at which the force sensor detects values greater than the background ("noise") levels to the point where maximum user-defined force is encountered. Thus, displacement is actually a measurement of the movement of the impactor rack as it touches the exposed spinal cord. The software displays a graph of changes in the force and the displacement as a function of time when the injury sequence was initiated. In this way, the experimenter can immediately evaluate the important parameters of the injury sequence. The operation of the device is extremely simple but requires that the spinal cord be exposed and the spinal column stabilized for a reproducible injury.

2 Materials

Adult Fischer (200–225 gm) or Sprague-Dawley (SD) (250–275 gm) rats are the animals of choice. Infinite Horizon 0400 impactor fully assembled and connected to a Windows®-based personal computer running the IH Impactor software.

3 Procedure

a. Preinjury sequence

1. According to the manufacturer's instructions, turn on the IH device and attach the rat tip to the force sensor and impactor rack (Fig. 1).
2. At the beginning of each injury session, the impactor function is tested by impacting a piece of ultrasoft polyurethane (supplied with the IH device). It is important to keep this polyurethane moist when not used.
3. The entire injury sequence is carried out five times at each force that is to be used during that day's experimental protocol.
4. The values for actual force, displacement, and velocity are recorded to determine instrument reliability between experimental surgery days. This procedure also helps familiarize the experimenter with the instrumentation software.

b. Surgery: Thoracic (T10) spinal cord contusion

1. These procedures should be performed according to local institutional animal use guidelines for aseptic surgery.
2. Anesthetize an adult rat with a mixture of ketamine (80 mg kg^{-1}, i.p.) and xylazine (10 mg kg^{-1}, i.p.). Alternatively, sodium pentoparbital (50 mg kg^{-1}, i.p.) or a mixture of isoflurane and oxygen can be used.
3. Apply Lacrilube ophthalmic ointment (Allergan Pharmaceuticals, Irvine, CA) to the eyes to prevent drying.
4. Shave the hairs on the back and prepare the surgical area with alcohol and betadine.
5. Place the animal on a heating pad to maintain body temperature in the surgical staging area.
6. Make a midline sagittal incision approximately 5–6 cm centered over thoracic vertebra T10. Dissect muscle and connective tissue to expose the underlying vertebral spinal column segments.
7. Perform a dorsal laminectomy of the T10 vertebra using a fine tipped rongeur (e.g. Tiemann 160-421-08) and remove the dorsal vertebral processes from T9 and T11. It is important that the laminectomy be larger than the diameter of the impactor tip (2.5 mm).
8. Use of a surgical or dissecting microscope is recommended to prevent bruising of the spinal cord during the laminectomy process.

9. Leave the dura intact but clear any boney protuberances so that at least 3.0 mm of the spinal cord is exposed.
10. Musculature around the transverse processes should be cleared with care taken not to remove any excess bone in this region.
11. Use a pair of Adson forceps (e.g., Tiemann 105-236-1) to test the location for clamping of the vertebral spinal column. Since clamping should be as close to the laminectomy as possible, it is important that enough tissue be removed from this area.
12. Control bleeding during the laminectomy process. Use of Ethicon surgifoam (Johnson & Johnson, New Jersey) or careful cauterization may be necessary.

c. Clamping the vertebral spinal column

1. The vertebral spinal column should be rigidly secured during the impact injury.
2. The vertebral spinal column is secured and rigidly stabilized by two pairs of Adson forceps attached to flexible armatures on the surgical plate (Fig. 2). These clamping forceps are supplied with the IH impactor.
3. Attach the Adson forceps to the flexible armatures on the surgical plate. Be sure the forceps are screwed on tightly to the flexible armature.
4. The surgical plate should be moved to the surgical staging area. **Do not** try to clamp the vertebral spinal column with the plate on the IH impactor.
5. Move the armatures with the attached Adson forceps laterally away from the middle of the plate in preparation for the animal. We have found that removing one of the segments on the rostral armature makes clamping easier (Fig. 3).
6. Gently move the rat with the exposed spinal cord to the middle of the surgical plate and under a dissecting or surgical microscope.
7. Lower either the rostral or caudal armature with the attached forceps toward the exposed spinal cord by loosening the tightening screws on the armature (Fig. 2).
8. It is helpful if the Adson forceps are screwed closed at the initiation of this process.
9. Once one pair of forceps is in the correct location, tighten the screws of the armature to hold it rigidly in place.
10. With an unattached pair of Adson forceps, lift the spinal column toward the attached Adson forceps on the rigid armature. Slowly open these forceps with the thumb screw and position them to secure the transverse processes of the vertebral spinal column (Fig. 4).
11. When the forceps lightly clamp the transverse processes, tighten the screw until the forceps holds the vertebral spinal column. Be careful not to close them too tightly. It is all right if some muscle is associated with this clamping process.
12. Test to ensure that the forceps clamps securely hold the transverse processes and rigidly support the spinal column.
13. The opposite clamping forceps can now be attached using the same process as above.
14. The vertebral spinal column should be slightly suspended by the armatures but not enough to raise the rat's abdomen off the surgical plate.

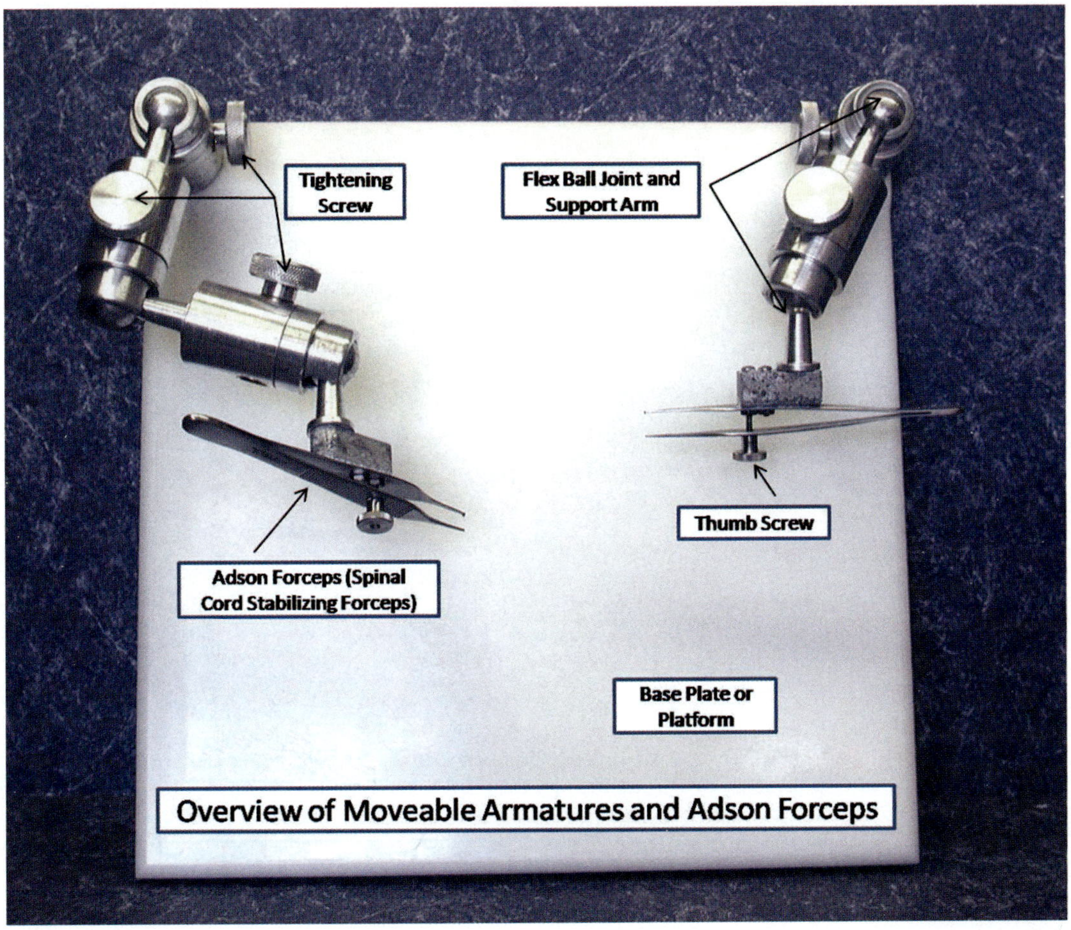

Fig. 2 Surgical plate with attached flexible armatures and Adson forceps. This removable surgical plate comes as part of the IH device and is used to stabilize the vertebral spinal column as part of the preinjury procedure. Each armature can be positioned to accommodate the exposed vertebral spinal column. The position of the Adson forceps can be set in place with tightening screws on the different segments of the flexible armatures. Thumb screws are used to open and close the Adson forceps

15. When appropriately clamped, the Adson forceps are approximately 1 mm rostral and 1 mm caudal to the laminectomy and the exposed spinal cord is level (horizontally) when viewed laterally (Fig. 4). The dorsal spinal vein should not be bleached.
16. The vertebral spinal column should also be straight and not curved or "twisted". The two sides of the forceps should be parallel to the vertebral column.
17. If the spinal cord is not level, adjustments can be made by loosening one of the armatures and adjusting the vertebral spinal column. It is not necessary to reclamp the animal.
18. Under the dissecting or surgical microscope, view the laminectomy site and remove any miscellaneous debris that may have found its way into this space.
19. Be sure to control any bleeding.

Fig. 3 The flexible armature supplied with the IH impactor comes with three segments and can be easily modified by the user. Some find that clamping the rostral end of the vertebral spinal column is easier with one of the segments removed. Often the rostral end is more difficult to clamp and having a less complicated armature makes the process easier

d. Injury sequence

1. Carefully move the animal on the surgical plate to the IH impactor and center the laminectomy site under the impactor tip.
2. Using a plastic quick grip clamp, secure the surgical plate to the base of the impactor (Fig. 5).
3. Use the X and Y axial control wheels for fine alignment of the impactor tip over the exposed spinal cord (Fig. 6).
4. Lower the impactor tip manually with the top *Z*-axis control knob to within 2–3 mm above the exposed cord. Make fine adjustments in the X and Y axis with the control wheels.
5. View the positioning of the tip from different angles to check whether or not the exposed cord is level. The impactor tip should be positioned over the middle of the spinal cord. The large spinal cord blood vessel does not necessarily indicate the midline. Make adjustments as needed.
6. Following careful alignment of the exposed spinal cord, move the impactor tip close to the exposed spinal cord without touching it.
7. Raise the impactor tip by turning the *Z*-axis control wheel two complete rotations *Counter Clockwise*. This should raise the tip to about 4 mm above the exposed spinal cord.
8. Using the interfacing computer define the value for the Injury **Force** (Fig. 7). For a mild injury use 100 kdyne, moderate injury use 150–175 kdyne, and

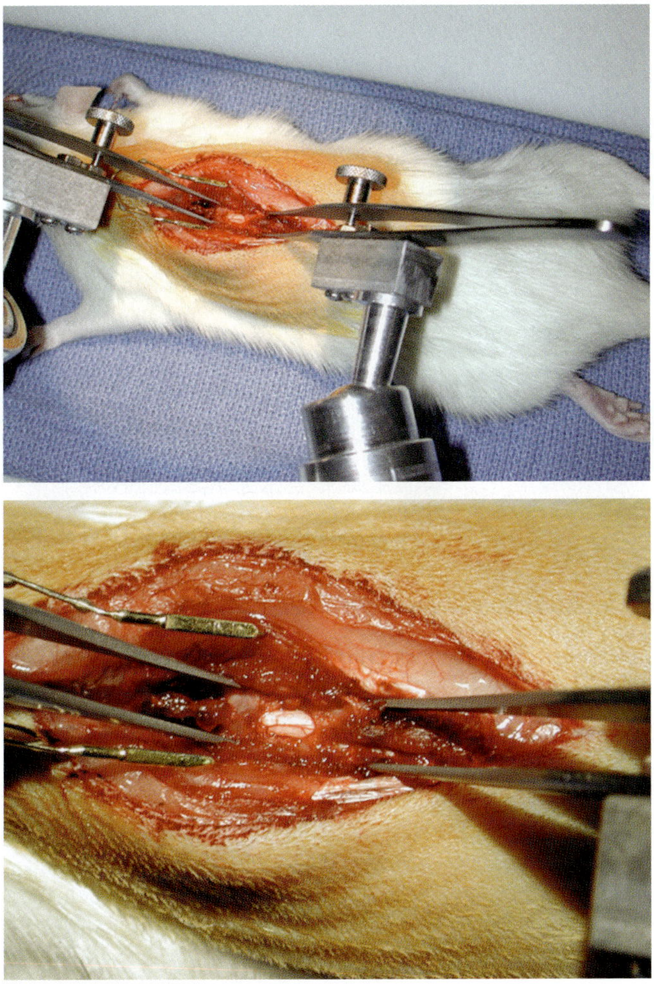

Fig. 4 Following exposure of the spinal cord, Adson forceps are used to clamp the vertebral spinal column both rostral and caudal to the laminectomy. It is important that the forceps clamp the transverse processes and not the dorsal spinal processes

for a severe injury use 200–225 kdyne. The displacement and velocity settings cannot be changed from their default values.

9. Although it can be changed, it is recommended that the **Dwell Time** setting be left at 0 (default value).
10. After the appropriate **force** value has been entered, use the computer mouse to activate (left click) the "**Start Experiment**" sequence.
11. Once activated the impactor tip will move slightly as a calibration check by the instrument.
12. A new pop-up "**Impact Now**" window will become available. Use the computer mouse to activate (left click) the actual injury sequence.
13. Following the injury, there is a short delay before the actual injury parameters are displayed on the computer screen. Determine whether those are appropriate for the experimental paradigm.

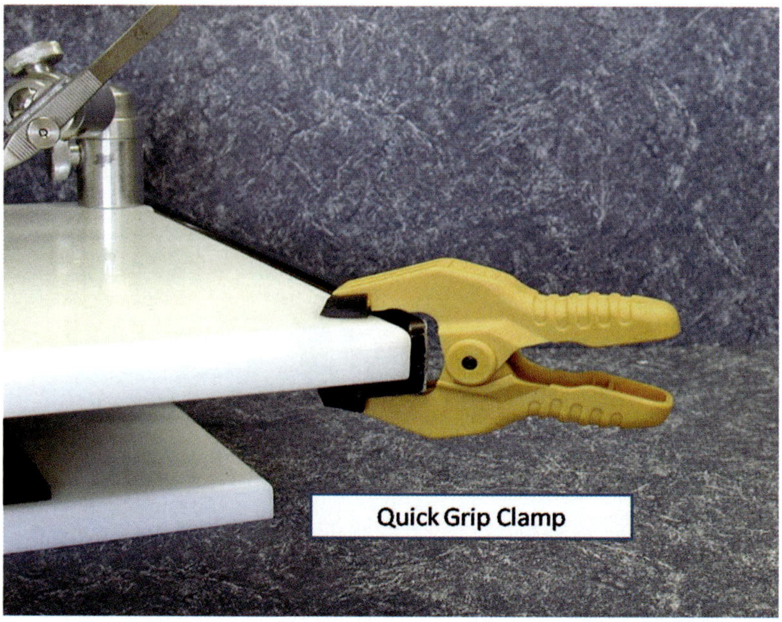

Fig. 5 An inexpensive plastic quick grip clamp holds the surgical plate in position on the base of the IH impactor. This type of clamp can be purchased from any local hardware store but it is important to protect the surgical plate from moving if accidentally bumped during the alignment process

14. Move the impactor tip away from the injury site by turning the Z-axis control knob *Counter Clockwise*.
15. Remove the surgical plate from the IH device and loosen the screws of the Adson forceps in preparation of removing the animal for closing and suturing of the incision.
16. Visually inspect the spinal cord for bruising.
17. Move the cursor on the computer screen into the "**Experiment Comments**" box and left click the computer mouse. The date and time of the experiment are automatically recorded.
18. Type any particular comments connected with the surgery or injury that should be attached to the animal's record. Notes can be made about visual inspection of bilateral bruising, etc.
19. Move the cursor to **File** and highlight **Save-As**
20. Locate the appropriate folder and save the data using the animal's unique identification number.

e. **Postoperative wound closure**
   1. Irrigate the wound with 2–3 cc warm sterile saline.
   2. Remove saline with sterile gauze or cotton-tipped applicator.
   3. Appose the dissected muscle beginning posterior to the laminectomy site and suture the muscle continuing to just rostral of the laminectomy, with Ethicon 4–0 coated Vicryl suture.
   4. Be careful not to suture any fat to the muscle at this time.
   5. The surface incision can be closed with sterile surgical staples (Stoelting). These should be removed 7–10 days post-op fi the incision is healing.
   6. The incision can be cleaned with hydrogen peroxide, and a topical antibiotic is applied to the wound.

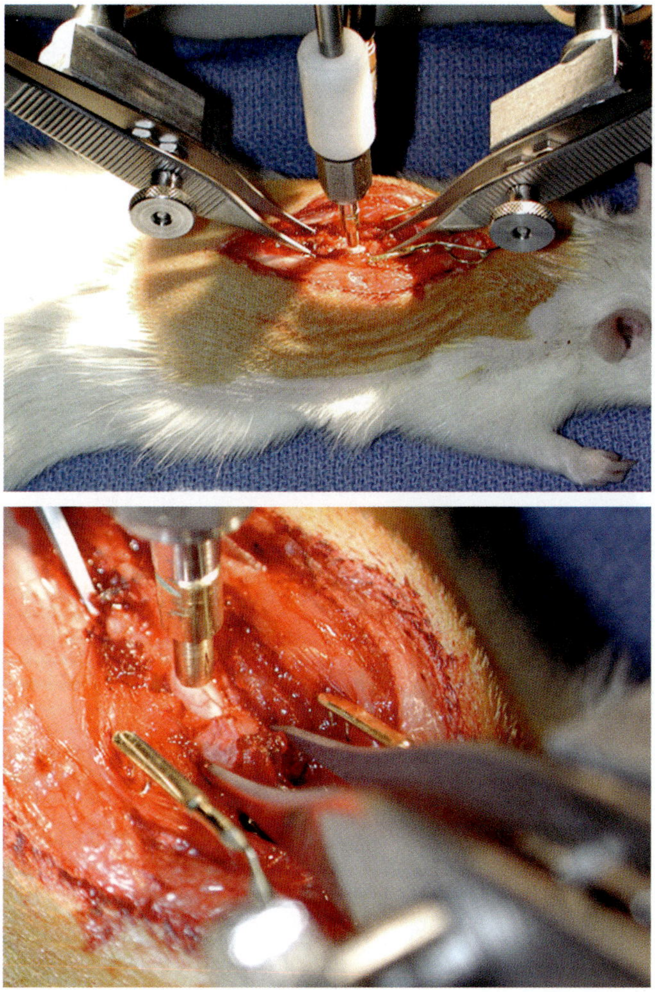

Fig. 6 The impactor tip is positioned over the exposed spinal cord and then lowered to within 2–3 mm above the tissue. It is important to view the relationship between the impactor tip and laminectomy to be sure the tip is centered and will not hit bone or other tissues during the injury sequence. Raise the tip to about 4 mm above the spinal cord prior to the injury sequence

7. Each animal is treated with a general antibiotic (e.g. ceflazolin)
8. Each animal receives 10 cc of saline injected s.c.

4 Equipment Maintenance

1. After each injury, the attached Adson forceps should be removed from the surgical plate, cleaned, and sterilized.
2. The surgical instruments should be cleaned and sterilized. Dry heat sterilization can be used between animals during the same day. Steam heat should be used between surgery days.
3. The impactor tip should be carefully cleaned with 95% alcohol.
4. Do not touch the encoder strip.
5. Keep ball joint sockets of the moveable armatures clean.
6. Keep the instrument covered when not in use.

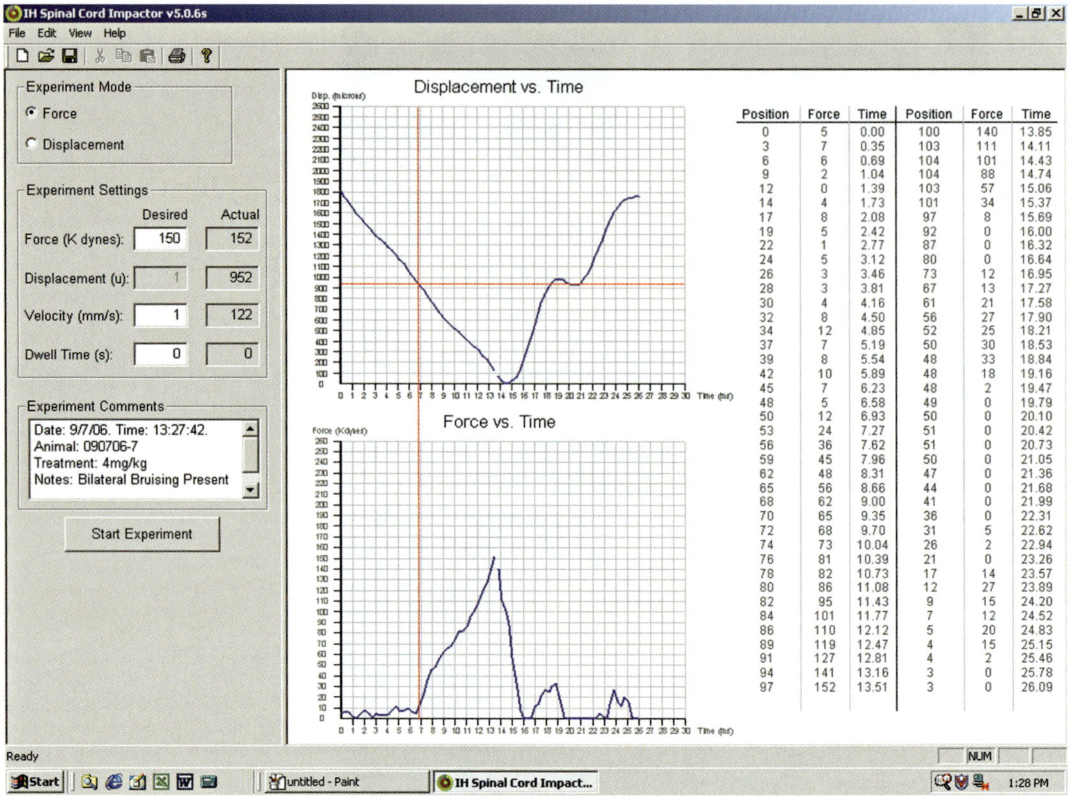

Fig. 7 The computer software is extremely user-friendly. It allows the investigator to define the force to be used during the injury sequence. The display following the injury shows the actual force encountered by the spinal cord as well as a graphic display of the injury sequence. Included is a box that can be used to comment on the surgical manipulation and treatment of the animal

Reference

1. Scheff SW, Rabchevsky AG, Fugaccia I et al. Experimental modeling of spinal cord injury: characterization of a force-defined injury device. J Neurotrauma. 2003;20:179–193

Chapter 37

The Ohio State University ESCID Spinal Cord Contusion Model

Lyn B. Jakeman, Dana M. McTigue, Patricia Walters, and Bradford T. Stokes

Abstract Approximately half of the clinical cases of spinal cord injury are caused by a rapid contusive impact to the soft tissue of the spinal cord. Reproducible models of contusion injury are produced in rodents, using a variety of approaches following the exposure of the intact dura with a dorsal laminectomy. In the contusion model developed at The Ohio State University (OSU), an impact to the spinal cord is mediated by the rapid and calibrated displacement of a vertical shaft, using a sensitive electromagnetic shaker controlled by interactive software. Key elements of the technology are the initiation of the impact from a fixed and measurable starting force or "touch" feature, investigator control of the slope, duration and amplitude of peak displacement, rapid retraction of the impactor after a single impact to prevent bounce effects, and the direct and real time measurement of displacement and force of impact upon the spinal compartment. In this chapter, we briefly describe the history of the design and the procedures that are used to calibrate, perform, and evaluate controlled contusion injuries using the OSU electromagnetic spinal cord injury device (ESCID).

Keywords: Contusion, Mouse, Rat, Vertebral column, Electromagnet, Laminectomy

1 The History and Evolution of the OSU-Controlled Impactor

1.1 The Use of the Electromagnetic Impounder

Several models of spinal cord contusion injury have been developed in small animals following the initial work of Allen.[1] Using a simple design with a vertical tube and free-falling mass, a rapid impact to the canine spinal cord was produced, using gravity as the primary force of impact to the surface of the exposed spinal cord following a dorsal laminectomy of a single vertebra. The initial studies demonstrated that mild and moderate contusion injury pathology that closely resembles the clinical condition could be produced in animal models.

From: *Animal Models of Acute Neurological Injuries.*
Edited by: J. Chen (eds.) New York, Totowa, NJ

In work done several decades later, the concept of a gravity-driven impact was adapted to rat models, where contusion injury was produced by dropping a calibrated mass from a defined distance above the exposed spinal cord, using a variety of vertical tube designs.[2,3] In these models, a weight is rapidly imparted either directly onto the surface of the exposed dura or onto a platform that rests on the dura, providing a fairly standardized impact force to the spinal compartment. Important modifications of the experimental procedures further improved reliability by standardizing the laminectomy size, stabilizing the spinal cord vertebrae to reduce respiratory artifacts, and standardizing of the impact device components across laboratories.[4–6] The most widely distributed of these designs was initially developed at New York University (NYU) and was adopted for the Multi Animal Spinal Cord Injury Studies (MASCIS) program, which expanded spinal cord injury research through a decade of collaborative research[4,7,8] (Chap. 35). The NYU device incorporated a transducer to record the effect of the impact on the vertebral column, providing feedback to the user regarding the biomechanical characteristics of the impact. Weight-drop injury devices have been shown to produce fairly reliable and graded injuries in rats and mice over the last two decades. Many basic mechanistic and preclinical studies have been pursued with these devices, which reproduce the basic hemorrhagic pathology observed in human contusion injuries. However, the desire to control the parameters of impact beyond height and mass has supported interest in alternative approaches.

At Ohio State University (OSU), an unique device was designed in the late 1980s with a goal to better control some of the mechanical variables of contusion impact.[9–13] The "OSU Injury Device" incorporated an electromagnetic driver capable of rapid and forceful vibration (Ling Shaker) and a pattern generator and amplifier to allow compression of the dorsal surface of the spinal cord with a single impact and retraction (Fig. 1). User control of the compression pattern was incorporated to reduce variability in the mechanics of displacement. In addition, the elimination of the long vertical tube that is required for a weight-drop design helped to decrease concerns with maintenance and consistency over time associated with friction and angle of descent. Displacement and force/acceleration transducers (impedance head) were incorporated into the vertical shaft of the OSU device to provide direct and instant information about the precise dynamics of the impact on the spinal cord. Initially, a closed loop feedback-based design was developed to permit dynamic control of the shaker displacement during the course of the impact,[11,14] so that impact force parameters could be incorporated into the signal generation. However, the feedback features proved difficult to control in the short time frame necessary for a contusion mimicking the clinical condition, and so the amplitude of displacement was chosen as the independent variable for producing injuries of graded severity, and the design modified to an open-feedback system.[15] The OSU device was also modified to incorporate a mechanical *stop plate* that effectively prevented electronic overshoot and better control of the slope and duration of the displacement.[12,13,16] The resulting device was thoroughly validated and shown to produce reliable graded injuries in rats with reproducible biomechanical, behavioral, and histological outcomes predicted by the choice of displacement distance.[17,18] The OSU Injury Device was thus used exclusively at Ohio State for several years.[19–28]

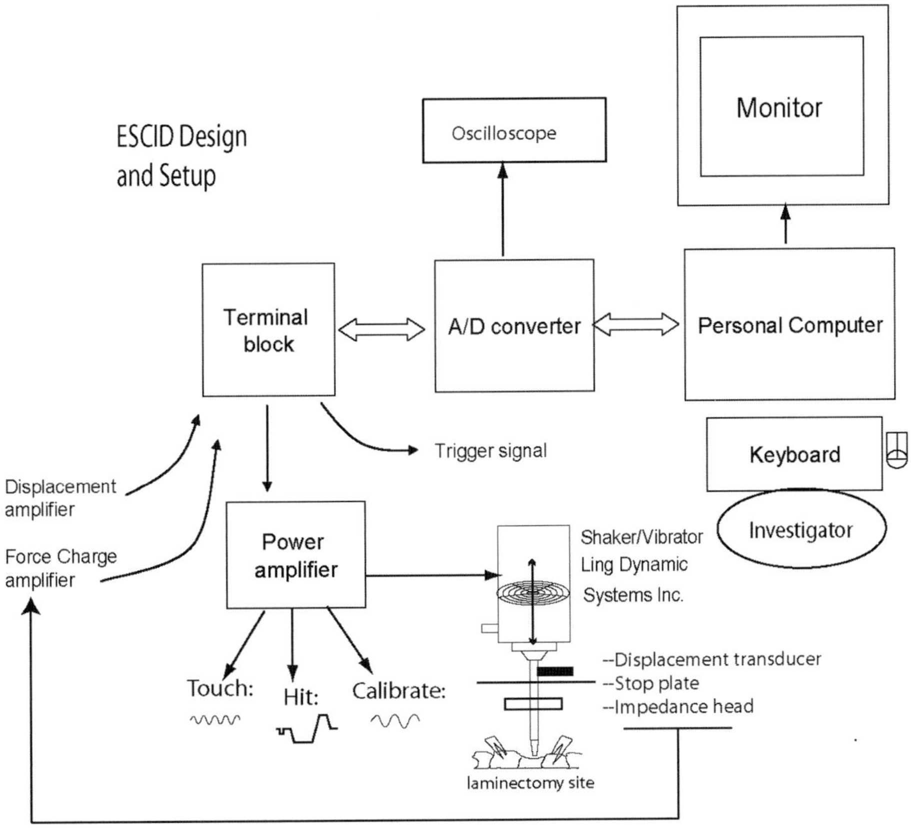

Fig. 1 The OSU injury device consists of a personal computer (PC), data acquisition board, terminal block, power amplifier, and the shaker/vibrator (Ling Dynamic Systems). The personal computer runs a control program, using software that was created using LabView™ programming language. The control program generates one of three different signals, sends them through the terminal block via an interface cable, to a power amplifier. The resulting signal from the power amplifier causes the coil and armature to move axially in the field of the magnet. The control program also initiates a trigger pulse that is used in recording and measuring output data from the transducers for displacement, force and acceleration of the moving probe. Three transducers are located in line with the vertical shaft of the impactor. The displacement is tightly controlled and protected from electronic overshoot by a mechanical Stop plate in line with the impactor

1.2 Revision and Updating the OSU Device: The ESCID 2000

The advantages of an user-controlled and single impact design were evident to many investigators from the initial descriptions, but reproduction of the OSU device was complicated by the unique design of the pattern generator, which had been created by using carefully selected electronic components and an unique engineered design.[14] In the late 1990s, three major developments came together that inspired further revision of the Ling Shaker-based OSU Injury Device. The first was the evolution and affordability of interface boards and user-friendly programming software for personal computers, such as those made available commercially from National Instruments, Inc. This permitted the production of a controlled analog output signal with unlimited flexibility and reproduction, and simultaneous data collection from transducers to a standard ASCII data file. The second development was the increased need for reproducible mouse models of contusion injury, and a need to easily alter user-defined parameters for adaptation to this small species. Finally, the financial backing for revision of the impactor was provided by a nonprofit organization, the International Spinal Research Trust. The objective of the project was to establish a device that could be reproduced by academic machine shops and programming staffs anywhere in the world, using commercially available components and software. The resulting design was renamed the Electromagnetic Spinal Cord Injury Device 2000 (ESCID 2000) to reflect its design, standardization, and era of development.[29] The ESCID successfully reproduced the essential features of the OSU Impactor while providing the flexibility for establishment of the model in a number of other institutions, including several major spinal cord injury research centers.[30–32]

1.3 OSU/ESCID Features

The key design features and components that are used for the ESCID impactor are illustrated in Fig. 1. The user interface is composed of a PC-based computer with an analog/digital processor card and LabView™ programming software. The resulting analog signal is forwarded to a terminal block and amplifier needed to drive the electromagnetic impact pattern.

1.3.1 "Touch" Signal

The principle characteristic of the OSU injury device design is the use of three distinct electronic signal patterns to ensure biomechanical reproducibility of the impact (Fig. 1). The first of these is a "Touch" signal, composed of a small (0.4 V) sinusoidal displacement (±15 μm) at 60 Hz. This is used in conjunction with the force transducer to obtain real-time measures of the dynamic force and acceleration from the impact probe as changes in static impedance. As the impact probe is slowly lowered onto the exposed dura, the recorded force is used to indicate when a precise amount of initial resistance is encountered, consistently indicating the location of the cord surface. Thus, the "Touch" signal is used to establish the displacement start point (or 0 distance) from which to initiate the injury sequence. The ESCID is the only current model that incorporates this starting point feature. By starting the injury sequence with a small and reproducible dimpling of the cord surface, the initial force is also standardized. The effect of this small force on the shape of the sinusoidal pattern measured from the force transducer provides immediate feedback regarding the initial mechanical properties of the specimen. This slight pressure on the surface of the spinal cord causes no detectable behavioral deficits or histological damage in rats or mice, although a very mild microglial hypertrophy can be observed in the dorsal columns at 2–3 days postsurgery with either laminectomy alone or laminectomy plus the touch pattern.

1.3.2 "Hit" Sequence

The second computer generated signal sequence produced by the ESCID is a "Hit" pattern, with the slope and dwell time dictated by the investigator through a software interface. Control of this sequence provides the necessary voltage ramp settings for the impact and retraction phases of impact to cause a single, rapid contusion injury. Reduction in the slope of impact or increasing the duration of dwell can be used to modify the total energy or impulse imparted upon the spinal cord with injury. Using these parameters, the extent of hemorrhage and ischemic insult may be independently manipulated.

1.3.3 "Calibration" Signal

The dependence of the user interface on the conversion of force and displacement measures to voltages requires a means to exclude daily variation from the intended injury characteristics. To account for this required precision, a third signal sequence, composed of a 0.64 V sinusoidal "Calibration" pattern, is provided to permit calibration of the voltage output from the force/acceleration transducer and ensure consistent measures over time.

The ability to easily generate these standardized signal patterns and transmit them to the injury probe on the shaft of the electromagnet provides exquisite and precise control of the independent displacement variable and flexibility necessary to incorporate variations in the sequence of injury.

1.4 Recording the Biomechanical Variables at Impact

The use of sensitive monitoring devices to permit evaluation of biomechanical variables during the injury process has proven to be important for developing and standardizing injuries across animals. With the compression rate and amplitude of displacement tightly controlled, inter-animal variation is principally a function of differences within the spinal cord compartment.[33] The OSU Injury Device allows the investigator to record the actual displacement and force imparted to the spinal cord during the injury (Fig. 2). Displacement of the vertical probe shaft, by definition, reflects the independent input by the investigator, and shows little to no variation under experimental conditions. In contrast, variations between animals are reflected primarily in the peak force measurement and in the shape of the recorded force curve. In the OSU device design, this parameter is a precise force measurement that is corrected for acceleration of the mass below the probe.[11] Evaluation of the changes in force over time gives very sensitive feedback regarding biological variability that can be used to assess subtle characteristics of the injury event.[29] Additional biomechanical parameters are derived from displacement and force parameters, including impulse, velocity, power, and energy (Fig. 3). The calculation and display of these biomechanical measurements provide the investigator with immediate feedback that can be used to identify and exclude subjects with injury conditions that are predicted to lie outside of a predetermined range of variation (see later).

2 Device Setup and Validation

2.1 Device Components

A qualified electronics lab can assemble the ESCID from commercially available components, according to specifications available from the authors. The device is controlled using a standard personal computer with data acquisition board (National Instruments, Inc., Austin, TX), terminal block, and LabView application builder programming software installed (National Instruments, Inc.). A LabView control program and documentation developed by Dr. Rodger Dwonczyk and updated by Dr. John Buford in collaboration with the authors is also available on request (© 1997, 2006 The Ohio State University). The impactor unit assembly consists of the Shaker unit and lateral

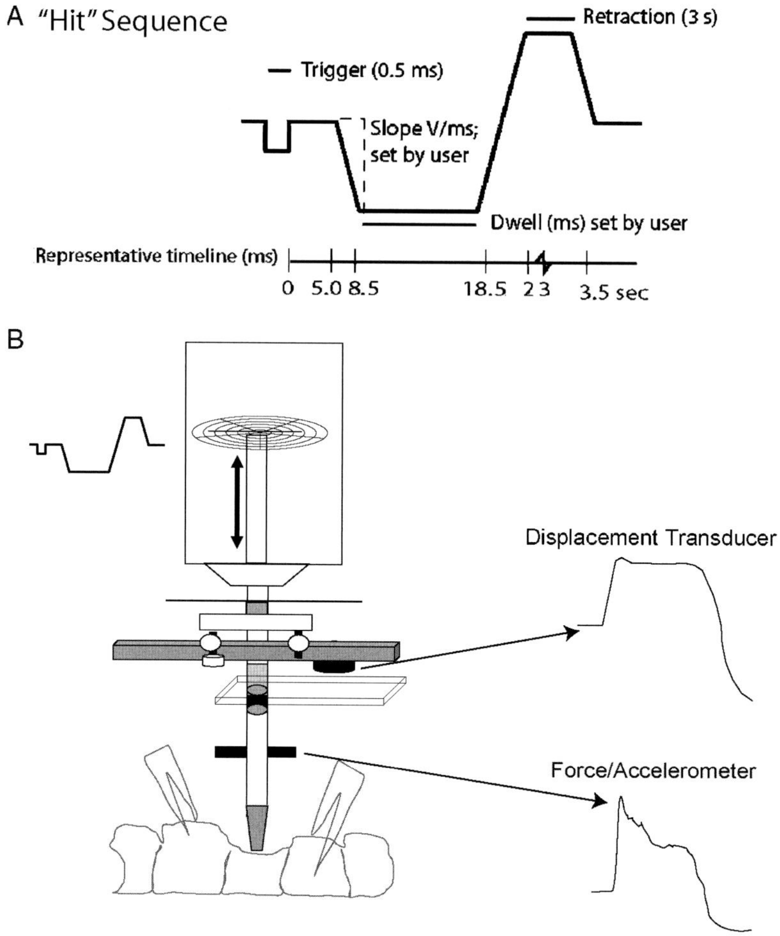

Fig. 2 (**A**) The "Hit" sequence is defined by the user in the control program menu. The sequence consists of a trigger pulse followed by downward displacement with user-defined slope and dwell at maximal displacement. After a single impact, the probe is retracted beyond the start point and held for 3 full seconds, which allows the investigator to remove the subject from the injury table. (**B**). Illustration of the "Hit" sequence as defined by the user and the impactor components. The voltage signal is converted to an electromagnet, causing vertical displacement of the central shaft. In its current configuration, the displacement transducer is located below the stop plate with the reference plate just below it. The impedance head (force and acceleration transducer) is mounted directly on the shaft. Traces on the right illustrate typical readings from the displacement and force transducers, respectively. Scales for the recordings are included in Fig. 3

holder (V203/S, T-200; Ling Dynamics Systems, Los Angeles, CA), power supply and power amplifier, a steel vertical threaded rod, reference plate and transducer mount for the displacement transducer (KDM-7200 4SB, Kaman Instrumentation Corp., Colorado Springs, CO), impedance head mount and force/acceleration transducer (Wilcoxin Research, Bethesda, MD), impedance charge amplifier (Model 104, Endevco, Pasadena, CA), displacement transducer pre-amp (Endevco), aluminum displacement manual adjustment disk assembly, steel mechanical stop plate, and a Lucite impactor tip sized for rat or mouse injuries (Fig. 4g).

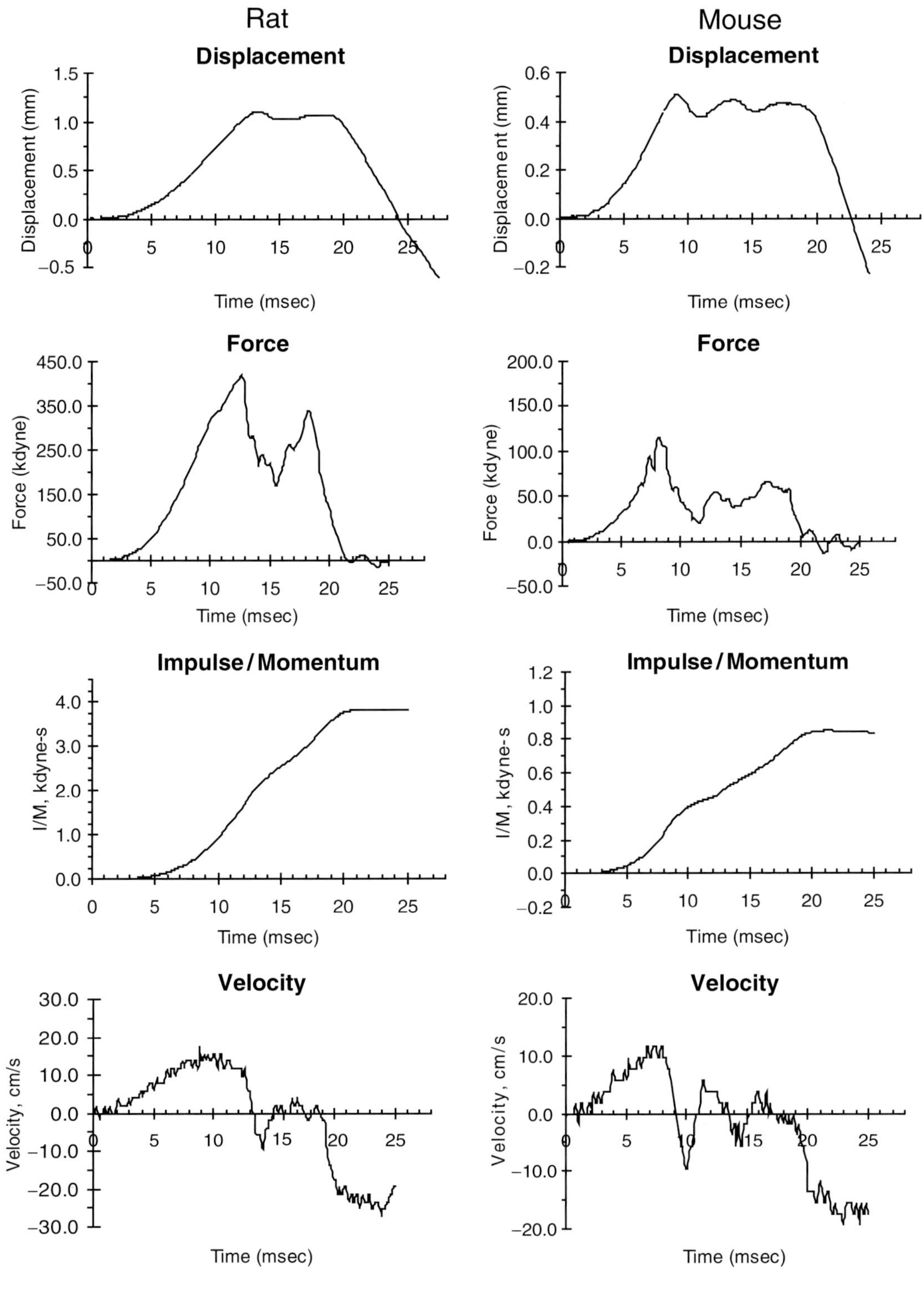

Fig. 3 Direct measures of displacement (*top row*) and force (*second row*) were obtained from representative moderate injuries performed in rat (*left*; 1.0 mm displacement) and mouse (*right*; 0.5 mm displacement) subjects. Impulse/momentum represents the calculated area under the force curve (*third row*), and velocity amplitude and direction were calculated as displacement per unit time (*bottom row*). Note the initial acceleration of the impact head to a maximum of 10–20 cm s^{-1} at the time of impact

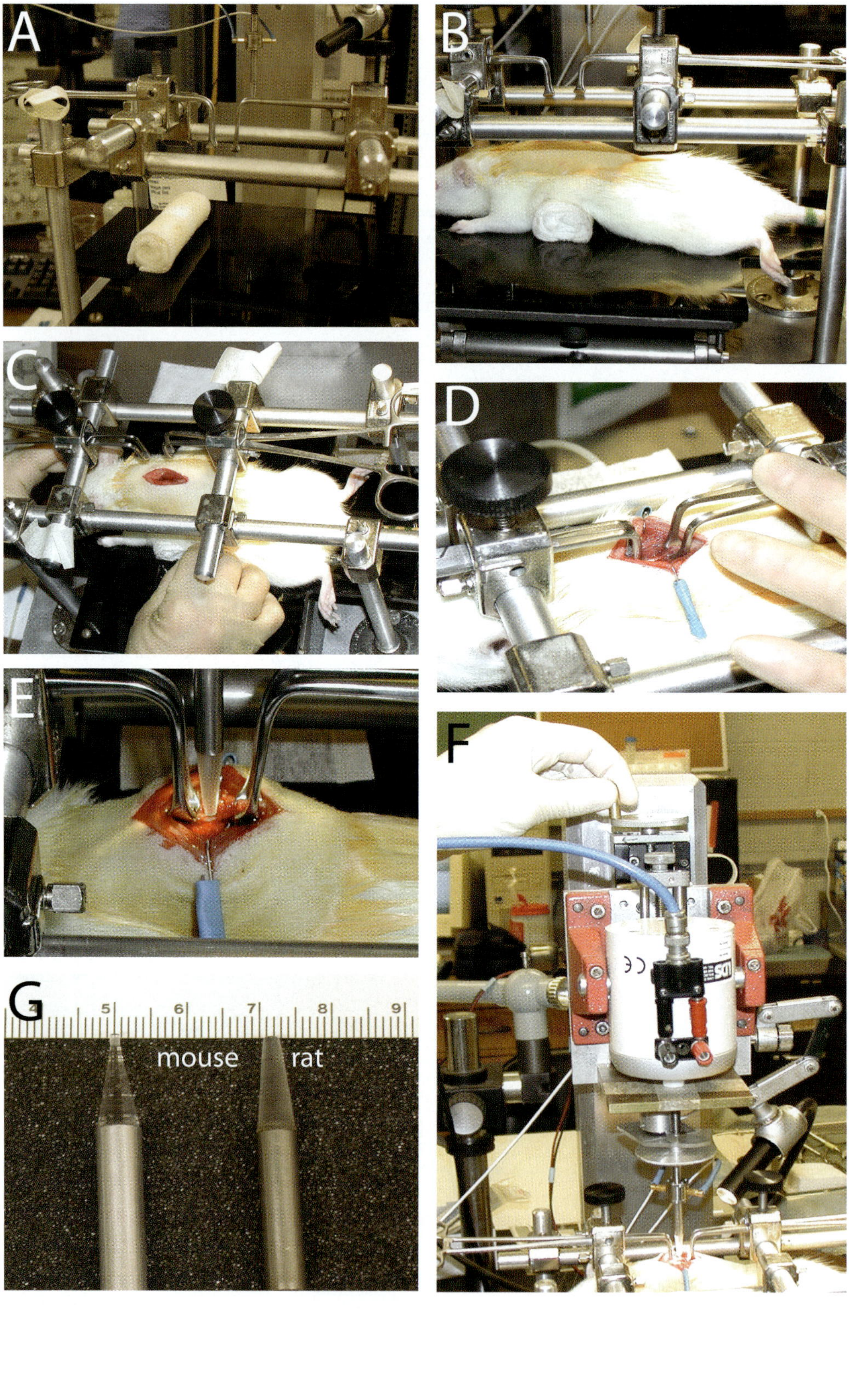
A
B
C
D
E
F
G
5
6
7
8
9
mouse
rat

2.2 Accessories

The investigators must devise and assemble appropriate equipment to support the instrument components (above), as well as an oscilloscope to record the transducer output in real time, a lab jack or adjustable table to raise and lower the experimental animal in a prone position, and a frame and clamping mechanism to stabilize the vertebral column of the experimental animal before, during, and immediately after the impact. Finally, a dial displacement indicator is required for routine calibration of the displacement/voltage conversion (Model 2046-11; Mitutoyo, City of Industry, CA). With few modifications, the support system and accessories will permit reproducible injuries in rats and mice, and potentially translate to larger animals as well.

2.3 Familiarization and Validation of Output Signals

Once the components have been assembled and the power amplifier calibrated and adjusted, several steps must be completed to establish the reliability and reproducibility of the injury device. The investigators should first familiarize themselves with the dialog screens and graphics manipulations of the control program. Using the documentation provided with the control program software and prepared sham tubing materials prepared and provided by the author's laboratory, sample touch, air hit, and injury sequences should be initiated and the resulting output data recorded for comparison with examples in the documentation. It is important to ensure that the output signals are converted to achieve the anticipated magnitude and direction of impactor displacement.

3 The Calibration and Use of the ESCID

3.1 Displacement Voltage Calibration

The independent variable of impact displacement is achieved with precision by applying a defined voltage ramp to the impactor head. Likewise, the precise measurement of actual displacement in time is recorded as a voltage and converted to millimeter by the LabView software. Variations in voltage readings are known to occur over time, and the direct conversion from millimeter to voltage is not precisely linear. Therefore, it is important to calibrate displacement on a regular basis. The following steps are used to measure and establish the conversion factor for the displacement transducer. These steps are performed

Fig. 4 Photographs of the procedures used to stabilize and impart injury to the rodent spinal cord. (**A**) The labjack is set below the impactor with a gauze pillow positioned to support the subject during positioning. Allis clamps with 90° bends are used to stabilize the rostral and caudal dorsal vertebral processes in the rat. (**B**) A rat is positioned directly under the clamps and impactor. (**C**) Dorsal view of the positioning of the rat. (**D**) The rostral and caudal vertebral processes are secured with modified Allis Clamps, while the skin is retracted laterally. (**E**) The impactor probe is lowered down to within 2 mm of the spinal cord surface at the laminectomy site. Fiber optic lights are used to help guide the investigator during this fine positioning step. (**F**) With the control program running a sinusoidal "touch" pattern, the entire impactor device is slowly lowered with the top wheel assembly while viewing the output from the force transducer on an oscilloscope. At a fixed starting force that just barely dimples the surface of the spinal cord (3,000 dynes for rat; 1,500 dynes for mouse), the "touch" pattern is stopped and the "Hit" pattern initiated. (**G**) Tips of the mouse and rat ESCID impactor probes shown to scale. The rat impactor is ~1.85 mm in diameter; the mouse probe is ~1.3 mm in diameter each time the device and transducers are powered on, and whenever the desired displacement setting is established for a group of subjects. The voltage calibration is also checked over the course of the day and adjusted if needed. The voltage calibration is performed as follows:

1. Place the displacement dial indicator directly below the impactor probe tip and lower the impactor to approximately 1 mm above the top of the indicator.
2. Using the Displacement Calibration screen of the LabView interface, initiate the "touch" signal profile to establish a recording from the impedance head/force transducer.
3. Slowly lower the impactor device using the adjustment handle until the force transducer reading on an oscilloscope reveals spikes that indicate the 0 or starting point of the indicator plate. Select the "Stop Touch" button on the LabView screen to stop the vibration. Then, select the "Calibrate" button to record the calibration procedure.
4. Manually depress the impactor ring until the indicator reaches the desired displacement setting for the study that will be performed (0.30–1.300 mm). The central shaft must descend precisely vertically. This is accomplished by using the thumbs of both hands, each thumb placed on opposing sides of the impactor ring. Hold the ring at the desired displacement for 3–5 s and release.
5. Using the cursors on the Displacement Calibration screen, determine the voltage difference that corresponds to the manual displacement profile. Adjust the transducer output until the voltage difference corresponds to the manual displacement (1 mm = 1 mV). Repeat steps 1–5 until 5 consecutive voltage readings are within 0.005 mV. Select "Return" to the main control program screen.
6. Make final adjustments to the mechanical stop plate. Confirm that there is nothing in the path of the impactor to prepare for air hits. On the LabView main control screen, select "Hit", and record actual displacement. Using the fine adjustment set the position of the stop plate until actual displacement corresponds to the desired value. The ESCID is now ready for use.

3.2 Vertebral Suspension and Impact Procedures

Once the injury device is operational and displacement has been calibrated, the procedures for obtaining reproducible injuries are straightforward. These steps assume that the investigator has anesthetized the animal appropriately and prepared a laminectomy at the desired spinal level. The laminectomy site must be sufficiently large to allow the impact probe tip to freely contact the spinal surface. Residual perisoteal membrane on the surface of the dura should be removed prior to injury, and the dura must be kept moist with sterile saline to prevent changes in elasticity and turgor. A suitable suspension apparatus will be chosen, and the surgical site prepared to secure the vertebrae by the dorsal or lateral processes as chosen by the investigator. The steps used to proceed with the injury procedure are as follows:

1. Prior to positioning the animal at the site of the injury device, check the anesthesia level and manually void the animal's bladder. The Lab-jack, chest pillow, retractor hooks, and vertebral clamps should be in position (Fig. 4a).

2. Place the animal on the Lab-jack or adjustable table with a chest pillow if needed (Fig. 4b, c). Be sure the laminectomy incision is parallel to long axis of the animal and the table. Use retractors or hooks to pull overlying muscles laterally and open the laminectomy incision for maximal visibility. Confirm hemostasis and hydration, and ensure that no muscle, bone chips, or other objects will obstruct a clear displacement trajectory for the probe.
3. Using the desired suspension and stabilization apparatus, clamp the vertebrae rostral and caudal to the laminectomy site (Fig. 4d). Again, maintain alignment, hemostasis, and hydration. Do not stretch or compress the surrounding vertebrae in the steps used to suspend and stabilize the vertebrae.
4. When the vertebrae are secure and stable, lower the support table and remove any pillows. This is important for reducing variations due to respiratory artifacts and to keep the airway clear.
5. Position the laminectomy site under the probe. Once the animal is in position, lower the probe until it is 2–3 mm above the exposed spinal cord surface and well centered beneath the probe (Fig. 4f, g).
6. Initiate the "touch" signal to the impactor probe and manually lower the probe toward the cord. By observing the sinusoidal output of the force transducer, determine when the probe touches the surface of the spinal cord. A brief horizontal reading of the force transducer, followed by a continued steady sinusoidal wave, will reflect this. If the sine wave appears jagged or skewed to one side, the probe is not evenly apposed to spinal cord tissue or fluid. Raise the impactor head and reposition the vertebrae if needed to center the probe, or remove any obstructing tissues. In some cases, the animal will need to be removed to the surgical table and the laminectomy site enlarged.
7. Once a clear path is obtained, continue descending slowly, to just cause a small dimpling of the dural surface. When the amplitude of touch wave increases to 3 kdyne (for rat), or 1.5 kdyne (for mice), stop the "touch" signal and initiate the "hit" sequence without delay. The sequence will end with the retraction of the probe to a position approximately 1.5 mm above the start, or 0 displacement position.
8. Immediately following the hit, use the impactor crank arm to raise the probe approximately 1 cm to easily clear the top of the incision site.
9. Raise the Lab-Jack or table and release the vertebral clamps. The investigator must carefully hold the animal's hindquarters and tail while the animal is released. Frequently, reflex kicking will occur as an immediate response to the injury.
10. When the subject is clear of all obstructions and any reflex movements have ceased, support the body and return the animal to the surgery table.
11. Observe the injury site under a surgical microscope; record the appearance of the dorsal cord surface. Bruising is typically rated as mild (slight discoloration or surface vessel damage), moderate (discoloration or hemorrhage visible and bilateral), or severe (extensive subdural hemorrhage, occasionally accompanied by swelling).
12. Close the incision site and perform postoperative procedures as dictated by an institutionally approved survival surgery and care protocol.

4 Discussion

The OSU/ESCID spinal cord injury device produces a reliable and reproducible spinal contusion injury in rats and mice. The choice of displacement as the independent variable for the design is supported by early data indicating that displacement can be fixed with low variability, and shows a high correlation with anatomical sparing and functional recovery.[18,20,27,34,35] The initiation of the injury sequence from a standardized "Touch" force recorded directly from the impact probe ensures the reliability of the displacement/force relationship across animals.

The user-controlled impact or "Hit" sequence serves to strike a balance between the goals of producing highly reproducible injury pathophysiology of injuries of graded severity while maintaining biological conditions that closely mimic the biomechanical consequences of rapid contusive injury in humans. Using the software interface, the user can dictate the slope of the displacement and the dwell time of compression. Recent work with a force-driven spinal cord injury device (IH Impactor) has demonstrated that different pathological and functional events can be obtained by increasing the dwell time of a single impact. Although the mechanisms for this are not clear, the longer dwell time may increase the ischemic component of the injury, but this has not yet been studied in detail.

Immediate feedback is available to the user, regarding the biomechanical characteristics of the impact event. The greatest advantage of this feedback is the ability to analyze the impact parameters in comparison with historical and experimental controls. In most studies, the output data can be quickly evaluated to determine whether abnormal physical events may contribute to nonstandard biological outcome.[29] For example, errors of surgical preparation such as bone or muscle impact effects, dural adhesion, vertebral slipping from the suspension apparatus, or a very dehydrated tissue compartment would be evident as abnormalities in the shape of the recorded waveforms during the "Touch" or "Hit" sequences. Furthermore, in those studies where very low variability is essential for testing a hypothesis regarding injury severity, analysis of the biomechanical conditions within the injury compartment can be used to carefully select a narrow range of variation.[35] A recent study discusses the advantages of setting narrow limits for peak force or displacement parameters to ensure consistent injury outcomes and reduce structural and functional variations with injury.[36]

The ESCID has been successfully used for producing reliable injuries in rats and mice with the simple replacement of the impact probe. Although most of these contusion models have been developed for thoracic spinal cord levels, this device is also useful for preparing contusive injuries in the cervical spinal cord.[37] To date, the contusion injury studies using this device have employed a standard single impact with a displacement ranging from 0.3–1.5 mm, a slope of 0.3–0.5 V ms^{-1}, and dwell times of 8–10 ms duration. However, the control program can be adjusted to accommodate more prolonged slope or longer dwell times, allowing the development of more complex models of contusion/compression injury.

There are many advantages to the use of the OSU/ESCID model, including the use of a fixed starting point through the "touch" feature, the ability to adjust impact parameters with a user interface, the precision of the hit sequence that is possible with the mechanical stop, and the immediate feedback available from the displacement and force transducer readings. However, any validated and reproducible contusion injury device will create a lesion with secondary pathology that closely resembles that of contusion injuries in the clinical theatre. Despite its advantages, the one limitation of the OSU/ESCID for modern laboratories to date has been the requirement for extensive assembly and testing of the components and device and the limited technical assistance available. Core and individual laboratories with a long substantive commitment to spinal cord injury research have been very successful in establishing this model. One day in the future, it may be feasible to obtain commercial investment in sales, assembly, and technical support for the injury device. This type of commitment would greatly facilitate the growth of reproducible spinal cord injury models.

Acknowledgments: The early discussions, documentation, and first prototype ESCID was developed in conjunction with Drs. Philip Horner and Fred Gage at the Salk Institute. Development of the ESCID was possible through contributions by Roger Dzwonczyk and Ernie Gresh. User methods and portions of Fig. 4 were adapted with permission from the Spinal Cord Injury Training Manual (unpublished), NINDS, Facilities of Research Excellence in Spinal Cord Injury, © The Ohio State University, 2004–2008 with technical help from Zhen Guan and Todd Lash. Collaborating laboratories and consultations regarding ESCID device assembly have included Dr. Joseph Springer (University of Kentucky) and Dr. John Bethea (University of Miami). Funding is provided by the International Spinal Research Trust, N01-NS-3–2350, NS045758, and NIH NS043246.

References

1. Allen AR. Surgery of experimental lesion of spinal cord equivalent to crush injury or fracture dislocation of spinal column. A preliminary report. JAMA 1911;57:870–880.
2. Ford RW. A reproducible spinal cord injury model in the cat. J Neurosurg 1983;59:268–275.
3. Noble LJ, Wrathall JR. An inexpensive apparatus for producing graded spinal cord contusive injury in the rat. Exp Neurol 1987;95:530–533.
4. Young W, Decrescito V, Flamm ES, Blight AR, Gruner JA. Pharmacological therapy of acute spinal cord injury: Studies of high dose methylprednisolone and naloxone. Clin.Neurosurg. 1988;34:675–697.
5. Gruner JA. A monitored contusion model of spinal cord injury in the rat. J Neurotrauma 1992;9:123–128.
6. Basso DM, Beattie MS, Bresnahan JC. Graded histological and locomotor outcomes after spinal cord contusion using the NYU weight-drop device versus transection. Exp Neurol 1996;139:244–256.
7. Basso DM, Beattie MS, Bresnahan JC, Anderson DK, Faden AI, Gruner JA, Holford TR, Hsu CY, Noble LJ, Nockels R, Perot PL, Salzman SK, Young W. MASCIS evaluation of open field locomotor scores: Effects of experience and teamwork on reliability. J Neurotrauma 1996;13:343–359.
8. Beattie MS, Bresnahan JC, Komon J, Tovar CA, Van Meter M, Anderson DK, Faden AI, Hsu CY, Noble LJ, Salzman SK, Young W. Endogenous repair after spinal cord contusion injuries in the rat. Exp Neurol 1997;148:453–463.

9. Somerson SK, Stokes BT. Functional analysis of an electromechanical spinal cord injury device. Exp Neurol 1987;96:82–96.
10. Stokes BT, Somerson SK. Spinal cord microenvironment: Can the changes resulting from trauma be graded? Neurochem Pathol 1978;7:47–55.
11. Noyes DH. Electromechanical impactor for producing experimental spinal cord injury in animals. Med Biol Eng Comp 1987;25:335–340.
12. Stokes BT. Experimental spinal cord injury: A dynamic and verifiable injury device. J Neurotrauma 1992;9:129–134.
13. Stokes BT, Noyes DH, Behrmann DL. An electromechanical spinal injury device with dynamic sensitivity. J Neurotrauma 1992;9:187–195.
14. Noyes DH. Correlation between parameters of spinal cord impact and resultant injury. Exp Neurol 1987;95:535–547.
15. Stokes BT, Horner PJ. Spinal cord injury modeling and outcome assessment. In: Narayan RK, Wilberger JE, Povlishock JT, eds. Neurotrauma. New York: McGraw-Hill Co, 1996:1395–1403.
16. Anderson TE, Stokes BT. Experimental models for spinal cord injury research: Physical and physiological considerations. J Neurotrauma 1992;9 Suppl 1: S135–S142.
17. Bresnahan JC, Beattie MS, Todd FD3, Noyes DH. A behavioral and anatomical analysis of spinal cord injury produced by a feedback-controlled impaction device. Exp Neurol 1987;95:548–570.
18. Behrmann DL, Bresnahan JC, Beattie MS, Shah BR. Spinal cord injury produced by consistent mechanical displacement of the cord in rats: Behavioral and histologic analysis. J Neurotrauma 1992;9:197–217.
19. Behrmann DL, Bresnahan JC, Beattie MS. A comparison of YM-14673, U-50488H, and Nalmefene after spinal cord injury in the rat. Exp Neurol 1993;119:258–267.
20. Basso DM, Beattie MS, Bresnahan JC. A sensitive and reliable locomotor rating scale for open field testing in rats. J Neurotrauma 1995;12:1–21.
21. Chen XY, Wolpaw JR, Jakeman LB, Stokes BT. Operant conditioning of H-reflex in spinal cord-injured rats. J Neurotrauma 1996;13:755–766.
22. Popovich PG, Wei P, Stokes BT. The cellular inflammatory response after spinal cord injury in Sprague-Dawley and Lewis rats. J Comp Neurol 1997;377:443–464.
23. Jakeman LB, Wei P, Guan Z, Stokes BT. Brain-derived neurotrophic factor stimulates hindlimb stepping and sprouting of cholinergic fibers after spinal cord injury. Exp Neurol 1998;154:170–184.
24. McTigue DM, Horner PJ, Stokes BT, Gage FH. Neurotrophin-3 and brain-derived neurotrophic factor induce oligodendrocyte proliferation and myelination of regenerating axons in the contused adult rat spinal cord. J Neurosci 1998;18:5354–5365.
25. McTigue DM, Popovich PG, Morgan TE, Stokes BT. Localization of transforming growth factor-β1 and receptor mRNA after experimental spinal cord injury. Exp Neurol. 2000;63:220–230.
26. McTigue DM, Popovich PG, Jakeman LB, Stokes BT. Strategies for spinal cord injury repair. Prog Brain Res 2000;128:3–8.
27. Ma M, Basso DM, Walters P, Stokes BT, Jakeman LB. Behavioral and histological outcome following graded contusion injury in C57Bl/6 mice. Exp Neurol 2001;169: 239–254.
28. Stokes BT, Jakeman LB. Experimental modelling of human spinal cord injury: A model that crosses the species barrier and mimics the spectrum of human cytopathology. Spinal Cord 2002;40:101–109.
29. Jakeman LB, Guan Z, Wei P, Ponnappan R, Dzwonczyk R, Popovich PG, Stokes BT. Traumatic spinal cord injury produced by controlled contusion in mouse. J Neurotrauma 2000;17:299–319.
30. Brambilla R, Bracchi-Ricard V, Hu WH, Frydel B, Bramwell A, Karmally S, Green EJ, Bethea JR. Inhibition of astroglial nuclear factor kappaB reduces inflammation

and improves functional recovery after spinal cord injury. J Exp Med 2005;202: 145–156.
31. McEwen ML, Springer JE. Quantification of locomotor recovery following spinal cord contusion in adult rats. J Neurotrauma 2006;23:1632–1653.
32. Pearse DD, Sanchez AR, Pereira FC, Andrade CM, Puzis R, Pressman Y, Golden K, Kitay BM, Blits B, Wood PM, Bunge MB. Transplantation of Schwann cells and/or olfactory ensheathing glia into the contused spinal cord: Survival, migration, axon association, and functional recovery. GLIA 2007;55:976–1000.
33. Blight A. Mechanical factors in experimental spinal cord injury. J Am Paraplegia Soc 1988;11:26–34.
34. Ma M, Wei P, Wei T, Ransohoff RM, Jakeman LB. Enhanced axonal growth into a spinal cord contusion injury site in a strain of mouse (129X1/SvJ) with a diminished inflammatory response. J Comp Neurol 2004;474:469–486.
35. Kloos AD, Fisher LC, Detloff MR, Hassenzahl DL, Basso DM. Stepwise motor and all-or-none sensory recovery is associated with nonlinear sparing after incremental spinal cord injury in rats. Exp Neurol 2005;191:251–265.
36. Ghasemlou N, Kerr BJ, David S. Tissue displacement and impact force are important contributors to outcome after spinal cord contusion injury. Exp Neurol 2005;196:9–17.
37. Pearse DD, Lo TP, Jr., Cho KS, Lynch MP, Garg MS, Marcillo AE, Sanchez AR, Cruz Y, Dietrich WD. Histopathological and behavioral characterization of a novel cervical spinal cord displacement contusion injury in the rat. J Neurotrauma 2005;22:680–702.

Chapter 38

Acute Clip Impact-Compression Model

Charles H. Tator and Peter Poon

Abstract In human spinal cord injury, the most common mechanism is the combination of acute impact and continuing compression. To simulate combined impact-compression, we developed in the 1970s the acute clip impact-compression model, one of the first nontransection models in the rodent. Subsequently, we characterized the relationships between clip strength, duration of compression, and neurological recovery, and established dose-response relationships between the forces of clip compression injury, axon-evoked potentials, spinal cord blood flow, neurological function, axon counts, and retrograde labeling of supraspinal neurons with axonal tracers. Clip injury is useful for in vitro and in vivo SCI studies in rats and mice for cervical, thoracic, and lumbar injuries, and is consistent, reliable, and relatively inexpensive.

Keywords: Spinal cord injury, Clip impact-compression model, Injury mechanisms

1 Introduction

There are many mechanisms involved in producing spinal cord injury (SCI) in humans, although the most common are compression, contusion, laceration, transaction, and traction of the spinal cord.[1] No single experimental model can simulate all these diverse mechanisms, and therefore, there is a need for more than one model of SCI, especially for screening potential therapies for use in human SCI.[2,3] There have been several useful reviews of experimental models for SCI,[4–6] and in our view, there are several reliable and consistent animal models some of which replicate human injuries reasonably well.[6,7]

The most common mechanism of SCI in humans is a combination of acute impact followed by persisting compression,[1,8] and this combination occurs in fracture-dislocations and burst fractures of the spine, the two most common mechanisms in human SCI. In anterior fracture-dislocation, the most common type, the anterior surface of the spinal cord is compressed between the posterior aspect of the rostral vertebra while the posterior surface of the spinal cord is compressed by the anterior aspect of the lamina of the caudal vertebra. In burst fractures, the anterior surface of the spinal cord is compressed by

From: *Animal Models of Acute Neurological Injuries.*
Edited by: J. Chen (eds.) New York, Totowa, NJ

the posterior aspect of the vertebral body while the posterior surface of the spinal cord is compressed by the anterior aspect of the lamina of the same vertebra. To simulate these mechanisms of combined impact-compression, we developed in the 1970s the acute clip impact-compression model, one of the first to simulate the sequence of acute impact followed by persisting compression.[9] Sudden impact is produced by rapid release of the clip. This model was one of the first nontransection models in the rodent.[7] Subsequently, in my laboratory, we characterized the relationships between clip strength, duration of compression, and neurological recovery.[10,11] We also established dose response-curves for the relationships between the force of clip compression injury, axon-evoked potentials, and spinal cord blood flow,[12] and for the relationships between the force of clip compression injury, neurological function, axon counts, and retrograde labeling of supraspinal neurons with retrograde axonal tracers.[13] Since then, most of the studies in the Tator and Fehlings laboratories at the University of Toronto have utilized this model in in vitro and in vivo SCI studies in rats and mice in cervical, thoracic, and lumbar injuries.

In the initial phase of the development of this model of experimental SCI, we found that an one-minute duration of clip compression was the minimum time required to produce a consistent pathological and clinical effect in rats with clips varying in force from 2–178 g. With clips of this range of forces, we varied the duration of compression from 3 s to as long as 4 h, and found a complex correlation between both the force and duration of compression in rats at C7-T1 or T7-8 and several histopathological and clinical outcome measures. Recently, we showed similar findings in injuries at T2-3 in the rat.[14] Of major importance is that the clip impact-compression SCI can be used to model injuries at all levels of the spinal cord, at a range of severities, and also for acute, subacute, and chronic SCI.[15]

In the Fehlings laboratory, the clip was adapted for producing SCI in the mouse, and the mouse clip was found to produce consistent SCI of graded severity similar to what had been previously established in the rat.[16,17] However, the pathological changes in the mouse spinal cord differ significantly from what has been established in rats and humans. In both rats and humans, major SCI causes ischemia, necrosis, and cavitation, whereas in the mouse the lesions are generally without cavitation.

The range of clip forces can be varied to represent the range of injury severities associated with incomplete injuries in humans graded by the system of the American Spinal Injury Association (ASIA) as ASIA grades B to D.[18] More severe injuries such as a 50 g injury at T4 in the rat produces the clinical and histological features of a complete SCI,[19] graded as ASIA A in the human. In a recent study in our laboratory with the clip compression model,[14] we demonstrated the relative increase in severity of thoracic injuries compared with cervical injuries, which is the case in humans.[8,20]

It should be noted that many other investigators have used clip compression SCI, some with the same modified Kerr-Lougheed aneurysm clip that we have used.[19,21] However, many other types of clips have been used in experimental SCI.[22] For example, von Euler et al. used bulldog clips of varying forces, and placed them on the cord vertically, so that the compression force was from side-to-side or lateral rather than in the antero-posterior direction.[23] In many reports, the authors did not state whether the clips were slowly released to produce compression or rapidly released to produce impact-compression.

2 Advantages and Disadvantages of the Acute Clip Impact-Compression Model

The main advantages and disadvantages of the clip impact-compression model are as follows:

2.1 Advantages

1. Inexpensive
2. Durable. The same clip can be used for an infinite number of times, although the springs will weaken after multiple openings and closings.
3. Useful for injuries at all levels of the spinal cord-cervical, thoracic, and lumbar.
4. Useful for modeling acute, subacute, and chronic injuries.
5. Useful in both rats and mice.
6. Does not require opening of the dura.
7. Produces anterior-posterior compression of the spinal cord.
8. Rapid release of the clip produces sudden impact of the spinal cord.
9. Both the force and duration of compression can be altered to model injuries of low to high severity, and reliable dose-response curves can be generated for both force and duration of compression.
10. Stabilization of the spine during injury is not required.
11. Precise positioning of the spine during injury is not required.
12. The same clip can be mounted with different rings to produce injuries of varying severity.
13. Very low mortality and morbidity rates.

2.2 Disadvantages

1. After hundreds of openings and closings, the springs upon which the force of the clips depends will weaken. Thus, it is wise to calibrate the force of compression before and after each experiment. Calibration is not very complicated and the method of calibration, including a description of the calibration device has been published.[24] Alternatively, the clips can be sent to the Tator or Fehlings laboratories for calibration.
2. Practice and some surgical expertise are required for optimal and consistent placement of the clip around the cord with minimal cord trauma.
3. Laminectomy is required for clip placement.

3 Other Impact-Compression Models

Several other impact-compression models were developed after the acute clip impact-compression model. Although the New York University (NYU) weight-drop impactor[25] is an improvement over the original Allen[26] weight-drop model, none of the weight-drop models produce persisting compression, and thus, they do not simulate the impact-compression of the most common human SCIs. Also, there is inherent variation and wide standard deviations in the results probably because of the randomness with which the weight contacts the walls of the cylinder and the impounder placed on the surface of the cord. The Ohio State University electromagnetic SCI device[27] is based on cord displacement, which can be accurately measured. The same is true of the more recent Infinite Horizons impactor,[28,29] but it is expensive, impacts the spinal cord from the dorsal surface only, has limited dwell-times (i.e., persisting compression for only up to 60 s), and requires both stabilization by forceps and horizontal positioning of the spinal column at the time of SCI. However, the Infinite Horizons device has the advantages of being able to produce SCI of varying displacement and force in rats[29] and mice.[30] The clip compression model is easily adaptable for use at the cervical level, which is not true for several of the contusion models where the adjacent and overlying spinal musculature can interfere with lesion making.[31]

The temporal effect of impact-compression injury is not achieved in most other compression models, including the Infinite Horizons model, which has a limited dwell time. In our previous studies with the clips, we examined these temporal effects after cervical cord SCI in the rat.[10,11] The ability to study varying compression times is an advantage of the clip technique as it can be used to examine the duration of cord compression and the value of early decompression which are important issues in human SCI.[2] The clip can remain in place compressing the spinal cord for clinically relevant durations of compression.

4 Nonimpact Models of SCI

There are many nonimpact models, and only a fraction of them will be discussed. A large number of complete and incomplete transection models have been used to great advantage to examine finite pathophysiological events, but laceration and transection are rare in human SCI. Photochemical injury delivered by irradiation of the spinal cord after intravenous administration of the photosensitive rose bengal dye[32–34] produces ischemia and graded infarction of the spinal cord and is useful to examine posttraumatic ischemia and infarction, which are common secondary mechanisms after SCI.[1,35,36] Also, this model is less invasive because irradiation can reach the spinal cord without laminectomy. Some compression models involve "placement" of a static weight on the cord, and thus do not cause sudden impact,[37] and therefore, these have been considered nonimpact.

Balloon compression models were first introduced byTarlov[38] in the 1950s, and there have been many subsequent versions. Most are slowly compressing and nonimpact, although the more recent balloon materials can be rapidly inflated without bursting. In general, balloon compression models suffer from a lack of reproducibility due to the uncertainty of the balloon-cord interface.

5 Methods for Clip Impact-Compression Injury

The following description applies specifically to rats.

5.1 Anesthesia

Rats can be anesthetized using a variety of techniques. We have found that the combination of 1.5–2% halothane and 2:1 mixture of nitrous oxide and oxygen produces rapid induction, good analgesia, and lack of movement during SCI.

A 3-ml injection of saline is given subcutaneously before surgery, and the rats are placed on a heating pad at 36°C during surgery.

5.2 Microscopy

After the skin incision is made the entire dissection, clip placement, and removal are performed with a surgical operating microscope.

5.3 Laminectomy

A laminectomy is performed at the desired cervical, thoracic, or lumbar level. To facilitate the laminectomy a self-retaining retractor is essential for holding laterally the paraspinal muscles (Fig. 1a). Failure to retract the muscles sufficiently far laterally impairs accurate and atraumatic clip placement. We usually remove two laminae to facilitate dissection of the extradural plane. The laminae are removed as far laterally as possible (Fig. 1b). Usually, this involves removal of the pedicles as far anteriorly as possible so that the stumps of the pedicles will not obstruct the passage of the clip (Fig. 1b). Bone removal is performed with a small bone-cutting instrument or rongeur with slender tapering ends (Fig. 2). A nail cutting instrument is ideal.

5.4 Exploration of the Extradural Plane

A dissecting hook, (Figs. 1c and 2) with similar curvature and thickness as the clip, is used to dissect the extradural plane between the dura and the adjacent vertebrae. The clips, dissecting hook, and clip applicators are manufactured to specification (Fig. 2). The hook is especially important for locating the nerve roots, and for clearing away any extradural adhesions at the optimal site for clip placement, which is at the mid-point between adjacent roots. It is important to pass the hook extradurally from both the right and left sides to determine the optimal trajectory for clip placement. Bleeding will often occur at this stage from the epidural veins and can be controlled with application of small cotton pledgets and with gelfoam sponge. Bleeding must be controlled to have optimal visualization of the cord during clip placement.

5.5 Placement of the Clip

With the clip held in the applicator in its fully opened position (Fig. 1e), the anterior blade of the clip is passed extradurally anterior to the cord with avoidance of damage to the adjacent nerve roots, and with as little cord displacement as possible (Fig. 1e). Some displacement is necessary, but it should be done very slowly to avoid any traction or impact of the cord. The clip held in the applicator can be used to clear away any remaining adhesions that are encountered, but only minimal force can be used, otherwise the clip will be unintentionally and prematurely released from the applicator.

5.6 Rapid Release of the Clip to Deliver the Impact-Compression Injury

The clip is then rapidly released from the applicator to produce the acute impact-compression injury (Fig. 1f). This is done by allowing the blades of the applicator to spring open at maximum velocity. The clip is then left compressing the spinal cord for the desired duration of compression.

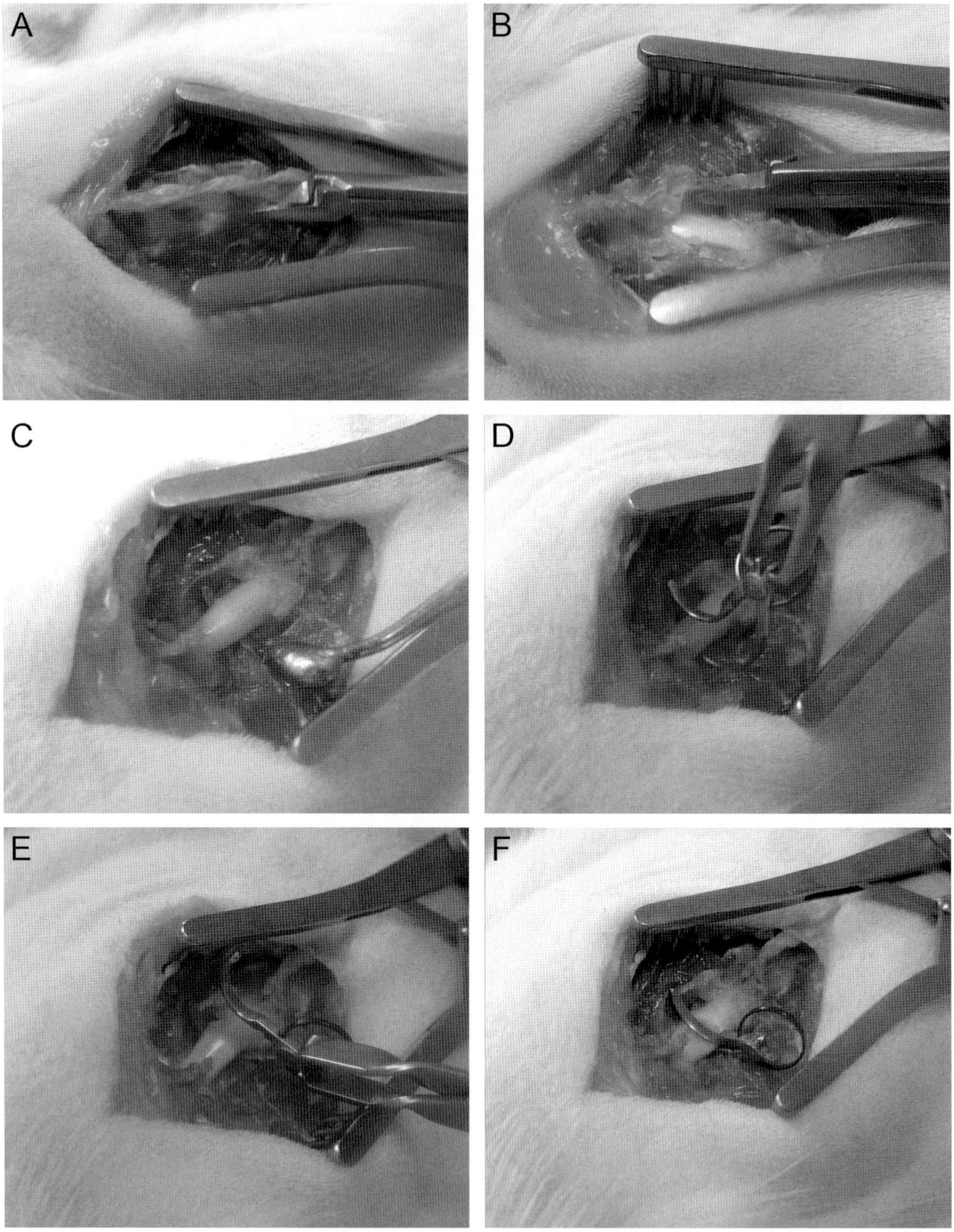

Fig. 1 The operative procedure: The steps are shown in a rat that has been perfused. (**A**) This shows the self-retaining retractor in place holding back the paraspinal muscles. (**B**) This shows the rongeur removing portions of the pedicle far laterally and anteriorly. (**C**) This shows the cord exposed after the two-level laminectomy, and with the passage of the probe anteriorly and extradurally. (**D**) This shows the clip held in the clip applicator, with the clip in the fully opened position, just prior to passage of the blade of the clip extradurally. (**E**) shows the blades of the clip anterior and posterior to the cord in the extradural space just prior to release of the clip from the applicator. The clip is still in the fully opened position. (**F**) The clip has been released from the applicator and is compressing the cord

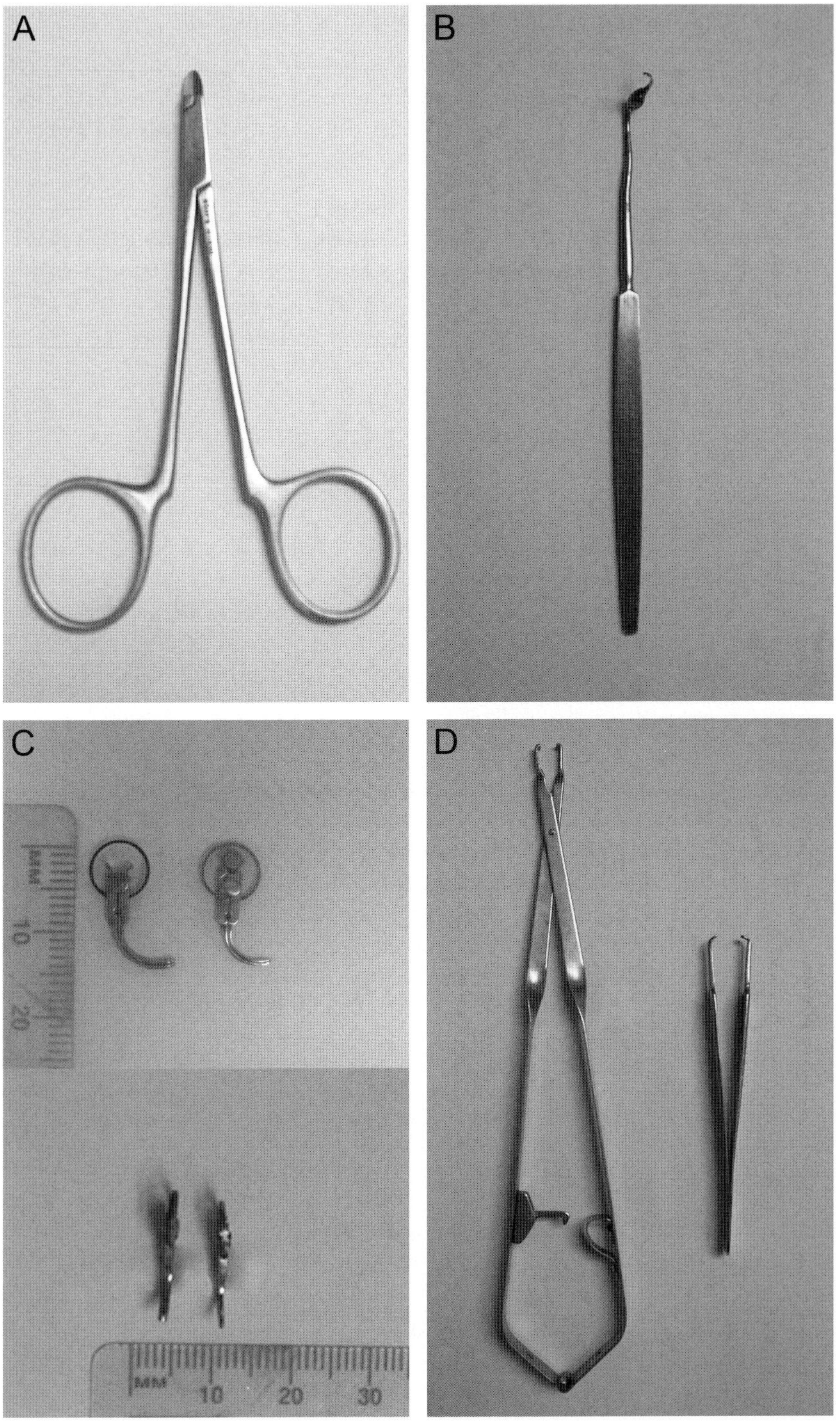

Fig. 2 The instruments for performing clip compression and the clips: (**A**) The bone cutting instrument for removing the laminae and pedicles (rongeur). (**B**) The probe for exploring the extradural plane is a dental hook bent to conform to the shape of a clip blade. (**C**) The *upper panel* is a side view of the clips and the *lower panel* is a top view. In both the upper and lower panels, a rat clip is on the left and a mouse clip is on the right. The length of the blades of the mouse clip is smaller, and the blades and spring of the mouse clip are thinner. (**D**) Two styles of clip applicators. The applicators have ledges at the ends that fit into the grooves at the end of the clip blades to permit firm grasping of the clips

5.7 Removal of the Clip with the Applicator

After the desired interval, the applicator is reintroduced and used to grasp and then open the clip maximally without any further manipulation of the cord. The clip is then removed with as little cord displacement as possible.

5.8 Suturing of the Wound

After injury, the muscles are sutured in one layer using 3–0 vicryl suture (Ethicon, Somerville, New Jersey) and the skin closed with Michel clips (Fine Science Tool, B.C., Canada). Buprenorphine (0.03 mg kg^{-1}, Temgesic, Schering-Plough, UK) is given subcutaneously before the animals awake, and then every 12 h for 48 h.

5.9 Postoperative Care

The rats are housed singly in a temperature-controlled room at 26°C for the duration of the experiment with a 12 h light/dark cycle. Bladders are expressed three times daily until spontaneous voiding occurs, and any hematuria or urinary tract infection is treated with ampicillin (100 mg kg^{-1}, s.c., Novopharm, Toronto, Canada) twice daily for 5 days. Water and food are provided *ad libitum*.

5.10 Ordering Information

The clips and applicators for rats and mice can be ordered directly from the Tator or Fehlings laboratories. The rat clips are approximately $2,000 each and the mouse clips are approximately $1,300 each. They can be supplied in a variety of forces of clip closing strength.

The clips for mice known as Fejota (Fehlings-Joshi-Tator) clips can be ordered from the Fehlings laboratory, and a signed Research Agreement is required. The University Health Network, the group of hospitals in which the Tator and Fehlings laboratories are located is the holder of the patent and trademark on the mouse clip.

5.11 Calibration of Clips

Investigators can perform their own force of clip closure calibrations or can request this service from the Tator or Fehlings laboratories.

5.12 Contact Information for Authors

Charles Tator charles.tator@uhn.on.ca
Peter Poon ppoon@chem-eng.utoronto.ca

5.13 Outcome Evaluations

A full range of histological (Fig. 3), imaging, and functional outcome measures has been used successfully after clip impact-compression injury.

6 Summary

Our group has used the clip impact-compression injury model to injure the spinal cord at the cervical,[39–41] thoracic,[15,42] and lumbar[43] levels since the 1970s, and this model has given reliable and consistent results for studying

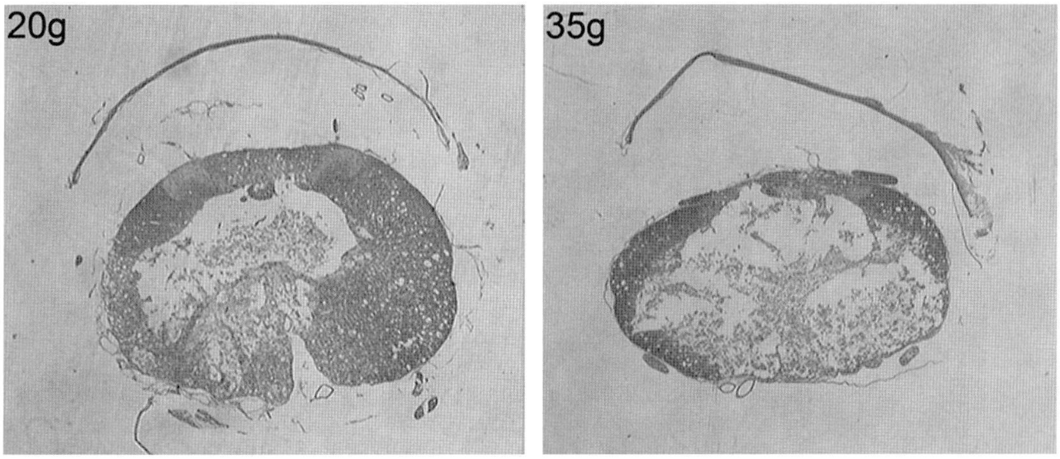

Fig. 3 Histological sections of the rat thoracic cord several weeks after clip compression injury. Hematoxylin and eosin and luxol fast blue staining. The central necrosis and cavitation are much greater after 35 g injury than after 20 g injury

acute, subacute, and chronic SCI in rodents. Other groups have also successfully used our clips for experimental SCI in rodents.[19,21,44] Several other clip designs were subsequently published and appear to give reliable results.[45–48] The clips have been used for examining a variety of injury mechanisms including inflammatory cytokines,[49] secondary injury antagonists,[22] and ischemia.[12] The clips have also been used for studying axonal physiology and morphology,[50] and neuroprotective and regenerative measures including localized drug delivery,[51,52] and cell transplantation.[48,53,54]

References

1. De Girolami U, Frosch MP, Tator CH. Regional Neuropathology: Diseases of the Spinal Cord and Vertebral Column. In: Graham DI, Lantos PL, eds. Greenfield's Neuropathology. 7th edn. London, England: Arnold; 2002:1063–101.
2. Tator CH. Review of treatment trials in human spinal cord injury: issues, difficulties, and recommendations. Neurosurgery 2006;59(5):957–82; discussion 82–7.
3. Anderson TE, Stokes BT. Experimental models for spinal cord injury research: physical and physiological considerations. J Neurotrauma 1992;9 (Suppl 1):S135–S142.
4. Kwon BK, Oxland TR, Tetzlaff W. Animal models used in spinal cord regeneration research. Spine 2002;27(14):1504–10.
5. Grill RJ. User-defined variables that affect outcome in spinal cord contusion/compression models. Exp Neurol 2005;196(1):1–5.
6. Fehlings MG, Tator CH. A review of experimental models of acute spinal cord injury. In: Illis LS, ed. Spinal Cord Dysfunction:Assessment. Oxford: Oxford University; 1988:3–43.
7. Wrathall JR. Spinal cord injury models. J Neurotrauma 1992;9 (Suppl 1): S129–S134.
8. Tator CH. Spine-spinal cord relationships in spinal cord trauma. Clin Neurosurg 1983;30:479–94.
9. Rivlin AS, Tator CH. Effect of duration of acute spinal cord compression in a new acute cord injury model in the rat. Surg Neurol 1978;10(1):38–43.
10. Dolan EJ, Tator CH, Endrenyi L. The value of decompression for acute experimental spinal cord compression injury. J Neurosurg 1980;53(6):749–55.

11. Guha A, Tator CH, Endrenyi L, Piper I. Decompression of the spinal cord improves recovery after acute experimental spinal cord compression injury. Paraplegia 1987;25(4):324–39.
12. Fehlings MG, Tator CH, Linden RD. The relationships among the severity of spinal cord injury, motor and somatosensory evoked potentials and spinal cord blood flow. Electroencephalogr Clin Neurophysiol 1989;74(4):241–59.
13. Fehlings MG, Tator CH. The relationships among the severity of spinal cord injury, residual neurological function, axon counts, and counts of retrogradely labeled neurons after experimental spinal cord injury. Exp Neurol 1995;132(2): 220–8.
14. Poon PC, Gupta D, Shoichet MS, Tator CH. Clip compression model is useful for thoracic spinal cord injuries: histological and functional correlates. Spine 2007; 32(25):2853–2859.
15. Parr AM, Kulbatski I, Tator CH. Transplantation of adult rat spinal cord stem/ progenitor cells for spinal cord injury. J Neurotrauma 2007;24(5):835–45.
16. Joshi M, Fehlings MG. Development and characterization of a novel, graded model of clip compressive spinal cord injury in the mouse: Part 2. Quantitative neuroanatomical assessment and analysis of the relationships between axonal tracts, residual tissue, and locomotor recovery. J Neurotrauma 2002;19(2):191–203.
17. Joshi M, Fehlings MG. Development and characterization of a novel, graded model of clip compressive spinal cord injury in the mouse: Part 1. Clip design, behavioral outcomes, and histopathology. J Neurotrauma 2002;19(2):175–90.
18. Geisler FH, Coleman WP, Grieco G, Poonian D. Measurements and recovery patterns in a multicenter study of acute spinal cord injury. Spine 2001;26 (24 Suppl):S68–S86.
19. Weaver LC, Verghese P, Bruce JC, Fehlings MG, Krenz NR, Marsh DR. Autonomic dysreflexia and primary afferent sprouting after clip-compression injury of the rat spinal cord. J Neurotrauma 2001;18(10):1107–19.
20. Sekhon LH, Fehlings MG. Epidemiology, demographics, and pathophysiology of acute spinal cord injury. Spine 2001;26(24 Suppl):S2–S12.
21. Khan M, Griebel R, Rozdilsky B, Politis M. Hemorrhagic changes in experimental spinal cord injury models. Can J Neurol Sci 1985;12(3):259–62.
22. Gorio A, Gokmen N, Erbayraktar S, et al. Recombinant human erythropoietin counteracts secondary injury and markedly enhances neurological recovery from experimental spinal cord trauma. Proc Natl Acad Sci USA 2002;99(14):9450–5.
23. von Euler M, Seiger A, Sundstrom E. Clip compression injury in the spinal cord: a correlative study of neurological and morphological alterations. Exp Neurol 1997;145(2 Pt 1):502–10.
24. Dolan EJ, Tator CH. A new method for testing the force of clips for aneurysms or experimental spinal cord compression. J Neurosurg 1979;51(2):229–33.
25. Gruner JA. A monitored contusion model of spinal cord injury in the rat. J Neurotrauma 1992;9(2):123–6; discussion 6–8.
26. Allen AR. Surgery of experimental lesions of the spinal cord equivalent to crush injury of fracture dislocation of the spinal column. A preliminary report. JAMA 1911;57:878–80.
27. Behrmann DL, Bresnahan JC, Beattie MS, Shah BR. Spinal cord injury produced by consistent mechanical displacement of the cord in rats: behavioral and histologic analysis. J Neurotrauma 1992;9(3):197–217.
28. Cao Q, Zhang YP, Iannotti C, et al. Functional and electrophysiological changes after graded traumatic spinal cord injury in adult rat. Exp Neurol 2005;191 (Suppl 1): S3–S16.
29. Scheff SW, Rabchevsky AG, Fugaccia I, Main JA, Lumpp JE, Jr. Experimental modeling of spinal cord injury: characterization of a force-defined injury device. J Neurotrauma 2003;20(2):179–93.

30. Ghasemlou N, Kerr BJ, David S. Tissue displacement and impact force are important contributors to outcome after spinal cord contusion injury. Exp Neurol 2005;196(1):9–17.
31. Pearse DD, Lo TP, Jr., Cho KS, et al. Histopathological and behavioral characterization of a novel cervical spinal cord displacement contusion injury in the rat. J Neurotrauma 2005;22(6):680–702.
32. Watson BD, Prado R, Dietrich WD, Ginsberg MD, Green BA. Photochemically induced spinal cord injury in the rat. Brain Res 1986;367(1–2):296–300.
33. Bunge MB, Holets VR, Bates ML, Clarke TS, Watson BD. Characterization of photochemically induced spinal cord injury in the rat by light and electron microscopy. Exp Neurol 1994;127(1):76–93.
34. Verdu E, Garcia-Alias G, Fores J, et al. Morphological characterization of photochemical graded spinal cord injury in the rat. J Neurotrauma 2003;20(5):483–99.
35. Tator CH. Strategies for recovery and regeneration after brain and spinal cord injury. Inj Prev 2002;8 (Suppl 4):IV33–6.
36. Tator CH. Update on the pathophysiology and pathology of acute spinal cord injury. Brain Pathol 1995;5(4):407–13.
37. Koda M, Nishio Y, Hashimoto M, et al. Up-regulation of macrophage migration-inhibitory factor expression after compression-induced spinal cord injury in rats. Acta Neuropathol (Berl) 2004;108(1):31–6.
38. Tarlov IM. Spinal cord compression: mechanisms of paralysis and Treatment. Springfield: Thomas; 1957.
39. Casha S, Yu WR, Fehlings MG. Oligodendroglial apoptosis occurs along degenerating axons and is associated with FAS and p75 expression following spinal cord injury in the rat. Neuroscience 2001;103(1):203–18.
40. Park YK, Tator CH. Failure of topical DMSO to improve blood flow or evoked potentials in rat spinal cord injury. J Korean Med Sci 1998;13(6):638–44.
41. Midha R, Fehlings MG, Tator CH, Saint-Cyr JA, Guha A. Assessment of spinal cord injury by counting corticospinal and rubrospinal neurons. Brain Res 1987;410(2):299–308.
42. Namiki J, Kojima A, Tator CH. Effect of brain-derived neurotrophic factor, nerve growth factor, and neurotrophin-3 on functional recovery and regeneration after spinal cord injury in adult rats. J Neurotrauma 2000;17(12):1219–31.
43. Park YK, Tator CH. Prevention of arachnoiditis and postoperative tethering of the spinal cord with Gore-Tex surgical membrane: an experimental study with rats. Neurosurgery 1998;42(4):813–23; discussion 23–4.
44. Schultke E, Kendall E, Kamencic H, Ghong Z, Griebel RW, Juurlink BH. Quercetin promotes functional recovery following acute spinal cord injury. J Neurotrauma 2003;20(6):583–91.
45. Baffour R, Achanta K, Kaufman J, et al. Synergistic effect of basic fibroblast growth factor and methylprednisolone on neurological function after experimental spinal cord injury. J Neurosurg 1995;83(1):105–10.
46. Kureshi IU, Ho SY, Onyiuke HC, Wakefield AE, D'Arrigo JS, Simon RH. The affinity of lipid-coated microbubbles to maturing spinal cord injury sites. Neurosurgery 1999;44(5):1047–53.
47. Saporta S, Kim JJ, Willing AE, Fu ES, Davis CD, Sanberg PR. Human umbilical cord blood stem cells infusion in spinal cord injury: engraftment and beneficial influence on behavior. J Hematother Stem Cell Res 2003;12(3):271–8.
48. Boyd JG, Lee J, Skihar V, Doucette R, Kawaja MD. LacZ-expressing olfactory ensheathing cells do not associate with myelinated axons after implantation into the compressed spinal cord. Proc Natl Acad Sci USA 2004;101(7):2162–6.
49. Fu ES, Saporta S. Methylprednisolone inhibits production of interleukin-1beta and interleukin-6 in the spinal cord following compression injury in rats. J Neurosurg Anesthesiol 2005;17(2):82–5.

50. Nashmi R, Fehlings MG. Changes in axonal physiology and morphology after chronic compressive injury of the rat thoracic spinal cord. Neuroscience 2001;104(1):235–51.
51. Jimenez Hamann MC, Tator CH, Shoichet MS. Injectable intrathecal delivery system for localized administration of EGF and FGF-2 to the injured rat spinal cord. Exp Neurol 2005;194(1):106–19.
52. Gupta D, Tator CH, Shoichet MS. Fast-gelling injectable blend of hyaluronan and methylcellulose for intrathecal, localized delivery to the injured spinal cord. Biomaterials 2006;27(11):2370–9.
53. Karimi-Abdolrezaee S, Eftekharpour E, Wang J, Morshead CM, Fehlings MG. Delayed transplantation of adult neural precursor cells promotes remyelination and functional neurological recovery after spinal cord injury. J Neurosci 2006;26(13):3377–89.
54. Roussos I, Rodriguez M, Villan D, Ariza A, Rodriguez L, Garcia J. Development of a Rat Model of Spinal Cord Injury and Cellular Transplantation. Transplant Proc 2005;37(9):4127–30.

Chapter 39

Experimental Laceration Spinal Cord Injury Model

Yi Ping Zhang, Lisa B.E. Shields, and Christopher B. Shields

Abstract An experimental laceration spinal cord injury (L-SCI) is an animal model designed to study mechanisms of axonal regeneration following SCI. Current techniques for creating L-SCI are performed freehand using visual estimation that cause significant errors in both the lesion depth and shape. In this chapter, we describe an unique method to create precise L-SCI in rodents using the Louisville Injury System Apparatus. We will explain the step-by-step procedure and describe the conditions that affect the outcomes following laceration injury.

Keywords: Spinal cord injury, LISA, Laceration injury, Experimental model, Mouse SCI model, Rat SCI model, Graded injury severity

1 Introduction

Contusion spinal cord injuries (C-SCI) occur in 25–40% of all spinal cord injuries followed by laceration spinal cord injuries (L-SCI) that occur in 25% of all SCI.[1–3] The latter injuries are caused by bone fragments, gunshot wounds, and stab wounds. In experimental SCI research, the L-SCI model is frequently used to trace ascending and descending pathways of the spinal cord and their respective functions.[4] In addition, the L-SCI model has become frequently utilized in research for axonal regeneration of the spinal cord. Recent discoveries have indicated that CNS axons have an intrinsic capacity for regeneration but are inhibited by their own hostile environment, including chondroitin sulfate proteoglycans,[5,6] CNS myelin derived inhibitory molecules,[7–10] and semaphorin 5A or Ephrin-3B/Eph-A4.[11–13] The L-SCI model in rodents has played a large role in understanding mechanisms of regeneration and inhibition of axon growth as well as in designing corresponding therapeutic strategies in SCI repair.

The International Spinal Research Trust (ISRT) published guidelines for SCI models indicating that the nature and magnitude of the lesion should be precisely defined and that the conclusions of the study would be misleading if an imprecise lesion was created.[14] A precise L-SCI model is essential in experimental SCI research to study mechanisms of injury and to screen novel

From: *Animal Models of Acute Neurological Injuries.*
Edited by: J. Chen (eds.) New York, Totowa, NJ

therapeutic agents. Criteria for an optimal L-SCI model includes: (1) the designated target undergoing laceration should not spare any axons to avoid false interpretation of regeneration caused by spared axons; (2) the laceration lesion should be uniform to avoid other types of lesion such as compression or contusion. The laceration should avoid repetitive cutting of the same site which may result in multiple cuts if the cutting instrument fails to cut along the same track. This error would cause an imprecise lesion; (3) the need for a precise laceration lesion would be critical when working on small rodents such as the mouse spinal cord. Use of genetically modified mice has greatly increased as they serve as precise tools to study specific genes or proteins that regulate axon regrowth.[15]

A unique rodent L-SCI model requires the ability to create a definitive depth and shape of injury, to avoid concomitant damage (contusion, compression), and be easily used in the mouse.

2 Current Status of Experimental L-SCI Models

A L-SCI is the easiest lesion to create experimentally yet is the most difficult to execute with a high degree of precision. A reliable L-SCI model must be highly accurate. An example of this need for precision is underscored by contradictory results from three independent research laboratories that studied the Nogo pathway for axon regeneration on gene knockout mice.[16–18] Their disparate results may be explained by varied strategies used to delete the Nogo gene or differences in the creation of SCI lesion models.

Currently available laceration models are created manually using microscissors, scalpel, razor blade, needle tip, or a wire knife.[15,16,19–21] A L-SCI lesion is often performed freehand under a surgical microscope guided solely by visual assessment. To enhance laceration accuracy, the tip of a razor blade or scissors may be marked to control the lesion depth when incising the spinal cord. However, the lesion quality is completely independent on the experience of the surgeon. The accuracy can only be confirmed by postmortem analysis. Although a complete spinal cord transection can be reliably produced using these techniques, they are incapable of creating precise lesions requiring an accurate depth and shape. Several inaccuracies arise in the creation of precise rodent L-SCI models such as: (1) spine flexibility that prevents precise surgical manipulation of the rodent spinal cord; (2) yield and deformity of spinal cord tissue caused by pressure of the tools used to create the laceration and tissue recoil following removal of the tool. Tissue yield and recoil may produce significant variability in lesion shapes and depths, thereby sparing fibers targeted for laceration; and (3) bleeding and swelling of the spinal cord during laceration that obscures visual landmarks. Hemostasis may be obtained by the topical application of cold saline. Thus, the creation of a L-SCI using visual landmarks to guide the laceration depth and shape is highly subjective.

A lesion executed manually often needs to be repeated several times in order to reach the required depth and/or shape.[22] This process demands that each pass must enter the same site otherwise shredding of the lesion will occur. Furthermore, creation of a L-SCI with micro-scissors is associated with a variable degree of contusion and/or compression on the spinal cord. This effect cannot be avoided as scissor blades will deform tissue between them before creating the laceration. Thus, use of visually generated lesions are incapable of creating accurate and reproducible L-SCI (Fig. 1).

3 Precision L-SCI Model

Improvements have been introduced to create a L-SCI based on precise cutting distances on a stabilized spinal cord using the Louisville Injury System Apparatus (LISA) device (Louisville Impactor System Apparatus) (Louisville Impactor System Inc., Louisville, KY). The LISA is multifunctional and is capable of producing a variety of CNS lesions. One attacment of the LISA creates a L-SCI using the laceration attachment which is attached to the spine stabilizer. Using a highly precise micromanipulator, the cord is elevated towards an oscillating blade that creates the L-SCI.

3.1 Importance of Spine Stabilization

Flexibility of the vertebral joints creates a fulcrum effect during spinal cord surgery on rodents, particularly if the spine is fixed by clamps attached to the spinous processes. The spinal skeleton is immobilized using the spine stabilizer (Louisville Impactor System, Louisville, KY) attached to the LISA at the level of the planned L-SCI. Our spine stabilizer provides fixation using a pair of stainless steel arms that attach to the vertebral facets in either the cervical or thoracic spine (Fig. 2a). Spine stabilizers have been designed for use in the mouse and rat to maximally reduce target movement (Figs. 2b, c). The stabilized spine is slightly suspended to prevent respiratory compromise.

3.2 Laceration Lesion Using the LISA

The laceration attachment of the LISA was formerly referred to as the Vibraknife.{Iannotti, 2006 59/id;Onifer, 2005 153/id;Sivasankaran, 2004 170/id;Zhang, 2004 189/id} This method creates the lesion depth with a mechanical accuracy of ±0.005 mm (Fig. 3) without dependence on visual or manual parameters. The LISA is capable of making precise lacerations on the spinal cord compared to microscissors which are manually controlled (Fig. 1). Figure 4shows a lateral view of the LISA. Graded precise dorsal hemisection lesions are created at a depth of 0.5, 0.8, 1.1, and 1.4 mm on the mouse (Fig. 3).

3.3 Creation of a L-SCI Using the LISA

The spine stabilizer is placed on top of the stage with the rodent in place. A high frequency oscillating blade is aimed at the spinal cord in preparation for the injury. The vibration amplitude can be varied by adjusting a control knob. A step-by-step procedure consists of:

1) Skin and subcutaneous soft tissue are opened, and paraspinal muscles are retracted bilaterally to expose the spine.
2) The rodent (rat or mouse) is placed on the U-shaped rodent spine stabilizer. The cervical or thoracic stainless steel arms are selected depending on the target segment (the cervical arms are slightly longer than the thoracic arms). The tips of the stabilizing arms are inserted and locked bilaterally under the lateral facets. The spine stabilizer is locked firmly to the base of the LISA.

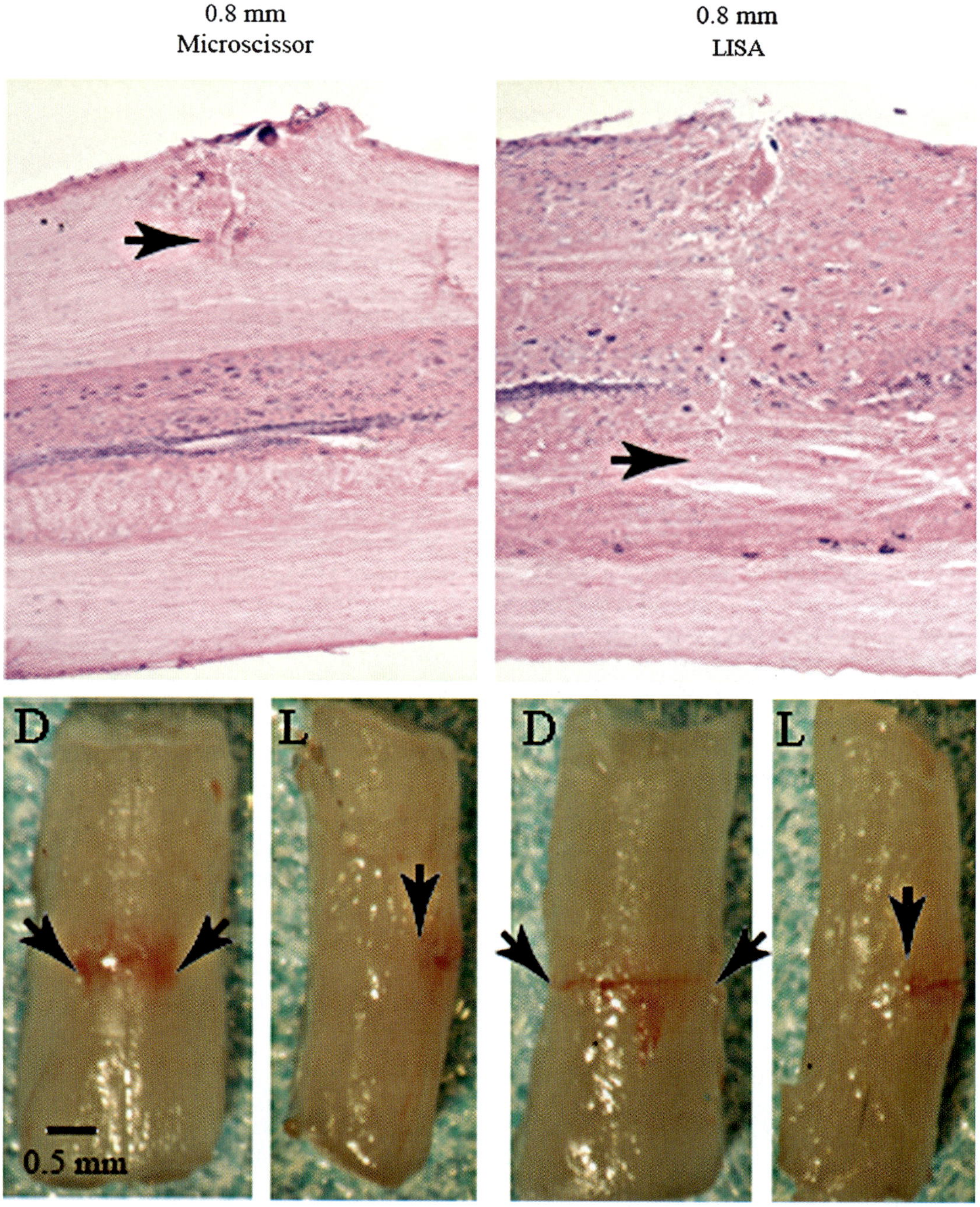

Fig. 1 Comparison of a 0.8 mm deep dorsal hemisection lesion in mice created by hand (*top left*), and top left using the LISA. (**D**) Dorsal and (**L**) lateral views of the spinal cord subjected to the 8 mm lesion. (*Lower left* two photos) A L-SCI created manually that demonstrates the imprecise and shallow lesions than expected due to tissue yield under the microscissor blades. (*Lower right* two photos) A L-SCI produced by the LISA that created a precise laceration without tissue deformation

3) If a single laceration of the cervical spinal cord is planned (without dural closure), a laminectomy is not performed. Removal of the interspinous ligament and ligamentum flavum are adequate to expose the dura mater.

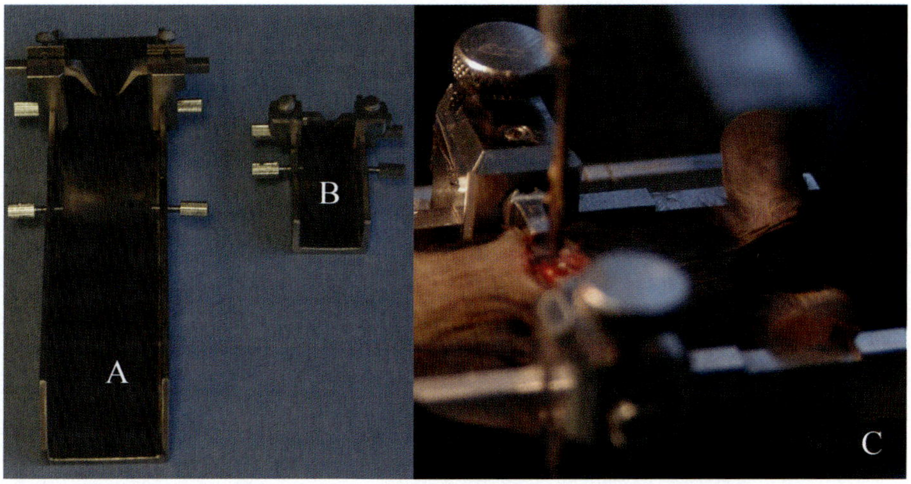

Fig. 2 (**A**) and (**B**) demonstrate the spine stabilizer designed for the rat and mouse, respectively. (**C**) The stabilized mouse undergoing a L-SCI by the vertically oriented vibrating blade

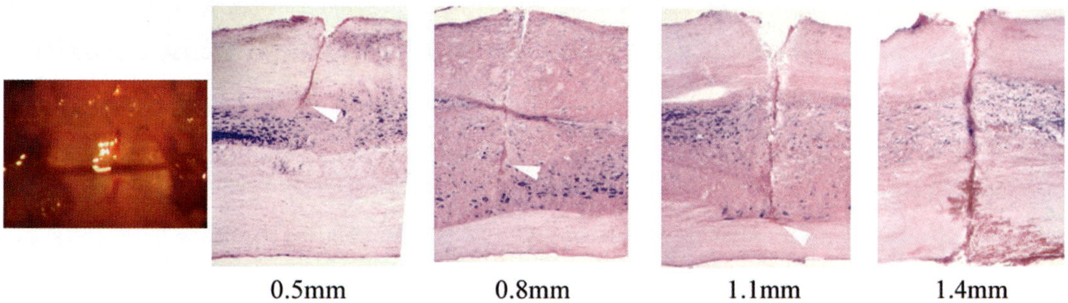

Fig. 3 A graded laceration can be reliably created in the mouse spinal cord. (*Left*) surgical view of a T8 L-SCI. (*Right* four photos) Graded depths of a L-SCI created using the LISA. Note the sharply outlined lesion in each lesion. The hemorrhage in the 1.4 mm lesion arose from the damaged anterior spinal artery

4) Dura is opened using the tip of a 30 G needle, and a durotomy is performed using microscissors. If dural closure is required following the L-SCI, a single-level laminectomy should be performed before opening dura.
5) The spine stabilizer (with the stabilized animal) is attached on the LISA stage that is adjusted until the spinal cord is positioned to lie perpendicular to the laceration blade. The shape of the spinal cord lesion is determined by the width and shape of the blade.
6) The oscillating knob is turned to elevate the LISA stage. When the blade touches the spinal cord, the "0" surface is reached which is characterized by hemorrhage arising from the severed dorsal vein. The blade is lubricated with a few drops of cool saline. The stage is raised to the desired predetermined level controlled by the microdrivers that cause the laceration. The vibration is terminated.
7) The quality of the lesion is checked before the knife is withdrawn. The spinal cord is irrigated with several rounds of cool saline. The blade should remain in the gap without evidence of tissue distortion.

Fig. 4 Picture of the LISA setup. (1) *Stage* that can be accurately moved in three dimensions attaining a mechanical accuracy of 0.005 mm, using three microdrivers. The angle of the spine stabilizer can be altered using a ball joint. (2) LISA *body frame*. (3) *Vibrating attachment* of the oscillating blade. The amplitude of the oscillations can be adjusted by turning (4) *knob*. (5) The mouse *spine stabilizer* is attached to the LISA stage. After placing the target spinal cord beneath the oscillating blade, the stage is gradually raised by turning the vertical microdriver that creates the laceration

8) Any remaining blood clot should be removed. If dural repair is necessary, it is performed with 10–0 nylon suture. The stainless steel arms of the spine stabilizer are detached from the vertebra.
9) Muscles and soft tissue are closed in layers using 5–0 (rat) or 7–0 (mice) braided silk suture (Ethicon Inc., Piscataway, NJ), and 9 mm stainless steel wound clips (Clay Adams, Parsippany, NJ).

3.4 Blade Design and Blade Vibration Characteristics

Large vibration amplitudes increase the linear velocity of the blade which creates the laceration more rapidly. However, small vibration amplitudes maintain the lesion shape similar to that of the blade, but completion of the laceration takes longer and requires using a very sharp blade.

The vibration amplitude of the LISA can be adjusted from 0–1.3 mm. A minimum of 0.5 mm vibration amplitude is recommended to facilitate laceration. The blades are made from breakable razor-blades that can be fashioned into a square or diamond shape. Figure 5 demonstrates three blades and their respective lesions. The diamond shaped blade is ideal to create a dorsal funicular lesion The L-SCI can be designed to create selective precise lesions of the dorsal hemisection, dorsal funiculus, and dorsolateral funiculus.[23,24]

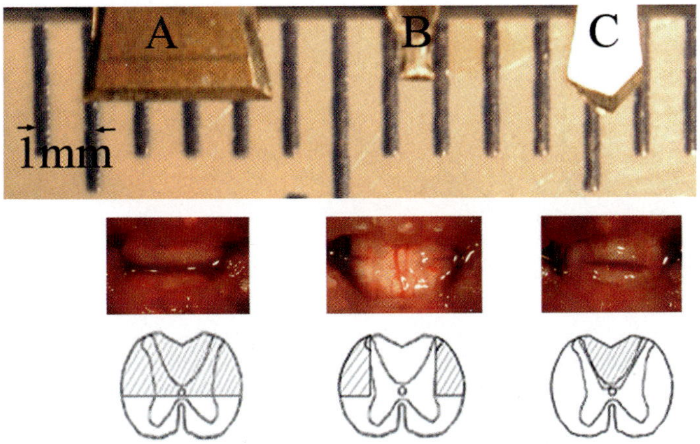

Fig. 5 The blades are made from breakable razor blades that can be fashioned into a square or diamond shape (*top row*). Three blades and their respective lesions (*second and third rows*). The diamond shaped blade (*top row* **C**) is ideal to create a dorsal funiculus lesion. The LISA can be adjusted to create selective precise lesions of the dorsal hemisection (*left column*), dorsolateral funiculus (*middle column*), and the dorsal funiculus (*right column*)

3.5 Reduction of the Lesion Gap Following a L-SCI

The injury type and location in experimental L-SCI are parameters that are often overlooked in experimental lacerations. The spinal cord is constantly subjected to a longitudinal tension due to traction caused by nerve roots and filum terminale. Following an incomplete L-SCI, a gap develops between the two stumps following the laceration. This problem does not exist in the dorsal column lesions because of the intact dorsolateral funiculus. A large gap may generate a cavity at the lesion site where it may pose a major physical barrier that impedes axon regeneration. There are two ways to minimize the physical gap: (1) suture the dura that approximates the stumps {Zhang, 2004 189/id} or (2) create the L-SCI in a location that has less tension on the spinal cord. Dural closure is feasible in the rat but becomes problematic in the mouse. The choice of the segmental level to create a L-SCI to reduce the gap is the obvious choice. For example, the T1-3 segments are concave in the thoracic segment of the rodent spine, and the curve is kyphotic at T9-11. The convex and kyphotic curve of the spine affects the tension on the dorsal half of the spinal cord. Thus, the spinal cord dorsal surface tension at the T9-10 segment will be higher than at T1-3 (Fig. 6). Wound healing may be impaired as more scar may develop in conjunction with spinal cord gap which usually occurs more frequently at T9-11 than at T1-3 following spinal cord dorsal hemisection laceration. The cord warp caused by tissue tension during the healing process may disturb the original orientations of the axons (Fig. 7). The T1 lesion in the mouse may have a greater mobility than at T9 during 1.1 mm dorsal hemisection lesion (unpublished data). In addition this issue is not only anatomical but is also affected by the cervical segment containing more fibers to and from the brain.

In vivo experimental SCI models are necessary to reliably evaluate the safety and efficacy of therapeutic strategies before being tested in humans. The

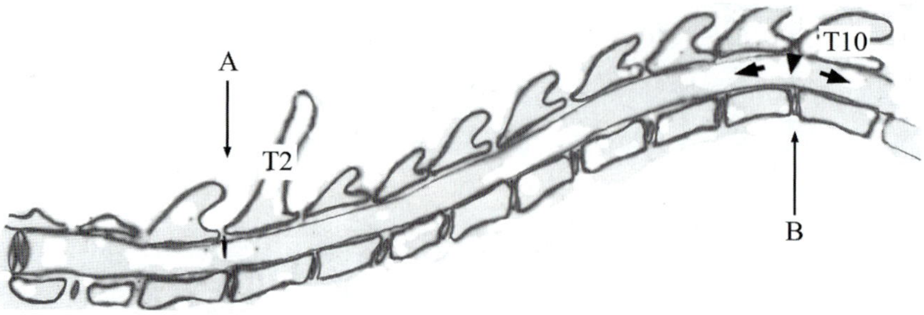

Fig. 6 Cartoon showing the rodent's spine curvature. The tension of the dorsal surface of the spinal cord is high on the convex segment (arrow B) compared with the concave segment (arrow A). Because of the different dorsal surface tensions following dorsal hemisection, the gap between the opposing surfaces of the spinal cord is larger following a T9-10 L-SCI compared with a T1-2 L-SCI

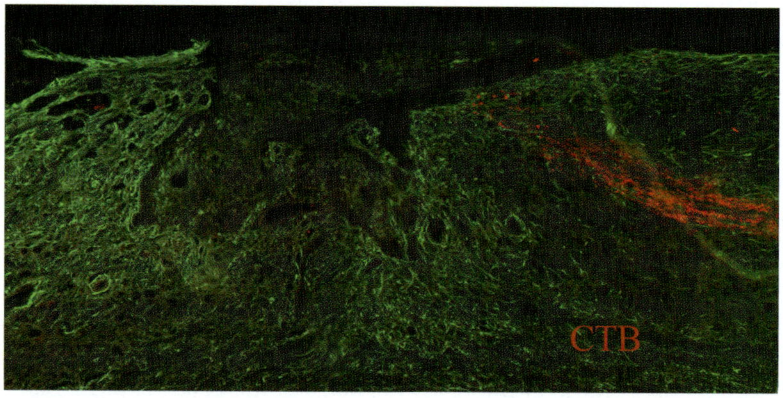

Fig. 7 This longitudinal section of spinal cord demonstrates cord warping as a result of tissue tension during the healing process of a mouse dorsal spinal cord after a T9-10 dorsal hemisection L-SCI. The gap is wide, and the orientation of fibers near the gap is distorted. Warping of the gracile funiculus seems to be migrating toward the dorsal surface of the spinal cord; however, this appearance is due to the gaping of the stumps of the lacerated spinal cord (Red = CTB tracing of posterior funiculus of the spinal cord)

reproducibility of the animal model is critical. We have introduced the method of creating an accurate, reproducible, and reliable L-SCI in rodents.

References

1. Bunge RP, Puckett WR, Hiester ED. Observations on the pathology of several types of human spinal cord injury, with emphasis on the astrocyte response to penetrating injuries. Adv Neurol 1997;72:305–315.
2. Quencer RM, Bunge RP. The injured spinal cord: imaging, histopathologic clinical correlates, and basic science approaches to enhancing neural function after spinal cord injury. Spine 1996;21(18):2064–2066.
3. Liverman CT, Altevogt BM, Joy JE, Johnson RT, eds. Progression of spinal cord injury. In: Spinal Cord Injury: Progress, Promise, and Priorities. Washington, D.C.: The National Academies Press, 2005:30–63.

4. Tracey D. Ascending and descending pathways in the spinal cord. In: Paxinos G, ed. The Rat Nervous System. 3rd ed. San Diego: Elsevier Academic Press, 2004:149–164.
5. Bradbury EJ, Moon LD, Popat RJ, et al. Chondroitinase ABC promotes functional recovery after spinal cord injury. Nature 2002;416(6881):636–640.
6. Silver J, Miller JH. Regeneration beyond the glial scar. Nat Rev Neurosci 2004;5(2):146–156.
7. Chen MS, Huber AB, van der Haar ME, et al. Nogo-A is a myelin-associated neurite outgrowth inhibitor and an antigen for monoclonal antibody IN-1. Nature 2000;403(6768):434–439.
8. Liu BP, Fournier A, GrandPre T, Strittmatter SM. Myelin-associated glycoprotein as a functional ligand for the Nogo-66 receptor. Science 2002;297(5584):1190–1193.
9. McGee AW, Strittmatter SM. The Nogo-66 receptor: focusing myelin inhibition of axon regeneration. Trends Neurosci 2003;26(4):193–198.
10. Wang KC, Koprivica V, Kim JA, et al. Oligodendrocyte-myelin glycoprotein is a Nogo receptor ligand that inhibits neurite outgrowth. Nature 2002;417(6892):941–944.
11. Benson MD, Romero MI, Lush ME, Lu QR, Henkemeyer M, Parada LF. Ephrin-B3 is a myelin-based inhibitor of neurite outgrowth. Proc Natl Acad Sci USA 2005;102(30):10694–10699.
12. Goldberg JL, Vargas ME, Wang JT, et al. An oligodendrocyte lineage-specific semaphorin, Sema5A, inhibits axon growth by retinal ganglion cells. J Neurosci 2004;24(21):4989–4999.
13. Goldshmit Y, Galea MP, Wise G, Bartlett PF, Turnley AM. Axonal regeneration and lack of astrocytic gliosis in EphA4-deficient mice. J Neurosci 2004;24(45):10064–10073.
14. Ramer MS, Harper GP, Bradbury EJ. Progress in spinal cord research – a refined strategy for the International Spinal Research Trust. Spinal Cord 2000;38(8):449–472.
15. Inman D, Guth L, Steward O. Genetic influences on secondary degeneration and wound healing following spinal cord injury in various strains of mice. J Comp Neurol 2002;451(3):225–235.
16. Kim JE, Li S, GrandPre T, Qiu D, Strittmatter SM. Axon regeneration in young adult mice lacking Nogo-A/B. Neuron 2003;38(2):187–199.
17. Simonen M, Pedersen V, Weinmann O, et al. Systemic deletion of the myelin-associated outgrowth inhibitor Nogo-A improves regenerative and plastic responses after spinal cord injury. Neuron 2003;38(2):201–211.
18. Zheng B, Ho C, Li S, Keirstead H, Steward O, Tessier-Lavigne M. Lack of enhanced spinal regeneration in Nogo-deficient mice. Neuron 2003;38(2):213–224.
19. Hermanns S, Reiprich P, Muller HW. A reliable method to reduce collagen scar formation in the lesioned rat spinal cord. J Neurosci Methods 2001;110(1–2):141–146.
20. Seitz A, Aglow E, Heber-Katz E. Recovery from spinal cord injury: a new transection model in the C57Bl/6 mouse. J Neurosci Res 2002;67(3):337–345.
21. Frisen J, Fried K, Sjogren AM, Risling M. Growth of ascending spinal axons in CNS scar tissue. Int J Dev Neurosci 1993;11(4):461–475.
22. Dyer JK, Bourque JA, Steeves JD. Regeneration of brainstem-spinal axons after lesion and immunological disruption of myelin in adult rat. Exp Neurol 1998;154(1):12–22.
23. Onifer SM, Zhang YP, Burke DA, et al. Adult rat forelimb dysfunction after dorsal cervical spinal cord injury. Exp Neurol 2005;192(1):25–38.
24. Zhang YP, Shields LBE, Zhang Y, et al. Use of magnetic stimulation to elicit motor evoked potentials, somatosensory evoked potentials, and H-reflexes in non-sedated rodents. J Neurosci Methods 2007;165(1):9–17.

25. Iannotti C, Zhang YP, Shields LBE, et al. Dural repair reduces connective tissue scar invasion and cystic cavity formation after acute spinal cord laceration injury in adult rats. J Neurotrauma 2006;23(6):853–865.
26. Sivasankaran R, Pei J, Wang KC, et al. PKC mediates inhibitory effects of myelin and chondroitin sulfate proteoglycans on axonal regeneration. Nat Neurosci 2004;7(3):261–268.
27. Zhang YP, Iannotti C, Shields LBE, et al. Dural closure, cord approximation, and clot removal: enhancement of tissue sparing in a novel laceration spinal cord injury model. J Neurosurg 2004;100(4 Suppl Spine):343–352.

Chapter 40

Rodent Spinal Cord Demyelination Models

Sarah C. Jernigan, Yi Ping Zhang, Christopher B. Shields, and Scott R. Whittemore

Abstract Demyelination is a significant component of contusive spinal cord injury as well as in multiple sclerosis. As such, remyelination, either by endogenous or grafted exogenous myelinating cells is a viable therapeutic target to restore function. To assess specific approaches to facilitate functional remyelination in vivo, appropriate injury models are needed. This chapter will discuss the strengths and weaknesses of a number of demyelinating lesions of the spinal cord and provide guidelines for choosing which model best suits which experimental condition. Step by step procedures for both creating and assessing the lesion will be provided.

Keywords: Spinal cord injury, Demyelination, Ethidium bromide, Lysolethicin, Cuprizone

1 Model Selection

There are a number of methods that can be used to demyelinate axons. These include injection of ethidium bromide (EB), lysolethicin, murine hepatitis virus (MHV), a cocktail of myelin-specific antibodies and complement, or by giving cuprizone in the diet. When injected into white matter, EB is a gliotoxic compound that intercalcates in the small groove of DNA and kills both oligodendrocytes and astrocytes; at higher doses, endothelial cells are also killed. Lysolethicin (lysophosphatidyl choline) rapidly and specifically disrupts myelin sheaths.[1] MHV lytically infects glial cells[2] and cuprizone is a copper chelator that specifically kills oligodendrocytes because they are metabolically compromised to support large areas of myelin and die by apoptosis.[3] Antibodies to galactocerebroside C (GalC), an oligodendrocyte glycolipid, specifically target oligodendrocytes for complement-mediated lysis.[4] Without further intervention, all of these models will spontaneously remyelinate, albeit with different time courses.[5] Methodologically, all approaches that require direct injection into spinal cord parenchyma utilize the exact same protocol, and only the volumes and concentrations differ. Cuprizone is provided in the diet, but it does not efficiently or reproducibly demyelinate the spinal cord and will not be discussed further. Importantly, all of these approaches can be used in both mice and rats.

From: *Animal Models of Acute Neurological Injuries.*
Edited by: J. Chen (eds.) New York, Totowa, NJ

All of these lesions will produce inflammatory reactions, the extent of which is dependent on the volumes and concentrations injected. Virus injection or complement lysis will obviously elicit more extensive inflammatory responses than the other lesion methods. Similarly, high doses of glial toxins can also result in severe axonal pathology, making them incompetent for remyelination. Depending on the region that is being demyelinated, axons in those areas may be differentially sensitive to these treatments and may require initial titration to establish optimally effective doses. Lastly, EB targets both neurons and astrocytes, and this aglial environment presents a very different landscape for remyelination than does one that contains astrocytes. Figure 1a–d shows an example of an EB lesion of the ventrolateral funiculus (VLF) of an adult rat that is devoid of myelin (B), astrocytes (C), and oligodendrocytes (D). Figure 1e shows a bilateral lysolethicin demyelination of the VLF in the adult mouse spinal cord. Note the presence of more astrocytes in this lesion compared with that seen following EB demyelination. One concern that must be kept in mind when using these lesions is that chronic demyelination leads to axonal pathology (Fig. 2). We routinely make our therapeutic interventions within one week postdemyelination.

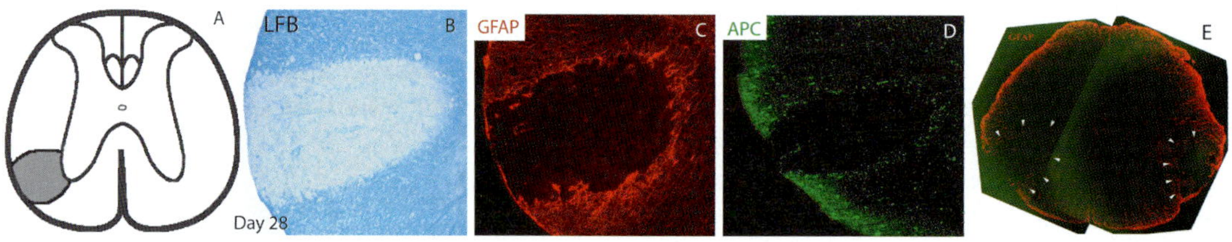

Fig. 1 Stereotactic injection of 1.0 μl of a 0.1 mg ml^{-1} EB solution into the VLF consistently resulted in a defined lesion, schematically depicted in (**A**). Luxol fast blue (LFB) staining in (**B**) shows the lack of myelin throughout the lesion 1 month post-EB. The lack of both astrocytes (**C**) and oligodendrocytes (**D**) in the lesion is shown by glial fibrillary acidic protein (GFAP) and APC immunoreactivity, respectively. (**E**) Adult mouse spinal cord with bilateral lysolethicin lesions of the VLF (outlined by *arrows*) immunostained with GFAP (*red*). Note the increased GFAP in the lesion core compared with EB lesions (**C**)

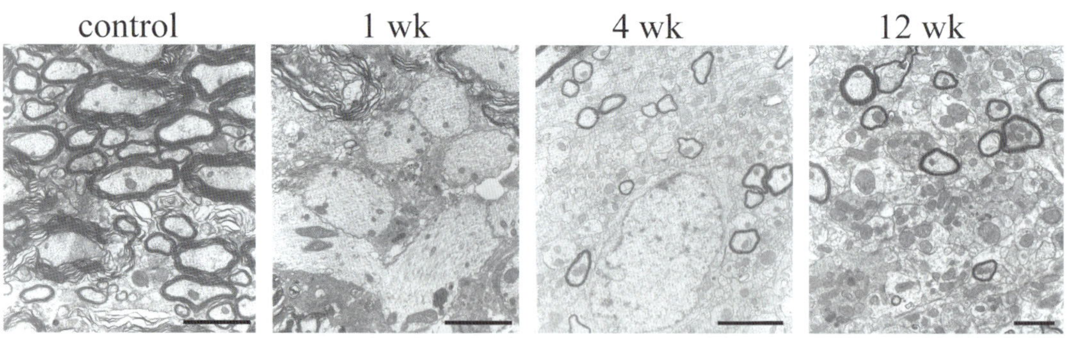

Fig. 2 Mouse demyelinating spinal cord lesion after 0.1 mg ml^{-1} unilateral EB injection into the VLF. Compared with the contralateral VLF region as depicted in (**A**), note the demyelinated but otherwise intact axons 1 week (**B**), 4 weeks (**C**), and 12 weeks (**D**) postinjection. At the 3 month time point, the demyelinated axons are swollen (**D**), and the lesion area are often sealed off by the perpendicular arrangement of fibers (data not shown). Magnification bars represent 1 μm

2 Materials

1. 7–0 monofilament
2. Spring scissors with fine blades (F.S.T. 15013-12)
3. Spring scissors with extra fine blades (Morica 160 MC 50)
4. Dural scissors (F.S.T. 15002-08)
5. Graefe Forceps, 0.8 mm (F.S.T. 11053-10)
6. Graefe forceps with extra fine tips, 0.5 mm (F.S.T. 11153-10)
7. Jeweler's forceps (F.S.T. 11254-20)
8. Friedman-Pearson Micro Rongeurs, curved tips (F.S.T.16221-14)
9. Scalpel handle (F.S.T. 10003-12) and #15 blades (F.S.T. 10015-00)
10. Adson Brown forceps (F.S.T. 11627-12)
11. Castroviejo needle holder (F.S.T. 12565-14)
12. Staple applicator (F.S.T. 12031-09)
13. Staples (F.S.T. 12032-07)
14. Cotton tip applicators (Henry Schein 100-6015)
15. Surgical microscope
16. Glass micropipette tips
17. 30 gauge needle
18. Ketamine (20 mg ml^{-1})/xylazine (2 mg ml^{-1})
19. Lacrilube
20. Bacitracin
21. Gentamicin 1 mg ml^{-1} (or equivalent antibiotic)
22. EB 0.1–0.3 mg ml^{-1} (Sigma E8751)
23. Lysolethicin (1%) (Sigma L1381)
24. Mouse vertebral stabilizer (Louisville Injury Systems or its equivalent)
25. Stereotaxis frame
26. Picoinjector or nanoinjector system
27. Micropipette puller

3 Procedures

3.1 Glass Micropipette Preparation

The selection of the size of the capillary glass depends on the volume of the demyelination solution, preferably 1.12 mm internal diameter (ID) capillary glass with the Picospritzer (Parker Company Picospritzer II or WPI PV800 Picopump) for the rat spinal cord and 0.53 mm ID capillary glass with the Nanojector II (Drummond Scientific Company Nanoject II Auto-nanoliter injector) for the mouse spinal cord. These micropipettes can be attached to a Hamilton syringe as well, but precise delivery of small volumes is more difficult. The fine tips of the glass micropipette need to be prepared using a micropipette puller (Sutter Instrument Co. Model P-80 PC or its equivalent). High temperature (870°C) and small force (20 psi) is recommended to produce a relatively long tapered tip for deep tissue injection.

With the aid of a micrometer, the tip is trimmed to 20–30 μm outer diameter (OD) with #5 forceps. To facilitate tissue penetration, we bevel the pipette tip to 45°. Always prepare 5–10 pipettes for each operative day to replace clogged or broken ones. Sterilize the glass pipettes by laying them on adhesive tape inside a petri dish exposing them to UV light for an hour.

No additional preparation of the micropipette needles is needed prior to loading and injecting the glial-toxin with the Picospritzer, which uses a vacuum and compressed air. However, for the Nanojector II, the micropipette needle must be backfilled with mineral oil. Once the oil has completely filled the micropipette, avoiding any air bubbles, it can be attached onto the device. The glial-toxin is loaded into the pipette – drawing it into pipette slowly and carefully to avoid the introduction of air bubbles.

3.2 Stereotaxis

To create the desired demyelinating lesion, stereotaxis is used to align the depth of the injection in the ventral-dorsal plane as well as the location of the injection in the medial-lateral plane. The lesions depend not only on the target but also on the volume and concentration of the solution, each of which has a specific pattern of diffusion. We generate thoracic VLF demyelination lesions to study the descending functional conduction using the following parameters:

For VLF lesions in rats, inject at 1.6 and 1.3 mm depths and 0.7 mm lateral to midline.

For VLF lesions in mice, inject at 1.1 and 0.9 mm depths and 0.5 mm lateral to midline.

3.3 Thoracic Demyelination Surgery

Adult mice weighing 15–25 g or adult rats weighing 180–220 g are used. Because of the similarities of the procedures, details on the mouse surgery are presented below (see [6], for a description of the rat surgery).

1. Anesthetize mice with a ketamine/xylazine solution.
2. Shave the animal along the dorsal spinal surface, clean the surgical area with betadine solution, and apply lacrilube ointment to the eyes.
3. Inject 1 cc normal saline bolus subcutaneously to prevent dehydration.
4. Make a sagittal incision in the skin from T8-T11.
5. Mobilize the fat pad anteriorly. This fat pad will be reapproximated at the end of surgery to cover the surgical site.
6. Using the extra fine scissors (F.S.T 15370-50 or F.S.T. 15023-10), separate the muscles from the lamina, exposing the lateral aspects of the facets bilaterally. It is important that the facets be adequately exposed to allow the vertebral column to be grasped and lifted onto the stabilizer using the facets. We use the spine stabilizer developed for the Louisville Injury System Aparatus[7] (LISA) (Fig. 3). However, other devices will work well provided the spine is securely immobilized.
7. Use the extra fine Graefe forceps to grasp the lateral facets and place the toothed blades of the stabilizer on either side of the facets, thereby securing the vertebral column in the stabilizer by tightening both sides of the stabilizer and then removing the forceps. The vertebral column must be aligned parallel with the base of the stabilizer (Fig. 4).
8. Use a scalpel with a #15 blade to remove any excess tissue from the lamina.
9. Once the T9 lamina has been identified, expose the underlying spinal cord between the interspinous spaces at T8-9 and T9-10, without performing a

Fig. 3 Mouse vertebral stabilizer developed by Louisville Injury Systems (LIS; Louisville, KY)

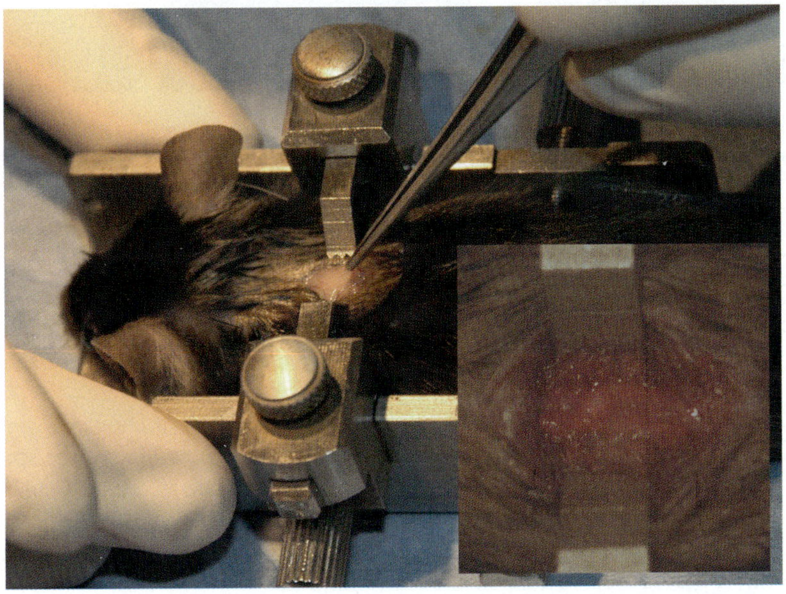

Fig. 4 Mouse on the LIS stabilizer. The inset shows the blades placed under facets of the spinal column

laminectomy. In rats, a partial laminectomy may need to be performed to expose an adequate space to insert the micropipette.

10. After cleaning the interlaminar soft tissue, open the dura using the tip of a 30 gauge needle. The dural opening is extended using dural scissors (F.S.T. 15002-08) to accommodate the glass pipette tip. The dural opening should only be large enough to allow the micropipette tip to enter the spinal cord.

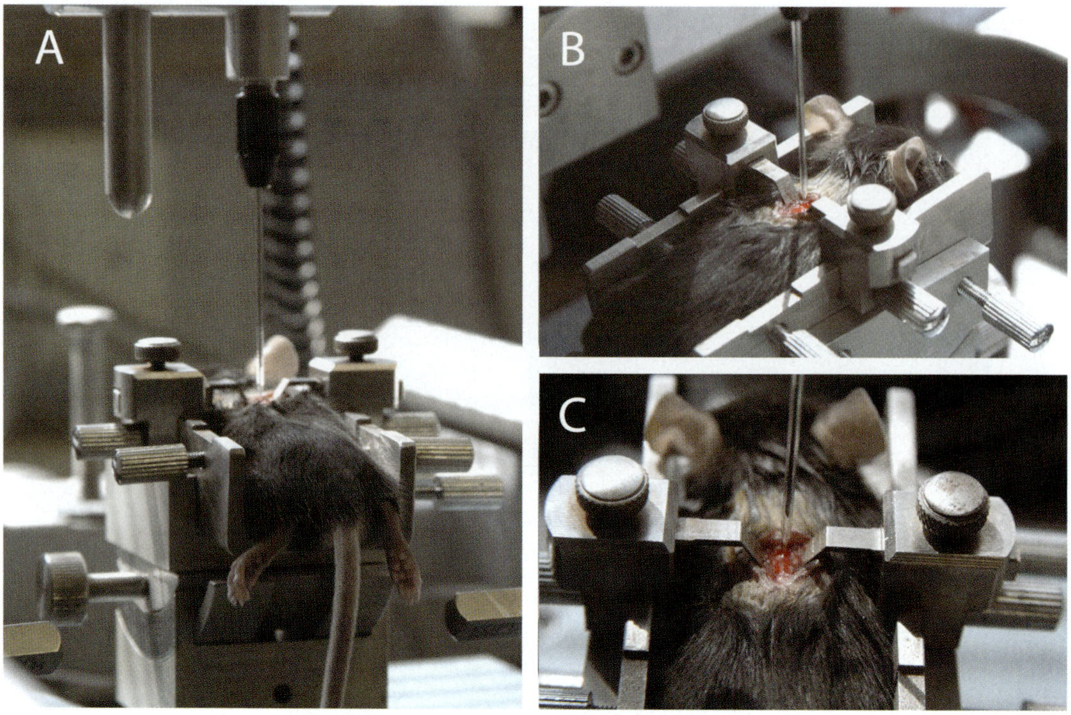

Fig. 5 Injection of demyelinating EB into the mouse spinal cord. Note that the spine of the mouse on the Nanoinjector system is exactly perpendicular to the plane of the injector, which is attached to a stereotaxic delivery frame

11. Transfer the mouse in the spine stabilizer to the injection system (Fig. 5). The base of the injection system can be leveled to make the spinal cord perpendicular to the tip of the micropipette.
12. Load the glass pipette with the demyelinating solution of choice.
13. Align the tip with the dural opening and bring the tip to edge of the spinal cord surface, making sure that the spinal cord surface is perpendicular to the needle (Fig. 5).
14. Penetrate the spinal cord surface with the glass microelectrode tip and inject the desired volume of the solution at the desired depth.
15. Leave the pipette in place for 2 min then remove it from the spinal cord or if two injection points are being made along the same trajectory, withdraw the micropipette to the second depth and inject again. Wait another 2 min and then remove the tip from the spinal cord.
16. Between each use, the micropipette tip should be kept submerged in liquid.
17. Once the injections are complete, loosen the pins of the facet blades and remove the mouse from the vertebral stabilizer.
18. Using 7–0 monofilament, reapproximate the muscles over the laminectomy site and then reattach the fat pad to the midline.
19. Using the staple applicator, apply two skin staples to close the incision then apply bacitracin.

4 Outcome Evaluations

The optimal approach to assess remyelination is to combine multiple outcome measures. We routinely use electophysiological methods in combination with terminal histological assessment of remyelination. What electrophysiological approach one takes depends on where anatomically in the spinal cord the demyelination lesion was performed. Conduction is measured by somatosensory evoked potentials (SSEPs)[8] for the ascending dorsal columns, transcranial magnetic motor evoked potentials (tcMMEPs)[9] for the medial aspects of the VLF, and magnetic inter-enlargement reflexes (MIER)[10] for the mediolateral white matter. A detailed description of these methods is beyond the scope of this chapter, but changes in conduction can be measured by electrophysiological assessment. Histologically, one must quantify both spared white matter[9] and the extent of remyelination and by which cells remyelination occurs.[11] It is imperative that the latter assessment utilize unbiased stereological methods.

5 Applications, Complications, and Limitations

5.1 When injecting larger volumes into the spinal cord, the volume of fluid itself can compress the surrounding spinal cord or the force of injection can cause the fluid to leak out through the needle trajectory. It is recommended that two depths be injected so that the volume at each depth is half the total. This prevents undesirable fluid leakage. Each glial-toxin has different diffusion properties as well so the two injection sites also provide a more consistent concentration of toxin throughout the desired target area. The concentration, volume, and stereotaxic coordinates for any desired lesion must be optimized.

5.2 All of these lesions can be used to follow the spatiotemporal progression of remyelination by both endogenous or engrafted myelinating cells, e.g. oligodendrocyte precursor cells (OPCs). If one wants only to look at the contribution of engrafted myelinating cells, the lesion site can be X-irradiated either before or after spinal cord injury (SCI) to prevent endogenous OPCs and Schwann cells from proliferating and migrating to the lesioned area and remyelinating. If endogenous remyelination is to be evaluated, the relative contributions of oligodendrocyte and Schwann remyelination must be quantified as both cells will migrate to the lesion site.[11] Lesions of the dorsal columns remyelinate more efficiently than do lesions of the lateral or ventral white matter.

5.3 One must be extremely careful to monitor axonal loss with these models. This can be done by immunostaining for neuron-specific antigens such as the neurofilament proteins. Silver staining can also be used. It is important that longitudinal sections be examined as it is difficult to assess axon continuity on cross sections. The extent and progression of the inflammatory response, which can affect both axonal pathology and remyelination, can be monitored by immunostaining for microglia/macrophage markers, OX42 or ED1 for the rat or Iba1 for the mouse.

5.4 When interpreting data from any of these lesion models, it is important to recognize that while these are demyelinating models, they are not the same as what occurs in vivo after SCI or in multiple sclerosis. The activation state of astrocytes has profound effects on the extent of remyelination and the aglial environment of the EB lesion is a particularly artificial environment. Nonetheless, important information about potential mechanism(s) regulating remyelination or successful therapeutic approaches can be obtained in these model systems, which can subsequently be evaluated in more clinically relevant lesion paradigms.

5.5 If axons are demyelinated chronically, they can degenerate and/or become not patent for remyelination (Fig. 2). In all these lesion models, demyelination is complete within 7 days, and it is recommended that therapeutic treatments are not delayed more than one month postdemyelination.

References

1. Hall SM, Gregson NA. The in vivo and ultrastructural effects of injection of lysophosphatidyl choline into myelinated peripheral nerve fibres of the adult mouse. J Cell Sci 1971; 9:769–789.
2. Dubois-Dalcq ME, Doller EW, Haspel MV, Holmes KV. Cell tropism and expression of mouse hepatitis viruses (MHV) in mouse spinal cord cultures. Virology 1982; 119:317–331.
3. Matsushima GK, Morell P. The neurotoxicant, cuprizone, as a model to study demyelination and remyelination in the central nervous system. Brain Pathol 2001; 11:107–116.
4. Keirstead HS, Blakemore WF. Identification of post-mitotic oligodendrocytes incapable of remyelination within the demyelinated adult spinal cord. J Neuropathol Exp Neurol 1997; 56:1191–1201.
5. Woodruff RH, Franklin RJ. Demyelination and remyelination of the caudal cerebellar peduncle of adult rats following stereotaxic injections of lysolecithin, ethidium bromide, and complement/anti-galactocerebroside: a comparative study. Glia 1999; 25:216–228.
6. Loy DN, Magnuson DSK, Zhang YP et al. Functional redundancy of ventral spinal locomotor pathways. J Neurosci 2002; 22:315–323.
7. Zhang YP, Shields LB, Zhang Y et al. Use of magnetic stimulation to elicit motor evoked potentials, somatosensory evoked potentials, and H-reflexes in non-sedated rodents. J Neurosci Methods 2007; 165:9–17.
8. Zhang YP, Burke DA, Shields LBE et al. Combined assessment to discriminate small functional differences following spinal cord contusion based on precise vertebral stabilization and tissue displacement. J Neurotrauma (submitted).
9. Cao QL, Zhang YP, Iannotti C et al. Functional and electrophysiological consequences of graded contusive spinal cord injury in the adult rat. Exp Neurol 2005; 191:S3–S16.
10. Beaumont E, Onifer SM, Reed WR, Magnuson DSK. Magnetically-evoked inter-enlargement response: An assessment of ascending propriospinal fibers following spinal cord injury. Exp Neurol (in press).
11. Totoiu MO, Keirstead HS. Spinal cord injury is accompanied by chronic progressive demyelination. J Comp Neurol 2005; 486:373–383.

Chapter 41

Spinal Cord Lateral Hemisection and Implantation of Guidance Channels

Xiaofei Wang and Xiao-Ming Xu

Abstract Spinal cord laceration injuries such as complete transection, lateral and dorsal hemisection, over hemisection, and dorsal funiculotomy in rodents have been widely used in the studies of neuronal responses to axotomy as well as axonal regeneration through and beyond the site of the injury. The latter approach has utilized transplantation of fetal cord tissues, pieces of peripheral nerves, and biosynthetic guidance channels seeded with different cell types, trophic factors, and matrix molecules. In this chapter, we introduce hands-on procedures to perform a spinal cord lateral hemisection, create a lesion cavity on the same side, and transplant a piece of guidance channel that bridges the two ends of the cord stumps. By suturing the dura, cerebrospinal fluid circulation is restored and connective tissue invasion into the lesion site is prevented. With modification, these procedures should also apply to creating above-mentioned laceration injuries when microscissors are used.

Keywords: Spinal cord injury, Laceration, Hemisection, Transection, Transplantation, Guidance channel

1 Model Selection

Spinal cord lateral hemisection and guidance channel transplantation is a useful model for studying axonal regeneration following spinal cord injury (SCI). This model was developed on the basis of an earlier model in which the spinal cord was completely transected followed by implantation of either a blind-ended[1] or open-ended[2] semipermeable guidance channel, seeded with Schwann cells, into a 6- to 8-mm-long gap created between the two stumps of the transected spinal cord. The lateral hemisection and guidance channel implantation model was first reported in 1999,[3] and is characterized by a lateral hemisection on one side of the spinal cord, creation of a lesion gap of desired length, implantation of a hollow channel seeded with Schwann cells (or other cell types) into the lesion gap, and restoration of cerebrospinal fluid (CSF) flow by suturing the dura. Major advantages of this model include the following:

From: *Animal Models of Acute Neurological Injuries.*
Edited by: J. Chen (eds.) New York, Totowa, NJ

1. Studying regeneration of specific axonal pathways that project unilaterally in the spinal cord, such as the rubrospinal tract, while their counterparts on the intact side serve as internal controls.
2. Allowing the lesion and transplantation to be made at all levels of the spinal cord, including the cervical level when compared with a complete spinal cord transaction, which is limited by animal survival.
3. Restoring CSF circulation after dura repair ensuring the delivery of nutrients to the graft tissue and prevention of connective tissue invasion to the graft region.
4. Creating more stable cord-graft interfaces than a complete spinal cord transection due to the function of intact hemi-cord.
5. Providing a modifiable channel environment with unlimited potential to test different trophic factors, matrix molecules, and cell types in their ability to promote axonal regeneration, modify the graft-host interface, and protect axotomized host neurons from injury induced cell death.
6. Providing a relatively isolated graft environment which is excellent for axonal tract tracing to identify origins of neurons, whose axons have regenerated into the graft environment.

2 Materials

Adult Fisher or Sprague Dawley (SD) rats weighing 160–180 gm are the animals of choice. For the surgical procedures, the following materials are recommended (Fig. 1).

1. Rat vertebral stabilizer (Developed by YP Zhang and CB Shields, University of Louisville)
2. 10–0 black monofilament nylon (Ethilon 2888)
3. Retractor (Agricola, F.S.T. 17005-04)
4. Spring scissors with especially fine, small blades (F.S.T. 15002-08)
5. Spring scissors with especially fine, small Blades (F.S.T. 15005-08)
6. Moria forceps (F.S.T. 11154-10)
7. Forceps (Adson, F.S.T. 11027-12)
8. Dumont angled forceps (F.S.T. 11253-20)
9. Dumont forceps (F.S.T. 11251-35)
10. Needle holder (F.S.T. 12500-12)
11. Guidance channel (outer diameter – 1.25 mm; length – 3 mm)
12. Scalpel handle (F.S.T. 10003-12) and scalpel blades (F.S.T. 10015-00)
13. Spring scissors with especially fine, small blades (F.S.T. 15013-12)
14. Spring scissors with especially fine, small blades (F.S.T. 15370-26)
15. Friedman-Pearson micro rongeur (F.S.T. 16021-14)
16. MikRon 9 mm autoclip
17. Fine iris forceps (F.S.T. 14017-14)
18. All-purpose sponges (Kendall 9022)
19. Cotton tipped applicators (Henry Schein 100-6015)
20. A surgical microscope

3 Procedures

1. Anaesthetize a rat with sodium pentobarbital (50 mg kg^{-1}; i.p.).
2. Inject an antibiotic (such as Ampicillin, 150 mg kg^{-1}, i.m.) to prevent infection.

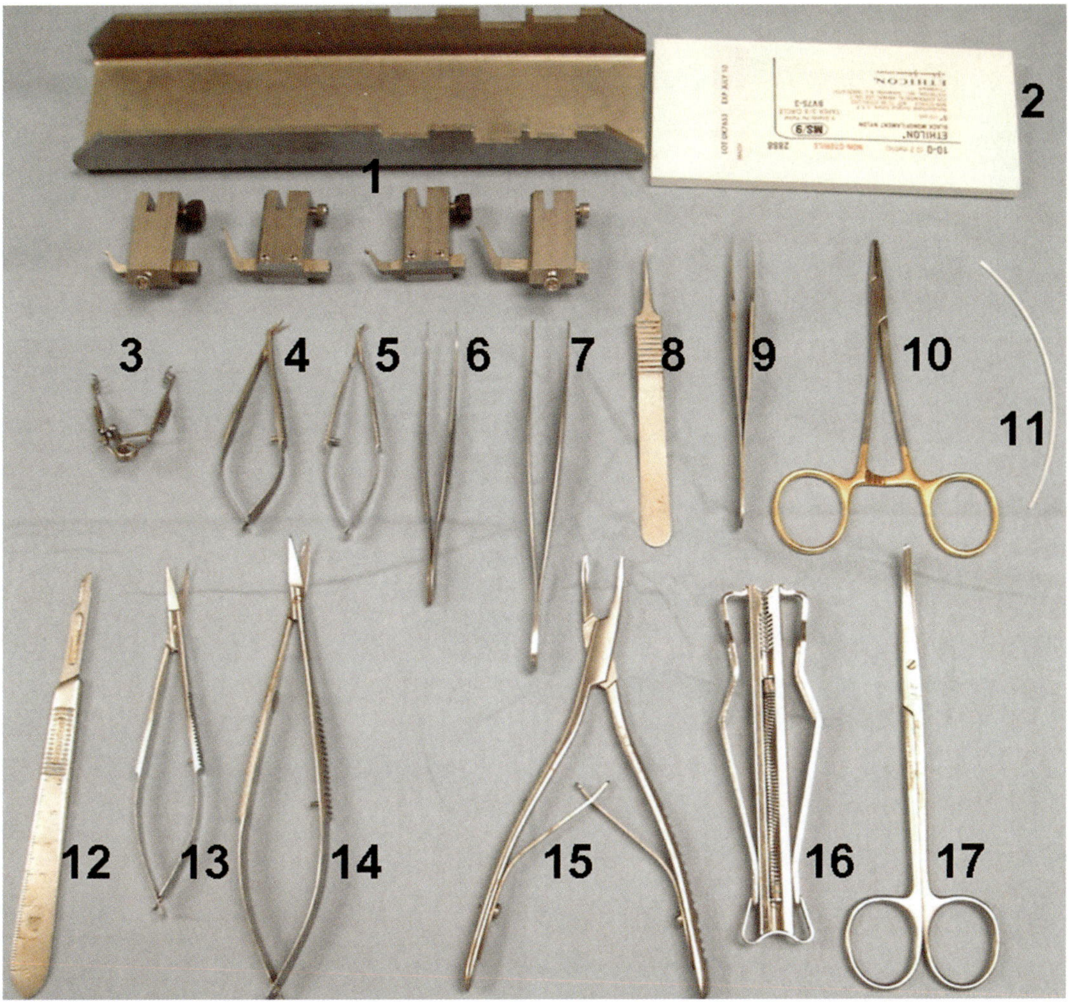

Fig. 1 List of tools (1–17) used for this model

3. Apply Lacrilube ophthalmic ointment (Allergan Pharmaceuticals, Irvine, CA) to eyes to prevent drying.
4. Place the animal on a heating pad to maintain body temperature.
5. Expose the injury site by incising the skin sagittally over the 9th and 10th thoracic vertebrae (T9-10).
6. Perform a laminectomy of the T9-10 vertebrae to expose the underlying cord segments. Remove more of the lamina on both sides of the cord (Fig. 2A) using the micro rongeur (Fig. 1, Item 15).
7. Incise the dura longitudinally along the midline (Fig. 2B) using spring scissors with especially fine blades (Fig. 1, Item 4).
8. Pull the dura laterally, stitch a 10–0 suture thread on each side of the incised dura at both rostral and caudal ends of the incision and leave the ends of the sutures free (Fig. 2B).

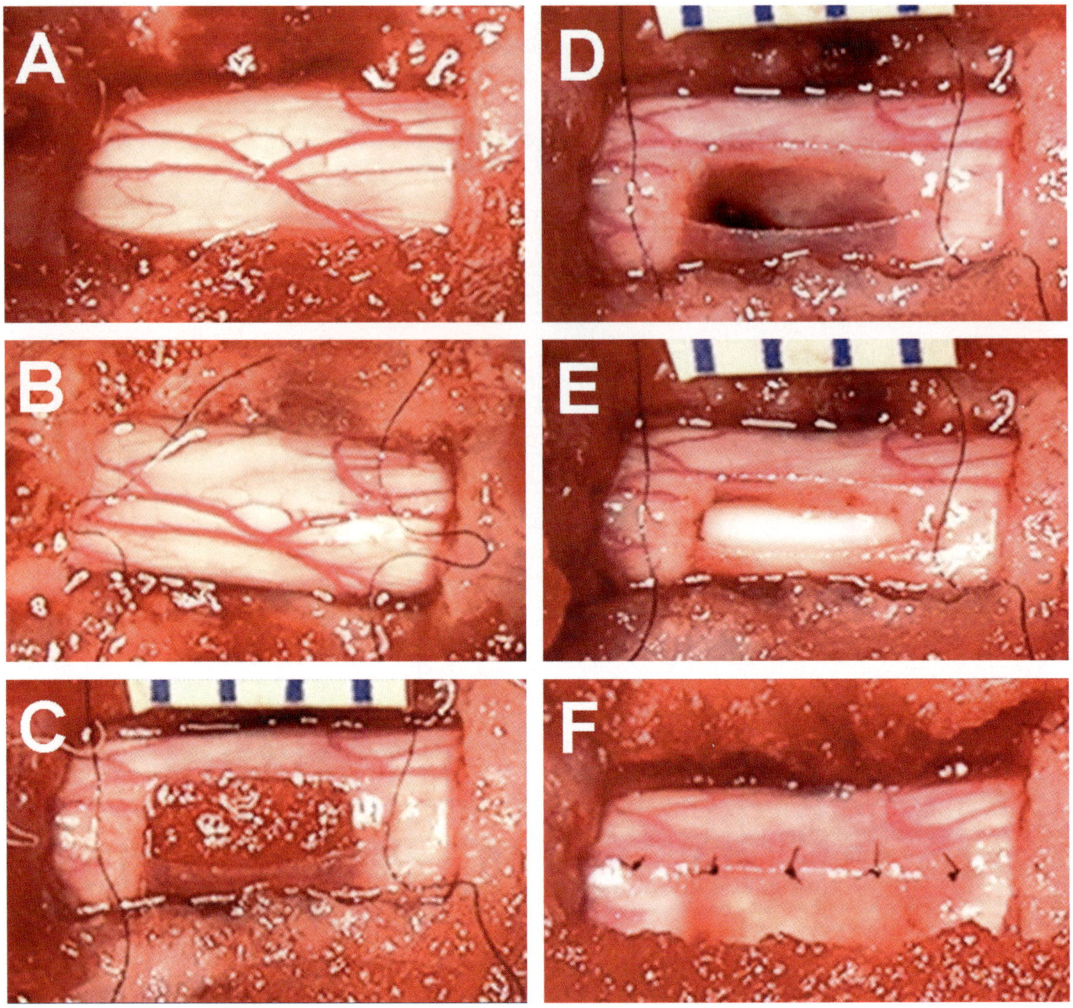

Fig. 2 Step by step procedures: (**A**) Exposure of the T9-10 spinal cord. (**B**) Incision of the dura. (**C**) Hemisection, creation of a lesion cavity, and insertion of a hemostatic sponge to stop bleeding. (**D**) Removal of the sponge to show the lesion cavity. Note that the cutting edge of two cord stumps was sharp and the lateral edge of the incised dura was well-preserved. (**E**) Implantation of a piece of guidance channel. Note the close apposition between the graft and host cord stumps at both ends. (**F**) Suture the dura. The channel and blood vessels underneath the dura can clearly be seen

9. Make a 3-mm long sagittal cut with a spring scissors (Fig. 1, Item 4) along the midline of the spinal cord. The incision should be made through the entire depth of the spinal cord, and the tip of scissors should touch the vertebral canal while cutting.
10. Pull the edge of the cut dura laterally with one angled forceps (Fig. 1, Item 8) and make two transverse cuts, one rostrally and one caudally using a spring scissors (Fig. 1, Item 5), of the hemi-cord to create a gap of 2.5–2.8 mm (Fig. 2C).
11. Remove the hemisected cord tissue by vacuum aspiration and stop bleeding by inserting a piece of absorbable collagen hemostatic sponge of desired size into the lesion cavity (Fig. 2C).

12. Remove the sponge and make sure that there is no bleeding (Fig. 2D).
13. Trim the guidance channel into a 3-mm-long segment. Insert one end of the channel into the lesion cavity in such a way that the caudal stump of the cord is positioned slightly within the caudal opening of the channel. Push the rostral end of the channel down to the bottom of the cavity; care should be taken not to damage the rostral stump of the cord during this process. Slightly slide the channel rostrally so that the rostral stump of the hemi-cord is also tightly apposed to the rostral opening of the channel (Fig. 2E).
14. Close the dura with 5–6 stitches of 10–0 nylon suture threads (Fig. 2F).
15. Suture the muscle and skin in layers.

4 Outcome Evaluations

1. Histology: To evaluate whether the spinal cord hemisection is complete, toluidine blue-stained semithin transverse sections through the lesion/graft mid-point should be prepared (Fig. 3). Quantitative measures including the number of myelinated axons, blood vessels, and regenerative tissue cable surface areas can be conducted at the light microscopic level. Ultrathin sections of the same tissue block can be examined at the electron microscopic level, and the ratio between myelinated and unmyelinated axons as well as the total number of regenerated axons can be estimated. Correlations between these parameters and functional assessments after various treatments may also be established.
2. Immunohistochemistry: Immunohistochemistry or immunofluorescence double-labeling can be used to examine the survival of grafted cells, the association between grafted cells and regenerated axons, the astrocytic response to the injury and transplantation, the regeneration of supraspinal

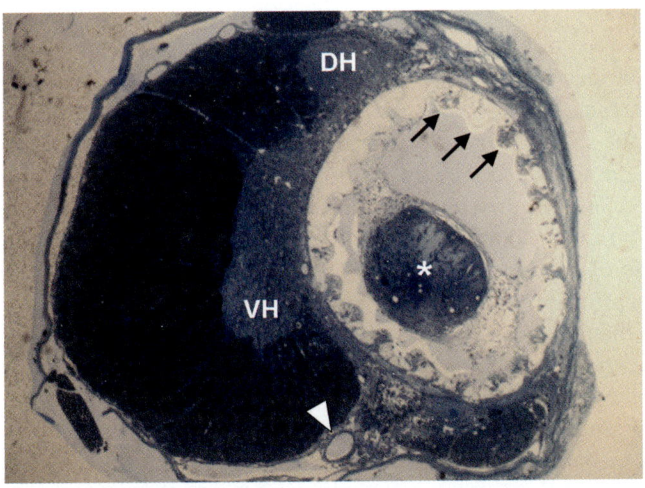

Fig. 3 A Toluidine Blue-stained plastic embedded transverse section through the graft mid-point shows the guidance channel (*arrows*) and regenerated tissue cable (*asterisk*) within the channel. Note that the ventral (VH) and dorsal (DH) horns are easily identified in the intact left hemi-cord. An *arrowhead* indicates the ventral spinal artery.

axons that express tract-specific transmitters, such as serotonin (5-HT) for the raphe-spinal pathway, and dopamine β-hydroxylase (DBH) for the ceroleo-spinal pathway, and the cavity formation at the graft-host interface.[1,4] These parameters can also be correlated with functional assessments after various treatments.

3. Retrograde tracing: Retrograde tracers such as Fast Blue (FB) or Fluoro-Gold (FG) may be injected into the graft mid-point or beyond the graft to label neuronal cell bodies whose axons regenerated into or beyond the graft, respectively.[1] If the injection is made distal to the graft, the intact hemicord should be completely transected prior to the tracer injection to prevent the labeling of neurons whose axons course around, rather than thorough, the graft.
4. Anterograde tracing: Axonal regeneration through and beyond the bridge transplant can be examined using anterograde tracers such as Phaseolus Valgaris-Leucoagglutinin (PHA-L) or Biotinylated-Dextran Amine (BDA) by injecting the tracer into specific regions of the cortex, brainstem, spinal cord, or dorsal root ganglia to label corticospinal, brainstem-spinal, propriospinal, or dorsal ascending tracts, respectively. Prior to the injection, the intact hemicord should be completely transected to prevent the labeling of axons course around, rather than thorough, the graft to the distal host spinal cord.[5]
5. Behavior assessments: According to our experience, grid walking and footprint analysis are sensitive measures for spinal cord hemisection injuries, whereas Basso, Beattie, and Bresnahan (BBB) open-filed locomotor rating scale and Louisville Swimming Scale (LSS) are less sensitive (unpublished observations).

5 Applications, Complications, and Limitations

1. The model is ideal for testing axonal regeneration of proprio- and supraspinal pathways. Importantly, it is appropriate for testing the growth-promoting potential of different trophic factors, matrix molecules, and cell types, since the channel environment can be modified. This model is not appropriate for testing tissue sparing at the site of injury.
2. When the spinal cord is completely transected, a large channel (inner diameter: 2.6 mm) can be placed into the lesion cavity with two cord stumps inserted into the openings of the channel.[2] Evidence for regeneration observed in the complete transection model is more convincing. However, because of the inability to repair the dura and instability of the host-graft interface in the complete transection model, axonal regeneration is less robust than the hemisection model.
3. The creation of spinal cord hemi-cavity on one side of the cord could also be used for transplantation of other tissues such as segments of peripheral nerves[6] and fetal spinal cord tissues.[7]
4. The procedures for creating a spinal cord hemisection without inducing a lesion cavity by using spring scissors or a VibraKnife device (see Chapter 39 of this book) can also be used for creation of a complete transection,[8] lateral over-hemisection,[9] dorsal hemisection (transection of the dorsal half of the spinal cord),[10,11] and dorsal funiculotomy (transection of the dorsal funiculus that contains the corticospinal tract and dorsal ascending tracts in rodents).[12]

5. Keep the midline: To ensure that the hemisection is made exactly through the midline, the spring scissors should be vertically inserted to the vertebral canal and the tips aligned along the longitudinal axis of the spinal cord. This is particularly important if functional assessments are to be conducted, which require consistent amounts of spared hemi-cord on the contralateral side. In many cases, dorsal vertebral veins are running irregularly on the surface of the cord. These vessels cannot be used as a landmark for the midline and should be cut if necessary. According to our experience, cutting such vessels does not cause much bleeding and, when bleeding occurs, it could be stopped easily.
6. Prevent cord retraction: After hemisection, the two stumps of the hemisected cord may retract for a short distance. Therefore, the cutting distance between the two stumps should be between 2.5 and 2.8 mm to ensure a tight apposition between the two stumps and the open ends of a 3-mm-long channel.
7. Avoid damaging the hemisected cord stumps: We found that cutting the cord stumps with spring scissors is better than using scalpel blades because the former induces a sharp laceration. Multiple cuts should be avoided. While introducing guidance channels, care should be taken not to push the channel wall against the cord stumps. If the cord stumps are damaged, they will swell and the regenerative outcomes will be poor.
8. Stop bleeding completely: Bleeding may occur during the surgical procedures. Stop it completely during the procedures, including prior to the transplantation.
9. Suture the dura: We feel that suturing the dura is critical for creating a growth-promoting environment and avoiding connective tissue invasion and scar formation. Extra care should be taken to not damage the dura during the surgical procedures and to suture it carefully after transplantation.
10. The rat vertebral column stabilizer we use was developed by Drs. Christopher B. Shields (cbshields1@gmail.com) and Y. Ping Zhang (ypzhan01@louisville.edu) at University of Louisville. If not available, other rat stabilizers may be used.
11. The semipermeable guidance channels that we use are composed of a 60:40 acrylonitrile/vinylchloride (PAN/PVC) with a nominal molecular weight cut off of 50 kDa. These channels are provided by Dr. Xuejun Wen (xjwen@clemson.edu) at Clemson University. Other biosynthetic channels or membranes can readily be tested using this model.

References

1. Xu XM, Guenard V, Kleitman N, Bunge MB. Axonal regeneration into Schwann cell-seeded guidance channels grafted into transected adult rat spinal cord. J Comp Neurol 1995;351:145–60.
2. Xu XM, Chen A, Guenard V, Kleitman N, Bunge MB. Bridging Schwann cell transplants promote axonal regeneration from both the rostral and caudal stumps of transected adult rat spinal cord. J Neurocytol 1997;26:1–16.
3. Xu XM, Zhang SX, Li H, Aebischer P, Bunge MB. Regrowth of axons into the distal spinal cord through a Schwann-cell-seeded mini-channel implanted into hemisected adult rat spinal cord. Eur J of Neurosci 1999;11:1723–40.
4. Iannotti C, Li H, Yan P, Lu X, Wirthlin L, Xu X-M. Glial cell line-derived neurotrophic factor-enriched bridging transplants promote propriospinal axonal regeneration and enhance myelination after spinal cord injury. Exp Neurol 2003;183:379–93.

5. Bamber NI, Li H, Lu X, Oudega M, Aebischer P, Xu XM. Neurotrophins BDNF and NT-3 promote axonal re-entry into the distal host spinal cord through Schwann cell-seeded mini-channels. Eur J Neurosci 2001;13:257–68.
6. Houle JD. Demonstration of the potential for chronically injured neurons to regenerate axons into intraspinal peripheral nerve grafts. Exp Neurol 1991;113:1–9.
7. Bregman BS, Kunkel-Bagden E, Reier PJ, Dai HN, McAtee M, Gao D. Recovery of function after spinal cord injury: mechanisms underlying transplant-mediated recovery of function differ after spinal cord injury in newborn and adult rats. Exp Neurol 1993;123:3–16.
8. Rabchevsky AG. Segmental organization of spinal reflexes mediating autonomic dysreflexia after spinal cord injury. Prog Brain Res 2006;152:265–74.
9. Kunkel-Bagden E, Dai HN, Bregman BS. Recovery of function after spinal cord hemisection in newborn and adult rats: differential effects on reflex and locomotor function. Exp Neurol 1992;116:40–51.
10. Iannotti C, Zhang YP, Shields LB, et al. Dural repair reduces connective tissue scar invasion and cystic cavity formation after acute spinal cord laceration injury in adult rats. J Neurotrauma 2006;23(6):853–65.
11. Onifer SM, Zhang YP, Burke DA, et al. Adult rat forelimb dysfunction after dorsal cervical spinal cord injury. Exp Neurol 2005;192(1):25–38.
12. Warden P, Bamber NI, Li H, et al. Delayed glial cell death following Wallerian degeneration in white matter tracts after spinal cord dorsal column cordotomy in adult rats. Exp Neurol 2001;168:213–24.

Chapter 42

Spinal Root Avulsion and Repair Model

Tak-Ho Chu and Wutian Wu

Abstract Motoneurons undergo retrograde degeneration following axonal injuries. The degree of degeneration varies with types of injuries, proximity of injury site to the cell body, species, and age of the animal. Unlike other injuries, spinal root avulsion induces a rapid and massive cell loss of the motoneurons. It not only provides a mean to study the degenerative changes of spinal motoneurons but also resembles the brachial plexus injury in human. Despite being situated in the central nervous system, motoneurons have a strong ability to regenerate under a suitable environment. We herein present step-by-step procedures for performing extravertebral and intravertebral avulsion at seventh cervical segment in rat spinal cord and peripheral nerve implantation to allow regeneration of the motoneurons.

Keywords: Spinal cord injury, Root avulsion, Peripheral nerve implantation, Brachial plexus, Motoneurons

1 Model Selection

Spinal root avulsion is a model to study degenerative changes of motoneurons responding to axonal injuries.[1] A traction force is applied on the ventral root, which is then torn off from the surface of the spinal cord. The animal model resembles the most severe type of brachial plexus injuries, which could occur during difficult delivery of newborns, sport injuries, or car accidents in clinical cases. Unlike peripheral nerve injuries such as axotomy, crush, or ligation, root avulsion is a central nervous system injury. Besides, proximity of injury site to the cell body, mechanical strain resulted from traction and deprivation of peripheral trophic support contribute to the rapid degeneration of the motoneurons and lead to massive cell loss.[1,2] Implantation of a peripheral nerve (PN)[3,4] or avulsed ventral root[5–7] can rescue the injured motoneurons from death and allow regeneration to occur in animal studies and human patients. In this chapter, we will describe in details both intravertebral and extravertebral spinal root avulsion in the seventh cervical (C7) segment in adult rat. We will also present a protocol on implanting a PN to allow regeneration of the motoneurons. The advantages of the model are that (1) The C7 root avulsion will not

From: *Animal Models of Acute Neurological Injuries.*
Edited by: J. Chen (eds.) New York, Totowa, NJ

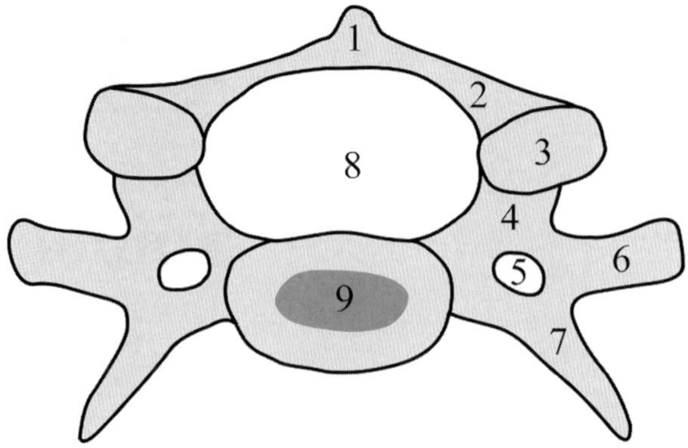

Fig. 1 Caudal surface of the sixth vertebra. 1, spinuos process; 2, lamina; 3, articular process; 4, pedicle; 5, transverse foramen; 6, posterior tubercle of transverse process; 7, anterior tubercle of transverse process; 8, vertebral foramen; 9, vertebral body

cause any observable functional disturbances in the experimental animals; (2) The surgery can be performed at different levels of spinal segments, most commonly cervical and lumbar segments; and (3) It can be done with or without removal of the vertebra[1] (Fig. 1), the latter two considerations give flexibility to the experimental design.

2 Materials

Surgical instruments common for extra and intravertebral avulsion are listed below, they could be obtained from various suppliers and some recommendations are given:

2.1 Basic Instruments

1. Surgical microscope (6×–40×)
2. Small animal retraction system (FST 18200-50)*
3. 12-cm tissue forceps
4. Scalpel handle 3 with blade
5. 2 Dumont 5 forceps
6. Vannas-Tübingen spring scissors (FST 15003-08)
7. Straight fine iris scissors
8. Crile hemostat
9. Michel suture clips
10. Gel foam (Johnson & Johnson)
11. 3–0 suture (Johnson & Johnson)
12. Autoclaved cotton and 70% ethanol

*A simple and economic retraction system can be made by wiring a small metal hook to a pin with a rubber band. Then the animal is placed on a cork board.

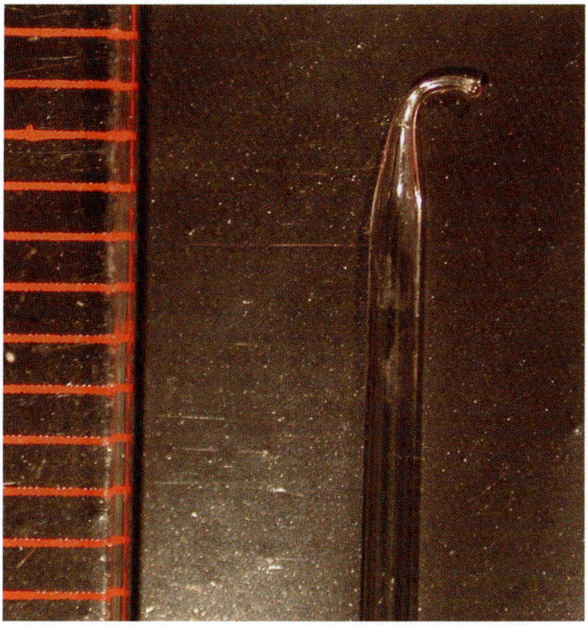

Fig. 2 A fire polished hook made of capillary tube. Scale: 1 division = 1 mm

2.2 Additional Tool for Extravertebral Avulsion

Serrated Halsted-mosquito hemostats (FST13008-12) or equivalent

2.3 Additional Tools for Intravertebral Avulsion

1. Curved Friedman-Pearson Rongeur with 1-mm cup (FST 16021-14 or Roboz RS-8305)
2. Fire polished avulsion hook (Fig. 2) or a hooked 5 forceps can be used

2.4 Additional Tools Needed for Peripheral Nerve Implantation

1. 1.6-ml microfuge tube with normal saline to temporarily store the nerve graft
2. 10–0 and 11–0 sutures (Johnson & Johnson)
3. Micro knife (FST10056-12) or an angled 27 gauge needle

3 Procedure

3.1 Anesthesia

Adult male Sprague-Dawley rats weighing 280–300 g are used. Anesthetize rats with ketamine (80 mg kg^{-1}) and xylazine (8 mg kg^{-1}) intramuscularly or intraperitoneally

3.2 Extravertebral Avulsion (Fig. 3)

1. Shave the hair on the right chest
2. Lay the animal on surgical platform in supine position

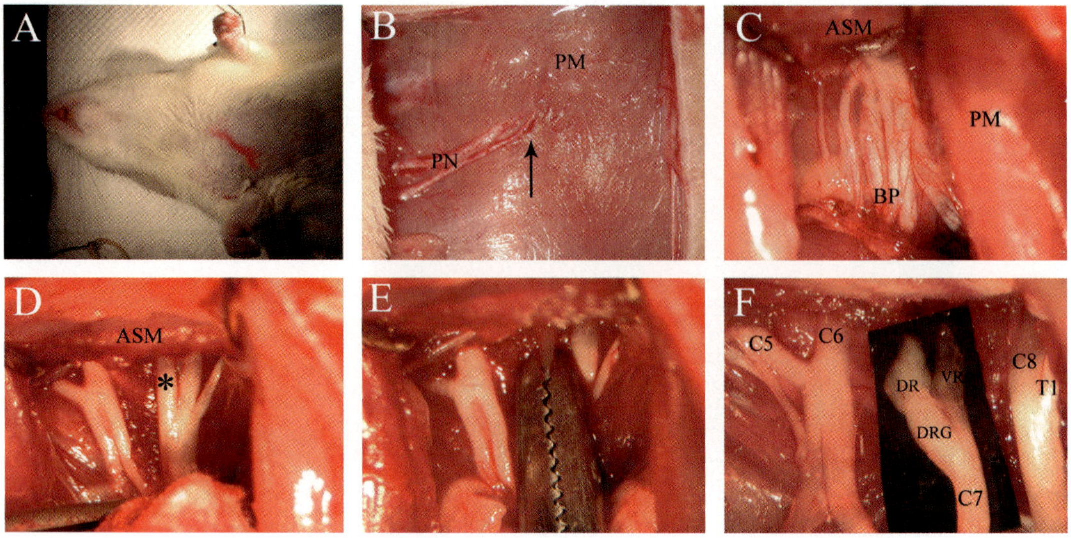

Fig. 3 Extravertebral avulsion: (**A**) An incision line on the right chest. (**B**) Pectoral nerve landmark (*arrow*). (**C**) Exposure of the brachial plexus. (**D**) Exposure of C5-T1 spinal nerves. Note the C7 spinal nerve (*asterisk*). (**E**) Avulsion by a hemostat. (**F**) Examination of the avulsed root. C5-C8 and T1: C5-C8 and T1 spinal nerves; *ASM* anterior scalene muscle; *BP* brachial plexus; *DR* dorsal root; *DRG* dorsal root ganglion; *PM* pectoral muscle; *PN* pectoral nerves, *VR* ventral root

3. Disinfect the skin over the shaved area
4. Make a 2-cm hockey stick incision on the right chest (Fig. 3a)
5. The pectoral nerves can be seen (Fig. 3b), blunt separate the greater pectoral muscle where the nerve pierces
6. By tracing the pectoral nerve, further separate the underlying ascending pectoral muscle, retract the muscles laterally to expose the brachial plexus (Fig. 3c)
7. Identify the trunks of the brachial plexus, the C7 spinal nerve is in the middle trunk
8. Dissect between the anterior and middle scalene muscles where the nerve passes through. Retract the anterior scalene muscle to identify the C7 spinal nerve (Fig. 3d)
9. Dissect the nerve free from the surrounding tissue and as close to the vertebra as possible
10. Use a Halsted-mosquito hemostat to grab along a segment of the spinal nerve. While using one forceps to push against the vertebra, slowly pull the spinal nerve out from the intervertebral foramen (Fig. 3e)
11. Stop bleeding by applying a gel foam to the foramen
12. Cut and remove the spinal nerve with the dorsal root ganglion, dorsal and ventral roots to prevent reinnervation
13. Evaluate the success of avulsion by examining the length of the avulsed ventral root (Fig. 3f)
14. Suture close the muscles and skin in layers with 3–0 suture and Michel suture clips

3.3 Intravertebral Avulsion (Fig. 4)

1. Locate a prominent protrusion on the center of the back caudal to the neck. It is the spinous process of second thoracic vertebra (T2), shave the hair from the neck to the protrusion

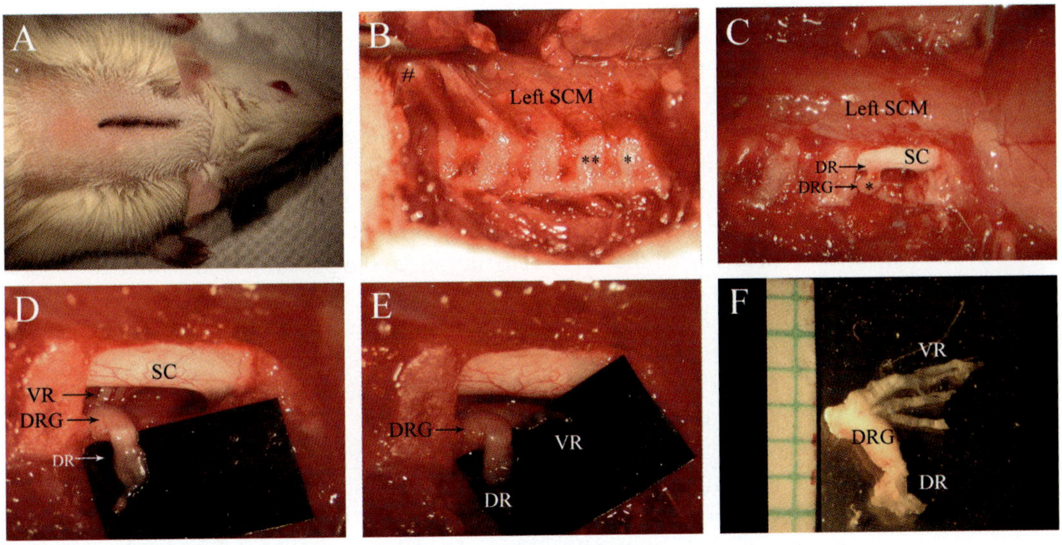

Fig. 4 Intravertebral avulsion: (**A**) An incision line on the neck. (**B**) Exposure of the C5-T2 vertebrae. Note the T2 spinous process (*hash*). C5 (*asterisk*) and C6 (*double asterisks*) vertebrae are indicated. (**C**) Exposure of dorsal root ganglion (*asterisk*) and dorsal root. (**D**) After dorsal root transection, ventral root (*arrow*) is exposed. (**E**) Avulsion of ventral root. (**F**) Avulsed spinal nerve. Scale: 1 division = 1 mm. *DR* dorsal root; *DRG* dorsal root ganglion; *Left SCM* left semispinalis cervicis muscle; *SC* spinal cord; *VR* ventral root

2. Put the rat on the surgical platform and disinfect the shaved area with 70% ethanol
3. Make a 2.5–3 cm midline skin incision (Fig. 4a)
4. Retract the muscles laterally to expose the semispinalis cervicis muscles inserted to the cervical vertebrae
5. Using the T2 protrusion as a bony landmark, identify the sixth cervical (C6) vertebra which encloses the C7 spinal segment (Fig. 4b)
6. While using a tissue forceps to grab the T2 protrusion to stabilize the vertebrae, remove the right-sided C6 lamina and articular process by a bone ronguer, in practice, the C5 lamina can also be removed to provide more space for manipulation
7. The dorsal root and its ganglia can clearly be seen beneath the dura, carefully pick up and incise the dura with #5 forceps and spring scissors. Cerebral spinal fluid (CSF) flows out, absorb the fluid before proceeding (Fig. 4c)
8. Incise the dorsal root where it attaches to the spinal cord to expose the ventral root (Fig. 4d)
9. Use a fine hook to avulse the ventral root, exert a gentle traction almost parallel to the spinal cord (Fig. 4e)
10. Stop bleeding by a gel foam if necessary
11. Together with the ventral root, dorsal root and its ganglion, about 2–3 mm of the spinal nerve is severed and removed (Fig. 4f), a gel foam can be put in the intervertebral foramen if bleeding occurs
12. (Optional) Chemicals such as inhibitors, growth factors, or others can be soaked in gel foam and topically applied to the avulsed site
13. (Optional) Suture the dura mater with 10–0 suture
14. Suture the muscles and skin in layers with 3–0 suture and Michel suture clips if no extra procedure is required

3.4 PN Graft Implantation (Fig. 5)

1. Shave the medial surface of the right hind leg of the animal and disinfect with 70% ethanol
2. Lay the animal on surgical platform in supine position
3. Saphenous vein should be visible, make a 2 cm skin incision (Fig. 5a)
4. Carefully incise the deep fascia along the nerve and separate the nerve from adjacent saphenous artery (Fig. 5b), remove a 2-cm segment of the nerve, label the nerve at 10 mm from the proximal end with a 10–0 suture on the epineurium (can be done in later step)
5. Temporarily store the nerve in chilled normal saline before implantation
6. Suture close the skin with Michel suture clips
7. Following step 11 in intravertebral avulsion, take out the nerve from the saline solution
8. Suture the epineurium (1 mm to proximal) of the nerve graft on the lateral funiculus of the spinal cord with 11–0 suture
9. Cut the lateral funiculus longitudinally with an angled micro knife or an angled 27 gauge needle of about 1 mm wide and 0.5 mm deep near the 11–0 suture
10. Use a sharp #5 forceps to gently spread apart the cut to create a myelotomy groove, the forceps should not be inserted deeper than 1.5 mm
11. Introduce the peripheral nerve graft almost perpendicularly into the myelotomy groove (Fig. 5c)
12. Suture the distal end of the nerve graft into the juxtaposed muscle with 10–0 suture
13. Suture close muscles and skin in layers
14. To access the regeneration of motoneurons, 2 days before the designated period of survival, 1 μl of 5% Fluorogold or other neuronal tracers can be injected into the peripheral nerve graft at the 10 mm position

4 Outcome Evaluations

For avulsion injury, the whole avulsed segment can be harvested and serially cut into 40-μm thick cross or longitudinal sections after fixation with 4% formaldehyde. Cresyl violet or neutral red can be used to counterstain the sections. Motoneuron can be easily identified by its location in Rexed's lamina IX in ventral horn, its large soma size, and granular appearance of Nissl bodies in the cytoplasm under high magnifications. Number of surviving motoneurons and size of soma can be assessed and compared with ipsilateral side. Typically, 600–900 motoneurons are found at one side of the C7 segment in adult rats. After avulsion, 20% of motoneurons dies at 2 weeks and half of them are lost 4 weeks postinjury.

Regeneration of motoneurons can be assessed by counting the number of fluorescent-labeled motoneurons (Fig. 6). Most labeled cells are motoneurons and from the injured segment; however, a few from neighboring segments can be found. Axon number, density, and size can be assessed in the cross section of the implanted nerve. Functional recovery can also be assessed after implantation of avulsed roots if a suitable set of ventral roots are avulsed, and grooming test and grasping test can be used according to the location of avulsed site and groups of muscle affected.

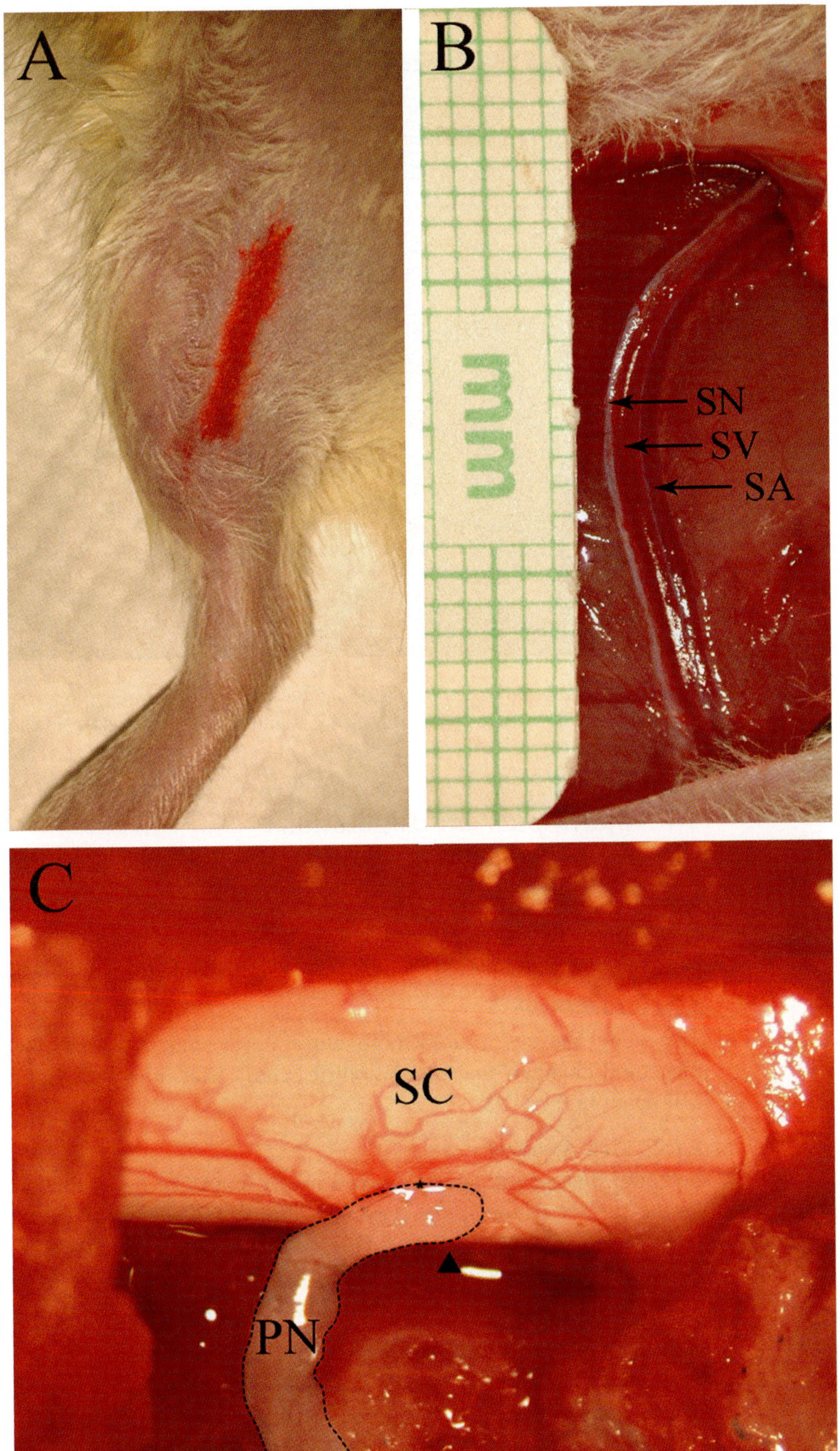

Fig. 5 Peripheral nerve graft implantation: (**A**) An incision line on the right hind leg. (**B**) Exposure of saphenous nerve and its artery and vein. (**C**) Implantation of the nerve graft (*dashed line*) into the spinal cord. Note the 11–0 suture point (*asterisk*) and the nerve insertion (*arrow head*). *PN* peripheral nerve; *SA* saphenous artery; *SC* spinal cord; *SN* saphenous nerve; *SV* saphenous vein

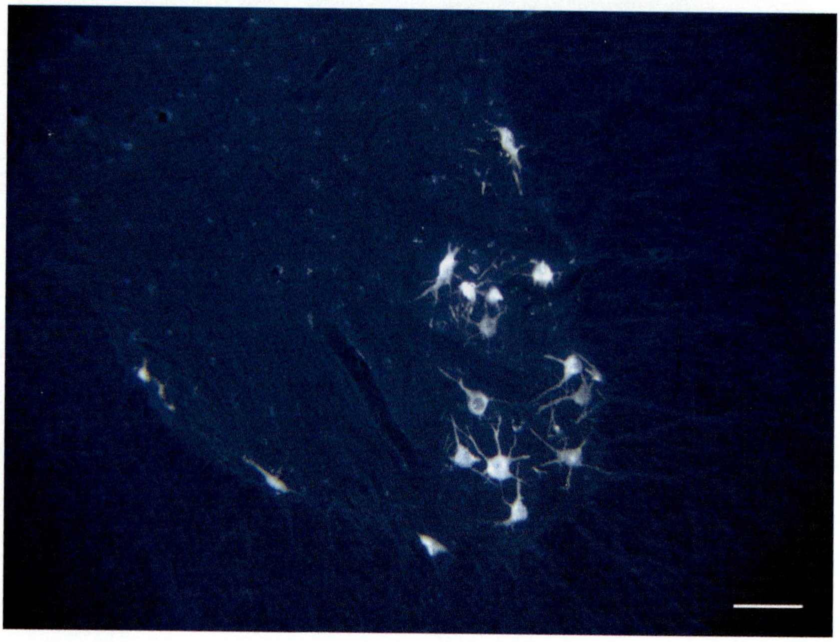

Fig. 6 Regeneration of motoneurons can be examined by application of neuronal tracers. Fluorogold positive motoneurons are found in the ventral horn of the spinal cord. Scale bar = 100 μm

5 Applications and Limitations

5.1 Applications

Spinal root avulsion is a useful and reproducible model in studying cell death, survival, and regeneration of motoneurons in small and large animals.

5.1.1 Cell Death

The model induces a rapid motoneuron death evident at 2 weeks postinjury. Structural changes[8] and regulation of genes[9] after injury provide insight into cell death in the CNS.

5.1.2 Cell Survival

Potential therapeutic agents soaked in gel foam can be applied conveniently on the surface of the spinal cord near the ventral root exit zone.[10,11] Massive cell loss if left untreated can provide a sharp contrast to the survival promoting effect of the drug under testing.

5.1.3 Regeneration

Regeneration of motoneurons under different conditions can be tested in a fairly short observation period in this animal model. Observable functional deficit can be induced by combination of avulsion and ligation, thus allowing functional assessment after implantation.[7,12] More importantly, studying avulsion and regeneration can provide useful cues to develop effective treatment to avulsion injuries in human.[13]

5.1.4 Choice of Animals

Experimental avulsion has been performed on larger mammals such as sheeps, monkeys, and cats, but a larger pool of data are from small rodents as in rats

and mice. Rat is in a reasonable size to perform complicated surgeries such as implantation, whereas knockout mice are helpful in studying the underlying molecular pathway. However, with the advance of RNA interference, cell type and site-specific gene knockdown can be achieved by viral infection,[14] thus making rat an ideal animal model to work with.
Limitations

1. If extravertebral avulsion fails, it is impossible to remove the remaining ventral root. If complete avulsion is a concern, intravertebral avulsion should be considered. Implantation cannot be done in extravertebral avulsion.
2. Both intravertebral and extravertebral avulsions damage the integrity of blood–spinal cord barrier and result in leakage of CSF.
3. Continuous flow of CSF will dilute the drug applied topically on the avulsed segment, a higher concentration of drug could be used, alternatively, repeated renewal of the drug or implantation of an osmotic pump are justified.
4. Attention has to be paid on laminectomy, which could damage the spinal cord if the procedure is not done carefully. Keep the bone ronguer away from the surface of the spinal cord, and do not twist the bone inward toward the spinal cord.

References

1. Koliatsos VE, Price WL, Pardo CA, Price DL. Ventral root avulsion: an experimental model of death of adult motor neurons. J Comp Neurol 1994; 342(1):35–44.
2. Wu W. Expression of nitric-oxide synthase (NOS) in injured CNS neurons as shown by NADPH diaphorase histochemistry. Exp Neurol 1993; 120(2):153–159.
3. Wu W, Han K, Li L, Schinco FP. Implantation of PNS graft inhibits the induction of neuronal nitric oxide synthase and enhances the survival of spinal motoneurons following root avulsion. Exp Neurol 1994; 129(2):335–339.
4. Holtzer CA, Marani E, van Dijk GJ, Thomeer RT. Repair of ventral root avulsion using autologous nerve grafts in cats. J Peripher Nerv Syst 2003; 8(1):17–22.
5. Carlstedt T, Grane P, Hallin RG, Noren G. Return of function after spinal cord implantation of avulsed spinal nerve roots. Lancet 1995; 346(8986):1323–1325.
6. Hoffmann CF, Marani E, van Dijk JG, vd KW, Thomeer RT. Reinnervation of avulsed and reimplanted ventral rootlets in the cervical spinal cord of the cat. J Neurosurg 1996; 84(2):234–243.
7. Gu HY, Chai H, Zhang JY et al. Survival, regeneration and functional recovery of motoneurons in adult rats by reimplantation of ventral root following spinal root avulsion. Eur J Neurosci 2004; 19(8):2123–2131.
8. Li L, Houenou LJ, Wu W, Lei M, Prevette DM, Oppenheim RW. Characterization of spinal motoneuron degeneration following different types of peripheral nerve injury in neonatal and adult mice. J Comp Neurol 1998; 396(2):158–168.
9. Yang Y, Xie Y, Chai H et al. Microarray analysis of gene expression patterns in adult spinal motoneurons after different types of axonal injuries. Brain Res 2006; 1075(1):1–12.
10. Wu W, Li L, Yick LW et al. GDNF and BDNF alter the expression of neuronal NOS, c-Jun, and p75 and prevent motoneuron death following spinal root avulsion in adult rats. J Neurotrauma 2003; 20(6):603–612.
11. Zhou L, Wu W. Antisense oligos to neuronal nitric oxide synthase aggravate motoneuron death induced by spinal root avulsion in adult rat. Exp Neurol 2006; 197(1):84–92.
12. Wu W, Chai H, Zhang J, Gu H, Xie Y, Zhou L. Delayed implantation of a peripheral nerve graft reduces motoneuron survival but does not affect regeneration following spinal root avulsion in adult rats. J Neurotrauma 2004; 21(8):1050–1058.

13. Carlstedt T. Approaches permitting and enhancing motoneuron regeneration after spinal cord, ventral root, plexus and peripheral nerve injuries. Curr Opin Neurol 2000; 13(6):683–686.
14. Ralph GS, Radcliffe PA, Day DM et al. Silencing mutant SOD1 using RNAi protects against neurodegeneration and extends survival in an ALS model. Nat Med 2005; 11(4):429–433.

Index

Printed in the United States of America